AF353101

# Computation in Complex Networks

# Computation in Complex Networks

Editors

**Clara Pizzuti**
**Annalisa Socievole**

MDPI • Basel • Beijing • Wuhan • Barcelona • Belgrade • Manchester • Tokyo • Cluj • Tianjin

*Editors*
Clara Pizzuti
Institute for High Performance
Computing and Networking
(ICAR)
Italy

Annalisa Socievole
Institute for High Performance
Computing and Networking
(ICAR)
Italy

*Editorial Office*
MDPI
St. Alban-Anlage 66
4052 Basel, Switzerland

This is a reprint of articles from the Special Issue published online in the open access journal *Entropy* (ISSN 1099-4300) (available at: https://www.mdpi.com/journal/entropy/special_issues/Computation_Complex_Networks).

For citation purposes, cite each article independently as indicated on the article page online and as indicated below:

LastName, A.A.; LastName, B.B.; LastName, C.C. Article Title. *Journal Name* **Year**, *Volume Number*, Page Range.

**ISBN 978-3-0365-0682-1 (Hbk)**
**ISBN 978-3-0365-0683-8 (PDF)**

# Contents

# About the Editors

**Clara Pizzuti**, Prof., Ph.D., received her master's degree in Mathematics from University of Calabria, Italy, and a Ph.D. in Science from the Radboud Universiteit Nijmegen, NL. She is Research Director at the Institute of High Performance Computing and Networking (ICAR) of the Italian National Research Council (CNR), where she leads the Social and Complex Data Intelligence research laboratory. Her research interests include knowledge discovery in databases, data mining, data streams, bioinformatics, social network analysis, evolutionary computation. She is serving as program committee member of international conferences, and as reviewer for several international journals.

**Annalisa Socievole**, Eng. Ph.D., is a Researcher at the National Research Council of Italy (CNR), Institute for High Performance Computing and Networking (ICAR). She received a Ph.D. in Systems and Computer Science Engineering in February 2013 and a master's degree in Telecommunications Engineering in July 2009, both from the University of Calabria. From October 2011 to April 2012, she has been a visiting Ph.D. student in the Systems Research Group of Cambridge Computer Laboratory (UK) working on multi-layer opportunistic social networks. From October 2013 to June 2014, she also spent a post-doc period in NAS Group at TU Delft (The Netherlands) to carry out the research project CONTACTO on opportunistic networking and community-based epidemic spreading. Her research interests include opportunistic networks, epidemic models, evolutionary techniques and community detection.

*Editorial*

# Computation in Complex Networks

**Clara Pizzuti * and Annalisa Socievole ***

National Research Council of Italy (CNR), Institute for High Performance Computing and Networking (ICAR), Via Pietro Bucci, 8-9C, 87036 Rende, Italy

* Correspondence: clara.pizzuti@icar.cnr.it (C.P.); annalisa.socievole@icar.cnr.it (A.S.)

Received: 27 January 2021; Accepted: 2 February 2021; Published: 5 February 2021

The Special Issue on "Computation in Complex Networks" focused on gathering highly original papers in the field of current complex network research. Due to their ability to model a wide variety of daily-life systems—including the Internet, communication, chemical, neural, social, political and financial networks—complex network systems and their behavior need to be deeply understood. As such, the focus of this Special Issue has been highlighting and promoting current interdisciplinary contributions on the various fields of complex networks, thus providing a collection of high-quality research papers that capture the challenges recently posed by these networks. We selected 20 manuscripts, which are described below.

In the paper "Active Learning for Node Classification: An Evaluation" by Madhawa and Murata [1], the active learning framework was used as a method to make node classification on attributed graphs by representing data instances as nodes of the graph. The authors performed an empirical evaluation of different state-of-the-art active learning algorithms proposed for graph neural networks, as well as other data types, such as images and text, on several real-world attributed graphs. The results showed that active learning algorithms designed for other data types do not perform well on graph-structured data, highlighting the importance of complementing uncertainty-based active learning models with an exploration term.

In the paper "Spreading Control in Two-Layer Multiplex Networks" by Jaquez et al. [2], the problem of controlling an SIS (Susceptible-Infected-Susceptible) epidemic spreading over a network with two layers was addressed. The stabilization of the extinction state for the nonlinear discrete-time model was obtained by properly tuning system parameters, such as intralayer and interlayer transmission rates, for a limited number of nodes characterized by a parametric threshold condition. The sufficient conditions for the choice of the subset of nodes and the parameters to be controlled were established through a rigorous mathematical analysis guaranteeing the exponential stability of the extinction state globally, with respect to the set of all possible probability states.

In the paper "Investigating the Influence of Inverse Preferential Attachment on Network Development" by Siew and Vitevitch [3], the growth mechanism of phonological language networks, in terms of the acquisition of new words that are phonologically similar to existing ones, was explored. Specifically, the authors analyzed the network structure and the degree distributions of networks synthetically generated through preferential attachment, an inverse variant of the classical version where new nodes are connected to existing nodes with fewer edges, or combinations of both network growth mechanisms. The simulation results showed that preferential attachment—followed by inverse preferential attachment—in the network growth resulted in densely connected network structures.

In the paper "Classification of Literary Works: Fractality and Complexity of the Narrative, Essay, and Research Article" by Ramirez-Arellano [4], the problem of the classification of literary works was tackled. This research analyzed the node degree, betweenness, shortest path length, clustering coefficient, nearest neighborhoods' degree, fractal dimension, complexity, area under box-covering, and area under robustness curve of the complex networks. The literary works of Mexican writers were analyzed, with the aim of classifying them according to their genre. The results of this analysis classified 87% of the full word co-occurrence networks as a fractal.

In the paper "Detecting Overlapping Communities in Modularity Optimization by Reweighting Vertices" by Tsung et al. [5], the community detection problem was considered, specifically focusing on overlapping community sets of nodes. By first introducing a node weight allocation problem to formulate the overlapping property, the authors proposed a genetic algorithm, exploiting an extension of the modularity function for solving the node weight allocation problem and detecting the overlapping communities. Moreover, three refinement strategies for improving the quality of results were added. On both real-world and synthetic networks, the proposed algorithm was able to better detect nontrivial overlapping nodes, compared to other contestant algorithms.

In the paper "Modelling and Recognition of Protein Contact Networks by Multiple Kernel Learning and Dissimilarity Representations" by Martino et al. [6], the authors focused on predicting the proteins' functional role, proposing a hybrid classification system based on a linear combination of multiple kernels defined over multiple dissimilarity spaces. Here, the training procedure jointly optimized the kernel weights and the representatives' selection in the dissimilarity spaces. The classification system was thus characterized by a double knowledge discovery phase in which the analysis of the weights allowed the authors to check which representations were better for solving the classification problem—whereas the pivotal patterns selected as representatives give further insight into the modelled system. Experimental results showed how the proposed classification system was able to reliably analyze the considered protein contact networks.

In the paper "Cross-Domain Recommendation Based on Sentiment Analysis and Latent Feature Mapping" by Wang et al. [7], a cross-domain recommendation algorithm (CDR-SAFM) based on sentiment analysis and latent feature mapping was proposed. This algorithm specifically combined the sentiment information extracted from different domains of users' ratings. The sentiment is categorized into (1) positive, (2) negative and (3) neutral. Moreover, the latent Dirichlet allocation (LDA) was used to model the users' semantic orientation to generate the latent sentiment review features. Finally, by applying multilayer perceptron (MLP), the CDR-SAFM was able to obtain the cross-domain nonlinear mapping function to transfer the users' sentiment review features. Tested on the Amazon dataset, the proposed recommendation algorithm outperformed other existing recommendation algorithms in the considered cross-domain scenario.

In the paper "Complex Contagion Features without Social Reinforcement in a Model of Social Information Flow" by Pond et al. [8], the problem of information spreading over social networks through a complex contagion model was considered. Focusing on the quoter model (a model of the social flow of written information copying or "quoting" short subsequences of text from neighbors), the authors showed how this model has features of complex contagion, including the weakness of long ties and the high network density that limits information flow rather than boosting it, despite lacking an explicit mechanism of social reinforcement that distinguishes complex contagion from epidemic spread.

In the paper "Optimizing Variational Graph Autoencoder for Community Detection with Dual Optimization" by Choong et al. [9], variational graph autoencoders for community detection were considered. The research underlined how variational autoencoder (VAE)-based approaches suffer from a deviation increase from the primary objective when minimizing loss using the stochastic gradient descent, resulting in suboptimal community structure. To smooth this effect, a dual optimization procedure was proposed to guide the optimization process toward better communities. The results of the experiments showed that the proposed community detection algorithm outperformed its predecessor.

In the paper "Properties of the Vascular Networks in Malignant Tumors" by Chimal-Eguìa et al. [10], both synthetic and real angiogenic vascular networks of patients with Hepato-Cellular Carcinoma (HCC), extracted from digital tomographies, were analyzed. From the measurements of network properties, such as average path length, clustering coefficient, degree of distribution and fractal dimension, the authors showed that there is a well-connected network (high clustering coefficient), different from previous related works. The network exhibited efficient communication. This was also reflected by the small average path length.

In the paper "Complex Network Construction of Univariate Chaotic Time Series Based on Maximum Mean Discrepancy" by Sun [11], the focus was on the analysis of chaotic time series; more specifically, on how measuring the similarity between time series affected construction of the corresponding network. Here, a method that first transforms univariate time series into high-dimensional phase space, then exploits a Gaussian mixture model (GMM) to represent time series, and finally introduces maximum mean discrepancy (MMD) to measure the similarity between GMMs was proposed. The introduced MMD was validated using the Lorenz system, showing that the similarity between GMMs can be measured more effectively.

In the paper "Analyzing Uncertainty in Complex Socio-Ecological Networks" by Maldonado et al. [12], the aim was to assess the impact of using the Bayesian network structure for modeling complex socio-ecological networks, whose behavior is often uncertain. The conducted analysis was two-fold. The first experiment assessed the impact of the Bayesian network structure on the entropy of the model. The second compared the entropy of the posterior distribution of the class variable obtained from the different structures. For the experiments, three types of Bayesian networks are analyzed: naive Bayes (NB), tree augmented networks (TAN) and networks with unrestricted structure (GSS). The results showed that GSS consistently outperformed both NB and TAN when evaluating the uncertainty of the entire model, while NB and TAN resulted in lower entropy values of the posterior distribution of the class variable, making them suitable for prediction tasks.

In the paper "Multi-Type Node Detection in Network Communities" by Ezeh et al. [13], a new community detection method—able to uncover disjoint clusters of nodes, clusters with overlapping nodes, and single isolated nodes forming a partition with a unique node—was proposed. Differing from previous state-of-the-art methods, the authors proposed an approach which iteratively computes the bridging centrality value of the nodes to find those with the highest bridging centrality value. Once a bridge node has been identified, the algorithm computes the node similarity between the bridge and its neighbors, and the neighbors with the least node similarity values are disconnected. This step is iterated until a stopping criterion condition is satisfied. Simulations on both real-world and synthetic networks demonstrated that the proposed method was able to efficiently classify multi-type nodes in network communities.

In the paper "Predicting the Evolution of Physics Research from a Complex Network Perspective" by Liu et al. [14], the problem of quantitative knowledge evolution in physics research was addressed through complex networks, built on bibliographic coupling and co-citation data extracted from the American Physical Society repository from 1981 to 2010. For each year, the topical clusters (TCs) were uncovered through the Louvain method and compared to subsequent years to assess their similarity. Once this information was gathered, a machine learning model was applied to predict the evolution of the clusters in terms of permanence, disappearance, merging or splitting. This research showed that the number of papers from certain journals, degree, closeness, and betweenness mostly drove the predictor.

In the paper "Uncovering the Dependence of Cascading Failures on Network Topology by Constructing Null Models" by Ding et al. [15], the problem of cascading failures in complex network infrastructures was taken on. The authors analyzed the impact that underlying network topology has on cascading failures in realistic Internet Autonomous System network scenarios by constructing different types of null models. By analyzing the shortest paths in different topological configurations, the results revealed the effects that microscale (e.g., degree distribution, assortativity, and transitivity) and mesoscale (e.g., rich-club and community structure) network properties have on cascade robustness when intentional node attacks are performed.

In the paper "Service-Oriented Model Encapsulation and Selection Method for Complex System Simulation Based on Cloud Architecture" by Xiong et al. [16], a service-oriented model encapsulation and selection method to construct complex system simulation applications was proposed. The method encapsulates models with large computational requirements in shared simulation services in the cloud architecture. It also allows the distributed scheduling of model services and a semantic search framework, useful for the users in searching the required models. An optimization selection algorithm based on quality of service (QoS) was proposed to support users in obtaining an ordered candidate model set satisfying a certain QoS. The performed experiment proved that the proposed method was able to effectively improve the execution efficiency of complex system simulation applications.

In the paper "Minimum Memory-Based Sign Adjustment in Signed Social Networks" by Qi et al. [17], the authors focused on signed social networks—and in particular, on the impact of limited memory on the convergence of the network. The research analyzed random and minimum memory-based sign adjustment rules. Under these rules, the impacts of an initial ratio of positive links, rewiring probability, network size, neighbor number and randomness upon structural balance are compared. The experimental results showed that the minimum memory-based sign adjustment can globally balance the network if the rewiring probability in the Newman–Watts small world model exceeds a critical value. When the rewiring probability is large, the resulting network is denser, and as a consequence, it is easier for the influence of each sign adjustment to spread to the whole network.

In the paper "A SOM-Based Membrane Optimization Algorithm for Community Detection" by Liu et al. [18], an evolutionary membrane community detection algorithm based on self-organizing maps (SOMs) was proposed. Initially, the community detection problem was formulated as a discrete optimization problem. Then, three features typical of the membrane algorithm—objects, reaction rules, and membrane structure—were designed to analyze the characteristics of the community structure. Here, an object was defined as a partition. Genetic algorithms and differential evolution were employed as two reaction rules, to let the objects evolve in different regions of the membrane. Finally, to choose the number of membranes by learning, and to mine the structure of the current objects in the decision space, the SOM was employed. To validate the algorithm, simulations were carried out on both synthetic and real-world networks. The experimental results showed that the proposed algorithm is highly accurate, stable and efficient in the execution when compared to other contestant algorithms.

In the paper "Image Entropy for the Identification of Chimera States of Spatiotemporal Divergence in Complex Coupled Maps of Matrices" by Smidtaite et al. [19], the complex networks of coupled maps of matrices (NCMM) are investigated. The authors proved that an NCMM can achieve two different steady states: quiet or divergence. The analysis of the regions around the boundary lines separating these two steady states showed the existence of chimera states of spatiotemporal divergence. This work demonstrated that for identifying such regions, digital image entropy can be exploited as an effective measure in different networks, including regular, feed-forward, random, and small-world NCMM.

In the paper "Evolution Model of Spatial Interaction Network in Online Social Networking Services" by Dong et al. [20], the research focused on modelling the evolution of spatial interactions between users of online social networks diffusing geospatial information at a city level. Through such interactions, a city interaction network was built. The proposed evolution model of the city interaction network takes into account two dynamics: the edge arrival time and the preferential attachment of the edge. More specifically, six preferential attachment models (Random-Random, Random-Degree, Degree-Random, Geographical distance, Degree-Degree, Degree-Degree-Geographical distance) were considered and compared. The authors found that the degree of the node and the geographic distance of the edge highly influenced the evolution of the spatial interaction network. Moreover, the experiments—comparing the optimal model with the real city interaction network, extracted from the information dissemination of WeChat users—revealed a good matching.

We hope that the selected papers described above will be of interest for the community of physicists, computer scientists and others working in the challenging field of complex networks.

**Acknowledgments:** We express our thanks to the authors of the above contributions, and to the journal Entropy and MDPI for their support during this work.

**Conflicts of Interest:** The authors declare no conflict of interest.

## References


1. Madhawa, K.; Murata, T. Active Learning for Node Classification: An Evaluation. *Entropy* **2020**, *22*, 1164. [CrossRef] [PubMed]
2. Bernal Jaquez, R.; Alarcón Ramos, L.A.; Schaum, A. Spreading Control in Two-Layer Multiplex Networks. *Entropy* **2020**, *22*, 1157. [CrossRef] [PubMed]
3. Siew, C.S.Q.; Vitevitch, M.S. Investigating the Influence of Inverse Preferential Attachment on Network Development. *Entropy* **2020**, *22*, 1029. [CrossRef] [PubMed]
4. Ramirez-Arellano, A. Classification of Literary Works: Fractality and Complexity of the Narrative, Essay, and Research Article. *Entropy* **2020**, *22*, 904. [CrossRef] [PubMed]
5. Tsung, C.-K.; Ho, H.-J.; Chen, C.-Y.; Chang, T.-W.; Lee, S.-L. Detecting Overlapping Communities in Modularity Optimization by Reweighting Vertices. *Entropy* **2020**, *22*, 819. [CrossRef] [PubMed]
6. Martino, A.; De Santis, E.; Giuliani, A.; Rizzi, A. Modelling and Recognition of Protein Contact Networks by Multiple Kernel Learning and Dissimilarity Representations. *Entropy* **2020**, *22*, 794. [CrossRef] [PubMed]
7. Wang, Y.; Yu, H.; Wang, G.; Xie, Y. Cross-Domain Recommendation Based on Sentiment Analysis and Latent Feature Mapping. *Entropy* **2020**, *22*, 473. [CrossRef] [PubMed]
8. Pond, T.; Magsarjav, S.; South, T.; Mitchell, L.; Bagrow, J.P. Complex Contagion Features without Social Reinforcement in a Model of Social Information Flow. *Entropy* **2020**, *22*, 265. [CrossRef]
9. Choong, J.J.; Liu, X.; Murata, T. Optimizing Variational Graph Autoencoder for Community Detection with Dual Optimization. *Entropy* **2020**, *22*, 197. [CrossRef] [PubMed]
10. Chimal-Eguía, J.C.; Castillo-Montiel, E.; Paez-Hernández, R.T. Properties of the Vascular Networks in Malignant Tumors. *Entropy* **2020**, *22*, 166. [CrossRef] [PubMed]
11. Sun, J. Complex Network Construction of Univariate Chaotic Time Series Based on Maximum Mean Discrepancy. *Entropy* **2020**, *22*, 142. [CrossRef] [PubMed]
12. Maldonado, A.D.; Morales, M.; Aguilera, P.A.; Salmerón, A. Analyzing Uncertainty in Complex Socio-Ecological Networks. *Entropy* **2020**, *22*, 123. [CrossRef] [PubMed]
13. Ezeh, C.; Tao, R.; Zhe, L.; Yiqun, W.; Ying, Q. Multi-Type Node Detection in Network Communities. *Entropy* **2019**, *21*, 1237. [CrossRef]
14. Liu, W.; Saganowski, S.; Kazienko, P.; Cheong, S.A. Predicting the Evolution of Physics Research from a Complex Network Perspective. *Entropy* **2019**, *21*, 1152. [CrossRef]
15. Ding, L.; Liu, S.-Y.; Yang, Q.; Xu, X.-K. Uncovering the Dependence of Cascading Failures on Network Topology by Constructing Null Models. *Entropy* **2019**, *21*, 1119. [CrossRef]
16. Xiong, S.; Zhu, F.; Yao, Y.; Tang, W.; Xiao, Y. Service-Oriented Model Encapsulation and Selection Method for Complex System Simulation Based on Cloud Architecture. *Entropy* **2019**, *21*, 891. [CrossRef]
17. Qi, M.; Deng, H.; Li, Y. Minimum Memory-Based Sign Adjustment in Signed Social Networks. *Entropy* **2019**, *21*, 728. [CrossRef] [PubMed]
18. Liu, C.; Du, Y.; Lei, J. A SOM-Based Membrane Optimization Algorithm for Community Detection. *Entropy* **2019**, *21*, 533. [CrossRef] [PubMed]
19. Smidtaite, R.; Lu, G.; Ragulskis, M. Image Entropy for the Identification of Chimera States of Spatiotemporal Divergence in Complex Coupled Maps of Matrices. *Entropy* **2019**, *21*, 523. [CrossRef] [PubMed]
20. Dong, J.; Chen, B.; Zhang, P.; Ai, C.; Zhang, F.; Guo, D.; Qiu, X. Evolution Model of Spatial Interaction Network in Online Social Networking Services. *Entropy* **2019**, *21*, 434. [CrossRef] [PubMed]

*Article*

# Active Learning for Node Classification: An Evaluation

**Kaushalya Madhawa * and Tsuyoshi Murata**

Department of Computer Science, Tokyo Institute of Technology, Tokyo 152-8552, Japan;
murata@c.titech.ac.jp
* Correspondence: kaushalya@net.c.titech.ac.jp

Received: 30 August 2020; Accepted: 12 October 2020; Published: 16 October 2020

**Abstract:** Current breakthroughs in the field of machine learning are fueled by the deployment of deep neural network models. Deep neural networks models are notorious for their dependence on large amounts of labeled data for training them. Active learning is being used as a solution to train classification models with less labeled instances by selecting only the most informative instances for labeling. This is especially important when the labeled data are scarce or the labeling process is expensive. In this paper, we study the application of active learning on attributed graphs. In this setting, the data instances are represented as nodes of an attributed graph. Graph neural networks achieve the current state-of-the-art classification performance on attributed graphs. The performance of graph neural networks relies on the careful tuning of their hyperparameters, usually performed using a validation set, an additional set of labeled instances. In label scarce problems, it is realistic to use all labeled instances for training the model. In this setting, we perform a fair comparison of the existing active learning algorithms proposed for graph neural networks as well as other data types such as images and text. With empirical results, we demonstrate that state-of-the-art active learning algorithms designed for other data types do not perform well on graph-structured data. We study the problem within the framework of the exploration-vs.-exploitation trade-off and propose a new count-based exploration term. With empirical evidence on multiple benchmark graphs, we highlight the importance of complementing uncertainty-based active learning models with an exploration term.

**Keywords:** machine learning; graph neural networks; node classification; active learning; graph representation learning

---

## 1. Introduction

Supervised learning is an important technique used to train machine learning models that are deployed in multiple real-world applications [1]. In a supervised classification problem, data instances with ground truth labels are used for training a model that can predict the labels of unseen data instances. Therefore, the performance of a supervised learning model depends on the quality and quantity of training data, often requiring a huge labeling effort. Usually, the labeling of data instances is done by humans. Labeling large amounts of data leads to a huge cost in both time and money. The labeling cost is significantly high when the labeling task needs to be done by domain experts. For example, potential tumors in medical images can be labeled only by qualified doctors [2,3].

With ever-increasing amounts of data, active learning (AL) is gaining the attention of researchers as well as practitioners as a way to reduce the effort spent on labeling data instances. Usually, a fraction of data instances are selected randomly and their labels are queried from an oracle (e.g., human labelers). This set of labeled instances are used for training the classifier. This process is known as *passive learning* [4] as the training data is selected at the beginning of the training process and it is assumed to stay fixed. An alternative approach is to iteratively select a small set of training instances, retrieve their labels, and update the training set. Then, the classification model is retrained using the acquired labeled instances and this process is repeated until a good level of performance (e.g., accuracy) is achieved. This process is known as *active learning* [5]. The objective of AL can be expressed as acquiring instances that maximize model performance. An *acquisition function* evaluates the informativeness of each unlabeled instance and selects the most informative ones. As quantifying the informativeness of an instance is not straightforward, a multitude of approaches have been proposed in AL literature [5]. For example, selecting the instance the model is most uncertain about is a commonly used acquisition function [6].

In this paper, we study the problem of applying AL for classifying nodes of an attributed graph (The term "network" is used as an alternative term in the literature. We use the term graph since the usage of the term network can be confused with neural networks in this paper.). This task is known as node classification. Reducing the number of labeled nodes required in node classification can benefit a variety of practical applications such as in recommender systems [7,8] and text classification [9] by selecting only the most informative nodes for labeling. Parisot et al. [3] demonstrated the importance of representing the associations between brain scan images of different subjects as a graph for the task of predicting if a subject has Alzheimer's disease. Features extracted from images are represented as node attributes. This is an example for a node classification problem where labeling is expensive as labeling a brain scan image is time-consuming and it can only be done by medical experts.

Node classification is an important task in learning from relational data. The objective of this problem is to predict the labels of unlabeled nodes given a partially labeled graph. Different approaches have been used for node classification including iterative classification algorithm (ICA) [10], label propagation [11], and Gaussian random fields (GRF) [12]. Approaching node classification as a semisupervised problem has contributed to state-of-the-art in classification performance [13–15]. In a semisupervised learning problem, the learning algorithm can utilize the features of all data instances including the unlabeled ones. Only the labels of unlabeled instances are not known. Semisupervised learning is a technique that utilizes unlabeled data to improve the label efficiency. Combining AL with semisupervised learning can increase the label efficiency further [16]. Graph neural network (GNN) models have achieved state-of-the-art performance in node classification [17].

Similar to other neural network-based models, GNN models are sensitive to the choice of hyperparameters. The common hyperparameters of a GNN model are learning rate, number of hidden layers, and the size of hidden units of each hidden layer. Unlike model parameters, the hyperparameters are not directly optimized to improve the model performance. Finding the most suitable set of values for hyperparameters is known as hyperparameter tuning. It is usually performed based on the performance of the model on a separate held-out labeled set known as the validation set. It is possible to leave a fraction of labeled data as the validation set when labeled data is abundant. However, in a label scarce setting, it is realistic to use all the available labeled instances for training the model. Therefore, we further reduce the size of the labeled set by not using a validation set and using fixed standard values for hyperparameters.

With the recent popularity of GNNs, several surveys on GNNs have been done [17–19]. These works provide a comprehensive overview of recent developments in graph representation learning and its applications. Surveys on AL research have been done separately [20,21]. However, as far as the authors know, a survey and a systematic comparison of existing AL approaches for the task of node classification have not been done yet. Moreover, only a handful of graph datasets are used for benchmarking such models. Most of the benchmark graphs are similar as they come from the same domain. In this paper, we study commonly used AL acquisition functions on the problem of node classification using a multitude of graph datasets belonging to different domains. As shown in previous work [22], the performance of AL algorithms is not consistent across different datasets.

Our main contributions are

1.  we discuss the importance of performing AL experiments in a more realistic setting where an additional labeled dataset is not used for hyperparameter tuning;
2.  we perform a thorough evaluation of existing AL algorithms on the task of node classification of attributed graphs in a more realistic setting; and
3.  with empirical evidence on an extensive set of graphs with different characteristics, we highlight that graph properties should be considered in selecting an AL approach.

## 2. Background

### 2.1. Node Classification

Node classification plays an important part in learning problems when the data is represented as a graph. A graph $G$ consists of $V$ nodes and $E$ edges connecting pairs of nodes. Edges of a graph can be directional as well. However, we limit our study to undirected graphs. Node classification is widely used in practical applications such as recommender systems [8,23], applied chemistry [24], and social network analysis [25]. In a node classification problem, an attributed graph $G = (V, E)$ with $N$ nodes is given as an adjacency matrix $A \in \mathbb{R}^{N \times N}$ and a node attribute matrix $X \in \mathbb{R}^{N \times F}$. Here, $F$ is the number of attributes. An element $a_{ij} \in A$ represents the edge weight between two nodes $v_i$ and $v_j$. If there is no edge connecting $v_i$ and $v_j$, $a_{ij} = 0$. If the graph is undirected, the adjacency matrix $A$ is symmetric. The degree matrix $D$ is a diagonal matrix defined as $D = \{d_1, \cdots, d_N\}$, where each diagonal element $d_i$ is the row-sum of the adjacency matrix such that $d_i = \sum_{j=1}^{N} a_{ij}$. Each node $v_i$ has a real-valued feature vector $x_i \in \mathbb{R}^{N \times F}$ and $v_i$ belongs to one of the $C$ class labels.

The objective of this problem is to predict the labels of unlabeled nodes $V_{\mathcal{U}}$ given a small set of labels $V_{\mathcal{L}}$. Earlier attempts for solving this problem relied on classifiers based on the assumption that nodes connected by an edge are likely to share the same label [26,27]. A major weakness of such classifiers is that this assumption restricts the modeling capacity and the node attributes are not used in the learning process. The use of node attributes of an attributed graph significantly improves the classification performance.

### 2.2. Graph Neural Networks (GNNs)

A GNN is a neural network architecture specifically designed for learning with attributed graphs. GNN models [14,28,29] achieve state-of-the-art performance on the node classification problem providing a significant improvement over previously used embedding algorithms [30,31]. What sets GNNs apart from previous models is their ability to jointly model both structural information and node attributes. In principle, all GNN models consist of a message passing scheme that propagates feature information of a node to its neighbors. Most GNN architectures use a learnable parameter matrix for projecting features to a different feature space. Usually, two or more of such layers are used along with a nonlinear function (e.g., ReLU). Let $G$ be an undirected attributed graph represented by an adjacency matrix $A$ and a feature matrix $X$. By adding self-loops to the adjacency matrix we get $\tilde{A} = A + I$ and its degree matrix $\tilde{D} = D + I$. Using this notation, the graph convolutional network (GCN) model [14] can be expressed as

$$\tilde{H}^{(k)} = \tilde{D}^{-1/2} \tilde{A} \tilde{D}^{-1/2} H^{(k-1)}, \tag{1}$$

where $\tilde{D}^{-1/2} \tilde{A} \tilde{D}^{-1/2}$ is the normalized adjacency matrix. Then, the hidden representation of a layer $H^{(k)}$ is obtained by multiplying the feature matrix $\tilde{H}^{(k)}$ with a parameter matrix $\theta$ and applying an activation function $\sigma$ as

$$H^{(k)} = \sigma(\tilde{H}^{(k)} \theta^{(k)}). \tag{2}$$

With normalized adjacency matrix $\hat{A} = \tilde{D}^{-1/2}\tilde{A}\tilde{D}^{-1/2}$ a two-layer GCN model [14] can be expressed as

$$Y_{\text{GCN}} = \text{softmax}\left(\hat{A}\,\text{ReLU}\left(\hat{A}X\theta^{(0)}\right)\theta^{(1)}\right), \tag{3}$$

where $X$ is the node attribute matrix and $\theta^{(0)}$ and $\theta^{(1)}$ are the parameter matrices of two neural layers. The softmax function defined as $\text{softmax}(x) = exp(x)/\sum_{c=1}^{C} exp(x_c)$ normalizes the output of the classifier across all classes. Rectified linear unit (ReLU) is a commonly used activation function where $\text{ReLU}(x) = max(0, x)$.

Wu et al. [29] showed that a simplified GNN model named SGC can achieve competitive performance on most attributed graphs at a significantly lower computational cost. They obtained this model by removing hidden layers and nonlinear activation functions in the GCN model. This model can be written as

$$Y_{\text{SGC}} = \text{softmax}\left(\hat{A}^k X\theta\right), \tag{4}$$

where $\hat{A}^k$ is the $k$th power of the adjacency matrix $A$. The parameter $k$ determines the number of hops the feature vectors are propagated to. This approach is similar to propagating node attributes over the $k$-hop neighborhood and then performing logistic regression. Using a 2-hop neighborhood ($k = 2$) often results in good performance.

*2.3. Active Learning*

In this paper, we consider the pool-based AL setting [5]. In a pool-based AL problem, the labeled dataset $\mathcal{L}$ is much smaller compared to a large pool of unlabeled items $\mathcal{U}$. We can acquire the label of any unlabeled item by querying an oracle (e.g., a human annotator) at a uniform cost per item. Suppose we are given a query budget $K$, such that we are allowed to query labels of a maximum of $K$ unlabeled items. We use the notation $f_\theta$ to denote a classification model with trainable parameters $\theta$. The probability of an instance $q$ belonging to class $c$ predicted by this model is written as $P_\theta(\hat{y}_q = c|x, \mathcal{D}_\mathcal{L})$. We calculate this likelihood as

$$P_\theta(\hat{y}_q = c|x, \mathcal{D}_\mathcal{L}) = \text{softmax}\left(f_\theta(x_q)\right)_{[q=c]}. \tag{5}$$

AL research has contributed to a multitude of approaches for training supervised learning models with less labeled data. We recommend the work in [5] as a detailed review of existing AL research. The objective of AL approaches is to select the most informative instance for labeling. This task is performed with the use of an acquisition function, where the acquisition function decides which unlabeled example should be labeled next. Existing acquisition functions can be grouped into a few general frameworks based on how they are formulated. In this section, we describe a few commonly used AL frameworks.

2.3.1. Uncertainty Sampling

Uncertainty sampling [32] is one of the most widely used AL approaches. The active learner selects the instance for which the classifier predicts a label with the least certainty. The information entropy of the label predictions is usually used to quantify the uncertainty of the model for a given instance $x_q$ such that

$$\mathbb{H}(y_q|x, \mathcal{D}_\mathcal{L}) = -\sum_{c=1}^{C} P_\theta(\hat{y}_q = c|x_q, \mathcal{D}_\mathcal{L})log\left(P_\theta(\hat{y}_q = c|x_q, \mathcal{D}_\mathcal{L})\right). \tag{6}$$

The instance corresponding to the maximum entropy is selected for querying

$$q^* = \arg\max_q \mathbb{H}(y_q|x_q, \mathcal{D}_\mathcal{L}). \tag{7}$$

The entropy computed over model predictions of a neural network does not correctly represent the model uncertainty for unseen instances. Even though Bayesian models are good at estimating the model uncertainty, Bayesian inference can be prohibitively time-consuming. Gal and Ghahramani [33] demonstrated that using dropout [34] at evaluation time is an approximation to a Bayesian neural network and this can be used to calculate the model uncertainty. Gal et al. [35] used this Bayesian approach to perform uncertainty sampling for active learning on image data with convolutional neural networks (CNN). Additionally, Gal et al. [35] performed a comparison of various acquisition functions proposed for quantifying the model uncertainty of CNN models. It is shown that uncertainty sampling is prone to select outliers [20].

Bayesian Active Learning by Disagreement (BALD) [6] is another uncertainty-based acquisition function used with Bayesian models. BALD algorithm selects the instance that maximizes the mutual information between the predictions and the model posterior. This can be written as

$$q^* = \arg\max_q \mathbb{H}(y_q|x_q, \mathcal{D}_\mathcal{L}) - \mathbb{E}_{\theta \sim p(\theta|\mathcal{D}_\mathcal{L})} \left[ \mathbb{H}(y_q|x_q, \theta, \mathcal{D}_\mathcal{L}) \right] . \tag{8}$$

The left side term of the Equation (8) is the entropy of the model prediction and the right side term is the expectation of the model prediction over the posterior of the model parameters. If the model is certain of its predictions for each draw of parameter values, the right side term becomes smaller. In this case the active learner selects the examples $x_q$ for which the model is most uncertain of its predictions (high $\mathbb{H}(y_q|x_q, \mathcal{D}_\mathcal{L})$), but the model is confident for individual parameter settings (low $\mathbb{E}_{\theta \sim p(\theta|\mathcal{D}_\mathcal{L})} \left[ \mathbb{H}(y_q|x_q, \theta, \mathcal{D}_\mathcal{L}) \right]$) .

### 2.3.2. Query by Committee (QBC)

Query by committee (QBC) [36] is a simple method that outperforms uncertainty sampling in many practical settings. This method maintains a committee of models trained on the same labeled dataset. Each model in the committee predicts the label of an unlabeled instance. The instance for which label predictions of the most number of committee members (models) disagrees is selected as the most informative instance. However, QBC is not a popular choice when AL is used with deep neural network (DNN) models since training a committee of DNN models is time-consuming.

### 2.3.3. Expected Error Reduction (EER)

Expected Error Reduction (EER) [37] is an AL approach that directly calculates the expected generalization error of a model trained on labeled instances including unlabeled instances $\mathcal{L} \cup (x_q, y_q)$. Then, the active learner queries the instance which minimizes the future generalization error. However, this approach involves the retraining of a model for each unlabeled instance $x_q$ with each label $c \in C$, making it one of the most time-consuming AL approaches. Therefore, the EER approach has been limited to simple classification algorithms such as Gaussian random fields (GRF) for which faster online retraining is possible.

### 3. Active Learning for Graph Classification Problems

Compared to application of AL on other types of data such as image and text data, only a limited number of AL models has been developed for graph data. Previous work on applying AL on graph data [38–40] is tightly coupled with earlier classification models such as Gaussian random fields, in which the features of nodes are not being used. Therefore, selecting query nodes uniformly in random coupled with a recent GNN model can easily outperform such AL models. AL models which utilize recent GNN architectures [41,42] are limited. Moreover, a comprehensive comparison of AL algorithms proposed for other domains of data has not been done yet.

In Table 1, we provide an extensive comparison of the literature on AL approaches proposed for node classification. We compare each work with respect to the following attributes.

- AL approach
- Classifier: Classification model used for predicting the label of a node
- Attributes: Whether the node classifier uses node attributes
- Adaptive: Whether the active learner is updated based on the newly labeled instances
- Labels: Whether the active learner uses node labels in making a decision

In addition to generic approaches proposed for AL, there have been a few works that are specifically designed for graph-structured data. These algorithms use graph-specific metrics for selecting nodes for labeling. In addition to the attributes of data instances, graph topology provides useful information. For example, the degree centrality of a node represents how a particular data instance is connected with others. Table 1 demonstrates that most of the previous AL approaches proposed for node classification do not use the node attribute information. Moreover, some works [40,43] ignore the label information as well.

**Table 1.** Summary of existing work for active node classification on attributed graphs. The work by Gadde et al. [43] does not use the labels of the nodes. Therefore, this method does not use a classifier. We use the following abbreviations in the table. LR—Logistic Regression, GRF—Gaussian Random Fields, LP—Label Propagation, SC—Spectral Clustering, NA—Not Applicable.

| Work | AL Approach | Classifier | Attributes | Adaptive | Labels | Year |
|---|---|---|---|---|---|---|
| Zhu et al. [26] | EER | GRF | No | No | Yes | 2003 |
| Macskassy [44] | EER + Heuristics | GRF | No | Yes | Yes | 2009 |
| Bilgic et al. [39] | QBC | LR | No | Yes | Yes | 2010 |
| Gu and Han [38] | EER | LP | No | No | Yes | 2012 |
| Ji and Han [40] | Variation Minimization | GRF | No | No | No | 2012 |
| Ma et al. [45] | Uncertainty | GRF | No | No | Yes | 2013 |
| Gadde et al. [43] | SC | NA | No | No | No | 2015 |
| Cai et al. [41] | Uncertainty + Heuristics | GCN | Yes | Yes | Yes | 2017 |

*3.1. Active Learning Framework*

In this problem, we start with an extremely small set of labeled instances. We are given a query budget $K$ such that we are allowed to query $K$ number of nodes to retrieve their labels. In each acquisition step, we select a node and retrieve its label from an oracle (e.g., a human labeler). The GNN model is retrained using the training set including the newly labeled instance. We repeat this process $K$ times. The basic framework is shown in Algorithm 1. Here, $f_\theta$ is any node classification algorithm with parameters $\theta$ and we can use different acquisition functions (e.g., uncertainty sampling or QBC) as $g$.

---

**Algorithm 1** Active learning for node classification.

---

**Input:** Graph $G = (A, X)$, Query budget $K$, Initial labels $Y_{\mathcal{L}}$
**Output:** An improved model $f_\theta$
**for** $i \leftarrow 1$ to $n_q = K$ **do**

    Select the best unlabeled instance $q^*$ with an acquisition function $g$
    Retrieve its label $Y_{q^*}$
    Update label set $Y_{\mathcal{L}} \leftarrow Y_{\mathcal{L}} \cup Y_{q^*}$
    Retrain the model $\theta \leftarrow \arg\min_\theta l(f_\theta(G), Y_{\mathcal{L}})$
**end for**
**Return** $\theta$

---

*3.2. The Importance of Exploration*

After each acquisition step, the classifier is trained on a limited number of labeled instances, which in turn are selected by the active learner. Therefore, the selected labeled instances tend to bias towards instances evaluated to be "informative" by the active learner. Therefore, the distribution of labeled instances is often different from the true underlying distribution. The active learner cannot observe the consequences of selecting an instance which has lower "informativeness". This leads the active learner to converge to policies that are not able to generalize for unlabeled data. This problem is amplified by the lack of hyperparameter tuning. A simple approach to overcome this limitation is to query a few instances in addition to the ones maximizing our selection criteria. This step is known as "exploration" while selecting the instance maximizing the criteria is "exploitation". For example, if our criterion is model entropy, the exploration step involves acquiring labels of a few instances which do not have the maximum entropy. Intuitively, an active learner should perform more exploration initially, so it can have a better view of the true distribution of data.

This problem is known as the *exploration* vs. *exploitation trade-off* in sequential decision-making problems. Solving this trade-off requires the learner to acquire potentially suboptimal instances (i.e., exploration) in addition to the optimal ones. This problem is studied under the framework of multi-armed bandits (MAB) problems [46]. In a MAB problem, a set of actions are given and selecting an action results in observing a reward drawn from a distribution that is unknown to the learner. The problem is to select a sequence of actions that maximize the cumulative reward. A multitude of approaches is used in solving online learning problems modeled as MAB problems. $\epsilon$-greedy, upper confidence bounds (UCB) [47], and Thompson sampling [48] are a few of the frequently used techniques.

We compare the performance of each active learner using two different exploration techniques: $\epsilon$-greedy and count-based exploration.

### 3.2.1. $\epsilon$-Greedy

$\epsilon$-greedy is used as the simplest method of introducing exploration into an MAB algorithm. In the case of AL, with probability $\epsilon$ the active learner randomly selects an unlabeled instance for querying its label. The most informative instance is selected by an acquisition function with probability $(1 - \epsilon)$. A key problem with this approach is that, as each unlabeled instance is selected with uniform probability, some of the labeled instances can be wasteful. This phenomena is known as *undirected exploration* [49].

### 3.2.2. Count-Based Exploration

In MAB problems, count-based exploration addresses the problem of undirected exploration by assigning a larger probability to actions that have been selected fewer times compared to the remaining actions. Based on the principle of optimism in the face of uncertainty, a count-based exploration algorithm computes an upper confidence bound (UCB) [47] and selects the action corresponding to the maximum UCB. We adopt the notion of count-based exploration as the number of labeled nodes in the neighborhood of an unlabeled node. We define the exploration term of an instance $i$ as the logarithm of the number of unlabeled neighboring nodes of $i$. This term encourages the learner to sample nodes from neighborhoods with less number of labeled nodes. As this term and the informative metric used in the acquisition function (e.g., entropy) are on different scales, we normalize both of these quantities into $[0, 1]$ range and get $\phi_{\exp}(i)$ and $\phi_{\inf}(i)$, respectively. We linearly combine these normalized quantities to get the criterion for acquiring nodes as

$$\phi(i) = (1 - \gamma_t) \cdot \phi_{\inf}(i) + \gamma_t \cdot \phi_{\exp}(i), \tag{9}$$

where the exploration coefficient $\gamma_t$ is a hyperparameter that balances exploration and exploitation. Setting $\gamma_t$ to 0 corresponds to pure exploration disregarding the feedback of the classifier. On the other hand, $\gamma_t = 1$ is equivalent to pure exploitation selecting a node based only on the uncertainty sampling (e.g., entropy).

## 4. Experiments

### 4.1. Data

We evaluate the performance of all algorithms on 11 real-world datasets belonging to different domains. as shown in Table 2. In Table 2, we list the datasets used in experiments with several graph properties. These datasets belong to different domains such as citation networks, product networks, co-author networks, biological networks, and social networks.

**Table 2.** Dataset statistics. Labeling rate as a percentage of total nodes is shown within brackets. Avg. deg.: Average degree, Avg. CC: Average clustering coefficient, $r_D$: Degree assortativity, $r_L$: Label assortativity.

| Dataset | Nodes | Classes | Avg. Deg. | Avg. CC | $r_D$ | $r_L$ | Features | Labels (%) |
|---|---|---|---|---|---|---|---|---|
| CiteSeer | 2110 | 6 | 2.84 | 0.17 | 0.007 | 0.67 | 3703 | 12 (0.56) |
| PubMed | 19,717 | 3 | 6.34 | 0.06 | −0.044 | 0.69 | 500 | 6 (0.03) |
| n CORA | 2485 | 7 | 4.00 | 0.24 | −0.071 | 0.76 | 1433 | 14 (0.56) |
| Amazon Comp. | 13,752 | 10 | 36.74 | 0.35 | −0.057 | 0.68 | 767 | 20 (0.14) |
| Co-author Phy | 34,493 | 5 | 14.38 | 0.38 | 0.201 | 0.87 | 8415 | 10 (0.03) |
| Co-author CS | 18,333 | 15 | 8.93 | 0.34 | 0.113 | 0.79 | 6805 | 30 (0.16) |
| Disease | 1044 | 2 | 2.00 | 0.0 | −0.544 | 0.68 | 1000 | 4 (0.38) |
| Wiki-CS | 11,701 | 10 | 36.94 | 0.47 | −0.065 | 0.58 | 300 | 20 (0.17) |
| PPI-Brain | 3480 | 121 | 31.94 | 0.17 | −0.064 | 0.09 | 50 | 35 (1.0) |
| PPI-Blood | 3312 | 121 | 32.91 | 0.18 | −0.061 | 0.09 | 50 | 33 (1.0) |
| PPI-Kidney | 3284 | 121 | 31.70 | 0.18 | −0.067 | 0.09 | 50 | 33 (1.0) |
| Github | 37,700 | 2 | 15.33 | 0.17 | −0.075 | 0.38 | 4005 | 4 (0.01) |

CiteSeer, PubMed, and CORA [50] are commonly used citation graphs. Each of these undirected graphs is made of documents as nodes and citations as edges between them. If one document cites another, they are linked by an edge. The bag-of-words features of the text content of a document correspond to the attributes of a node.

Co-author CS and Co-author Physics are co-authorship graphs constructed from Microsoft Academic Graph [51]. Authors are represented as nodes and two authors are linked by an edge if they have co-authored a paper. Node features correspond to the keywords of the papers authored by a particular author. An author's most active field of study is used as the node label.

Amazon Computers is a subgraph of the Amazon co-purchase graph [52]. Products are represented as nodes, and two nodes are connected by an edge of those two products that are frequently bought together. Node attributes correspond to product reviews encoded as bag-of-words features. The product category is used as the node label.

The disease dataset [53] simulates the SIR disease propagation model [54] on a graph. The label of a node indicates whether a node is infected or not and the features indicate the susceptibility to the disease.

The Wiki-CS dataset [55] is a graph constructed from Wikipedia articles corresponding to computer science. A Wikipedia article is a node of this graph and two nodes are connected by an edge if one article has a hyperlink to the other. GloVe word embeddings [56] obtained from the text content of an article is used as the feature vector of the node corresponding to that article.

Each protein–protein interaction (PPI) graph represents physical contacts between proteins in a human tissue (brain, blood, and kidney) [57,58]. Unlike other datasets, in PPI graphs a protein (node) can have multiple functions as its label, making this a multi-label classification problem. Learning the protein function (cellular function from gene ontology) involves learning about node roles. Several properties of a protein such as positional gene sets, motif gene sets and immunological signatures are used as node attributes in a PPI graph.

Github is a social network dataset constructed from developer profiles on Github who have at least 10 public repositories [59]. Details of a developer such as location, employee, and starred repositories are represented as node attributes. Two nodes are linked by an edge if those two developers mutually follow each other on Github. The label of a node indicates whether a developer is primarily working on machine learning or web development projects.

From each dataset, we randomly select two nodes belonging to each label as the initial labeled set $V_{\mathcal{L}}$. We use 5% of the rest of the unlabeled nodes as the test set. The set of remaining unlabeled nodes $V_{\mathcal{U}}$ qualify to be queried. The size of the initial labeled set and its size as a fraction of the total nodes (labeling rate) are shown in Table 2.

Graph Properties

In some real-world graphs, such as social and communication networks, nodes tend to cluster together creating tightly knit groups of nodes. This phenomenon is known as clustering and the clustering coefficient [60] quantifies the amount of clustering present in a graph. The local clustering coefficient of a node $i$ is calculated as

$$C_i = \frac{\text{number of triangles connected to node } i}{\text{number of triples centered around node } i}. \tag{10}$$

Average clustering coefficient is calculated as the average of local clustering coefficients of all nodes of a graph.

In real-world graphs, nodes tend to connect with other nodes with similar properties. In network science literature this phenomenon is known as "assortative mixing" [61]. Assortativity coefficient quantifies the amount of assortative mixing present in a graph. Assortativity coefficient can be calculated with respect to any node attribute. We calculate the label assortativity ($r_L$) with

$$r_L = \frac{\sum_i e_{ii} - \sum_i a_i b_i}{1 - \sum_i a_i b_i}, \tag{11}$$

where $e_{ij}$ denotes the fraction of edges connecting a node with label $i$ with a node with label $j$. For multi-label graphs, we calculate label assortativity for each label separately and take the average. A higher label associativity indicates that a node tends to connect with another node with the same label. As shown in Table 2, citation and co-author graphs exhibit high assortativity. It is easier to predict labels in a graph exhibiting high assortativity since neighbors of a node tend to have the same label as the node. Many node classification models are based on this assumption. However, the PPI graphs show low assortativity indicating that nodes with the same label are not necessarily in the same neighborhood. This is due to the fact that the function of a protein (i.e., node) depends on the role of a node in that graph rather than its neighboring proteins (i.e., nodes). Using degree centrality as a node attribute degree assortativity $r_D$ of each node can be computed in a similar manner. Average degree assortativity of a graph indicates whether a high degree node prefers to connect with other high degree nodes.

*4.2. Experimental Setup*

4.2.1. Node Classification Model

Recent studies demonstrated that GNN-based classifiers significantly outperform previous classifier algorithms such as GRFs. Therefore, we restrict our study of AL to GNN-based learning models. In our experiments, we consider two types of graph neural network architectures: GCN [14] and SGC [29]. SGC is a simplified GNN architecture that does not include a hidden layer and nonlinear activation functions. As the goal of AL is to reduce the number of labeled instances used for training, we do not use a separate validation set for fine-tuning the hyperparameters of a GNN model. In addition, it is shown that tuning hyperparameters while training a model with AL can lead to label inefficiency [62].

For all datasets, we use the default hyperparameters used in GNN literature (e.g., learning rate = 0.01). We use the following algorithms in our experiments.

- Random: Select an unlabeled node randomly,
- PageRank: Select the unlabeled node with the largest PageRank centrality,
- Degree: Select the unlabeled node with the largest degree centrality,
- Clustering coefficient: Select the unlabeled node with the largest clustering coefficient,
- Entropy: Calculate the entropy of predictions of the current model over unlabeled nodes and select the node corresponding to the largest entropy.,
- BALD [6,35]: Select the node which has the the largest mutual information value between predictions and model posterior, and
- AGE [41]: Select the node which maximizes a linear combination of three metrics: PageRank centrality, model entropy and information density.

Here, PageRank, degree, and clustering coefficient-based sampling do not use node attributes or the feedback from the classification model. On the other hand, entropy BALD are uncertainty-based acquisition functions that calculate an uncertainty metric using the performance of the classifier trained using the current training set. We acquire the label of an unlabeled node and retrain the GNN model by performing 50 steps of adam optimizer [63]. We perform 40 acquisition steps (query budget = 40) and repeat this process on 30 different randomly initialized training and test splits for each dataset. Test dataset is often unbalanced. Therefore, accuracy is not suitable to be used as the performance metric. We report the average F1 score (macro-averaged) over the test set in each experiment. F1-score is the harmonic mean of the precision and recall metrics. Macro-F1 score is calculated by first calculating F1-scores for each class separately and then taking the average of class-wise F1-scores.

### 4.2.2. Packages and Hardware

We use the NetworkX library [64] for representing and processing graphs. We use the Pytorch [65] implementations of GCN [14] and SGC [29] node classification models. All experiments are run on a computer running Ubuntu 18.04 OS on an Intel(R) Core i9-7900X CPU @ 3.30GHz processor with 64GB memory and a NVIDIA GTX 1080-Ti GPU.

## 5. Results and Discussion

### 5.1. Performance Comparison of AL Approaches

In this section, we compare the performance of acquisition functions which use only a single type of approach. Figures 1 and 2 show how the performance of the node classification model varies with the number of acquisitions.

As shown in previous works, AGE [41], the current state-of-the-art AL algorithm, performs well on citation networks (CiteSeer, CORA, and PubMed). However, the performance of this algorithm is suboptimal on other datasets such as Wiki-CS. The citation datasets possess similar characteristics. For example, average degree centrality of them is in the same range as shown in Table 2. Therefore, selecting AL algorithms based on their performance on a handful of graphs from the same domain may result in suboptimal algorithms.

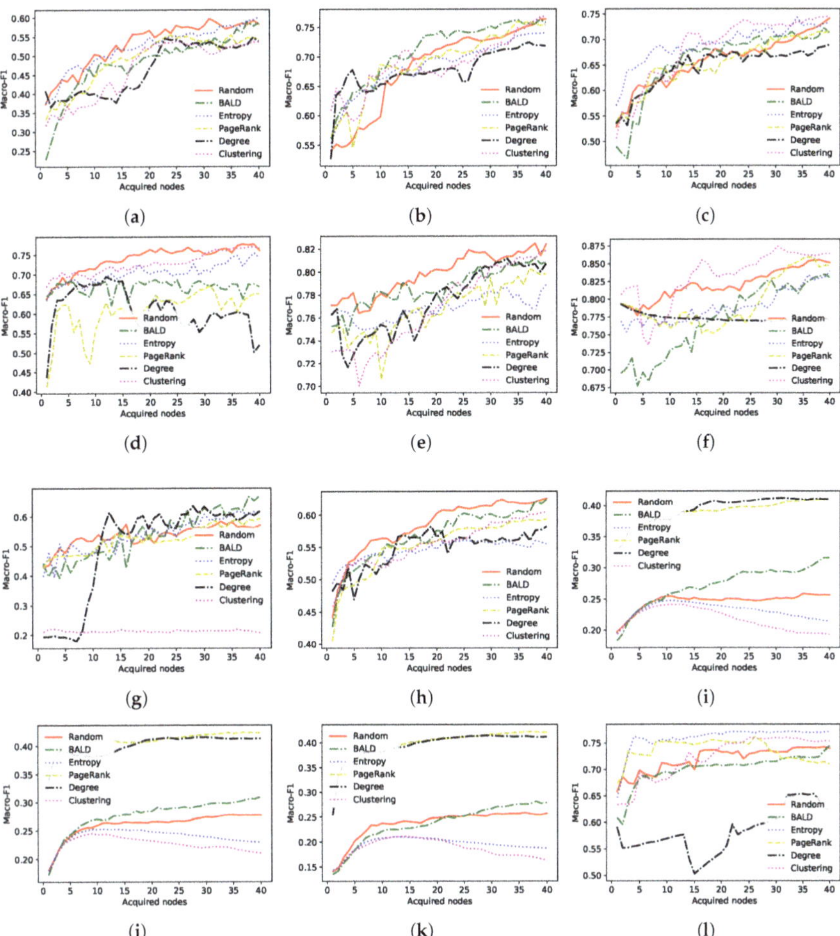

**Figure 1.** Macro-F1 score (test) of active learning algorithms with number of acquisitions. A two-layer graph convolutional network (GCN) is used as the graph neural network (GNN) model. (**a**) CiteSeer. (**b**) PubMed. (**c**) CORA. (**d**) Amazon Computers. (**e**) Co-author CS. (**f**) Co-author Physics. (**g**) Disease. (**h**) Wiki-CS. (**i**) PPI-Brain. (**j**) PPI-Blood. (**k**) PPI-Kidney. (**l**) Github.

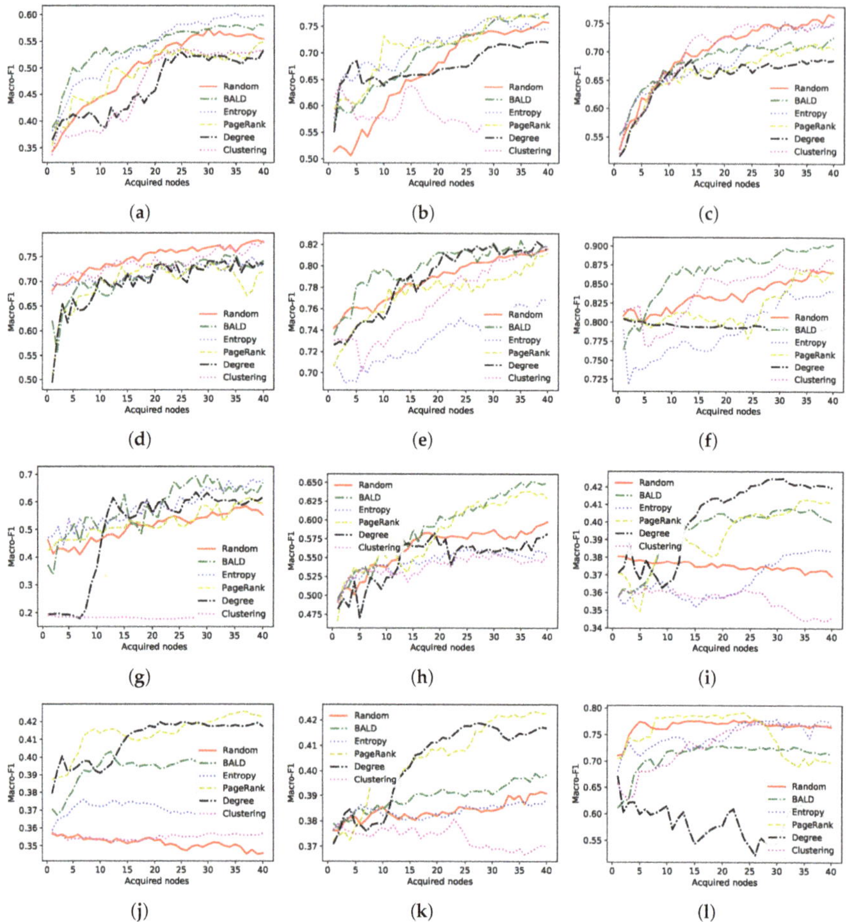

**Figure 2.** Macro-F1 score (test) of active learning algorithms with number of acquisitions. SGC model is used as the GNN model. (**a**) CiteSeer. (**b**) PubMed. (**c**) CORA. (**d**) Amazon Computers. (**e**) Co-author CS. (**f**) Co-author Physics. (**g**) Disease. (**h**) Wiki-CS. (**i**) PPI-Brain. (**j**) PPI-Blood. (**k**) PPI-Kidney. (**l**) Github.

### 5.2. Comparison of Exploration Strategies

In this experiment, we run uncertainty sampling algorithms: BALD and entropy with $\epsilon$-greedy and count-based exploration terms. In the count-based exploration policy, we set the exploration coefficient $\beta$ to 0.5. In Tables 3 and 4, we present the performance of GCN and SGC classifiers when 40 nodes are acquired using each of the acquisition functions. Entropy-Count and BALD-Count correspond to max entropy sampling and BALD policy combined with count-based exploration term. The values in bold indicate that the performance of an algorithm is significantly better (at 5% significance level) than the rest of the algorithms on that dataset. We calculate the statistical significance between the performance of two algorithms using paired t-test. If no single algorithm is significantly better than the rest, all statistically significant values are marked in bold. We summarize the results in Table 5 and show the best performing AL algorithm along with the classifier. Uncertainty-based acquisition functions, when combined with the count-based exploration term (Entropy-Count and BALD-Count), achieve the best performance on average on four datasets. It highlights that encouraging the active learner to select nodes in less explored neighborhoods is effective than selecting a node in random as the exploration step ($\epsilon$-greedy).

**Table 3.** Average F1-score of different acquisition functions. Forty query instances are selected (average of 30 runs). Standard deviation is shown underneath the macro-averaged F1-score. Classifier: GCN. Rand—Random, Ent—Entropy, PR—PageRank, Deg—Degree, CC: Clustering coefficient.

| Dataset | Rand | Ent | BALD | PR | Deg | CC | Ent $\epsilon$-Greedy | BALD $\epsilon$-Greedy | Ent Count | BALD Count | AGE |
|---|---|---|---|---|---|---|---|---|---|---|---|
| CiteSeer | 58.4 ± 6.9 | 60.1 ± 8.1 | 58.6 ± 5.1 | 54.4 ± 3.4 | 53.8 ± 4.3 | 53.6 ± 7.0 | 59.4 ± 4.2 | 52.0 ± 5.7 | 60.3 ± 4.2 | 59.1 ± 4.7 | **61.5** ± 3.7 |
| CORA | 74.1 ± 5.1 | 73.9 ± 6.6 | 71.4 ± 7.4 | 71.8 ± 5.1 | 70.2 ± 4.9 | **74.5** ± 5.5 | **75.1** ± 4.1 | 70.4 ± 6.4 | 73.3 ± 5.4 | 72.9 ± 3.6 | **74.5** ± 7.7 |
| PubMed | 76.4 ± 4.0 | 74.1 ± 3.0 | 75.7 ± 3.8 | 75.3 ± 3.3 | 71.8 ± 3.7 | 76.8 ± 1.4 | 74.2 ± 4.2 | 74.1 ± 4.1 | **77.3** ± 1.7 | 75.7 ± 4.4 | 74.5 ± 2.2 |
| Coauthor CS | 82.4 ± 2.3 | 78.9 ± 3.4 | 80.6 ± 3.6 | 79.7 ± 3.6 | 80.7 ± 2.5 | 81.9 ± 3.7 | 78.2 ± 4.4 | 81.2 ± 2.3 | 80.3 ± 4.9 | 82.1 ± 2.5 | **83.9** ± 2.2 |
| Coauthor Phy | 85.2 ± 3.3 | 84.1 ± 3.2 | 83.8 ± 2.8 | 85.2 ± 1.8 | 77.1 ± 2.1 | 86.4 ± 2.9 | 85.5 ± 2.7 | 83.4 ± 3.5 | 86.9 ± 2.9 | 83.7 ± 2.8 | **87.9** ± 2.6 |
| Amazon Comp. | **76.1** ± 5.4 | 74.2 ± 3.4 | 66.8 ± 7.6 | 65.2 ± 8.1 | 60.2 ± 15.6 | 76.7 ± 4.1 | 73.1 ± 6.0 | 70.8 ± 8.1 | 75.4 ± 3.8 | 73.3 ± 7.5 | 74.2 ± 5.9 |
| Disease | 57.1 ± 7.1 | **67.1** ± 8.7 | **67.2** ± 8.7 | 59.4 ± 8.8 | 53.2 ± 9.1 | 20.8 ± 5.1 | 61.0 ± 10.7 | **66.5** ± 9.4 | 65.8 ± 9.2 | **67.2** ± 7.2 | 63.3 ± 8.0 |
| Wiki-CS | 57.1 ± 7.1 | 55.0 ± 5.1 | **62.4** ± 2.5 | 59.4 ± 3.1 | 58.2 ± 2.2 | 60.5 ± 3.7 | 61.0 ± 10.7 | **63.3** ± 2.9 | 57.0 ± 3.3 | **62.1** ± 3.7 | 57.7 ± 4.9 |
| PPI Brain | 25.6 ± 6.5 | 21.4 ± 6.3 | 31.6 ± 6.1 | **41.1** ± 2.1 | **41.0** ± 2.4 | 19.3 ± 6.6 | 22.3 ± 6.0 | 30.0 ± 8.9 | 22.2 ± 5.0 | 35.3 ± 6.1 | 22.3 ± 6.2 |
| PPI Blood | 27.7 ± 3.2 | 22.9 ± 5.7 | 31.0 ± 6.5 | **42.4** ± 1.6 | 41.4 ± 1.9 | 21.0 ± 5.8 | 26.5 ± 4.9 | 36.9 ± 4.6 | 23.6 ± 5.4 | 37.4 ± 4.3 | 23.3 ± 5.6 |
| PPI Kidney | 25.7 ± 2.9 | 18.7 ± 6.8 | 27.9 ± 9.6 | **42.1** ± 1.6 | 41.1 ± 2.2 | 16.3 ± 5.9 | 18.8 ± 7.0 | 33.5 ± 3.3 | 29.2 ± 1.7 | 37.6 ± 3.4 | 19.4 ± 4.8 |
| Github | 74.0 ± 8.0 | **77.1** ± 1.9 | 74.5 ± 2.4 | 71.1 ± 2.9 | 62.3 ± 4.8 | 75.4 ± 1.8 | **77.3** ± 1.6 | 74.4 ± 2.2 | 76.4 ± 2.2 | 73.8 ± 2.3 | 73.9 ± 2.1 |

**Table 4.** Average F1-score of different acquisition functions. Forty query instances are selected (average of 30 runs). Standard deviation is shown underneath the macro-averaged F1-score. Classifier: SGC. Rand - Random, Ent—Entropy, PR—PageRank, Deg—Degree, CC: Clustering coefficient.

| Dataset | Rand | Ent | BALD | PR | Deg | CC | Ent $\epsilon$-Greedy | BALD $\epsilon$-Greedy | Ent Count | BALD Count | AGE |
|---|---|---|---|---|---|---|---|---|---|---|---|
| CiteSeer | 55.5 ± 4.6 | **59.9** ± 4.7 | 58.0 ± 4.0 | 55.0 ± 3.4 | 53.4 ± 5.3 | 53.4 ± 7.4 | 56.3 ± 5.6 | 56.0 ± 4.6 | **60.0** ± 6.3 | 56.6 ± 4.8 | **60.4** ± 5.6 |
| CORA | **76.1** ± 3.7 | 75.4 ± 4.0 | 71.4 ± 2.3 | 71.4 ± 5.1 | 69.3 ± 3.4 | 74.9 ± 6.1 | 73.9 ± 6.1 | 73.8 ± 4.4 | **76.7** ± 6.1 | 74.2 ± 3.1 | 74.7 ± 6.4 |
| PubMed | 75.8 ± 3.6 | 74.8 ± 2.3 | 77.5 ± 2.6 | 76.7 ± 2.5 | 72.3 ± 6.4 | 60.7 ± 7.9 | 75.3 ± 3.7 | 77.2 ± 2.4 | 76.6 ± 2.8 | **78.0** ± 1.7 | **77.7** ± 3.4 |
| Coauthor CS | 81.7 ± 2.9 | 76.8 ± 3.4 | 81.9 ± 3.9 | 81.3 ± 4.1 | 81.4 ± 4.0 | 81.9 ± 3.7 | 76.9 ± 4.1 | 82.6 ± 3.7 | 77.2 ± 4.7 | **82.7** ± 4.8 | **83.2** ± 2.9 |
| Coauthor Phy | 86.5 ± 3.3 | 84.1 ± 2.4 | **90.2** ± 0.9 | 86.7 ± 2.9 | 79.3 ± 3.7 | 88.1 ± 2.7 | 84.1 ± 3.1 | 89.6 ± 2.6 | 87.5 ± 3.6 | **90.4** ± 1.4 | 88.9 ± 2.1 |
| Amazon Comp. | 77.3 ± 4.1 | 73.4 ± 4.2 | 74.2 ± 5.3 | 71.9 ± 3.5 | 73.5 ± 6.1 | **78.3** ± 3.6 | 75.8 ± 5.4 | 74.5 ± 6.7 | 74.3 ± 3.2 | 74.9 ± 5.3 | 75.6 ± 3.8 |
| Disease | 55.4 ± 8.7 | **68.2** ± 6.1 | **67.2** ± 7.1 | 59.7 9.5 | 58.5 ± 8.9 | 17.8 ± 4.5 | 63.4 ± 7.5 | 67.4 ± 8.5 | 67.1 ± 9.7 | 66.2 ± 8.4 | 66.4 ± 11.1 |
| Wiki-CS | 59.8 ± 6.3 | 55.5 ± 3.6 | 64.7 ± 4.0 | 62.9 ± 3.6 | 61.3 ± 3.1 | 55.4 ± 6.6 | 57.5 ± 5.3 | 63.8 ± 2.4 | 56.5 ± 5.8 | **65.6** ± 3.1 | 50.4 ± 5.7 |
| PPI Brain | 36.9 ± 2.2 | 38.4 ± 2.4 | 40.0 ± 1.4 | 41.0 ± 1.4 | **41.8** ± 1.2 | 34.6 ± 3.6 | 38.2 ± 2.0 | 40.3 ± 1.4 | 40.6 ± 1.0 | **41.6** ± 1.2 | 33.2 ± 2.7 |
| PPI Blood | 34.6 ± 2.2 | 37.2 ± 3.7 | 39.5 ± 2.6 | **42.3** ± 2.3 | 41.7 ± 2.1 | 35.7 ± 1.7 | 37.0 ± 3.6 | 39.0 ± 2.9 | 39.8 ± 2.0 | 40.8 ± 2.1 | 39.4 ± 2.2 |
| PPI Kidney | 39.1 ± 1.8 | 38.8 ± 2.6 | 39.9 ± 1.4 | **42.3** ± 1.8 | 41.7 ± 2.0 | 37.0 ± 1.4 | 40.0 ± 1.7 | 39.9 ± 1.4 | 40.4 ± 2.1 | 41.0 ± 1.8 | 41.0 ± 1.7 |
| Github | 76.4 ± 2.5 | **77.4** ± 2.1 | 71.4 ± 2.5 | 69.7 ± 2.8 | 58.0 ± 5.6 | 76.8 ± 1.4 | **77.4** ± 2.2 | 72.8 ± 1.5 | 75.8 ± 2.7 | 72.9 ± 1.5 | 73.3 ± 4.0 |

**Table 5.** The best performing model according to Tables 3 and 4.

| Data | Without Exploration | | | With Exploration | | |
|---|---|---|---|---|---|---|
| | Macro-F1 | Model | Classifier | Macro-F1 | Model | Classifier |
| CiteSeer | 61.5 | AGE | GCN | 61.5 | AGE | GCN |
| CORA | 76.1 | Random | SGC | 76.7 | Entropy Count | SGC |
| PubMed | 77.7 | AGE | SGC | 78.0 | BALD Count | SGC |
| Coauthor CS | 83.9 | AGE | GCN | 83.9 | AGE | GCN |
| Coauthor Phy | 90.2 | BALD | SGC | 90.4 | BALD Count | SGC |
| Amazon Comp. | 78.3 | Clustering | SGC | 78.3 | Clustering | SGC |
| Disease | 68.2 | Entropy | SGC | 68.2 | Entropy | SGC |
| Wiki-CS | 64.7 | BALD | SGC | 65.6 | BALD Count | SGC |
| PPI Brain | 41.8 | Degree | SGC | 41.8 | Degree | SGC |
| PPI Blood | 42.4 | PageRank | GCN | 42.4 | PageRank | GCN |
| PPI Kidney | 42.3 | PageRank | SGC | 42.3 | PageRank | SGC |
| Github | 77.4 | Entropy | SGC | 77.4 | Entropy | SGC |

## 5.3. Running Time

Table 6 shows the execution time each algorithm spends to acquire a set of 40 unlabeled instances on average. AGE, the current state-of-the-art, is several magnitudes slower compared to the rest of the algorithms. The clustering step performed to compute the information gain is responsible for the additional time. The time complexity of this step grows $O(n^2)$ with the number of vertices $n$ of a graph making AGE not suitable for large attributed graphs. For example, the AGE algorithm is 80 times slower than random sampling for the Amazon Computers graph but achieves inferior performance. Additionally, the SGC model can be trained in a relatively less time compared to the GCN model and this difference is significant for larger graphs such as Wiki-CS and co-authorship graphs. However, in AL problems, the time spent for selecting an unlabeled example is a minor concern since the labeling time is more valued.

**Table 6.** Running time (seconds): average execution time to acquire 40 unlabeled instances. We run all experiments on a single NVIDIA GTX 1080-Ti GPU. PR: PageRank, CC: Clustering coefficient.

| Clf. | Dataset | Rand | Ent | PR | Deg | CC | AGE | BALD | $\epsilon$-Greedy | | Count | |
|---|---|---|---|---|---|---|---|---|---|---|---|---|
| | | | | | | | | | Ent | BALD | Ent | BALD |
| GCN | CiteSeer | 4.2 | 4.8 | 4.8 | 4.7 | 4.9 | 4.8 | 4.8 | 4.8 | 4.8 | 4.8 | 4.8 |
| | PubMed | 6.9 | 7.6 | 25.4 | 7.3 | 32 | 1125.9 | 7.9 | 7.5 | 7.8 | 7.6 | 7.9 |
| | CORA | 4.2 | 4.5 | 4.6 | 4.4 | 14.5 | 26.8 | 4.5 | 4.5 | 4.5 | 4.5 | 4.5 |
| | Coauthor CS | 20.4 | 22.3 | 40.8 | 21.9 | 39.3 | 2154 .2 | 23.7 | 22.3 | 23.6 | 22.4 | 23.6 |
| | Coauthor Phy | 46.1 | 50.5 | 116.4 | 48.5 | 98.6 | 2436.9 | 50.8 | 50.4 | 50.7 | 50.5 | 50.8 |
| | Amazon Comp. | 17.5 | 19.1 | 31.8 | 18.8 | 33.8 | 1688.9 | 19.2 | 19.1 | 19.1 | 19.1 | 19.2 |
| | Disease | 4.1 | 4.3 | 4.2 | 4.1 | 4.2 | 8.7 | 4.3 | 4.3 | 4.3 | 4.3 | 4.3 |
| | Wiki-CS | 15.3 | 16.6 | 30.0 | 28.3 | 33.0 | 410.8 | 16.7 | 16.6 | 16.6 | 16.7 | 16.7 |
| | PPI Brain | 8.3 | 8.9 | 11.5 | 10.2 | 10.9 | 133.3 | 9.0 | 8.4 | 8.6 | 8.4 | 8.7 |
| | PPI Blood | 7.9 | 8.2 | 10.4 | 9.4 | 9.9 | 130.2 | 8.4 | 8.2 | 8.4 | 8.3 | 8.5 |
| | PPI Kidney | 7.3 | 7.8 | 9.8 | 8.0 | 8.8 | 129.4 | 7.7 | 7.7 | 7.7 | 7.8 | 7.9 |
| | Github | 57.1 | 69.2 | 211.8 | 102.9 | 121.4 | 6810.0 | 72.1 | 69.6 | 71.1 | 70.5 | 73.2 |
| SGC | CiteSeer | 1.7 | 1.9 | 5.6 | 1.8 | 2.7 | 18.3 | 1.9 | 1.9 | 1.9 | 1.9 | 1.9 |
| | PubMed | 2.0 | 2.2 | 3.9 | 2.2 | 21.1 | 1229.2 | 2.2 | 2.2 | 2.2 | 2.2 | 2.2 |
| | CORA | 3.8 | 4.8 | 5.8 | 4.7 | 2.3 | 23.7 | 4.9 | 4.8 | 4.8 | 4.8 | 4.9 |
| | Coauthor CS | 16.8 | 19.8 | 33.2 | 19.3 | 37.9 | 2098.2 | 19.8 | 19.8 | 19.8 | 19.8 | 19.8 |
| | Coauthor Phy | 35.6 | 40.7 | 90.4 | 39.8 | 88.7 | 2232.3 | 40.8 | 40.4 | 40.5 | 40.7 | 40.7 |
| | Amazon Comp. | 12.2 | 14.7 | 17.2 | 16.9 | 17.1 | 1134.6 | 14.8 | 14.6 | 14.7 | 14.8 | 14.8 |
| | Disease | 1.4 | 1.4 | 1.5 | 1.4 | 1.4 | 6.0 | 1.4 | 1.4 | 1.4 | 1.4 | 1.4 |
| | Wiki-CS | 1.9 | 2.0 | 13.6 | 8.2 | 18.3 | 400.5 | 2.1 | 2.0 | 2.0 | 2.1 | 2.1 |
| | PPI Brain | 4.4 | 4.5 | 5.1 | 4.8 | 4.9 | 142.2 | 4.6 | 4.4 | 4.6 | 4.5 | 4.7 |
| | PPI Blood | 4.1 | 4.3 | 4.9 | 4.7 | 4.8 | 139.4 | 4.4 | 4.3 | 4.3 | 4.4 | 4.5 |
| | PPI Kidney | 3.9 | 4.1 | 4.4 | 4.3 | 4.5 | 135.6 | 4.1 | 4.1 | 4.1 | 4.1 | 4.2 |
| | Github | 22.3 | 24.5 | 166 | 78.3 | 106.2 | 4905.1 | 25.8 | 24.4 | 25.4 | 24.6 | 26.0 |

## 5.4. Discussion

As shown in Table 5, the performance of acquisition functions is diverse such that no single approach can be considered the best for all datasets. Sampling nodes based on graph properties leads to good performance depending on the graph structure. We make several key observations on how average clustering coefficient and label assortativity of a graph impact the performance of AL acquisition functions as following.

**Graphs with high level of clustering.** Amazon computers, co-authorship graphs, and Wiki-CS graphs have larger average clustering coefficients. For these datasets, sampling the node with the largest clustering coefficient outperforms sampling with other node centrality measures.

**Graphs with medium level of clustering.** CiteSeer, CORA, Github, and PPI graphs possess a medium level of average clustering in the range of 0.1 to 0.2. On CORA, CiteSeer, and Github datasets uncertainty-based acquisition functions and their variants obtain the best performance. However, the performance of PPI graphs is quite different since their label assortativity values are significantly low compared to all other datasets.

**Graphs with low level of clustering.** Pubmed and the disease graphs have the lowest average clustering coefficients. In most cases, the use of clustering coefficient to select the nodes for querying lead to suboptimal results. However, sampling with clustering coefficient on PubMed dataset obtained good performance when the GCN model was used as the node classifier.

**Graphs with low label assortativity.** Out of all graph datasets, PPI graphs exhibit the lowest label assortativity. As most of the graphs used in node classification literature exhibit high label assortativity, commonly used node classification models are build on the assumption that neighbors of a node may have the same label. Therefore, such models are not confident in predicting the labels of unlabeled nodes, specially when the training data is scarce. On PPI graphs, we observe that performing AL by sampling the query nodes based on PageRank and degree centrality contributes to the best performing models. However, one limitation in calculating the label assortativity is that node labels need to be known beforehand. When we are given an unlabeled graph, one way to overcome this problem is we can use similar labeled graphs belonging to the same domain to approximate the label assortativity.

## 6. Conclusions

In this paper, we studied the application of the active learning framework as a method to make node classification on attributed graphs label efficient. We have performed an empirical evaluation of state-of-the-art active learning algorithms on the node classification task using twelve real-world attributed graphs belonging to different domains. In our experiments, we initiate the active learner with an extremely small number of labeled instances. Additionally, we assumed a more realistic setting in which the learner does not use a separate validation set. Our results highlight that no single acquisition function can be performs consistently well on all datasets and the performance of acquisition functions depend on graph properties. We further show that selecting an acquisition function based on the performance on a handful of attributed graphs with similar characteristics result in suboptimal algorithms. Notably, our results point that SGC, a simpler variant of GNN performs better and efficiently on most datasets compared to more complex GNN models.

A key takeaway of this research is that AL is beneficial in reducing the labeling cost of semisupervised node classification models and the choice of an AL acquisition function depends on the properties of the graph data at hand. Using an extensive set of graph datasets with a wide variety of characteristics, we showed that there is no single algorithm that works across different graph datasets possessing different graph properties. We further made the observation that using node PageRank and degree centrality of nodes achieve the best performance on graphs with low label assortativity.

Moreover, the current state-of-the-art active learning algorithm (AGE) [41] uses a combination of multiple acquisition functions and it is several magnitudes slower than all other acquisition functions that were used in this paper. Therefore, it is not suitable for large real-world attributed graphs. Lack of hyperparameter tuning and a minuscule number of training instances lead to classifiers that cannot generalize well for unlabeled data. We expressed this problem as balancing the exploration-vs.-exploitation trade-off and propose introducing an exploration term into acquisition functions. We evaluated the performance of two exploration terms using multiple real-world graph datasets. The introduction of this exploration term into existing uncertainty-based acquisition functions make their performance competitive with the current state-of-the-art AL algorithm for node classification on some datasets. As future work, we would like to explore how AL can be utilized for other graph-related learning tasks.

**Author Contributions:** Conceptualization, K.M.; methodology, K.M.; software, K.M.; validation, K.M.; formal analysis, K.M.; investigation, K.M.; writing—original draft preparation, K.M.; writing—review and editing, T.M.; visualization, K.M.; supervision, T.M.; funding acquisition, T.M. All authors have read and agreed to the published version of the manuscript.

**Funding:** This work was funded by JSPS Grant-in-Aid for Scientific Research(B) (Grant Number 17H01785) and JST CREST (Grant Number JPMJCR1687).

**Conflicts of Interest:** The authors declare no conflicts of interest.

## Abbreviations

The following abbreviations are used in this manuscript.

| | |
|---|---|
| DNN | Deep neural network |
| GCN | Graph convolutional network |
| GNN | Graph neural network |
| SGC | Simplified graph convolution |
| AL | Active learning |
| CNN | Convolutional neural network |
| BALD | Bayesian Active Learning by Disagreement |
| QBC | Query by committee |
| EER | Expected error reduction |
| GRF | Gaussian random fields |
| AGE | Active graph embedding |
| PR | PageRank |
| UCB | Upper confidence bound |

## References


1. Mohri, M.; Rostamizadeh, A.; Talwalkar, A. *Foundations of Machine Learning*; MIT Press: Cambridge, MA, USA, 2018.
2. Hoi, S.C.; Jin, R.; Zhu, J.; Lyu, M.R. Batch mode active learning and its application to medical image classification. In Proceedings of the 23rd International Conference on Machine Learning, Pittsburgh, PA, USA, 25–29 June 2006; pp. 417–424.
3. Parisot, S.; Ktena, S.I.; Ferrante, E.; Lee, M.; Guerrero, R.; Glocker, B.; Rueckert, D. Disease prediction using graph convolutional networks: Application to Autism Spectrum Disorder and Alzheimer's disease. *Med. Image Anal.* **2018**, *48*, 117–130. [CrossRef]
4. Shalev-Shwartz, S.; Ben-David, S. *Understanding Machine Learning: From Theory to Algorithms*; Cambridge University Press: Cambridge, UK, 2014.
5. Settles, B. *Active Learning Literature Survey*; Technical Report; University of Wisconsin-Madison Department of Computer Sciences: Madison, WI, USA, 2009.
6. Houlsby, N.; Huszár, F.; Ghahramani, Z.; Lengyel, M. Bayesian Active Learning for Classification and Preference Learning. *arXiv* **2011**, arXiv:1112.5745.
7. Rubens, N.; Elahi, M.; Sugiyama, M.; Kaplan, D. Active Learning in Recommender Systems. In *Recommender Systems Handbook*; Springer: Berlin/Heidelberg, Germany, 2015; pp. 809–846.
8. Ying, R.; He, R.; Chen, K.; Eksombatchai, P.; Hamilton, W.L.; Leskovec, J. Graph convolutional neural networks for web-scale recommender systems. In Proceedings of the 24th ACM SIGKDD International Conference on Knowledge Discovery & Data Mining, London, UK, 19–23 August 2018; pp. 974–983.
9. Yao, L.; Mao, C.; Luo, Y. Graph convolutional networks for text classification. In Proceedings of the AAAI Conference on Artificial Intelligence, Honolulu, HI, USA, 27 January–1 February 2019; Volume 33, pp. 7370–7377.
10. Neville, J.; Jensen, D. Iterative classification in relational data. In Proceedings of the AAAI-2000 Workshop on Learning Statistical Models From Relational Data, Austin, TX, USA, 31 July 2000; pp. 13–20.
11. Zhu, X.; Ghahramani, Z. *Learning from Labeled and Unlabeled Data with Label Propagation*; Technical Report; Carnegie Mellon University: Cambridge, MA, USA, 2002.
12. Zhu, X.; Ghahramani, Z.; Lafferty, J.D. Semi-supervised learning using gaussian fields and harmonic functions. In Proceedings of the 20th International Conference on Machine Learning (ICML-03), Washington, DC, USA, 21–24 August 2003; pp. 912–919.
13. Defferrard, M.; Bresson, X.; Vandergheynst, P. Convolutional Neural Networks on Graphs with Fast Localized Spectral Filtering. In Proceedings of the 2016 Conference on Neural Information Processing Systems, Barcelona, Spain, 5–10 December 2016; pp. 1–14.

14. Kipf, T.N.; Welling, M. Semi-Supervised Classification with Graph Convolutional Networks. In Proceedings of the International Conference on Learning Representations (ICLR), Toulon, France, 24–26 April 2017.

15. Veličković, P.; Fedus, W.; Hamilton, W.L.; Liò, P.; Bengio, Y.; Hjelm, R.D. Deep Graph Infomax. *arXiv* **2018**, arXiv:1809.10341.

16. Fazakis, N.; Kanas, V.G.; Aridas, C.K.; Karlos, S.; Kotsiantis, S. Combination of Active Learning and Semi-Supervised Learning under a Self-Training Scheme. *Entropy* **2019**, *21*, 988. [CrossRef]

17. Wu, Z.; Pan, S.; Chen, F.; Long, G.; Zhang, C.; Philip, S.Y. A comprehensive survey on graph neural networks. *IEEE Trans. Neural Netw. Learn. Syst.* **2020**.

18. Zhou, J.; Cui, G.; Zhang, Z.; Yang, C.; Liu, Z.; Wang, L.; Li, C.; Sun, M. Graph neural networks: A review of methods and applications. *arXiv* **2018**, arXiv:1812.08434.

19. Zhang, Z.; Cui, P.; Zhu, W. Deep learning on graphs: A survey. *IEEE Trans. Knowl. Data Eng.* **2020**. [CrossRef]

20. Settles, B.; Craven, M. An analysis of active learning strategies for sequence labeling tasks. In Proceedings of the 2008 Conference on Empirical Methods in Natural Language Processing, Honolulu, HI, USA, 25–27 October 2008; pp. 1070–1079.

21. Fu, Y.; Zhu, X.; Li, B. A survey on instance selection for active learning. *Knowl. Inf. Syst.* **2013**, *35*, 249–283. [CrossRef]

22. Baram, Y.; Yaniv, R.E.; Luz, K. Online choice of active learning algorithms. *J. Mach. Learn. Res.* **2004**, *5*, 255–291.

23. Huang, Z.; Chung, W.; Ong, T.H.; Chen, H. A graph-based recommender system for digital library. In Proceedings of the 2nd ACM/IEEE-CS joint conference on Digital libraries, Portland, OR, USA, 14–18 July 2002; pp. 65–73.

24. Gilmer, J.; Schoenholz, S.S.; Riley, P.F.; Vinyals, O.; Dahl, G.E. Neural message passing for quantum chemistry. *arXiv* **2017**, arXiv:1704.01212.

25. Bhagat, S.; Cormode, G.; Muthukrishnan, S. Node classification in social networks. In *Social Network Data Analytics*; Springer: Berlin/Heidelberg, Germany, 2011; pp. 115–148.

26. Zhu, X.; Lafferty, J.; Ghahramani, Z. Combining Active Learning and Semi-supervised Learning using Gaussian Fields and Harmonic Functions. In Proceedings of the ICML 2003 Workshop on the Continuum from Labeled to Unlabeled Data in Machine Learning and Data mining, Washington, DC, USA, 21–24 August 2003; Volume 3.

27. Zhou, D.; Bousquet, O.; Lal, T.N.; Weston, J.; Schölkopf, B. Learning with local and global consistency. In Proceedings of the Advances in Neural Information Processing Systems, Vancouver, BC, Canada, 13–18 December 2004; pp. 321–328.

28. Li, Y.; Tarlow, D.; Brockschmidt, M.; Zemel, R. Gated Graph Sequence Neural Networks. In Proceedings of the International Conference on Learning Representations (ICLR), San Juan, Puerto Rico, 2–4 May 2016.

29. Wu, F.; Souza, A.; Zhang, T.; Fifty, C.; Yu, T.; Weinberger, K. Simplifying Graph Convolutional Networks. In Proceedings of the 36th International Conference on Machine Learning, PMLR, Beach, CA, USA, 10–15 June 2019; pp. 6861–6871.

30. Perozzi, B.; Al-Rfou, R.; Skiena, S. Deepwalk: Online Learning of Social Representations. In Proceedings of the 20th ACM SIGKDD International Conference on Knowledge Discovery and Data Mining, New York, NY, USA, 24–27 August 2014; pp. 701–710.

31. Yang, Z.; Cohen, W.; Salakhudinov, R. Revisiting Semi-Supervised Learning with Graph Embeddings. In Proceedings of the International conference on machine learning, New York, NY, USA, 14–19 June 2016; pp. 40–48.

32. Lewis, D.D.; Catlett, J. Heterogeneous uncertainty sampling for supervised learning. In *Machine Learning Proceedings 1994*; Elsevier: Amsterdam, The Netherlands, 1994; pp. 148–156.

33. Gal, Y.; Ghahramani, Z. Dropout as a Bayesian Approximation: Representing Model Uncertainty in Deep Learning. In Proceedings of the 33rd International Conference on Machine Learning, New York, NY, USA, 19–24 June 2016; Volume 48, pp. 1050–1059.

34. Srivastava, N.; Hinton, G.; Krizhevsky, A.; Sutskever, I.; Salakhutdinov, R. Dropout: A simple way to prevent neural networks from overfitting. *J. Mach. Learn. Res.* **2014**, *15*, 1929–1958.

35. Gal, Y.; Islam, R.; Ghahramani, Z. Deep Bayesian Active Learning with Image Data. In Proceedings of the 34th International Conference on Machine Learning, Sydney, Australia, 6–11 August 2017; Volume 70, pp. 1183–1192.

36. Seung, H.S.; Opper, M.; Sompolinsky, H. Query by committee. In Proceedings of the Fifth Annual Workshop on Computational Learning Theory, Pittsburgh, PA, USA, 27–29 July 1992; pp. 287–294.

37. Roy, N.; McCallum, A. Toward Optimal Active Learning through Monte Carlo Estimation of Error Reduction. In Proceedings of the 18th International Conference on Machine Learning, Williamstown, MA, USA, 28 June–1 July 2001; pp. 441–448.

38. Gu, Q.; Han, J. Towards Active Learning on Graphs: An Error Bound Minimization Approach. In Proceedings of the 2012 IEEE 12th International Conference on Data Mining, Brussels, Belgium, 10 December 2012; pp. 882–887.

39. Bilgic, M.; Mihalkova, L.; Getoor, L. Active Learning for Networked Data. In Proceedings of the 27th International Conference on Machine Learning, Haifa, Israel, 21–24 June 2010; pp. 79–86.

40. Ji, M.; Han, J. A Variance Minimization Criterion to Active Learning on Graphs. In Proceedings of the Artificial Intelligence and Statistics, La Palma, Canary Islands, 21–23 April 2012; pp. 556–564.

41. Cai, H.; Zheng, V.W.; Chang, K.C.C. Active Learning for Graph Embedding. *arXiv* **2017**, arXiv:1705.05085.

42. Gao, L.; Yang, H.; Zhou, C.; Wu, J.; Pan, S.; Hu, Y. Active Discriminative Network Representation Learning. In Proceedings of the 27th International Joint Conference on Artificial Intelligence, Stockholm, Sweden, 13–19 July 2018; AAAI Press: Menlo Park, CA, USA, 2018; pp. 2142–2148.

43. Gadde, A.; Anis, A.; Ortega, A. Active semisupervised learning using sampling theory for graph signals. In Proceedings of the 20th ACM SIGKDD International Conference on Knowledge Discovery and Data Mining, New York, NY, USA, 24–27 August 2014; pp. 492–501.

44. Macskassy, S.A. Using graph-based metrics with empirical risk minimization to speed up active learning on networked data. In Proceedings of the 15th ACM SIGKDD International Conference on Knowledge Discovery and Data Mining, Paris, France, 28 June–July 2009; pp. 597–606.

45. Ma, Y.; Garnett, R.; Schneider, J. $\sigma$-Optimality for active learning on gaussian random fields. In Proceedings of the Advances in Neural Information Processing Systems, Lake Tahoe, NV, USA, 5–8 December 2013; pp. 2751–2759.

46. Lattimore, T.; Szepesvári, C. *Bandit Algorithms*; Cambridge University Press: Cambridge, UK, 2020.

47. Auer, P. Using confidence bounds for exploitation-exploration trade-offs. *J. Mach. Learn. Res.* **2002**, *3*, 397–422.

48. Thompson, W.R. On the likelihood that one unknown probability exceeds another in view of the evidence of two samples. *Biometrika* **1933**, *25*, 285–294. [CrossRef]

49. Thrun, S.B. The role of exploration in learning control. In *Handbook of Intelligent Control: Neural, Fuzzy and Adaptive Approaches*; Van Nostrand Reinhold: New York, NY, USA, 1992; pp. 1–27.

50. Sen, P.; Namata, G.; Bilgic, M.; Getoor, L.; Galligher, B.; Eliassi-Rad, T. Collective Classification in Network Data. *AI Mag.* **2008**, *29*, 93–93. [CrossRef]

51. Shchur, O.; Mumme, M.; Bojchevski, A.; Günnemann, S. Pitfalls of Graph Neural Network Evaluation. In Proceedings of the Relational Representation Learning Workshop (NeurIPS 2018), Montréal, QC, Canada, 3–8 December 2018.

52. McAuley, J.; Targett, C.; Shi, Q.; Van Den Hengel, A. Image-based Recommendations on Styles and Substitutes. In Proceedings of the 38th International ACM SIGIR Conference on Research and Development in Information Retrieval, Santiago, Chile, 9–13 August 2015; pp. 43–52.

53. Chami, I.; Ying, Z.; Ré, C.; Leskovec, J. Hyperbolic Graph Convolutional Neural Networks. In *Advances in Neural Information Processing Systems 32*; Wallach, H., Larochelle, H., Beygelzimer, A., Alché-Buc, F.D., Fox, E., Garnett, R., Eds.; Curran Associates, Inc.: Red Hook, NY, USA, 2019; pp. 4868–4879.

54. Anderson, R.M.; Anderson, B.; May, R.M. *Infectious Diseases of Humans: Dynamics and Control*; Oxford University Press: Oxford, UK, 1992.

55. Mernyei, P.; Cangea, C. Wiki-CS: A Wikipedia-Based Benchmark for Graph Neural Networks. *arXiv* **2020**, arXiv:2007.02901.

56. Pennington, J.; Socher, R.; Manning, C. GloVe: Global Vectors for Word Representation. In Proceedings of the 2014 Conference on Empirical Methods in Natural Language Processing (EMNLP), Doha, Qatar, 25–29 October 2014; Association for Computational Linguistics: Stroudsburg, PA, USA, 2014; pp. 1532–1543. [CrossRef]

57. Zitnik, M.; Leskovec, J. Predicting multicellular function through multi-layer tissue networks. *Bioinformatics* **2017**, *33*, 190–198. [CrossRef]

*Entropy* **2020**, *22*, 1164

58. Oughtred, R.; Stark, C.; Breitkreutz, B.J.; Rust, J.; Boucher, L.; Chang, C.; Kolas, N.; O'Donnell, L.; Leung, G.; McAdam, R.; et al. The BioGRID interaction database: 2019 update. *Nucleic Acids Res.* **2019**, *47*, D529–D541. [CrossRef]
59. Rozemberczki, B.; Allen, C.; Sarkar, R. Multi-scale Attributed Node Embedding. *arXiv* **2019**, arXiv:1909.13021.
60. Watts, D.J.; Strogatz, S.H. Collective dynamics of 'small-world'networks. *Nature* **1998**, *393*, 440–442. [CrossRef]
61. Newman, M.E. Mixing patterns in networks. *Phys. Rev. E* **2003**, *67*, 026126. [CrossRef] [PubMed]
62. Ash, J.T.; Zhang, C.; Krishnamurthy, A.; Langford, J.; Agarwal, A. Deep Batch Active Learning by Diverse, Uncertain Gradient Lower Bounds. In Proceedings of the International Conference on Learning Representations (ICLR), Addis Ababa, Ethiopia, 26–30 April 2020.
63. Kingma, D.P.; Ba, J. Adam: A Method for Stochastic Optimization. *arXiv* **2014**, arXiv:1412.6980.
64. Hagberg, A.; Swart, P.; S Chult, D. *Exploring Network Structure, Dynamics, and Function Using NetworkX*; Technical Report; Los Alamos National Lab.(LANL): Los Alamos, NM, USA, 2008.
65. Paszke, A.; Gross, S.; Massa, F.; Lerer, A.; Bradbury, J.; Chanan, G.; Killeen, T.; Lin, Z.; Gimelshein, N.; Antiga, L.; et al. Pytorch: An imperative style, high-performance deep learning library. In Proceedings of the Advances in Neural Information Processing Systems, Vancouver, BC, Canada, 8–14 December 2019; pp. 8026–8037.

**Publisher's Note:** MDPI stays neutral with regard to jurisdictional claims in published maps and institutional affiliations.

*Article*

# Spreading Control in Two-Layer Multiplex Networks

**Roberto Bernal Jaquez [1], Luis Angel Alarcón Ramos [2],* and Alexander Schaum [3]**

[1]   Department of Applied Mathematics and Systems, Universidad Autónoma Metropolitana, Cuajimalpa, Mexico-City 05348, Mexico; rbernal@cua.uam.mx
[2]   Postgraduate in Natural Sciences and Engineering, Universidad Autónoma Metropolitana, Cuajimalpa, Mexico-City 05348, Mexico
[3]   Chair of Automatic Control, Kiel-University, 24143 Kiel, Germany; alsc@tf.uni-kiel.de
*   Correspondence: lalarcon@cua.uam.mx; Tel.: +52-55-5814-6500 (ext. 3884)

Received: 3 September 2020; Accepted: 8 October 2020; Published: 15 October 2020

**Abstract:** The problem of controlling a spreading process in a two-layer multiplex networks in such a way that the extinction state becomes a global attractor is addressed. The problem is formulated in terms of a Markov-chain based susceptible-infected-susceptible (SIS) dynamics in a complex multilayer network. The stabilization of the extinction state for the nonlinear discrete-time model by means of appropriate adaptation of system parameters like transition rates within layers and between layers is analyzed using a dominant linear dynamics yielding global stability results. An answer is provided for the central question about the essential changes in the step from a single to a multilayer network with respect to stability criteria and the number of nodes that need to be controlled. The results derived rigorously using mathematical analysis are verified using statical evaluations about the number of nodes to be controlled and by simulation studies that illustrate the stability property of the multilayer network induced by appropriate control action.

**Keywords:** multilayer complex networks; stability; spreading control

---

## 1. Introduction

Multiplex networks are a collection of coupled networks placed in different layers with each layer having the same set of nodes but not necessarily the same topology. Layer interactions are given via counterpart nodes of each network layer. Multilayer networks build key elements in the structure of many modern technological systems including social cyber and computer networks as well as in fundamental natural systems determining the functioning of gene regulation and brain dynamics [1–6]. A central advantage in comparison to single-layer networks is that each node can have different states in the different networks. This enables them e.g., to analyze the spreading of information or computational viruses among different social or cyber networks [7], thus enabling the identification, understanding and possibly the manipulation of the corresponding mechanisms associated to each layer and between layers.

Spreading processes in complex networks have attracted recent attention for the purpose of analyzing the intertwined dynamics of epidemics [8–13] or information transmission in [14–18]. The control of such problems has to address fundamental questions as (i) which parameters of the system are amenable to manipulation and (ii) which nodes must be actively controlled. The latter question goes in particular in hand with the aim to develop control strategies with minimum need of implementation costs. In multilayer networks the additional question arises if nodes need to be controlled in all layers or just in some of them or maybe only in one single layer, as long as the nodes to be controlled are defined accordingly.

The question of network control has been addressed on one side using classical control theory methods as controllability analysis [19–26] including statistic evaluations of the number of nodes to be controlled in networks of certain structures [11,27–33]. Given that nonlinear system controllability analysis is much more involved than for linear systems [34] controllability studies are typically focussing on linear models or the linearization about some equilibrium point. Only a few recent studies explicitly considered nonlinear controllability and control design approaches in complex networks [25,26]. It should be mentioned that even though network controllability ensures that a desired state can be reached or stabilized, it does not necessarily guide the way for the design of a decentralized control but typically leads to centralized control strategies. On the other hand, the control of networks has been explicitly addressed using stabilization and stability analysis leading the way to the choice of nodes to be controlled with implicit decentralized parametric control strategies [35–38]. In particular, the approach followed in [36–38] yields global stability assessments by means of the derivation of a global dominant linear dynamics. Furthermore, optimization based approaches for parameter adaptation and node or link removal have been widely discussed, as has been summarized in [39].

In the present study the control of a spreading process in a complex multilayer network is addressed on the basis of the classical Markov-based susceptible-infected-susceptible (SIS) dynamics [40–44] in a multilayer version that has been adapted from [7] in such a way that the unit polytope is an invariant set for the dynamics. Following the global stability analysis and parametric control design studies for SIS processes in homogenous and inhomogeneous single-layer complex networks [36,38] and extensions of it including quarantine [37,45] a decentralized parametric control strategy is developed providing sufficient conditions for global stability of the extinction state without altering the topology of the networks as is suggested in other studies related to adaptive networks [39,46]. Instead of involving computationally expensive optimization procedures, simple analytic measures are provided which can be quickly determined for a given network topology and parameter set. Accordingly, the present result provides (i) a solution to the problem of designing decentralized spreading control strategies with global stability assessment and without huge computational effort, which to the knowledge of the authors is still an open question, and (ii) presents an extension of the approaches in [36–38] to the case of two-layer multiplex networks. It turns out that the step from a single layer to a two-layer network allows to clearly identify some of the main challenges when considering multiplex networks. In particular, having in mind the nonlinear dynamics in each network and its non-trivial interplay between networks it is clear from the theory of input-to-state stability [47,48] that it is not sufficient that both nonlinear systems are asymptotically stable for their own but the specific interconnection needs to satisfy some additional, small-gain-like criteria. A sufficient criterion ensuring the asymptotic stability of the complete multiplex networks and its differentiation to the stability criteria for each network on its own is a central result that is derived. Based on this criterion it is highlighted how the number of nodes that need to be controlled changes when the interconnection of two networks is considered. Besides a rigorous mathematical derivation of the results some statistical analysis is provided to show the expected variation in the number of nodes that need to be controlled for some illustrative setups.

The paper is organized as follows: In Section 2 the problem formulation is stated, in Section 3 the system analysis is presented along with the main mathematical results of this work. Control design, a statistical analysis of the number of nodes to be controlled, and simulations to corroborate our results are presented in Sections 4 and 5, respectively. Finally, conclusions are presented in Section 6.

## 2. Problem Formulation

Consider a two layer network of any topology with adjacency matrices given by $\mathbf{A}$ and $\mathbf{B}$. Each network has the same set of $N$ nodes, and the adjacency matrix associated to network $A$ is defined as $\mathbf{A} = [\mathbf{a_{ij}}]$, where $a_{ij} = a_{ji} = 1$ if nodes $i$ and $j$ are connected and zero otherwise (that means, we consider non-directed graphs), the adjacency matrix associated to network $B$ is defined in the same way as $\mathbf{B} = [\mathbf{b_{ij}}]$. Any node $i$ in network $A$ is connected with node $i$ in the network $B$, as it is shown in Figure 1.

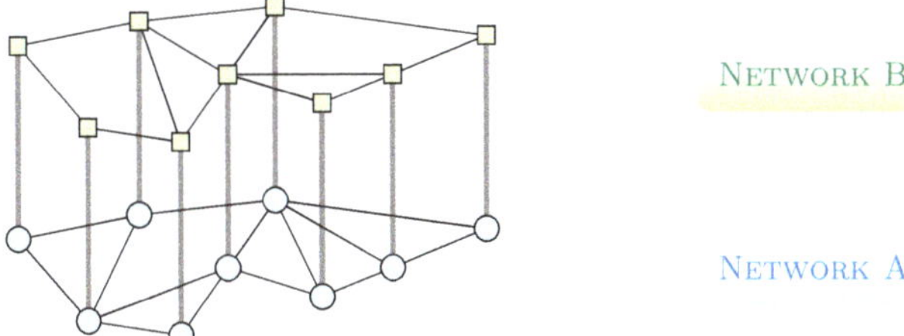


**Figure 1.** Networks $A$ and $B$ of arbitrary topology with each node $i$ in network $A$ being connected with its equivalent node $i$ in network $B$.

Using a slightly modified version of the model defined in [7,8], the underlying process for every node in both layers of the network is modeled as a discrete time SIS Markov process. A node $i$ can be in state $I$ (infected) with probability $p_{Ai}(t)$ (or $p_{Bi}(t)$) at time $t \in \mathbb{N}_0$, or in state $S$ (susceptible) with probability $1 - p_{Ai}(t)$ (or $1 - p_{Bi}(t)$). The probabilities $p_{Ai}(t)$ and $p_{Bi}(t)$ then correspond to the solutions of the following dynamical system:

$$
\begin{aligned}
p_{Ai}(t+1) &= (1 - \mu_{Ai})p_{Ai}(t) + (1 - q_{Ai}(t))(1 - p_{Ai}(t)), \\
p_{Bi}(t+1) &= (1 - \mu_{Bi})p_{Bi}(t) + (1 - q_{Bi}(t))(1 - p_{Bi}(t)), \\
p_{ki}(0) &= p_{ki0}, \quad k = \{A, B\}, \quad i = 1, 2, \ldots, N.
\end{aligned}
\tag{1}
$$

In the preceding Equations $\mu_{ki}$ is the recovery probability of node $i$ in the network $k \in \{A, B\}$, $q_{ki}(t)$ is the probability that node $i$ in network $k$ is not infected by some neighbor in network $A$ or $B$, which is given by

$$
\begin{aligned}
q_{Ai} = \varphi_{Ai}(P_A, P_B) &:= \prod_{j=1}^{N}(1 - \beta_{Ai}a_{ij}p_{Aj})(1 - \gamma_{Ai}p_{Bi}), \\
q_{Bi} = \varphi_{Bi}(P_A, P_B) &:= \prod_{j=1}^{N}(1 - \beta_{Bi}b_{ij}p_{Bj})(1 - \gamma_{Bi}p_{Ai}),
\end{aligned}
\tag{2}
$$

with $P_k = [p_{k1}, \ldots, p_{kN}]^T$ for $k = A, B$. The parameters $\beta_{Ai}$ and $\beta_{Bi}$ represent the transmission probabilities of the node $i$ in each layer-network, and $\gamma_{Ai}$ and $\gamma_{Bi}$ are the transmission probabilities of a node $i$ from $B$ to $A$ and from $A$ to $B$, respectively.

Note that in Equations (1) and (2)

$$
0 \leq p_{ki}(t), \mu_{ki}, q_{ki}(t), \gamma_{ki}, \beta_{ki}, p_{ki0} \leq 1, \quad k = \{A, B\}, \quad i = 1, 2, \ldots, N.
$$

Additionally, in order to propose a control mechanism, we consider that each node has a manipulable variable $u_{ki}(t)$ ($k \in \{A, B\}$), which is amenable for control. In the present study, we consider that the amenable variables are taken from the set $\{\gamma_{Ai}, \beta_{Ai}, \gamma_{Bi}, \beta_{Bi}; i = 1, \ldots, N\}$.

The problem addressed in the following consists in determining the $m \leq N$ nodes whose interaction parameters $\gamma_{ki}, \beta_{ki}$ have to be adapted in order to ensure the global exponential stability of the extinction state, i.e., such that for all $p_{ki0} \in [0, 1]$ there are constants $m_{ki} \geq 1, \alpha \in (0, 1)$ such that

$$
p_{ki}(t) \leq m_{ki}\alpha^t p_{ki0}.
\tag{3}
$$

## 3. System Analysis

The fixed points $p_{ki}^*$, $k = \{A, B\}$, $i = 1, \ldots, N$ associated with the dynamics (1) for some constant values $\mu_{ki}$, $\gamma_{ki}^*$, and $\beta_{ki}^*$ are determined by substituting $p_{ki}(t+1) = p_{ki}(t) = p_{ki}^*$ into (1). After some algebra it follows that

$$p_{ki}^* = \frac{1 - q_{ki}^*}{\mu_{ki} + 1 - q_{ki}^*}, \quad k = \{A, B\},\ i = 1, \ldots, N, \quad q_{ki}^* = \varphi_{ki}(P_A^*, P_B^*) \tag{4}$$

with $\varphi_{ki}$ defined in (2) and $P_k^* = [p_{k1}^*, \ldots, p_{kN}^*]^T$. Note that $p_{ki}^* = 0$ for all $k = \{A, B\}$, $i = 1, \ldots, N$ is a fixed point given that this condition implies that $q_{ki} = 1$. This fixed point is referred to as extinction state.

Given that model (1) represents the evolution of probabilities it is important to ensure that all solutions for $p_{ki}$ are contained in the unit hypercube $\mathcal{P} = [0, 1]^{2N}$. This is established in the following Lemma.

**Lemma 1.** *The set $\mathcal{P} = [0, 1]^{2N}$ is a positively invariant set for the dynamics (1).*

**Proof.** Let $p_{ki}(t) \in [0, 1]$, $k = \{A, B\}$, $i = 1, \ldots, N$. From (1) it follows that

$$p_{ki}(t+1) = (1 - \mu_{ki})p_{ki}(t) + (1 - q_{ki}(t))(1 - p_{ki}(t)) \leq p_{ki}(t) + (1 - p_{ki}(t)) = 1$$

and

$$p_{ki}(t+1) \geq (1 - \mu_{ki})p_{ki}(t) \geq 0.$$

□

Next, sufficient conditions for the (global in $\mathcal{P}$) exponential stability of the extinction state $(P_A, P_B)^T = (0, 0)^T$ are presented in the following Theorem.

**Theorem 1.** *Consider the dynamics (1) on a two-layer network with adjacency matrices $\mathbf{A}$ and $\mathbf{B}$. The extinction state $(P_A, P_B) = (0, 0)$ is globally exponentially stable in the hypercube $\mathcal{P}$ if*

$$\sigma(\mathbf{H}) < 1, \tag{5}$$

*where $\sigma(\cdot)$ is the spectral radius, and the matrix $\mathbf{H}$ is defined as follows*

$$\mathbf{H} = \begin{bmatrix} \mathbf{I} - \mathbf{M_A} + \mathbf{B_A}\mathbf{A} & \mathbf{G_A} \\ \mathbf{G_B} & \mathbf{I} - \mathbf{M_B} + \mathbf{B_B}\mathbf{B} \end{bmatrix},$$

*where $\mathbf{M_k} = diag(\mu_{ki})$, $\mathbf{B_k} = diag(\beta_{ki})$, $\mathbf{G_k} = diag(\gamma_{ki})$ ($k \in \{A, B\}$), and $\mathbf{I}$ is the identity matrix.*

**Proof.** The exponential stability is assessed through the determination of a linear dominant dynamics, whose stability features imply the desired result similar to the development in [36,37,45].

Note that for all $p_{ki} \in [0,1]$, $k = \{A, B\}$, $i = 1, \ldots, N$ it holds that

$$
\begin{aligned}
q_{Ai} &= \prod_{j=1}^{N}(1 - \beta_{Ai}a_{ij}p_{Aj})(1 - \gamma_{Ai}p_{Bi}) \\
&\geq \left(1 - \sum_{j}\beta_{Ai}a_{ij}p_{Aj}\right)(1 - \gamma_{Ai}p_{Bi}) \\
&= 1 - \sum_{j}\beta_{Ai}a_{ij}p_{Aj} - \gamma_{Ai}p_{Bi} + \sum_{j}\beta_{Ai}a_{ij}p_{Aj}\gamma_{Ai}p_{Bi}
\end{aligned}
$$

where in the second step the Weierstrass product inequality [49] has been employed. It follows that

$$
\begin{aligned}
1 - q_{Ai} &\leq \sum_{j}\beta_{Ai}a_{ij}p_{Aj} + \gamma_{Ai}p_{Bi} - \sum_{j}\beta_{Ai}a_{ij}p_{Aj}\gamma_{Ai}p_{Bi} \\
&\leq \sum_{j}\beta_{Ai}a_{ij}p_{Aj} + \gamma_{Ai}p_{Bi}.
\end{aligned}
$$

Equivalently it holds that

$$
1 - q_{Bi} \leq \sum_{j=1}^{N}\beta_{Bi}b_{ij}p_{Bj} + \gamma_{Bi}p_{Ai}.
$$

Substitution of these inequalities into Equations (1) and taking into account that $0 \leq 1 - p_{ki} \leq 1$ holds true it follows that

$$
\begin{aligned}
p_{Ai}(t+1) &\leq (1 - \mu_{Ai})p_{Ai}(t) + \sum_{j=1}^{N}\beta_{Ai}a_{ij}p_{Aj}(t) + \gamma_{Ai}p_{Bi}(t), \\
p_{Bi}(t+1) &\leq \gamma_{Bi}p_{Ai}(t) + (1 - \mu_{Bi})p_{Bi}(t) + \sum_{j=1}^{N}\beta_{Bi}b_{ij}p_{Bj}(t).
\end{aligned}
\tag{6}
$$

The preceding Equations can be written in matrix form as

$$
\begin{bmatrix} P_A(t+1) \\ P_B(t+1) \end{bmatrix} \leq \begin{bmatrix} \mathbf{I} - \mathbf{M_A} + \mathbf{B_A A} & \mathbf{G_A} \\ \mathbf{G_B} & \mathbf{I} - \mathbf{M_B} + \mathbf{B_B B} \end{bmatrix} \begin{bmatrix} P_A(t) \\ P_B(t) \end{bmatrix} \leq \mathbf{H} \begin{bmatrix} P_A(t) \\ P_B(t) \end{bmatrix}
\tag{7}
$$

with $\mathbf{I}$, $\mathbf{M_k}$, $\mathbf{B_k}$ and $\mathbf{G_k}$, $k = A, B$ defined in the statement of Lemma 1. In virtue of (5) it follows that there exists a constant $\alpha = \sigma(\mathbf{H}) \in (0,1)$ so that

$$
\left\| \begin{bmatrix} P_A(t+1) \\ P_B(t+1) \end{bmatrix} \right\| < \alpha \left\| \begin{bmatrix} P_A(t) \\ P_B(t) \end{bmatrix} \right\|
$$

implying the exponential stability (4) of the extinction state. $\square$

**Remark 1.** *It should be noted at this place that according to the dynamics in (7) for the asymptotic stability of the origin $[P_a^T, P_B^T]^T = 0$ it is not sufficient to ensure the asymptotic stability in both sub-networks, what would be ensured by analyzing the diagonal sub-matrices $\mathbf{I} - \mathbf{M_K} + \mathbf{B_K K}$ separately for $\mathbf{K} = \mathbf{A}, \mathbf{B}$, but that it is required to account explicitly for the particular interconnection structure and the associated transition probabilities between sub-networks. This establishes a significant difference to the case of single-layer networks as considered e.g., in [36–38]. Given that the solutions of the linear dynamics (7) bound the one for the nonlinear dynamics, Theorem 1 is intrinsically connected with the input-to-state stability and the small-gain condition [47,48] for the interconnection (1).*

## 4. Control Design

The next question to be addressed is how the sufficient condition established in Theorem 1 can be used to design an efficient control strategy, and how the number of nodes to be controlled varies when considering the interconnection of two networks. This question is addressed in the following Lemma.

**Lemma 2.** *Let $N_{ki}$, $k = \{A, B\}$, $i = 1, \ldots, N$ denote the number of neighbors of node $i$ in network $k$. For constant values $\mu_{ki}$, $\gamma^*_{ki}$, and $\beta^*_{ki}$, the extinction state is (globally in $\mathcal{P} = [0, 1]^{2N}$) exponentially stable if for every node $i$ in $A$ and $B$ it holds that*

$$\mu_{Ai} > \gamma^*_{Ai} + \beta^*_{Ai} N_{Ai}, \tag{8a}$$

$$\mu_{Bi} > \gamma^*_{Bi} + \beta^*_{Bi} N_{Bi}. \tag{8b}$$

**Proof.** In virtue of Lemma 1, it is sufficient to show that the conditions (8) ensure that $\sigma(\mathbf{H}) < 1$. This is achieved by applying Geršgorin's theorem [50] to the matrix $\mathbf{H}$ using an upper-bound estimate for the spectral radius.

Let $\lambda$ be an arbitrary eigenvalue of $\mathbf{H}$. Recalling that all entries of the matrices $\mathbf{A}$ and $\mathbf{B}$ are non-negative, Geršgorin's theorem [50] implies the following inequalities

$$|\lambda| \leq \gamma^*_{Ai} + \sum_{j=1}^{N} \beta^*_{Ai} a_{ij} + 1 - \mu_{Ai},$$

$$|\lambda| \leq \gamma^*_{Bi} + \sum_{j=1}^{N} \beta^*_{Bi} b_{ij} + 1 - \mu_{Bi}.$$

Thus $|\lambda| < 1$ is satisfied if

$$|\lambda| < \gamma^*_{Ai} + \sum_{j=1}^{N} \beta^*_{Ai} a_{ij} + 1 - \mu_{Ai} < 1,$$

$$|\lambda| < \gamma^*_{Bi} + \sum_{j=1}^{N} \beta^*_{Bi} b_{ij} + 1 - \mu_{Bi} < 1.$$

Rearranging and taking into account that the numbers of neighbors of node $i$ in network $\mathbf{A}$ and $\mathbf{B}$ is given by $N_{Ai} = \sum_{j=1}^{N} a_{ij}$, $N_{Bi} = \sum_{j=1}^{N} b_{ij}$, respectively, it follows that this condition is satisfied if

$$\gamma^*_{Ai} + \sum_{j=1}^{N} \beta^*_{Ai} a_{ij} = \gamma^*_{Ai} + \beta^*_{Ai} N_{Ai} < \mu_{Ai},$$

$$\gamma^*_{Bi} + \sum_{j=1}^{N} \beta^*_{Bi} b_{ij} = \gamma^*_{Bi} + \beta^*_{Bi} N_{Bi} < \mu_{Bi},$$

for $i = 1, 2, \ldots, N$. These inequalities correspond to the ones stated in (8). $\square$

**Remark 2.** *The stability conditions (8) of the system basically state that the recovery rate of each node must be higher than the rate with which it potentially receives infected messages or has contact with infected neighbors, measured by the total amount of intra-layer contacts in each network $k = \{A, B\}$ during one time interval, i.e., $\beta_{ki} N_{ki}$ plus the inter-layer contacts $\gamma_{ki}$ during the same time interval.*

Condition (8) can be used to determine which nodes should be controlled, i.e., for which nodes $i$ inequalities (8) are not satisfied in either of the networks $A$ and/ or $B$ and thus either of the rates $\gamma_{ki}$ or $\beta_{ki}$ should be adapted in such a way that $\gamma_{ki} < \gamma^*_{ki}$ and/ or $\beta_{ki} < \beta^*_{ki}$ with $\gamma^*_{ki}, \beta^*_{ki}$ chosen so that (8) holds. This is summarized in the following corollary.

**Corollary 1.** *The extinction state is (globally in $\mathcal{P}$) exponentially stable if for all nodes $i$ for which either of the conditions in (8) does not hold the parameter $\gamma_{ki}$ and/ or $\beta_{ki}$ are adapted so that the inequalities (8) are satisfied.*

**Remark 3.** *It should be noted that the conditions of Corollary 1 are only sufficient and not necessary. Actually, in specific scenarios the number of nodes for which the transmission parameters have to be adapted can be smaller. Alternative (non-analytic) approaches to determine the nodes to be controlled would be e.g., using optimization or genetic algorithms.*

**Remark 4.** *In comparison with the single-layer setup considered, e.g., in [36–38] the additional dependency on $\gamma_{ki}$, $k = A, B$ introduces stronger conditions. This will most probably imply a higher number of nodes to be controlled in the case of interconnecting the network with another one, i.e., the number of nodes that need to be controlled to ensure an asymptotically stable interconnection will be larger then the sum of the numbers of nodes that need to be controlled in each sub-network to achieve individual asymptotic stability. This is a particularly important point highlighting a consequence of the complex interplay of two nonlinear dynamical systems pointed out in Remark 1.*

**Remark 5.** *Conditions (8), as alternative to Corollary 1, also suggests as sufficient condition, to adapt the parameters $N_{Ai}$ and/or $N_{Bi}$. This adaptation requires disconnecting links from those nodes that do not satisfy condition (8) in order to reach the extinction state, resulting in an equivalent method as the one proposed in Adaptive Networks [39,46]. However, our approach keeps the network structure, modifying the parameters associated with the interaction probabilities of the model, avoiding disconnecting nodes.*

According to inequalities (8), a set of all possible scenarios for adaptation of parameters in every layer and for every node is presented in Table 1. That means that every node could have a different set of parameter to be controlled as shown in the Table, with the exception of those nodes that satisfied the condition (8) that do not need to be controlled as is shown in scenario 1. We can notice that in scenario 2 the critical parameter (i.e., the parameter to be controlled) of node $i$, situated in layer $k = \{A, B\}$, is given by $\gamma_{ki}$. For the scenario 5 we have several options and the criterion to be selected will depend on the specific implementation costs varying with the particular case example at hand.

Note from Table 1 that it is not necessary for the nodes of any layer to be acquainted of the structure and properties of the nodes of the other layer in order to control and eventually reach the extinction state. This constitutes one of the virtues of non centralized control.

**Table 1.** Amenable control parameters for the nodes of every layer $k = \{A, B\}$.

| Scenario | Critical Parameter | Satisfied | Not Satisfied |
|----------|--------------------|-----------|---------------|
| 1 | - | $\mu_{ki} > \gamma_{ki} + \beta_{ki} N_{ki}$ | - |
| 2 | $\gamma_{ki}$ | $\mu_{ki} - \beta_{ki} N_{ki} \geq 0$ | $\mu_{ki} > \gamma_{ki} + \beta_{ki} N_{ki}$ |
| 3 | $\beta_{ki}$ | $\mu_{ki} - \gamma_{ki} \geq 0$ | $\mu_{ki} > \gamma_{ki} + \beta_{ki} N_{ki}$ |
| 4 | $\gamma_{ki}$ and $\beta_{ki}$ | - | $\mu_{ki} > \gamma_{ki} + \beta_{ki} N_{ki}$ |
| 5 | $\gamma_{ki}$ or $\beta_{ki}$ | $\mu_{ki} - \beta_{ki} N_{ki} \geq 0$ <br> $\mu_{ki} - \gamma_{ki} \geq 0$ | $\mu_{ki} > \gamma_{ki} + \beta_{ki} N_{ki}$ |

## 5. Simulations

To corroborate the theoretical results, numerical simulations have been performed considering a spreading process in a two-layer network with $N = 10^5$ nodes in each layer. In the simulations performed, in order to verify that our results are independent of the topology, we have selected three different types of networks: Barabási–Albert scale-free (BA type), Regular nearest-neighbor (R type) and Small-World (WS type). Every network was built according to the methods discussed in [51], and as it is stated in this reference, the WS network was constructed randomly rewiring a Regular network with parameters shown in Table 2. As stated above, each node in layer $A$ is connected to its counterpart in layer $B$.

**Table 2.** Construction parameters for networks Barábasi-Albert ($BA$), Regular ($R$) and Small-World (WS).

| Network | Parameters |
|---|---|
| $BA_1$ | $m_0 = 10, m = 2$ |
| $BA_2$ | $m_0 = 5, m = 3$ |
| $R_1$ | Every node is connected with 20 nearest neighbors. |
| $R_2$ | Every node is connected with 10 nearest neighbors. |
| $WS_1$ | Every node in $R_1$ network was randomly rewired with probability 0.2. |
| $WS_2$ | Every node in $R_2$ network was randomly rewired with probability 0.3. |

For the subsequent analysis the parameter intervals shown in Table 3 were selected for $\mu_i$, $\gamma_i$ and $\beta_i$ and every type of network in Table 2 and for every node $i = 1, 2, \ldots, N$ in such a way that a considerable endemic response can be observed when the network parameters are uniformly distributed over these intervals.

**Table 3.** Simulation parameters for each node $i = 1, 2, \ldots, N$, in every network in Table 2.

| Network | $\mu_i$ | $\gamma_i$ | $\beta_i$ |
|---|---|---|---|
| $BA_1, R_1, WS_1$ | $(0.60, 0.80)$ | $(0.40, 0.80)$ | $(0.01, 0.03)$ |
| $BA_2, R_2, WS_2$ | $(0.50, 0.70)$ | $(0.20, 0.35)$ | $(0.02, 0.06)$ |

Considering the parameters shown in Table 2, six network layers were built (two networks for each network $BA$, $R$ and $WS$) that were combined to form six different two-layer networks as listed in Table 4. The parameters of each node in each layer were assigned randomly according to the intervals given in Table 3. Based on these scenarios the nodes to be controlled were identified and classified according to Table 1 to establish a control criteria. The results are summarized in Table 5 showing the number of nodes for which $\gamma$ needs to be adjusted, those for which $\beta$ needs to be adjusted, those for which either of both needs to be adjusted and those for which both need to be adjusted. Accordingly, the total number of nodes to be controlled is given in the last column.

**Table 4.** Amenable parameters chosen to control every two layer network. Compare this with data shown in Table 5.

| No. | Layer $A$ | Layer $B$ | Amenable Parameters Chosen | Figure |
|---|---|---|---|---|
| 1 | $R_1$ | $R_2$ | $\beta_{Ai}$ and $\beta_{Bi}$ | 2 |
| 2 | $BA_1$ | $BA_2$ | $\gamma_{Ai}, \beta_{Ai}$ and $\beta_{Bi}$ | 3 |
| 3 | $WS_1$ | $WS_2$ | $\gamma_{Ai}, \beta_{Ai}$ and $\beta_{Bi}$ | 4 |
| 4 | $R_2$ | $BA_2$ | $\beta_{Ai}$ and $\beta_{Bi}$ | 5 |
| 5 | $BA_1$ | $WS_2$ | $\gamma_{Ai}, \beta_{Ai}$ and $\beta_{Bi}$ | 6 |
| 6 | $R_1$ | $WS_2$ | $\gamma_{Ai}$ and $\beta_{Bi}$ | 7 |

**Table 5.** Number of nodes and their parameters to control for every network.

| Network | $\gamma_i$ | $\beta_i$ | $\gamma_i$ or $\beta_i$ | $\gamma_i$ and $\beta_i$ | Nodes to Control |
|---|---|---|---|---|---|
| $BA_1$ | 12,178 | 25,716 | 52,080 | 7761 | 97,735 |
| $BA_2$ | 0 | 5082 | 12,191 | 0 | 17,273 |
| $R_1$ | 19,939 | 0 | 67,448 | 0 | 87,387 |
| $R_2$ | 0 | 6138 | 62,286 | 0 | 68,424 |
| $WS_1$ | 19,915 | 30 | 66,501 | 24 | 86,470 |
| $WS_2$ | 0 | 11,697 | 52,678 | 0 | 64,375 |

The difference between analyzing and controlling the networks in a single layer context to the two-layer one becomes clear when comparing the numbers in Table 4. Without interconnection of the two layers only the third column is relevant, i.e., the number of nodes for which $\beta$ must be adjusted. It can be clearly seen that due to the coupling with a second layer very drastic changes occur, independent of the choice of topology in the attached layer. In particular, consider an interconnection of $R_1$ and $WS_1$. In the isolated network $R_1$ no node needs to be controlled as the extinction point is globally asymptotically stable. The network $WS_1$, when isolated only requires 30 nodes to be controlled. When interconnecting both networks it becomes necessary to control 87,387 nodes in $R_1$ and 86,470 in $WS_1$.

Note further that according to Table 5 several scenarios could arise depending on the networks selected to build the two layer multiplex network, for example, if we propose a two layer multiplex network made up of $R_2$ (layer $A$) and $BA_2$ (layer $B$) then, according to Table 5, it is only necessary to control both networks taking $\beta$ as amenable parameter.

In order to show the effect of the proposed control law, we simulate several two layer networks as described in Table 4. The changes in the transmission parameters are applied at time 35. In these simulations, and following the above discussion, the specific values for the control parameters are chosen either as one of the following:

$$\beta_{ki}(t) \;=\; \begin{cases} \beta_{ki} & t < 35, \\ 0.99\frac{\mu ki - \gamma ki}{N_{ki}} & t \geq 35 \end{cases} \tag{9a}$$

$$\gamma_{ki}(t) \;=\; \begin{cases} \gamma_{ki} & t < 35, \\ 0.99(\mu ki - \beta ki N_{ki}) & t \geq 35 \end{cases} \tag{9b}$$

for $k = \{A, B\}$ and $i = 1, 2, \ldots, N$. Besides, in this case it is also possible to chose $\gamma$ and $\beta$ (at the same time) as control parameters (scenario 4 from Table 1). This is also the case of networks 2, 3 and 5 in Table 4, where an specific combination of control parameters are chosen as $\gamma_{ki} = 0.99\mu_{ki}$ and (9a).

In consequence of this control scenario, at the beginning the state converge to an endemic fixed point that disappears after applying the control strategy at $t = 35$, causing the states to exponentially converge to the extinction state, as shown in Figures 2–7. In the figures each line corresponds to the mean value (or probability density)

$$\rho_A(t) = \sum_{i=1}^{N} p_{Ai}(t) \quad (\text{red}) \quad \text{and} \quad \rho_B(t) = \sum_{i=1}^{N} p_{Bi}(t) \quad (\text{blue}), \tag{10}$$

in the respective network for the initial conditions $p_{Ai}(0), p_{Bi}(0) \in \{0.1, 0.3, 0.5, 0.7, 0.9\}$, $i = 1, \ldots, N$. For example, in Figure 3 around 28% of the nodes in layer $A$ are infected meanwhile in network $B$, around 17% of the nodes are. Once the control is activated, in all simulations, the state of the system exponentially converges to the extinction state according to the assertion of Corollary 1.

In order to analyze the dependency of the number of nodes to be controlled on the particular choice of network a statistical analysis has been carried out for the networks $BA_1, BA_2, R_1, R_2, WS_1, WS_2$ with construction specified in Table 2 by randomly assigning the seeds for the network generation and the parameters using a uniform distribution over the intervals provided in Table 3. For the $BA$-type networks a total of 481 networks were considered, for the $R$-type networks 600, and for the $WS$-type networks 464. The resulting sample distributions showing the number of times a certain number of nodes needs to be controlled are shown in Figure 8. For all six networks two scenarios are evaluated: (a) the isolated network and (b) the network in interconnection with another one. From the sub-figures it can be seen that (i) in all networks a very small variation is observed in the number of nodes to be controlled, and (ii) in the passage from the isolated to the interconnected network the number of nodes to be controlled increases considerably. This last fact illustrates again the substantial difference between controlling isolated and interconnected networks, as highlighted above at several places.

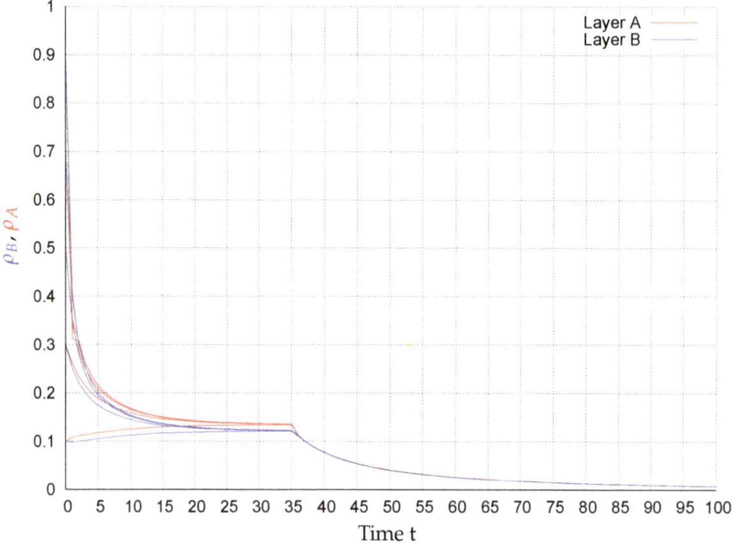


**Figure 2.** $\rho_A(t)$ (red) and $\rho_B(t)$ (blue) for several initial conditions in network $R_1$-$R_2$.

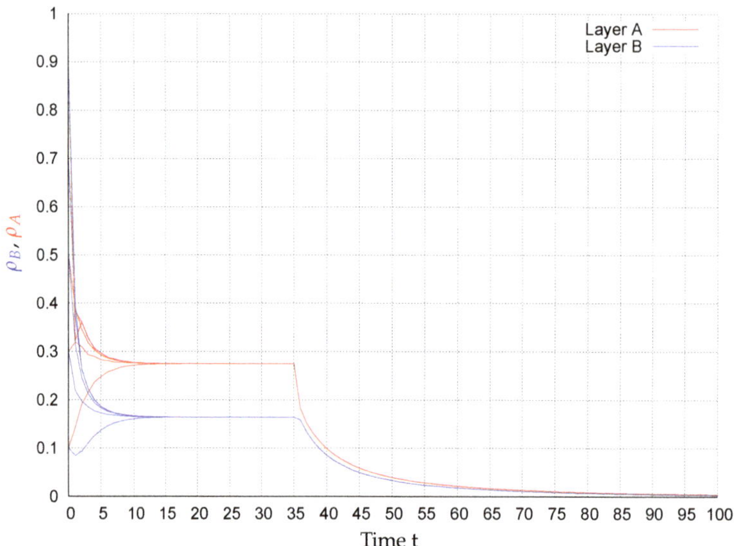


**Figure 3.** $\rho_A(t)$ (red) and $\rho_B(t)$ (blue) for several initial conditions in network $BA_1$-$BA_2$.

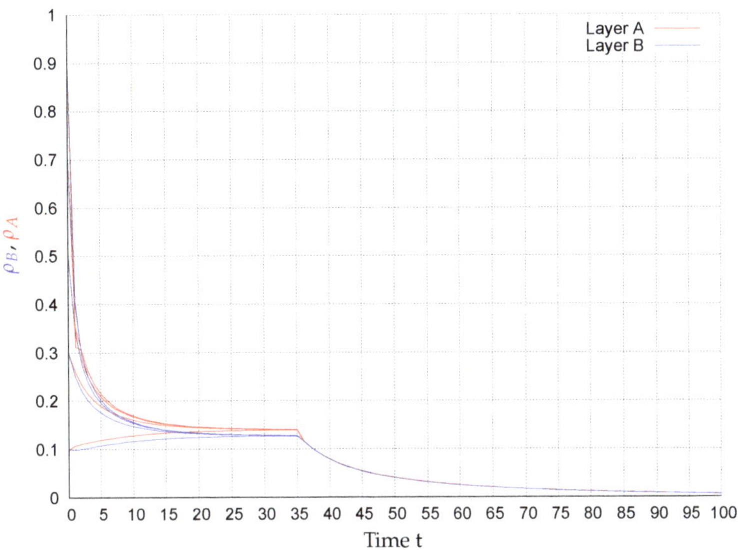


**Figure 4.** $\rho_A(t)$ (red) and $\rho_B(t)$ (blue) for several initial conditions in network $WS_1$-$WS_2$.

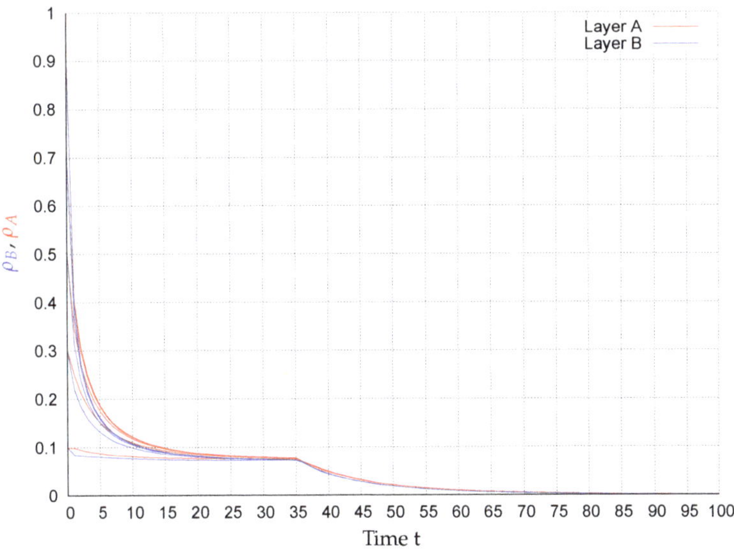


**Figure 5.** $\rho_A(t)$ (red) and $\rho_B(t)$ (blue) for several initial conditions in network $R_2$-$BA_2$.

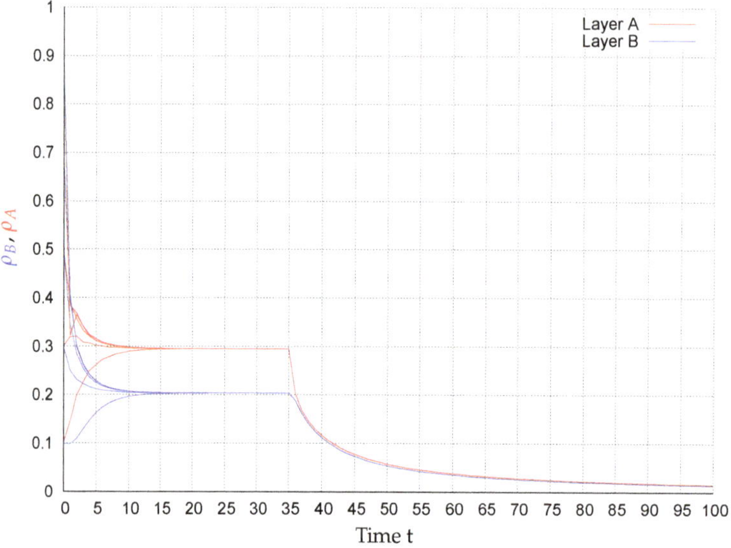

**Figure 6.** $\rho_A(t)$ (red) and $\rho_B(t)$ (blue) for several initial conditions in network $BA_1$-$WS_2$.

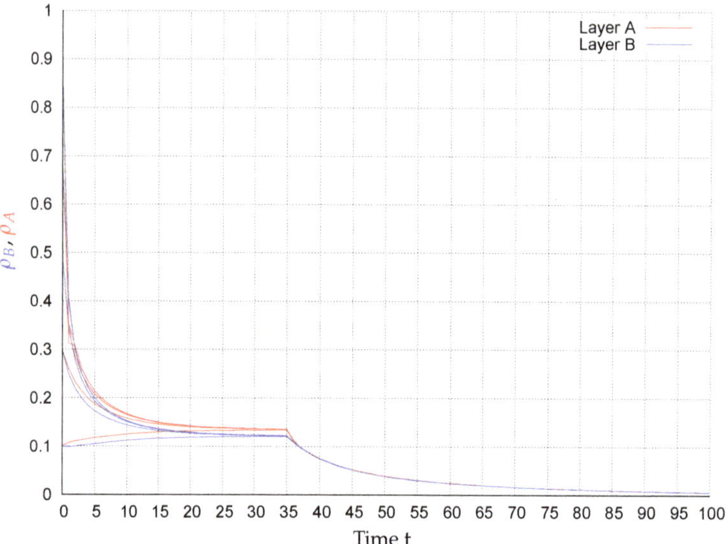

**Figure 7.** $\rho_A(t)$ (red) and $\rho_B(t)$ (blue) for several initial conditions in network $R_1$-$WS_2$.

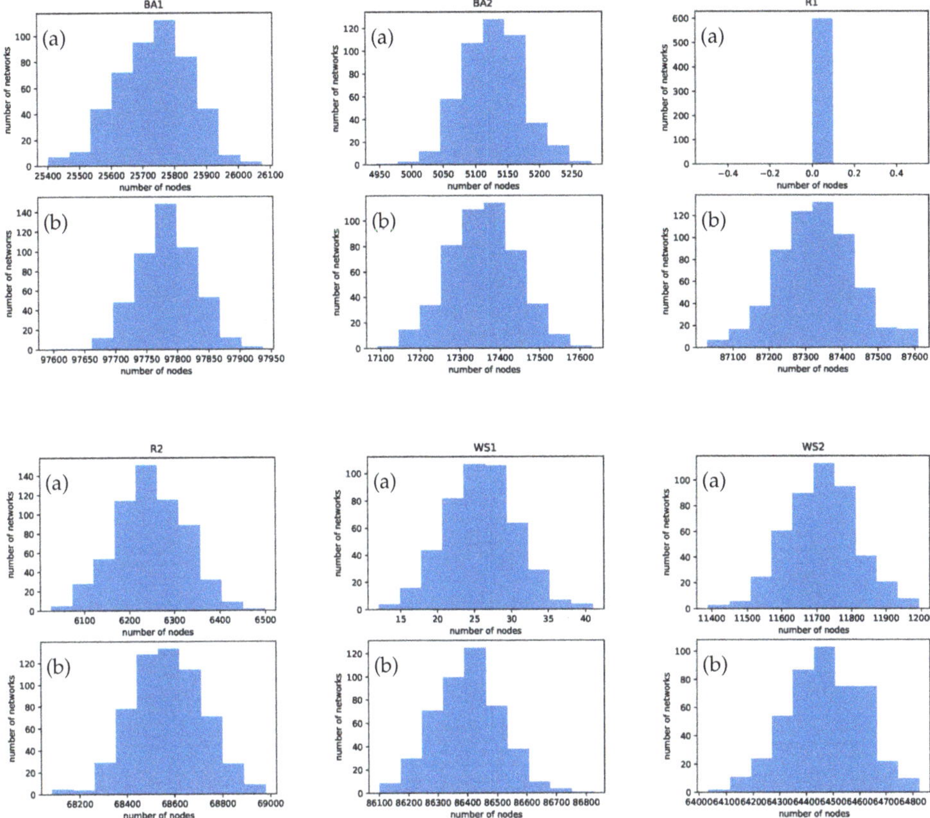

**Figure 8.** Sample distributions of the number of nodes to be controlled in the considered networks specified in Table 2: (**a**) isolated network and (**b**) interconnected network.

## 6. Conclusions

The control of a spreading process in a two-layer multiplex network with a parametric control strategy is analyzed. Sufficient conditions for the choice of nodes and parameters to be controlled are established using rigorous mathematical derivations ensuring the exponential stability of the extinction state globally with respect to the set of all possible probability states. The proposed control strategy consists in the adaptation of the parameters specifying the intra-layer and inter-layer transmission rates only for a limited number of nodes that are characterized by a parametric threshold condition. Particular emphasis is made on the substantial difference between controlling isolated and interconnected networks, showing intrinsic cnections with the individual input-to-state stability and the small-gain criterion. It results that in the passage from controlling isolated networks to interconnected ones, the number of nodes that need to be controlled significantly increases. The theoretical results are analyzed in multiplex networks with different representative topologies in each layer with $10^5$ nodes each. The corresponding simulation studies and statistical evaluations of the number of nodes to be controlled corroborate the theoretical findings.

Based on the presented results future studies will focus on the generalization of the discussed ideas to the case of *n*-layer multiplex networks, in order to further enlighten the expected challenges when adding additional layers. Furthermore, the model identification and testing of the presented approaches in real-world scenarios based on explicit data will be focused on in future studies.

**Author Contributions:** All authors designed and did the research, in particular the formal analysis; R.B.J. did the conceptualization; A.S. was in charge of the supervision and methodology and substantial formal analysis; L.A.A.R. performed the numerical experiments and wrote the simulation program; all authors analyzed the data; A.S. and R.B.J. wrote the paper; All authors reviewed the manuscript; all authors have read and approved the final manuscript. All authors have read and agreed to the published version of the manuscript.

**Funding:** This research was partially funded by UAM Cuajimalpa CNI Division's fund for internal academic groups.

**Conflicts of Interest:** The authors declare no conflict of interest.

## Abbreviations

The following abbreviations are used in this manuscript:

| | |
|---|---|
| BA | Barabási-Albert scale free network |
| R | Regular nearest-neighbor network |
| WS | Watts-Strogatz small-world network |
| MDPI | Multidisciplinary Digital Publishing Institute |

## References


1. Aleta, A.; Moreno, Y. Multilayer Networks in a Nutshell. *Annu. Rev. Condens. Matter Phys.* **2019**, *10*, 45–62. [CrossRef]
2. Liu, J.; Wu, X.; Lü, J.; Wei, X. Infection-Probability-Dependent Interlayer Interaction Propagation Processes in Multiplex Networks. *IEEE Trans. Syst.* **2019**, 1–12. [CrossRef]
3. Wei, X.; Chen, S.; Wu, X.; Feng, J.; Lu, J.-a. A unified framework of interplay between two spreading processes in multiplex networks. *Europhys. Lett.* **2016**, *114*, 26006. [CrossRef]
4. Tejedor, A.; Longjas, A.; Foufoula-Georgiou, E.; Georgiou, T.T.; Moreno, Y. Diffusion Dynamics and Optimal Coupling in Multiplex Networks with Directed Layers. *Phys. Rev. X* **2018**, *8*, 031071. [CrossRef]
5. Boccaletti, S.; Bianconi, G.; Criado, R.; del Genio, C.; nes, J.G.G.; Romance, M.; na Nadal, I.S.; Wang, Z.; Zanin, M. The structure and dynamics of multilayer networks. *Phys. Rep.* **2014**, *544*, 1–122. [CrossRef] [PubMed]
6. Buldú, J.M.; Porter, M.A. Frequency-based brain networks: From a multiplex framework to a full multilayer description. *Netw. Neurosci.* **2018**, *2*, 418–441. [CrossRef] [PubMed]
7. Jiang, J.; Zhou, T. Resource control of epidemic spreading through a multilayer network. *Sci. Rep.* **2018**, *8*, 1629. [CrossRef]
8. de Arruda, G.F.; Cozzo, E.; Peixoto, T.P.; Rodrigues, F.A.; Moreno, Y. Disease Localization in Multilayer Networks. *Phys. Rev. X* **2017**, *7*, 011014. [CrossRef]
9. Granell, C.; Gómez, S.; Arenas, A. Dynamical Interplay between Awareness and Epidemic Spreading in Multiplex Networks. *Phys. Rev. Lett.* **2013**, *111*, 128701. [CrossRef]
10. Granell, C.; Gómez, S.; Arenas, A. Competing spreading processes on multiplex networks: Awareness and epidemics. *Phys. Rev. E* **2014**, *90*, 012808. [CrossRef]
11. Schaum, A.; Jaquez, R.B. Estimating the state probability distribution for epidemic spreading in complex networks. *Appl. Math. Comput.* **2016**, *291*, 197–206. [CrossRef]
12. Stanoev, A.; Trpevski, D.; Kocarev, L. Modeling the Spread of Multiple Concurrent Contagions on Networks. *PLoS ONE* **2014**, *9*, 1–16. [CrossRef] [PubMed]
13. Chen, X.; Wang, R.; Tang, M.; Cai, S.; Stanley, H.E.; Braunstein, L.A. Suppressing epidemic spreading in multiplex networks with social-support. *New J. Phys.* **2018**, *20*, 013007. [CrossRef]
14. Pu, C.; Li, S.; Yang, X.; Yang, J.; Wang, K. Information transport in multiplex networks. *Phys. A* **2016**, *447*, 261–269. [CrossRef]

15. Freitas, C.G.S.; Aquino, A.L.L.; Ramos, H.S.; Frery, A.C.; Rosso, O.A. A detailed characterization of complex networks using Information Theory. *Sci. Rep.* **2019**, *9*, 16689. [CrossRef] [PubMed]

16. Wei, X.; Valler, N.C.; Prakash, B.A.; Neamtiu, I.; Faloutsos, M.; Faloutsos, C. Competing Memes Propagation on Networks: A Network Science Perspective. *IEEE J. Sel. Areas Commun.* **2013**, *31*, 1049–1060. [CrossRef]

17. Harush, U.; Barzel, B. Dynamic patterns of information flow in complex networks. *Nat. Commun.* **2017**, 2181. [CrossRef]

18. Baggio, G.; Rutten, V.; Hennequin, G.; Zampieri, S. Efficient communication over complex dynamical networks: The role of matrix non-normality. *Sci. Adv.* **2020**, *6*, doi:10.1126/sciadv.aba2282. [CrossRef]

19. Yuan, Z.; Zhao, C.; Di, Z.; Wang, W.X.; Lai, Y.C. Exact controllability of complex networks. *Nat. Commun.* **2013**, *4*, 2447. [CrossRef]

20. Nozari, E.; Pasqualetti, F.; Cortés, J. Time-invariant versus time-varying actuator scheduling in complex networks. In Proceedings of the 2017 American Control Conference (ACC), Seattle, WA, USA, 24–26 May 2017; pp. 4995–5000.

21. Lindmark, G.; Altafini, C. Minimum energy control for complex networks. *Sci. Rep.* **2018**, *8*, 3188. [CrossRef]

22. Chen, G. Pinning control and controllability of complex dynamical networks. *Int. J. Autom. Comput.* **2017**, *14*, 1–9. [CrossRef]

23. Pang, S.P.; Wang, W.X.; Hao, F.; Lai, Y.C. Universal framework for edge controllability of complex networks. *Sci. Rep.* **2017**, *7*, 4224. [CrossRef] [PubMed]

24. Song, K.; Li, G.; Chen, X.; Deng, L.; Xiao, G.; Zeng, F.; Pei, J. Target Controllability of Two-Layer Multiplex Networks Based on Network Flow Theory. *IEEE Trans. Cybern.* **2019**, 1–13. [CrossRef]

25. Structural Accessibility and Structural Observabilityof Nonlinear Networked Systems. *IEEE Trans. Netw. Sci. Eng.* **2020**, 1656–1666. [CrossRef]

26. Menara, T.; Baggio, G.; Bassett, D.S.; Pasqualetti, F. Conditions for Feedback Linearization of Network Systems. *IEEE Control Syst. Lett.* **2020**, *4*, 578–583. [CrossRef]

27. Menichetti, G.; Dall'Asta, L.; Bianconi, G. Control of Multilayer Networks. *Sci. Rep.* **2016**, *6*, 20706. [CrossRef] [PubMed]

28. Pósfai, M.; Gao, J.; Cornelius, S.P.; Barabási, A.L.; D'Souza, R.M. Controllability of multiplex, multi-time-scale networks. *Phys. Rev. E* **2016**, *94*, 032316. [CrossRef]

29. Li, G.; Ding, J.; Wen, C.; Pei, J. Optimal control of complex networks based on matrix differentiation. *Europhys. Lett.* **2016**, *115*, 68005. [CrossRef]

30. Watkins, N.J.; Nowzari, C.; Preciado, V.M.; Pappas, G.J. Optimal Resource Allocation for Competitive Spreading Processes on Bilayer Networks. *IEEE Trans. Control. Netw. Syst.* **2018**, *5*, 298–307. [CrossRef]

31. Zhang, X.; Wang, H.; Lv, T. Efficient target control of complex networks based on preferential matching. *PLoS ONE* **2017**, *12*, 1–10. [CrossRef]

32. Nicosia, V.; Criado, R.; Romance, M.; Russo, G.; Latora, V. Controlling centrality in complex networks. *Sci. Rep.* **2012**, *2*, 218. [CrossRef] [PubMed]

33. Nacher, J.C.; Ishitsuka, M.; Miyazaki, S.; Akutsu, T. Finding and analysing the minimum set of driver nodes required to control multilayer networks. *Sci. Rep.* **2019**, *9*, 576. [CrossRef] [PubMed]

34. Isidori, A. *Nonlinear Control Systems*; Springer: London, UK, 2000.

35. Wan, Y.; Roy, S.; Saberi, A. Designing spatially heterogeneous strategies for control of virus spread. *IET Syst. Biol.* **2008**, *2*, 184–201. [CrossRef]

36. Alarcón Ramos, L.A.; Bernal Jaquez, R.; Schaum, A. Output-Feedback Control for Discrete-Time Spreading Models in Complex Networks. *Entropy* **2018**, *20*, 204. [CrossRef]

37. Alarcón-Ramos, L.A.; Bernal Jaquez, R.; Schaum, A. Output-Feedback Control of Virus Spreading in Complex Networks With Quarantine. *Front. Appl. Math. Stat.* **2018**, *4*, 34. [CrossRef]

38. Alarcón-Ramos, L.A.; Schaum, A.; Lucatero, C.R.; Jaquez, R.B. Stability analysis for virus spreading in complex networks with quarantine and non-homogeneous transition rates. *J. Phys. Conf. Ser.* **2014**, *490*, 012011. [CrossRef]

39. Nowzari, C.; Preciado, V.M.; Pappas, G.J. Analysis and Control of Epidemics: A Survey of Spreading Processes on Complex Networks. *IEEE Control Syst.* **2016**, *36*, 26–46. [CrossRef]

40. Chakrabarti, D.; Wang, Y.; Wang, C.; Leskovec, J.; Faloutsos, C. Epidemic Thresholds in Real Networks. *ACM Trans. Inf. Syst. Secur.* **2008**, *10*, 1:1–1:26. [CrossRef]

41. Gómez, S.; Arenas, A.; Borge-Holthoefer, J.; Meloni, S.; Moreno, Y. Discrete-time Markov chain approach to contact-based disease spreading in complex networks. *Europhys. Lett.* **2010**, *89*, 38009. [CrossRef]
42. Wang, W.; Tang, M.; Stanley, H.E.; Braunstein, L.A. Unification of theoretical approaches for epidemic spreading on complex networks. *Rep. Prog. Phys.* **2017**, *80*, 036603. [CrossRef]
43. Gómez, S.; Arenas, A.; Borge-Holthoefer, J.; Meloni, S.; Moreno, Y. Probabilistic framework for epidemic spreading in complex networks. *Int. J. Complex Syst. Sci.* **2011**, *1*, 47–54.
44. Wang, Y.; Chakrabarti, D.; Wang, C.; Faloutsos, C. Epidemic Spreading in Real Networks: An Eigenvalue Viewpoint. In Proceedings of the 22nd International Symposium on Reliable Distributed Systems, Florence, Italy, 6–8 October 2003; pp. 23–25. [CrossRef]
45. Bernal Jaquez, R.; Schaum, A.; Alarcón, L.; Rodríguez, C. Stability analysis for virus spreading in complex networks with quarantine. *Publicaciones Matemáticas Del Urug.* **2013**, *14*, 221–233.
46. Achterberg, M.A.; Dubbeldam, J.L.A.; Stam, C.J.; Van Mieghem, P. Classification of link-breaking and link-creation updating rules in susceptible-infected-susceptible epidemics on adaptive networks. *Phys. Rev. E* **2020**, *101*, 052302. [CrossRef] [PubMed]
47. Sontag, E.D. On the Input-to-State Stability Property. *Eur. J. Control.* **1995**, *1*, 24–36. [CrossRef]
48. Jiang, Z.P.; Wang, Y. Input-to-state stability for discrete-time nonlinear systems. *Automatica* **2001**, *37*, 857–869. [CrossRef]
49. Bromwich, T.; Bromwich, T. *An Introduction to the Theory of Infinite Series*; American Mathematical Society: Providence, RI, USA, 2005.
50. Cullen, C.G. *Matrices and Linear Transformations*; Dover Publications: New York, NY, USA, 1990.
51. Wang, X.F.; Chen, G. Complex networks: Small-world, scale-free and beyond. *IEEE Circuits Syst. Mag.* **2003**, *3*, 6–20. [CrossRef]

**Publisher's Note:** MDPI stays neutral with regard to jurisdictional claims in published maps and institutional affiliations.

*Article*

# Investigating the Influence of Inverse Preferential Attachment on Network Development

Cynthia S. Q. Siew [1,*] and Michael S. Vitevitch [2]

[1] Department of Psychology, National University of Singapore, Singapore 117570, Singapore
[2] Department of Psychology, University of Kansas, Lawrence, KS 66045, USA; mvitevit@ku.edu
[*] Correspondence: cynthia@nus.edu.sg

Received: 22 July 2020; Accepted: 8 September 2020; Published: 15 September 2020

**Abstract:** Recent work investigating the development of the phonological lexicon, where edges between words represent phonological similarity, have suggested that phonological network growth may be partly driven by a process that favors the acquisition of new words that are phonologically similar to several existing words in the lexicon. To explore this growth mechanism, we conducted a simulation study to examine the properties of networks grown by inverse preferential attachment, where new nodes added to the network tend to connect to existing nodes with fewer edges. Specifically, we analyzed the network structure and degree distributions of artificial networks generated via either preferential attachment, an inverse variant of preferential attachment, or combinations of both network growth mechanisms. The simulations showed that network growth initially driven by preferential attachment followed by inverse preferential attachment led to densely-connected network structures (i.e., smaller diameters and average shortest path lengths), as well as degree distributions that could be characterized by non-power law distributions, analogous to the features of real-world phonological networks. These results provide converging evidence that inverse preferential attachment may play a role in the development of the phonological lexicon and reflect processing costs associated with a mature lexicon structure.

**Keywords:** network growth; preferential attachment; inverse preferential attachment; language networks; language development

---

## 1. Introduction

Many complex systems, such as the Internet, brain networks, and social networks, can be classified as networks—collections of entities connected to each other in a web-like fashion—permitting the application of network analysis to study these systems (see [1] for a review). A common feature across diverse complex networks is their scale-free degree distribution, whereby most nodes in the network have very few edges or links and a few nodes have many edges or links. Preferential attachment models of network growth, where new nodes that are added to the network tend to connect to existing nodes with many links (i.e., high degree nodes), have been prominent in the literature covering network growth and evolution, because such models describe a generic mechanism that provides an elegant account of the emergence of scale-free complex networks [2–5]. In this paper, we conducted a series of network simulations to specifically examine the properties of networks grown via a different mechanism, which we refer to as inverse preferential attachment, where new nodes added to the network tend to connect to existing nodes with fewer edges.

Our present approach of simulating network growth via inverse preferential attachment was directly motivated by recent research examining the development of language networks constructed from phonological similarity among words. In these language networks, nodes represent words, while edges are placed between words that share similar sounds [6]. Previous research has shown that the structure of the phonological lexicon has measurable influences on various language-related processes [7–9]. Research investigating the processes that facilitate the acquisition of the phonological form of a word indicate that phonological network growth may be driven by alternative network growth mechanisms other than the widely studied preferential attachment [10–12]. Central to the present study is a recent paper by Siew and Vitevitch [12], who conducted a longitudinal analysis of phonological networks of English and Dutch words and found that preferential attachment was a better predictor of acquisition than preferential acquisition. Furthermore, although the standard preferential attachment model was a significant predictor of acquisition at early stages of network growth (i.e., when the phonological network was "young"), there was a subsequent shift in the network growth mechanism, such that an inverse variant of preferential attachment became a significant predictor of acquisition at later stages of network growth (i.e., when the phonological network matured and contained many nodes and edges). To put it in another way, a network growth mechanism that prioritized the learning of words that were phonologically similar to words with many phonological neighbors (i.e., many edges) in the lexicon was important in the early stages of development, whereas a growth mechanism that prioritized the learning of words that were phonologically similar to words with few phonological neighbors (i.e., few edges) in the lexicon was important in the later stages of development. Siew and Vitevitch [12] provided further empirical support for inverse preferential attachment by conducting a word learning experiment, which found that people with mature lexicons (i.e., college students) were able to better learn made-up words that were phonologically similar to words with few phonological neighbors in the lexicon, as compared to made-up words that were phonologically similar to words with many phonological neighbors.

Given these intriguing patterns of phonological network growth observed in our prior work, the aim of the present paper was to conduct a computational exploration of these patterns. To this end, we conducted a series of network growth simulations to examine if networks generated by the preferential attachment growth algorithm and its inverse variant, as well as combinations of each algorithm, might lead to structurally different networks. Even though we examine a simple model of network growth here, this has potentially important theoretical implications for understanding how the large-scale development of the phonological lexicon could occur. For instance, the artificial randomly grown networks examined by Callaway and colleagues [13] exhibited many network characteristics that were also observed in real phonological networks [6], and we wanted to investigate if simulating network growth with typical or inverse preferential attachment mechanisms may also lead to networks with characteristics observed in real phonological networks. Computing network measures (such as average shortest path length, network diameter) and degree distributions is one way of evaluating the structure of simulated networks. Network measures such as the average shortest path length and network diameter provide an indication of the overall efficiency of the network (i.e., efficiency referring to a network's ability to quickly exchange information or for activation to spread in a network [14]), whereas degree distributions can be considered as structural signatures of the network, which can inform us about the growth processes that gave rise to the network [2]. If the overall network measures of simulated networks are qualitatively similar to real-world phonological language networks, this suggests that growth mechanisms that gave rise to the observed structure of the simulated networks might also contribute to the acquisition of phonological representations.

It is important to acknowledge that the approach taken here does not provide conclusive proof that either one of these network growth algorithms is entirely responsible for producing the structures observed in real-world phonological networks. Indeed, much research (e.g., [15]) has demonstrated that the famed scale-free network, for example, can be produced not only by the preferential attachment algorithm proposed by Barabási and Albert [2], but also by a number of other methods as well. In the absence of any other information, it would indeed be unwise to assert that a particular algorithm was responsible for producing a network with a particular set of characteristics. In the domain of psycholinguistics, however, there is a long and rich history of research that provides some guidance on which possible algorithms are unattested in the languages of the world, and therefore not plausible as a mechanism for the acquisition of words; or which algorithms have been observed with other research methods (e.g., case studies, archival analyses, laboratory-based experiments), and therefore might be plausible mechanisms for the acquisition of words and may also provide insight into certain language disorders (e.g., [16]). We performed the present simulation merely to offer an additional piece of evidence to complement the archival analyses and experiments in our earlier work [12], which might help to constrain the realm of possibility to the more restricted space of plausibility.

## 2. Materials and Methods

Each simulation began with a single node. The growth of the network was simulated by adding a new node and a single new link to the network at each iteration. Each simulation continued for 999 iterations, such that each resulting network consisted of 1000 nodes and 999 edges. To simulate the growth of the network via preferential attachment, the probability that a new node connected to a given existing node was proportional to the number of connections that the existing node had to other nodes in the network. Therefore, a new node was more likely to connect to an existing node with a high degree. To simulate the growth of the network via inverse preferential attachment, the probability that a new node connected to a given existing node was inversely proportional to the number of connections that the existing node had to other nodes in the network. In this case, a new node was more likely to connect to an existing node with a low degree. Finally, in random attachment, the new node had an equal probability of connecting to any existing node, regardless of its degree.

There was a total of 11 different network types, i.e., networks that were grown by different mechanisms and by various combinations of those mechanisms (see Figure 1 for a summary). Three network types were generated by a single mechanism, i.e., entirely via preferential attachment (PATT), entirely via inverse preferential attachment (iPATT), and via random attachment (Random). For ease of exposition, PATT refers to networks generated by preferential attachment and iPATT refers to networks generated by inverse preferential attachment. The remaining 8 network types were generated using a combination of preferential attachment and inverse preferential attachment, to explore how the "blending" of different growth models affected the development of the network, given that Siew and Vitevitch [12] found that preferential attachment was influential earlier in development but not later in development. Of these 8 network types, four were generated via preferential attachment first (for 200, 400, 600, and 800 iterations) followed by the inverse variant for the remainder of the iterations, and four were generated via inverse preferential attachment first (for 200, 400, 600, and 800 iterations) followed by the original preferential attachment model for the remainder of the iterations. The network growth simulations were repeated 100 times for each network type, resulting in a total of 1100 simulated networks. All simulations were conducted in R using the igraph library [17]. Analyses of the final network structure and their degree distributions were also conducted in R using the igraph and poweRlaw [18] libraries, respectively. The simulation and analysis R scripts, as well as the simulated network data, are available via the Supplementary Materials.

The characteristics of the 1100 simulated networks can be quantified in various ways to examine how the overall structures of these networks differ across different simulation conditions (i.e., network type). The following network measures will be computed: average shortest path length, network diameter, and degree distribution.

The shortest path length between two nodes refers to the fewest number of links that must be traversed to get from one node to another node in the network. The average shortest path length (ASPL) is the mean of the shortest path length obtained from every possible pairing of nodes in the network. A closely related measure is the diameter of the network; this is the longest shortest path length that exists in the network. The degree distribution refers to the probability distribution of node degrees in the network; in other words, how many nodes have a given number of connections in the network. Recall that degree refers to the number of connections incident to a node.

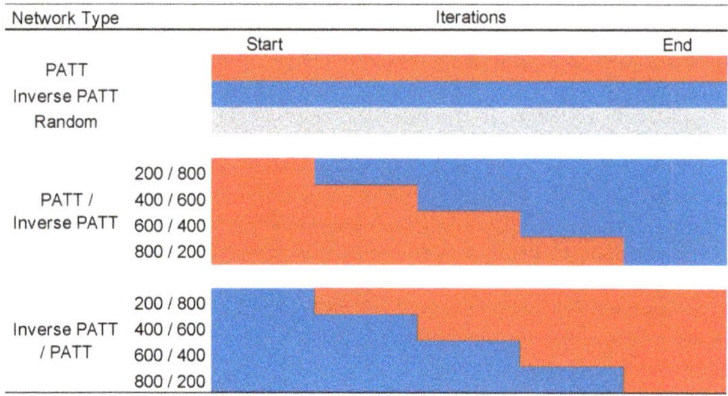


**Figure 1.** A summary of the 11 network growth conditions simulated in the present study. Red cells indicate growth by standard preferential attachment, blue cells indicate growth by inverse preferential attachment. PATT, preferential attachment.

## 3. Results

### 3.1. Overall Network Characteristics of Simulated Networks

For each of the simulated networks, the average shortest path length and network diameter was computed. Table 1 shows the mean ASPL and network diameter for the networks in each condition (i.e., network type) of the simulations.

**Table 1.** Means and standard deviations of network measures of simulated networks, summarized by each of the 11 simulation conditions. Note that all simulated networks had the same number of nodes and edges (1000 nodes and 999 edges).

| Network | | Nodes | Edges | ASPL | Diameter |
|---|---|---|---|---|---|
| PATT | M | 1000 | 999 | 8.34 | 11.55 |
| | SD | 0 | 0 | 0.50 | 1.42 |
| Inverse PATT | M | 1000 | 999 | 13.07 | 16.42 |
| | SD | 0 | 0 | 0.62 | 1.80 |
| Random | M | 1000 | 999 | 10.91 | 13.81 |
| | SD | 0 | 0 | 0.52 | 1.54 |
| PATT–Inverse PATT | | | | | |
| **200/800** | **M** | **1000** | **999** | **9.97** | **13.27** |
| | **SD** | **0** | **0** | **0.50** | **1.47** |
| 400/600 | M | 1000 | 999 | 9.22 | 12.58 |
| | SD | 0 | 0 | 0.49 | 1.58 |
| 600/400 | M | 1000 | 999 | 8.83 | 12.29 |
| | SD | 0 | 0 | 0.49 | 1.43 |
| 800/200 | M | 1000 | 999 | 8.55 | 11.93 |
| | SD | 0 | 0 | 0.49 | 1.50 |
| Inverse PATT–PATT | | | | | |
| 200/800 | M | 1000 | 999 | 11.25 | 14.28 |
| | SD | 0 | 0 | 0.59 | 1.60 |
| 400/600 | M | 1000 | 999 | 12.00 | 15.17 |
| | SD | 0 | 0 | 0.61 | 1.63 |
| 600/400 | M | 1000 | 999 | 12.48 | 15.59 |
| | SD | 0 | 0 | 0.59 | 1.76 |
| 800/200 | M | 1000 | 999 | 12.81 | 15.92 |
| | SD | 0 | 0 | 0.62 | 1.74 |

Legend: M = mean; SD = standard deviation; ASPL = average shortest path length; PATT = preferential attachment.

Independent samples *t*-tests comparing the average shortest path length and network diameter of PATT and iPATT networks showed that iPATT networks had larger diameters ($t(189.76) = 59.47$, $p < 0.001$) and longer ASPLs ($t(187.7) = 21.27$, $p < 0.001$) than PATT networks (see Figure 2). This suggests that networks generated by preferential attachment tend to be denser and more compact as compared to networks generated by inverse preferential attachment, despite having the same numbers of nodes and edges (see Figure 3 for network visualizations of the overall network structures).

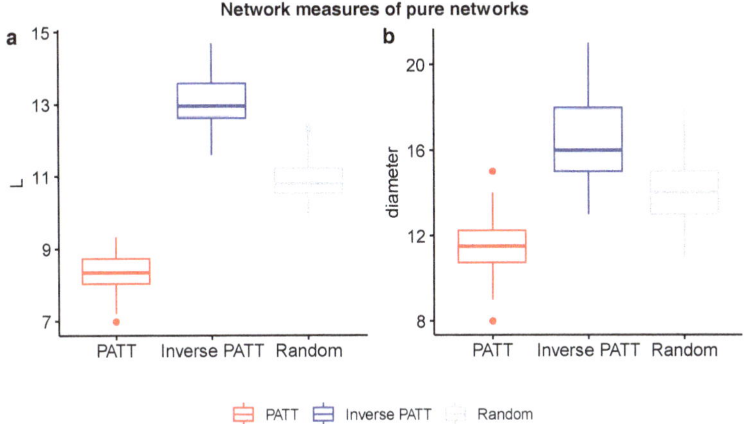



**Figure 2.** Boxplots of ASPL (**a**) and network diameter (**b**) values of networks grown via preferential attachment (PATT), inverse preferential attachment (Inverse PATT), and random attachment (Random).



**Figure 3.** Network visualizations of exemplar networks generated via pure preferential attachment (**left**), pure inverse preferential attachment (**center**), and preferential attachment followed by inverse preferential attachment (hybrid model; **right**). Each network consisted of 100 nodes. The size of each node reflects its degree.

Interestingly, the networks grown by PATT first followed by iPATT tended to have smaller diameters and shorter average shortest path lengths than the networks grown by iPATT first followed by PATT, regardless of when the growth model "switched" to the other growth model (see Figure 4). This observation is supported by significant the interaction effects of the network type (PATT–inverse PATT; inverse PATT–PATT) and time of switch (20%, 40%, 60%, 80% of nodes added) for ASPL ($F(1, 796) = 761.04$, $p < 0.001$) and diameter ($F(1, 796) = 91.68$, $p < 0.001$) in a between-group two-way analysis of variance. This result may suggest that networks generated by the preferential attachment growth algorithm at the initial stages (even for a short period) may be more navigable than networks that are generated by the inverse preferential attachment growth algorithm at the initial stages.

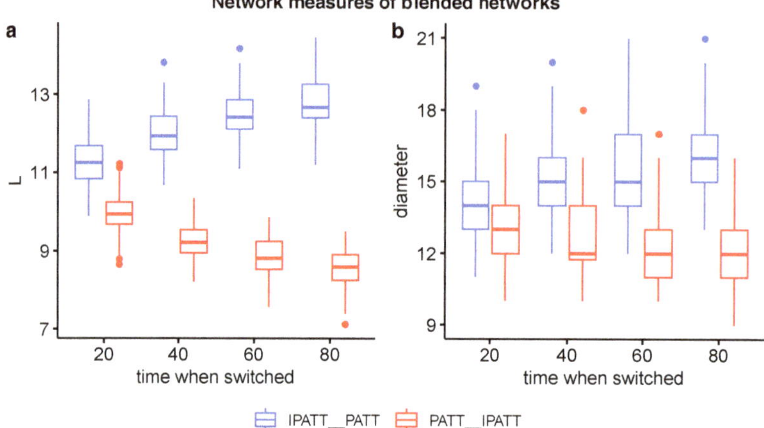



**Figure 4.** Boxplots of ASPL (**a**) and network diameter (**b**) values of networks grown via blends of preferential attachment and inverse preferential attachment. The x-axis indicates the percentage proportion of nodes added before the network algorithm was switched. Red bars indicate networks first grown by PATT followed by inverse PATT. Blue bars indicate networks first grown by inverse PATT followed by PATT.

*3.2. Degree Distributions*

In this section, we examine the degree distributions of networks generated in Section 2. Raw counts of the numbers of nodes with degrees of various values were obtained from each network. In part 1, a power law was first fitted to the degree distributions and the goodness-of-fit of the power law to the data was evaluated via a bootstrapping approach. In part 2, the data were fit to alternative distributions (log-normal, exponential, and Poisson distributions) and tests were conducted to assess the fit of the power law to the data as compared to alternative distributions. This sequence of analyses closely follows the recommendations of Clauset, Shalizi, and Newman [19] for analyzing power law-distributed data in a statistically rigorous manner (see [18] for more information on how to implement this analysis pipeline).

3.2.1. Test for Power Law Fits via Bootstrapping

A power law was fit to the degree distributions of each of the simulated networks. Specifically, a power law was fit to the data and the scaling parameter, $\alpha$ (i.e., the exponent of the power law), was computed for a given $x_{min}$ value (the minimum value for which the power law holds; see the $x_{min}$ and $\alpha$ columns in Table 2). Note that all exponents were <2, lower than what is usually observed in real-world networks, where $2 < \alpha < 3$ [19]. This may be due to the simplicity of the simulations conducted (i.e., only 1 node and 1 edge were added to the network at each iteration), which led to sparser networks.

As suggested by Clauset et al. [19], we evaluated whether the observed degree distributions actually followed a power law via a bootstrapping approach. Specifically, 1000 degree distributions were sampled from the empirical degree distribution of interest, a power law was fit to that degree distribution, and the exponent was computed. Mean $\alpha$ indicates the mean exponent of the 1000 bootstrapped networks and SD $\alpha$ indicates the standard deviation of the 1000 bootstrapped networks. A goodness-of-fit test was then conducted to determine if the exponent obtained from the original degree distribution was likely to have come from the bootstrapped "population" of exponents. As the point estimate *p*-values were not significant (all *p*-values > 0.05), this indicated that for all

11 network types, the power law distribution provided a plausible fit to the degree distributions (i.e., the exponent estimate is stable despite random fluctuations). Table 2 shows a summary of the results of the goodness-of-fit test for all 11 network types (see the mean $\alpha$, SD $\alpha$, Kolmogorov–Smirnov statistics, and $p$-value columns).

**Table 2.** Power law scaling parameter estimates and uncertainty estimates from the bootstrap procedure. Note that the bootstrap procedure was conducted for each simulated network; the means and standard deviations of estimates are shown in the table.

| Network | $x_{min}$ | | $\alpha$ | | Mean Bootstrapped $\alpha$ | | SD Bootstrapped $\alpha$ | | KS-Statistic | | $p$-Value | |
|---|---|---|---|---|---|---|---|---|---|---|---|---|
| | M | SD | M | SD | M | SD | M | SD | M | SD | M | SD |
| PATT | 2.09 | 2.27 | 1.45 | 0.06 | 1.55 | 0.08 | 0.36 | 0.14 | 0.11 | 0.02 | 0.74 | 0.16 |
| Inverse PATT | 9.98 | 10.97 | 1.42 | 0.15 | 1.70 | 0.24 | 0.74 | 0.26 | 0.25 | 0.02 | 0.34 | 0.08 |
| Random | 10.51 | 13.51 | 1.49 | 0.17 | 1.68 | 0.20 | 0.67 | 0.23 | 0.19 | 0.02 | 0.52 | 0.15 |
| PATT/Inverse PATT | | | | | | | | | | | | |
| 200/800 | 1.14 | 0.51 | 1.34 | 0.03 | 1.47 | 0.05 | 0.40 | 0.07 | 0.16 | 0.02 | 0.60 | 0.13 |
| 400/600 | 1.13 | 0.40 | 1.39 | 0.03 | 1.51 | 0.04 | 0.40 | 0.08 | 0.12 | 0.02 | 0.72 | 0.18 |
| 600/400 | 1.28 | 0.80 | 1.41 | 0.04 | 1.51 | 0.05 | 0.37 | 0.09 | 0.11 | 0.02 | 0.76 | 0.16 |
| 200/800 | 1.34 | 0.89 | 1.42 | 0.04 | 1.52 | 0.05 | 0.35 | 0.09 | 0.11 | 0.02 | 0.73 | 0.19 |
| Inverse PATT/PATT | | | | | | | | | | | | |
| 200/800 | 8.50 | 6.87 | 1.56 | 0.13 | 1.72 | 0.15 | 0.63 | 0.18 | 0.15 | 0.02 | 0.70 | 0.20 |
| 400/600 | 18.07 | 14.73 | 1.64 | 0.20 | 1.84 | 0.21 | 0.81 | 0.26 | 0.17 | 0.02 | 0.67 | 0.19 |
| 600/400 | 23.40 | 20.21 | 1.65 | 0.25 | 1.87 | 0.26 | 0.86 | 0.28 | 0.20 | 0.02 | 0.59 | 0.18 |
| 200/800 | 24.34 | 27.48 | 1.61 | 0.29 | 1.84 | 0.28 | 0.84 | 0.29 | 0.23 | 0.02 | 0.48 | 0.14 |

Legend: M = mean; SD = standard deviation; KS = Kolmogorov–Smirnov; PATT = preferential attachment.

Although the results of the bootstrap seem to suggest that both degree distributions from the PATT and iPATT networks followed a power law, a closer look at Table 2 indicates that the Kolmogorov–Smirnov statistic for the iPATT network ($D = 0.25$) was larger than the Kolmogorov–Smirnov statistic for the PATT network ($D = 0.11$). The magnitude of $D$ is an indicator of the "distance" between the fitted distribution and the actual data. In this case, the degree distribution of the network that was simulated via inverse preferential attachment deviated to a greater extent from a power law as compared to the network that was simulated via preferential attachment. This was confirmed by a visual inspection of the cumulative degree distributions (see Figure A3 in the Appendix A).

### 3.2.2. Statistical Comparison with Other Degree Distributions

As recommended by Clauset et al. [19], another way of investigating the nature of degree distributions in networks is to fit alternative distributions (exponential, log-normal, and Poisson distributions) to the degree distributions of all networks and conduct the relevant goodness-of-fit tests to compare the fit of these distributions to the fit of the power law to the data. The comparison of the power law and these distributions constitutes a non-nested model comparison, so Vuong's test of non-nested hypotheses was used instead of the likelihood ratio test (for details, please see [20]). Vuong's test computes a $V$-statistic, one-sided $p$-value, and two-sided $p$-value. The one-sided $p$-value indicates the probability of obtaining the particular value of log likelihood ratio if the power law is not true. In other words, a significant one-sided $p$-value indicates that the power law distribution is a good fit to the data (low probability that the alternate distribution could account for the data), whereas a non-significant one-sided $p$-value indicates that the power law distribution is a not good fit to the data (high probability that the alternate distribution could account for the data). The two-sided $p$-value indicates the probability that both distributions being compared are equally "distant" from the data. In other words, a significant two-sided $p$-value indicates that one distribution is a significantly better fit to the data than the other distribution, whereas a non-significant two-sided $p$-value indicates that neither distribution is preferred.

The results of these comparisons are summarized in Table 3 below, with more detailed statistics available in Table A1 of Appendix A. For the power law and Poisson comparison, the significant two-sided $p$-values and significant one-sided $p$-values for all 11 network types indicate that a power law distribution was a significantly better fit to the data than a Poisson distribution. For the power law and log-normal comparison, the non-significant two-sided $p$-values for all 11 network types indicate that one distribution cannot be favored over the other. See Figures A1 and A2 in the Appendix A for a visual depiction of these results.

**Table 3.** Summary of Vuong's tests of non-nested models comparing power law distributions to alternative distributions (exponential, log-normal, Poisson). The cell indicates the preferred distribution from the comparison; n.d. indicates that no distribution can be favored.

| Network | PL vs. Exp | PL vs. LN | PL vs. Pos |
|---|---|---|---|
| PATT | PL | n.d. | PL |
| Inverse PATT | n.d. | n.d. | PL |
| Random | n.d. | n.d. | PL |
| PATT/Inverse PATT | | | |
| **200/800** | **n.d.** | **n.d.** | **PL** |
| 400/600 | PL | n.d. | PL |
| 600/400 | PL | n.d. | PL |
| 200/800 | PL | n.d. | PL |
| Inverse PATT/PATT | | | |
| 200/800 | n.d. | n.d. | PL |
| 400/600 | n.d. | n.d. | PL |
| 600/400 | n.d. | n.d. | PL |
| 200/800 | n.d. | n.d. | PL |

Legend: PL = power law; LN = log-normal; Pos = Poisson; Exp = exponential; PATT = preferential attachment; n.d. = no difference.

The comparison between the power law and exponential distribution is more informative (see Figure 5). For the PATT network, the two-sided $p$-value was significant, indicating that the two distributions were not equivalent in terms of their fit to the data, with one distribution being a better fit. The results of the one-sided test indicate that the power law was a better fit for the degree distribution generated by the preferential attachment as compared to an exponential distribution. For the iPATT and random network, the two-sided $p$-value was not significant, indicating that the two distributions (power law and exponential) were equivalent in terms of their fit to the data.

Turning to the results of Vuong's test for the combination (i.e., blended) networks, we observe that for all network types generated with iPATT followed by PATT, the two-sided $p$-values for the power law and exponential comparison were non-significant, indicating that the two distributions were equivalent in terms of their fit to the data, similar to the iPATT-only and random networks. In contrast, the pattern of results varied for the networks generated with PATT followed by iPATT. The network where the first 200 iterations were based on the PATT model had two-sided $p$-values that were non-significant (similar to the iPATT-only network), whereas the other networks where the first 400, 600, and 800 iterations were based on the PATT model had two-sided and one-sided $p$-values that were significant, indicating that the power law was a better fit than the exponential distribution (similar to the PATT only network).

In summary, the key finding of the analyses of the network structure and degree distributions was that the blended network that was first generated by PATT followed by iPATT led to (i) a network structure with relatively low values for the ASPL and diameter (i.e., low values of ASPL and diameter in Figure 4a,b) and (ii) degree distributions that could not be exclusively classified as a power law (i.e., $p$-values > 0.05 in Figure 5b)—qualitatively resembling the properties of real-world phonological networks [21].

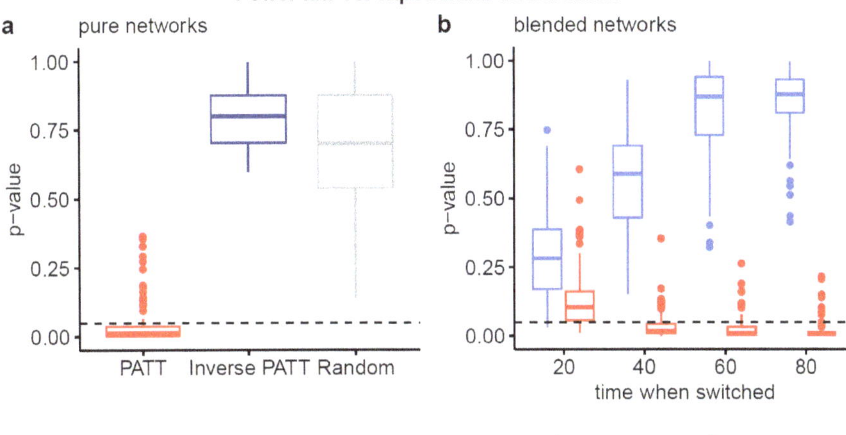



**Figure 5.** Boxplots of two-sided *p*-values from Vuong's test of non-nested models comparing the fits of power law and exponential distributions to the degree distributions of simulated networks. Panel (**a**) compares the degree distributions of pure networks and panel (**b**) compares the degree distributions of blended networks. Non-significant *p*-values (based on an alpha-level of 0.05) indicate that neither distribution is preferred. Significant *p*-values (based on an alpha-level of 0.05) indicate that one distribution fits the empirical data better. Based on the 1-sided *p*-values (see Table A1), the power law distribution provides a better fit than the exponential distribution for all networks grown by preferential attachment, except for when the switch to its inverse variant occurs early.

## 4. Discussion

The key finding from the simulations was that a model where the network was first generated by PATT for a short period (the first 200 out of 1000 iterations) before switching to the iPATT growth mechanism led to a network structure that was (i) more densely connected than if the growth models were reversed (i.e., smaller diameters and ASPL) and (ii) had a degree distribution that could be accounted for by alternative distributions (i.e., the exponential distribution that provided similar fits to the data as did the power law).

Recall that Siew and Vitevitch [12] found through an archival analysis and laboratory-based experiments that novel words that connected to existing words with few phonological neighbors in the lexicon were more likely to be learned than novel words that connected to existing words with many connections at later stages of development. We suggested that this switch may arise due to the increased processing costs associated with navigating a lexicon with a crowded phonological space [22], as well as the increased pressures on lexical representations to be better differentiated from each other in a more densely connected phonological lexicon (see the lexical restructuring hypothesis; [23,24]). We wished to explore these intriguing ideas computationally and simulated networks that were generated by a blend of different network growth mechanisms. Our results suggest that it is possible that the development of the phonological network may be better captured, at least partly, by an alternative network growth algorithm.

Overall, the simulations suggest that a particular combination of the PATT and iPATT network growth algorithms (i.e., the network that is initially "grown" by PATT followed by inverse PATT) led to the emergence of network characteristics that are suggestive of increased efficiencies in network navigation [25] (i.e., lower ASPL and smaller diameter) and degree distributions that are not necessarily best captured by a pure power law (i.e., not a purely scale-free degree distribution). We observed this in the case where the network was generated with PATT driving the initial stage for a short period (200 out of 1000 iterations) and iPATT driving the later stages of growth. This led to a network structure that was more densely connected than if the order of the growth models was reversed (i.e., iPATT followed by PATT) and a degree distribution that could be accounted for by alternative non-power law distributions, such as the exponential distribution, rather than if preferential attachment persisted for a longer period of time at the beginning (i.e., PATT continued for a longer period before the switch to iPATT occurred in the simulations).

Although small, simple networks were simulated in this study, the present findings nevertheless provide a proof-of-concept that the new growth principle that we proposed—inverse preferential attachment—can produce a degree distribution that is not necessarily captured by a power law and still lead to the emergence of network characteristics that facilitate efficient navigation (i.e., small diameter and low ASPL). These network features are qualitatively similar to the network features observed in real-world phonological networks [6,21]. In addition, we wish to highlight that the present analyses do not provide evidence that only the preferential attachment or inverse preferential attachment mechanisms are directly influencing the network structure of the phonological lexicon. What these results do suggest is that a countably infinite list of complicated and detailed constraints that capture the microscopic details of language may not be necessary to produce the structure observed in the phonological network. Rather, a simple assumption, such as the assumption examined mathematically by Callaway et al. [13]—stating that newly added nodes do not necessarily need to be attached to an existing node in the network—may lead to some of the structural features of the phonological network, such as the presence of lexical hermits in the phonological lexicon as observed by [6]. The results of the present simulation in conjunction with the long and rich history of research in psycholinguistics also allows us to constrain our search of possible algorithms involved in the acquisition of words to the space of plausible algorithms. Furthermore, the results of the present simulation lend credence to the idea that the principles that affect word learning may change over time as the lexicon becomes more "crowded" with similar sounding words or other cognitive constraints begin to exert an influence on acquisition (for similar influences on semantics, see [26]).

Finally, our results provide new avenues for research within the field of network science. First, although network scientists have previously examined the influence of constraints of costs on network growth (e.g., financial or space limitations on the expansion of air transportation networks [27]), the present findings suggest that it may also be important to consider how different costs introduced at different time-points of development shape future network growth. Second, network scientists commonly view network growth as operating via a process that maximizes node fitness [3,5]. In the case of preferential attachment and close variants of this model, the fitness of an individual node (i.e., its ability to gain new edges) is maximized by attaching to a high-degree node. The present findings suggest that understanding network growth requires a careful consideration of the functional purpose of each complex network. In the case of phonological network development, prioritizing the acquisition of new words that occupy sparser, peripheral areas of the phonological space at later stages of development when the core of the lexicon is already highly filled out may be especially important to increase the overall fitness and efficiency of the entire network. This provides accurate coverage of the entire phonological space in order to attain an overall network structure that is optimized for language processing. In other words, network growth may not be only about maximizing the fitness of individual nodes, but may also leverage on different types of network growth algorithms (such as inverse preferential attachment) to maximize the fitness of the network as a whole in order to facilitate the processes and operations that occur within that network.

**Supplementary Materials:** All relevant materials and programming scripts are available on the Open Science Framework (https://osf.io/e8gwr).

**Author Contributions:** Conceptualization, C.S.Q.S. and M.S.V.; methodology, C.S.Q.S.; formal analysis, C.S.Q.S.; writing—original draft preparation, C.S.Q.S.; writing—review and editing, C.S.Q.S. and M.S.V.; visualization, C.S.Q.S.; funding acquisition, M.S.V. All authors have read and agreed to the published version of the manuscript.

**Funding:** The APC was funded by the University of Kansas.

**Conflicts of Interest:** The authors declare no conflict of interest.

## Appendix A

The appendix contains additional detail in relation to the analyses of the degree distributions of the simulated networks.

**Table A1.** Degree distributions of simulated networks were fitted to exponential, log-normal, and Poisson distributions and compared against the fitted power law distribution using Vuong's test of mis-specified, non-nested hypotheses. Note that each set of model comparisons was conducted for each of the 100 simulated networks per condition or network type. The present table displays the mean and standard deviations of the test statistic and $p$-values.

| | \multicolumn{6}{c}{**Power Law Vs. Exponential**} |
|---|---|---|---|---|---|---|
| | \multicolumn{2}{c}{$V$-statistic} | \multicolumn{2}{c}{2-sided $p$} | \multicolumn{2}{c}{1-sided $p$} |
| **Network** | **M** | **SD** | **M** | **SD** | **M** | **SD** |
| PATT | 2.387 | 0.614 | 0.046 | 0.080 | 0.023 | 0.040 |
| Inverse PATT | −0.156 | 0.251 | 0.800 | 0.113 | 0.560 | 0.098 |
| Random | 0.420 | 0.359 | 0.682 | 0.222 | 0.347 | 0.120 |
| \multicolumn{7}{c}{PATT/Inverse PATT} |
| 200/800 | 1.614 | 0.399 | 0.135 | 0.111 | 0.067 | 0.056 |
| 400/600 | 2.320 | 0.483 | 0.037 | 0.048 | 0.018 | 0.024 |
| 600/400 | 2.597 | 0.578 | 0.025 | 0.041 | 0.012 | 0.021 |
| 200/800 | 2.626 | 0.511 | 0.021 | 0.039 | 0.011 | 0.020 |
| \multicolumn{7}{c}{Inverse PATT/PATT} |
| 200/800 | 1.116 | 0.357 | 0.293 | 0.151 | 0.147 | 0.076 |
| 400/600 | 0.593 | 0.288 | 0.570 | 0.176 | 0.285 | 0.088 |
| 600/400 | 0.221 | 0.241 | 0.817 | 0.161 | 0.416 | 0.088 |
| 200/800 | 0.009 | 0.253 | 0.849 | 0.120 | 0.498 | 0.097 |
| | \multicolumn{6}{c}{**Power law vs. Log-normal**} |
| | \multicolumn{2}{c}{$V$-statistic} | \multicolumn{2}{c}{2-sided $p$} | \multicolumn{2}{c}{1-sided $p$} |
| **Network** | **M** | **SD** | **M** | **SD** | **M** | **SD** |
| PATT | −0.735 | 0.350 | 0.488 | 0.171 | 0.756 | 0.085 |
| Inverse PATT | −0.695 | 0.227 | 0.498 | 0.137 | 0.751 | 0.069 |
| Random | −0.672 | 0.267 | 0.517 | 0.160 | 0.742 | 0.080 |
| \multicolumn{7}{c}{PATT/Inverse PATT} |
| 200/800 | −0.777 | 0.201 | 0.446 | 0.091 | 0.777 | 0.045 |
| 400/600 | −0.809 | 0.391 | 0.448 | 0.132 | 0.776 | 0.066 |
| 600/400 | −0.798 | 0.340 | 0.450 | 0.154 | 0.775 | 0.077 |
| 200/800 | −0.774 | 0.354 | 0.465 | 0.138 | 0.768 | 0.069 |
| \multicolumn{7}{c}{Inverse PATT/PATT} |
| 200/800 | −0.519 | 0.272 | 0.618 | 0.172 | 0.691 | 0.086 |
| 400/600 | −0.500 | 0.272 | 0.631 | 0.176 | 0.685 | 0.088 |
| 600/400 | −0.514 | 0.260 | 0.620 | 0.167 | 0.690 | 0.084 |
| 200/800 | −0.590 | 0.270 | 0.570 | 0.171 | 0.715 | 0.085 |
| | \multicolumn{6}{c}{**Power law vs. Possion**} |
| | \multicolumn{2}{c}{$V$-statistic} | \multicolumn{2}{c}{2-sided $p$} | \multicolumn{2}{c}{1-sided $p$} |
| **Network** | **M** | **SD** | **M** | **SD** | **M** | **SD** |
| PATT | 1.860 | 0.066 | 0.063 | 0.009 | 0.032 | 0.005 |
| Inverse PATT | 3.198 | 0.411 | 0.003 | 0.005 | 0.002 | 0.003 |
| Random | 2.453 | 0.151 | 0.015 | 0.006 | 0.008 | 0.003 |
| \multicolumn{7}{c}{PATT/Inverse PATT} |
| 200/800 | 3.390 | 0.161 | 0.001 | 0.001 | 0.000 | 0.000 |
| 400/600 | 2.926 | 0.138 | 0.004 | 0.002 | 0.002 | 0.001 |
| 600/400 | 2.554 | 0.118 | 0.011 | 0.003 | 0.006 | 0.002 |
| 200/800 | 2.184 | 0.075 | 0.029 | 0.005 | 0.015 | 0.003 |
| \multicolumn{7}{c}{Inverse PATT/PATT} |
| 200/800 | 1.899 | 0.072 | 0.058 | 0.009 | 0.029 | 0.005 |
| 400/600 | 1.967 | 0.084 | 0.050 | 0.009 | 0.025 | 0.005 |
| 600/400 | 2.103 | 0.118 | 0.037 | 0.010 | 0.018 | 0.005 |
| 200/800 | 2.460 | 0.240 | 0.017 | 0.012 | 0.008 | 0.006 |

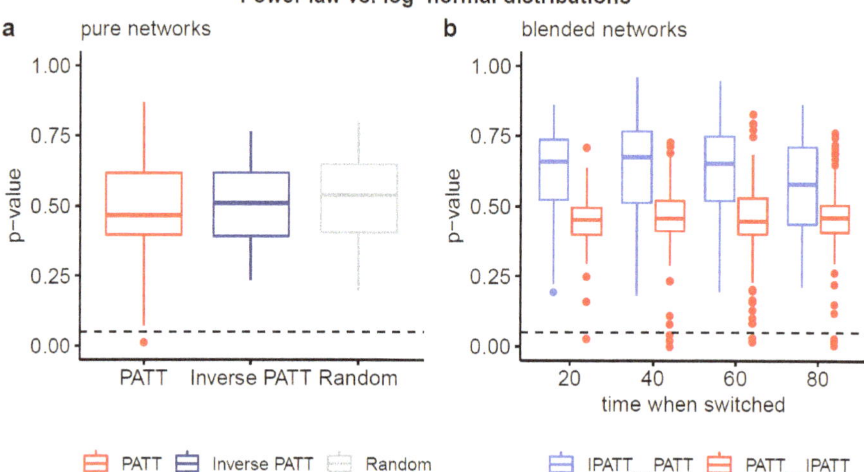


**Figure A1.** Boxplots of two-sided *p*-values from Vuong's test of non-nested models comparing the fits of power law and log-normal distributions to the degree distributions of simulated networks. Panel (**a**) compares the degree distributions of pure networks and panel (**b**) compares the degree distributions of blended networks. Non-significant *p*-values (based on an alpha-level of 0.05) indicate that neither distribution is preferred.

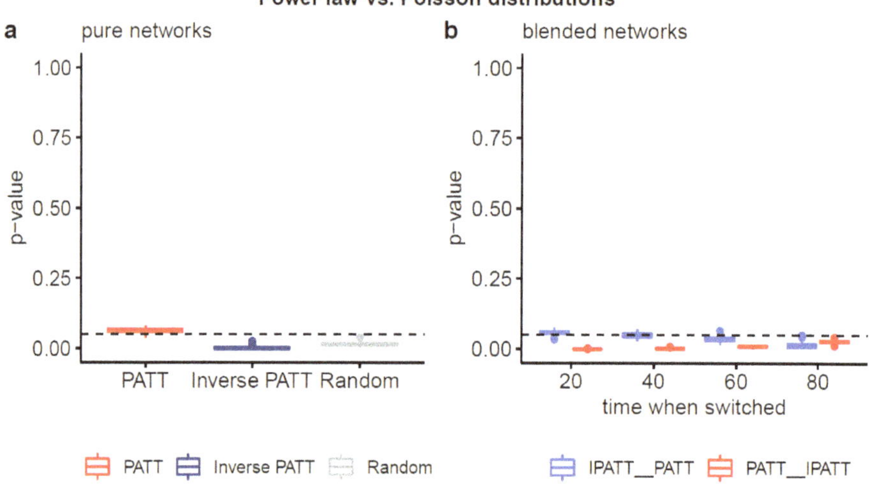


**Figure A2.** Boxplots of two-sided *p*-values from Vuong's test of non-nested models comparing the fits of power law and Poisson distributions to the degree distributions of simulated networks. Panel (**a**) compares the degree distributions of pure networks and panel (**b**) compares the degree distributions of blended networks. Significant *p*-values (based on an alpha-level of 0.05) indicate that one distribution fits the empirical data better. Based on the one-sided *p*-values (see Table A1), the power law distribution provides a better fit than the Poisson distribution.

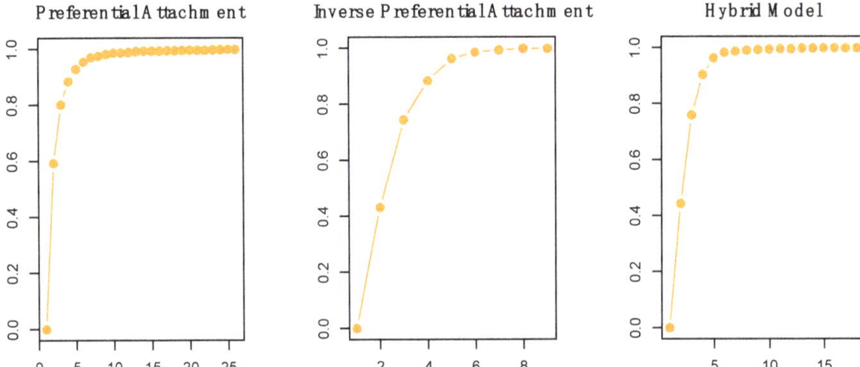


**Figure A3.** Cumulative degree distributions of exemplar networks generated via pure preferential attachment (**left**), pure inverse preferential attachment (**center**), and preferential attachment followed by inverse preferential attachment (hybrid model; **right**).

## References


1. Barabási, A.L. Scale-free networks: A decade and beyond. *Science* **2009**, *325*, 412–413. [CrossRef] [PubMed]
2. Barabási, A.L.; Albert, R. Emergence of scaling in random networks. *Science* **1999**, *286*, 509–512. [CrossRef] [PubMed]
3. Bell, M.; Perera, S.; Piraveenan, M.; Bliemer, M.; Latty, T.; Reid, C. Network growth models: A behavioural basis for attachment proportional to fitness. *Sci. Rep.* **2017**, *7*, 1–11. [CrossRef] [PubMed]
4. Bianconi, G.; Barabási, A.L. Competition and multiscaling in evolving networks. *EPL (Europhys. Lett.)* **2001**, *54*, 436. [CrossRef]
5. Mendes, G.A.; da Silva, L.R. Generating more realistic complex networks from power-law distribution of fitness. *Braz. J. Phys.* **2009**, *39*, 423–427. [CrossRef]
6. Vitevitch, M.S. What can graph theory tell us about word learning and lexical retrieval? *J. Speech Lang. Hear. Res.* **2008**, *51*, 408–422. [CrossRef]
7. Chan, K.Y.; Vitevitch, M.S. The influence of the phonological neighborhood clustering coefficient on spoken word recognition. *J. Exp. Psychol. Hum. Percept. Perform.* **2009**, *35*, 1934–1949. [CrossRef]
8. Goldstein, R.; Vitevitch, M.S. The influence of closeness centrality on lexical processing. *Front. Psychol.* **2017**, *8*, 1683. [CrossRef]
9. Siew, C.S.Q.; Vitevitch, M.S. The phonographic language network: Using network science to investigate the phonological and orthographic similarity structure of language. *J. Exp. Psychol. Gen.* **2019**, *148*, 475–500. [CrossRef]
10. Beckage, N.M.; Colunga, E. Network Growth Modeling to Capture Individual Lexical Learning. *Complexity* **2019**, *2019*, 7690869. [CrossRef]
11. Storkel, H.L.; Lee, S.Y. The independent effects of phonotactic probability and neighbourhood density on lexical acquisition by preschool children. *Lang. Cogn. Process.* **2011**, *26*, 191–211. [CrossRef] [PubMed]
12. Siew, C.S.Q.; Vitevitch, M.S. An investigation of network growth principles in the phonological language network. *J. Exp. Psychol. Gen.* **2020**. [CrossRef] [PubMed]
13. Callaway, D.S.; Hopcroft, J.E.; Kleinberg, J.M.; Newman, M.E.J.; Strogatz, S.H. Are randomly grown graphs really random? *Phys. Rev. E* **2001**, *64*. [CrossRef] [PubMed]
14. Latora, V.; Marchiori, M. Efficient behavior of small-world networks. *Phys. Rev. Lett.* **2001**, *87*, 198701. [CrossRef]
15. Fox Keller, E. Revisiting "scale-free" networks. *BioEssays* **2005**, *27*, 1060–1068. [CrossRef]
16. Beckage, N.; Smith, L.; Hills, T. Small worlds and semantic network growth in typical and late talkers. *PLoS ONE* **2011**, *6*, 19348. [CrossRef]
17. Csardi, G.; Nepusz, T. The igraph software package for complex network research. *Interj. Complex. Syst.* **2006**, *1695*, 1–9.

18. Gillespie, C.S. Fitting Heavy Tailed Distributions: The poweRlaw Package. *arXiv* **2015**, arXiv:1407.3492.
19. Clauset, A.; Shalizi, C.R.; Newman, M.E. Power-law distributions in empirical data. *SIAM Rev.* **2009**, *51*, 661–703. [CrossRef]
20. Vuong, Q.H. Likelihood ratio tests for model selection and non-nested hypotheses. *Econ. J. Econ. Soc.* **1989**, *57*, 307–333. [CrossRef]
21. Arbesman, S.; Strogatz, S.H.; Vitevitch, M.S. The structure of phonological networks across multiple languages. *Int. J. Bifurc. Chaos* **2010**, *20*, 679–685. [CrossRef]
22. Luce, P.A.; Pisoni, D.B. Recognizing spoken words: The neighborhood activation model. *Ear. Hear.* **1998**, *19*, 1–36. [CrossRef] [PubMed]
23. Garlock, V.M.; Walley, A.C.; Metsala, J.L. Age-of-acquisition, word frequency, and neighborhood density effects on spoken word recognition by children and adults. *J. Mem. Lang.* **2001**, *45*, 468–492. [CrossRef]
24. Metsala, J.L.; Walley, A.C. Spoken vocabulary growth and the segmental restructuring of lexical representations: Precursors to phonemic awareness and early reading ability. In *Word Recognition in Beginning Literacy*; Lawrence Erlbaum Associates Publishers: Mahwah, NJ, USA, 1998; pp. 89–120. ISBN 978-0-8058-2898-6.
25. Kleinberg, J.M. Navigation in a small world. *Nature* **2000**, *406*, 845. [CrossRef]
26. Samuelson, L.; Smith, L.B. Grounding development in cognitive processes. *Child. Dev.* **2000**, *71*, 98–106. [CrossRef]
27. Amaral, L.A.N.; Scala, A.; Barthelemy, M.; Stanley, H.E. Classes of small-world networks. *Proc. Natl. Acad. Sci. USA* **2000**, *97*, 11149–11152. [CrossRef]

*Article*

# Classification of Literary Works: Fractality and Complexity of the Narrative, Essay, and Research Article

Aldo Ramirez-Arellano

Sección de Estudios de Posgrado e Investigación, Unidad Profesional Interdisciplinaria de Ingeniería y Ciencias Sociales y Administrativas, Instituto Politécnico Nacional, Ciudad de México 07738, Mexico; aramirezar@ipn.mx; Tel.: +52-5557296000

Received: 22 July 2020; Accepted: 15 August 2020; Published: 17 August 2020

**Abstract:** A complex network as an abstraction of a language system has attracted much attention during the last decade. Linguistic typological research using quantitative measures is a current research topic based on the complex network approach. This research aims at showing the node degree, betweenness, shortest path length, clustering coefficient, and nearest neighbourhoods' degree, as well as more complex measures such as: the fractal dimension, the complexity of a given network, the Area Under Box-covering, and the Area Under the Robustness Curve. The literary works of Mexican writers were classify according to their genre. Precisely 87% of the full word co-occurrence networks were classified as a fractal. Also, empirical evidence is presented that supports the conjecture that lemmatisation of the original text is a renormalisation process of the networks that preserve their fractal property and reveal stylistic attributes by genre.

**Keywords:** complex networks; literary works; genre classification; stylistic attributes; lemmatization; renormalisation process

---

## 1. Introduction

A complex network as an abstraction of a language system has attracted attention in the last decade. The current linguistics research, based on the complex network approach, follows three major lines [1,2]: characterisation of human language as a multi-level system, linguistic typological research using quantitative measures, and the relationship between system-level complexity of human language and its microscopic features.

Word co-occurrence networks and their measures have been widely employed to analyse the syntactic features for multiple purposes, such as: identifying authors' writing styles [3–8] and evaluating machine translations [9]. Also, Ferraz de Arruda, Nascimento Silva [10], as well as F. de Arruda, Q. Marinho [11] built a complex network where the nodes are the representation of adjacent paragraphs that share a minimum semantical content to classify the text as real (written by an author) or randomly constructed (built from random blocks of real texts).

In most of the research mentioned above, well-known measures such as: node degree ($k$), shortest path length ($spl$), betweenness ($b$), clustering coefficient ($cc$), and the average of nearest neighbourhoods' degree ($nnd$) are applied to characterise the word co-occurrence networks. The $k$, $b$, and $nnd$ are centrality measures that characterise local properties of the network that are useful for authorship attribution [3–8]. However, these measures do not capture the global network structure that could give us insight into the literary genre. This research aims at showing that local and global measures of the word co-occurrence networks—of literary works of Mexican writers—let us classify them according to the genre. Thus, the following research questions are formulated:

1.   Are measures of the complex network useful to classify literary works by genre?

2. Is the full word co-occurrence network of literary works fractal?

3. Do pre-process tasks such as: deletion of number, punctuation, functional words, and lemmatisation generate fractal networks?

## 2. Measures of Complex Networks

Formally, a network is defined by $G = (V, E)$ where $V$ is the vertexes or nodes, and $E$ is the edges. The complex networks exhibit non-trivial topological features that do not occur in simple networks, such as: lattices or random graphs [12], and their overall behaviour cannot be predicted by observing the behaviour of their nodes [13]. Since the complex network theory has its root in graph theory, some measures are presented below.

The degree of a node $i$ is defined by:

$$k_i = \sum_j^N v_{ij}$$
(1)

where $j$ represents a given neighbour of the node $i$, and $N$ is the total neighbours. The value of $v_{ij}$ is defined as one, if there is a connection between nodes $i$ and $j$, and as 0 otherwise.

Similarly, the betweenness of a node is defined as:

$$b_i = \sum_{j,m \neq i} \frac{L_{jm}(i)}{L_{jm}}$$
(2)

where $L_{jm}$, is the number of shortest paths between nodes $j$ and $k$, and $L_{jm}(i)$ is the shortest paths between nodes $j$ and $m$ that go through $i$.

The average nearest neighbourhoods' degree ($nnd$) of a given node can be computed by:

$$nnd_i = \sum_{j \in V(i)} \frac{k_j}{k_i}$$
(3)

where $k_i$ is the degree of the node $i$, and the set $V(i)$ contains its nearest neighbours, and $k_j$ is the degree of a given neighbour.

A definition of network clustering is expressed by:

$$cc(G) = \frac{3\tau}{spl(2)}$$
(4)

where $\tau$ is the number of triangles of the network and $spl(2)$ is the shortest path of length two. A "triangle" is a set of three nodes in which each contacts the other two.

### 2.1. Fractality of Complex Networks

A fractal is an object that is similar to itself on all scales [14]. A network is a fractal network if its box-covering follows the power law given by:

$$N_b(l) \sim \beta l^{-d_b}$$
(5)

where $N_b(l)$ is the minimum number of boxes of diameter $l$ to cover the network—the procedure of box-covering that gives us this number is detailed later—$\beta$ is the scaling factor, and $d_b$ is the box dimension of a complex network that can be obtained as follows:

$$d_b = -\lim_{l \to 0} \frac{\ln N_b(l)}{\ln l}$$
(6)

On the other hand, a non-fractal network is characterised by a sharp decay of $N_b(l)$, with $l$ described by an exponential function as follows [15,16]:

$$N_b(l) \sim \beta e^{-d_b l} \tag{7}$$

*2.2. Complexity of Networks*

The complexity measure of a network proposed by Lei, Liu [17] is defined as:

$$c(G) = d(G)s(G) \tag{8}$$

where $d(G) = |E|/\left(4CR^3/3\Delta\right)$ is the absolute density [18]; $|E|$, $C$, $R$, and $\Delta$ are the number of edges, circumference, radius, and diameter of the network, respectively. $s(G) = -k\sum_{i=1}^{|V|}\left(p_i^{q_i} - p_i\right)/(1 - q_i)$ is known as structure entropy based on degree and betweenness [17], where $k$ is the Boltzmann constant, $|V|$ is the number of nodes, $p_i = \frac{k_i}{\sum_{i=1}^{|V|} k_i}$, $q_i = 1 + (b_{\max} - b_i)$, and $b_{max}$ is the maximum value of the betweenness computed by the Equation (2). This measure captures the topology of the networks, but it is not affected by scales and their types.

*2.3. Box-Covering of Complex Networks*

To obtain $N_b(l)$, consider the phrase "No one behind, no one ahead". Its word co-occurrence network is shown in Figure 1. The number of boxes to cover the network $N_b(l)$ for $l = 1$, and $l = \Delta + 1$—where $\Delta$ is the diameter of the network—is the number of nodes of the network and one, respectively. The $N_b(l)$ from 2 to $\Delta$ is not a trivial answer.

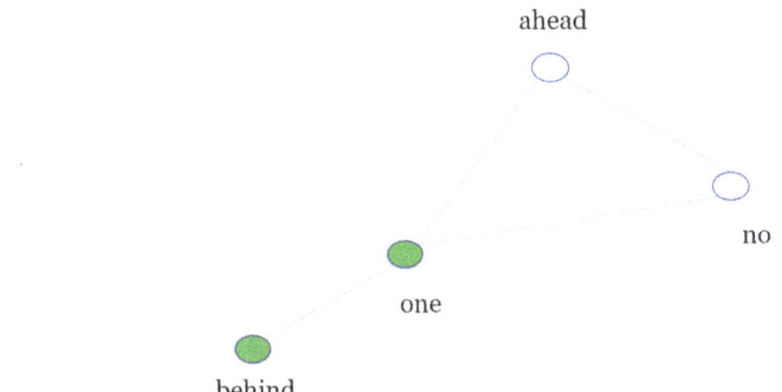


**Figure 1.** Word co-occurrence network of "No one behind, no one ahead"; the nodes in same colour belong to the same box.

For example, $N_b(l = 1) = 4$ and $N_b(l = \Delta + 1) = 1$ for the network of Figure 1. To obtain the $N_b(l = 2)$, we first compute a dual network ($G'$) from the original ($G$) as follows: given a distance $l$; two nodes $i, j$, in the dual network, are connected if the distance between $l_{ij}$ is greater than or equal to $l$. For example, we start the procedure from the node "no", see Figure 2; "no" and "behind" have a distance of two in $G$, thus, they will be connected in $G'$. Next, the node "ahead" as the starting node is chosen—notice that the distance from it to "behind" is two—thus, a connection in $G'$ will be drawn (see Figure 2).

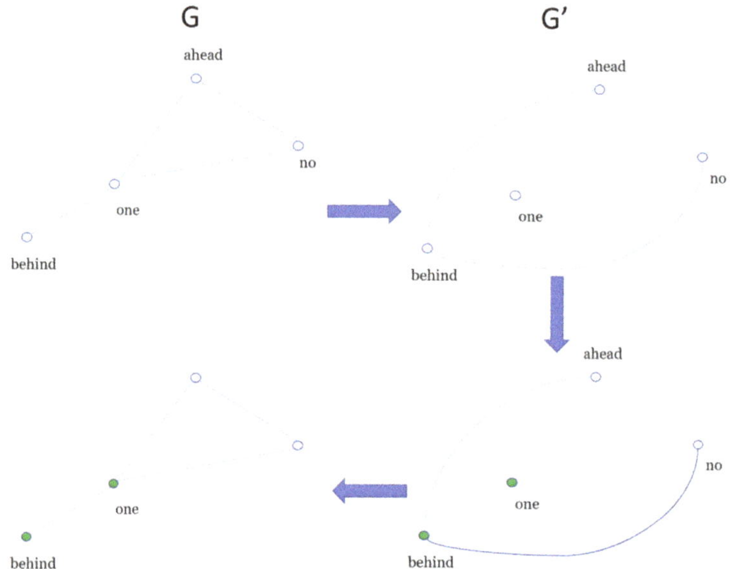


**Figure 2.** Covering of the network for a given box size ($l = 2$). The number of boxes in this network is $N_b(2) = 2$.

Then, the nodes of $G'$ must be coloured following a single rule: two nodes directly connected will be painted different colours. The nodes of the resulting coloured dual network $G'$ are mapped to original network $G$. The number of colours of $G'$ represents the minimum number of boxes $N_b(l)$ of a given value of $l$ to cover the network. The nodes of $G$, in the same colour, belong to the same box. The procedure described above is repeated until $l = \Delta + 1$. For more profound details of the box-covering algorithm, the reader is referred to the work of Chaoming, Lazaros [19]. Since $l$ vs. $N_b(l)$ characterises the topology of the network, the area under the box-covering curve, $l$ vs. $N_b(l)$ (AUB), was also included in the measures of the word co-occurrence network.

### 2.4. Robustness of Fractal Networks

Intentional network attacks are based on different centrality measures such as: the node degree or betweenness. They differ in the approach to compute those centrality measures such as: computing the global degree or betweenness, then performing the attack, or recomputing the centrality measure after a node is removed [20–23]. The fraction of nodes necessary to break down a fractal network ($p_c$) by a random attack are close to the total number of nodes; thus, these networks are extremely robust [24]. On the other hand, this robustness decreases drastically when the nodes with a high degree are selected to be removed [20,25]. This vulnerability to intentional attack relies on that a few nodes, with a high degree, maintain the connectivity of the network [26]. The robustness of each network is quantified by the size of the largest connected component $C_c$ after removing a fraction $p$ node from the network [20,24,26,27] when $C_c(p_c) \simeq 0$ the network has been disintegrated. The value of $p_c$ is low for fragile networks, and the opposite for robust networks.

Although the $p_c$ value is useful for measuring the overall damage caused by the attack strategy, it does not reflect the damage of an individual node removal; for example, Figure 3 shows the plot of $C_c$ vs. $p$, where the value of $p_c$ is 0.5 and 0.49 for networks one and two, respectively.

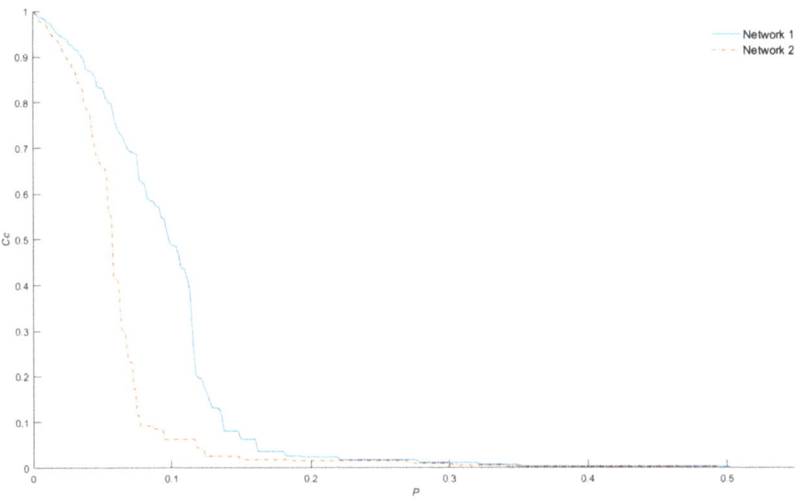

**Figure 3.** The damage of an individual node removal of network one and two. Although the $p_c$ of networks one and two are 0.5 and 0.49, respectively, the area under the robustness curve reflects more precisely the vulnerability of the networks (0.0956 for network one and 0.060 for network 2).

This means that for both networks, it is necessary to remove approximately 50% of the nodes to disintegrate them in components that contain at most one node. Moreover, based on Figure 3, the removal of the nodes from network two causes more damage than the removal of those from network one. This damage can be quantified by computing the Area Under the Robustness Curve (AURC)—0.0956 for network one and 0.060 for network 2—to a higher the value, the higher the robustness of the network. The AURC of the attack performed by node degree was included as a measure of network robustness instead of $p_c$.

## 3. Materials and Methods

From seven Mexican writers —Juan José Arreola Zúñiga, Carlos Fuentes Macías, Jorge Ibargüengoitia Antillón, Carlos Monsiváis Aceves, José Emilio Pacheco Berny, Octavio Irineo Paz Lozano, and Alfonso Reyes Ocha—21 essays, 21 narratives (15 tales and six novels), and 21 research articles were the corpus for this research (see Table 1). Noticeably, some authors wrote titles classified as essays, tales, or novels, such as Carlos Fuentes, Jorge Ibargüengoitia, and José Emilio Pacheco. The essays, narratives, and research articles were published between 1911 and 2019. All the titles were obtained in an electronic format such as pdf and then converted to plain text.

The node degree ($k$), betweenness ($b$), shortest path length ($spl$), clustering coefficient ($cc$), and nearest neighbourhoods' degree ($nnd$), as well as more complex measures such as: the fractal dimension ($d_b$) obtained by the Equation (6), the complexity of a network $c(G)$ given by the Equation (8), the Area Under Box-covering ($AUB$), and the Area Under the Robustness Curve (AURC), were computed for each network of each title. Statistical analysis was carried out to select those measures that have a significant difference by literary genres and produce a better classification.

Then the Support Vector Machine (SVM), Naïve Bayes (NB), Decision Tree (DT), and Neural Network (NN) implemented in Weka [28] and fourth data mining views—described later and based on the measures mentioned above—were employed to classify the literary works. The hyperparameter optimisation of the data mining techniques was conducted by sequential model-based algorithm configuration [29,30]. The hidden layers and the nodes learning function of NN were 28 and sigmoid, respectively. The polynomial kernel was used in SVM, and all measures of the networks were normalised before training and validating SVM and NN. The NN technique was used with a normal distribution to estimate the probabilities of the network measures. DT uses the C4.5 algorithm [31].

**Table 1.** The genre, number of titles, and primary author of the corpus.

| Genre | Number of Titles | Primary Author |
|---|---|---|
| Essay | 6 | Alfonso Reyes Ochoa |
| Essay | 3 | Carlos Fuentes Macías |
| Essay | 6 | Carlos Monsiváis Aceves |
| Essay | 6 | Octavio Irineo Paz Lozano |
| Narrative (Tale) | 2 | Carlos Fuentes Macías |
| Narrative (Tale) | 5 | José Emilio Pacheco Berny |
| Narrative (Tale) | 3 | Jorge Ibargüengoitia Antillón |
| Narrative (Tale) | 5 | Juan José Arreola Zúñiga |
| Narrative (Novel) | 1 | Carlos Fuentes Macías |
| Narrative (Novel) | 1 | José Emilio Pacheco Berny |
| Narrative (Novel) | 3 | Jorge Ibargüengoitia Antillón |
| Narrative (Novel) | 1 | Juan José Arreola Zúñiga |
| Research Article | 16 | Several authors |

The efficacy of each data mining technique and data mining views was validated by 5-fold cross-validation, comparing the Area under the Receiver Operating characteristic Curve (AROC). The AROC is useful to measure the performance of a data mining technique when the dataset is unbalanced [32]. Values of AROC closer to 1 mean a better classification than those closer to 0.5. This analysis shows the impact of data mining techniques and the measures on the classification of literary works. These results answer research question one (see Figure 4). Also, the accuracy of classification is presented as additional information that is computed as (TP Positive (TP) + False Positive (FP) + False Negative (FN) + True Negative (TN)). The computation of AROC and accuracy are well-known for a two-class problem. Furthermore, for a multi-class problem, for each time one class could be considered as positive, then all the others as negative. This means that TP, TN, FP, and FN are calculated for each class. Therefore, a confusion matrix and AROC curve is obtained for each class (see [33,34] for more details).

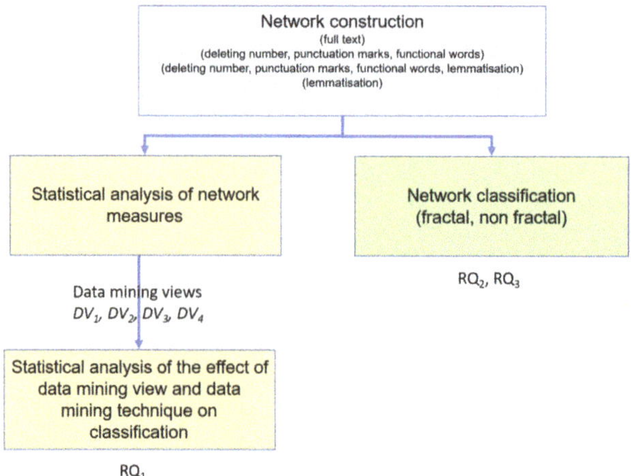


**Figure 4.** The experimental design followed to answer the research questions.

A set of word co-occurrence networks of each title was obtained and the first network was built using the full text. The second was obtained by deleting numbers and functional words.

A lemmatisation stage created the third after numbers and functional words deletion, and the fourth network was attained only through a lemmatisation stage (see Figure 4).

The networks were obtained by using the full text, by deleting numbers and functional words, by adding a lemmatisation stage after the numbers and functional words deletion, and through only a lemmatisation stage, are classified as fractal or non-fractal. Thus, research question two and three will be answered.

## 4. Results and Discussion

Tables 2–5 show the descriptive statistics by literary genre of the three types of networks—the first was built using the full text; the second was built by deleting numbers, punctuation marks, and functional words; the third was built by adding a lemmatisation stage; and the fourth was built through only a lemmatisation stage, denoted by subscripts $f$, $nf$, $l$ and $ol$, respectively.

**Table 2.** Mean and standard deviation by genre of node degree ($k$), betweenness ($b$), shortest path length ($spl$), clustering coefficient ($cc$), nearest neighbourhoods' degree ($nnd$), fractal dimension ($d_b$), complexity $c(G)$, the Area Under Box-covering ($AUB$), and the Area Under the Robustness Curve ($AURC$) of the networks built using the full text.

| Genre | $K_f(\mu-\sigma)$ | $b_f\ (\mu-\sigma)$ | $slp_f\ (\mu-\sigma)$ | $cc_f\ (\mu-\sigma)$ | $nnd_f\ (\mu-\sigma)$ | $d_{bf}\ (\mu-\sigma)$ | $c(G)_f\ (\mu-\sigma)$ | $AUB_f\ (\mu-\sigma)$ | $AURC_f\ (\mu-\sigma)$ |
|---|---|---|---|---|---|---|---|---|---|
| Essay | 5.39–0.94 | 8249.16–4869.58 | 2.9–0.81 | 0.427–0.107 | 282.51–234.12 | 6.07–0.772 | $7.84 \times 10^{-4}$–$3.83 \times 10^{-4}$ | 1510.61–969.37 | 0.0154–0.0031 |
| Narrative (Tale) | 4.76–0.536 | 3868.33–1709.2 | 3.02–0.94 | 0.311–0.082 | 99.57–55.59 | 5.06–0.801 | $7.22 \times 10^{-4}$–$3.62 \times 10^{-4}$ | 685.86–315.90 | 0.0231–0.0067 |
| Narrative (Novel) | 6.81–1.01 | 13667.19–4796.52 | 2.78–0.061 | 0.577–0.088 | 559.34–233.84 | 7.42–0.644 | $6.78 \times 10^{-4}$–$2.48 \times 10^{-4}$ | 2631.5–1013.36 | 0.01456–0.0017 |
| Research Article | 5.80–0.94 | 5995.1–2013.42 | 3.00–0.092 | 0.374–0.065 | 153.31–73.31 | 5.43–0.539 | $3.72 \times 10^{-4}$–$2.39 \times 10^{-4}$ | 1068.4–379.76 | 0.0284–0.0071 |

**Table 3.** Mean and standard deviation by genre of node degree ($k$), betweenness ($b$), shortest path length ($spl$), clustering coefficient ($cc$), nearest neighbourhoods' degree ($nnd$), fractal dimension ($d_b$), complexity $c(G)$, the Area Under Box-covering ($AUB$), and the Area Under the Robustness Curve ($AURC$) of the networks built by deleting numbers and functional words.

| Genre | $k_{nf}(\mu-\sigma)$ | $b_{nf}\ (\mu-\sigma)$ | $slp_{nf}\ (\mu-\sigma)$ | $cc_{nf}\ (\mu-\sigma)$ | $nnd_{nf}\ (\mu-\sigma)$ | $d_{bnf}\ (\mu-\sigma)$ | $c(G)_{nf}\ (\mu-\sigma)$ | $AUB_{nf}\ (\mu-\sigma)$ | $AURC_{nf}\ (\mu-\sigma)$ |
|---|---|---|---|---|---|---|---|---|---|
| Essay | 3.763–1.067 | 15092.572–8418.110 | 5.370–0.911 | 0.052–0.037 | 12.101–9.638 | 2.057–0.352 | $1.5 \times 10^{-5}$–$1.32 \times 10^{-5}$ | 2299.330–1416.600 | 0.077–0.010 |
| Narrative (Tale) | 3.998–0.861 | 13572.998–7935.093 | 5.109–0.922 | 0.710–0.042 | 12.319–5.054 | 2.141–0.302 | $2.1 \times 10^{-5}$–$1.61 \times 10^{-5}$ | 2022.467–1173.400 | 0.083–0.018 |
| Narrative (Novel) | 4.703–0.425 | 16327.442–11950.302 | 4.460–0.249 | 0.097–0.029 | 17.797–5.444 | 2.390–0.160 | $1.8 \times 10^{-5}$–$0.77 \times 10^{-5}$ | 2656.333–1971.806 | 0.087–0.006 |
| Research Article | 3.483–1.054 | 15538.651–7392.827 | 5.770–1.03 | 0.037–0.024 | 10.648–9.539 | 1.940–0.383 | $1.2 \times 10^{-5}$–$0.59 \times 10^{-5}$ | 2339.476–1314.540 | 0.070–0.012 |

**Table 4.** Mean and standard deviation by genre of node degree ($k$), betweenness ($b$), shortest path length ($spl$), clustering coefficient ($cc$), nearest neighbourhoods' degree ($nnd$), fractal dimension ($d_b$), complexity $c(G)$, the Area Under Box-covering ($AUB$), and the Area Under the Robustness Curve ($AURC$) of the networks built by deleting numbers, functional words, and lemmatisation stage.

| Genre | $k_l(\mu\text{–}\sigma)$ | $b_l\ (\mu\text{–}\sigma)$ | $slp_l\ (\mu\text{–}\sigma)$ | $cc_l\ (\mu\text{–}\sigma)$ | $nnd_l\ (\mu\text{–}\sigma)$ | $d_{bl}\ (\mu\text{–}\sigma)$ | $c(G)_l\ (\mu\text{–}\sigma)$ | $AUB_l\ (\mu\text{–}\sigma)$ | $AURC_l\ (\mu\text{–}\sigma)$ |
|---|---|---|---|---|---|---|---|---|---|
| Essay | 4.081–1.649 | 21871.411–7787.529 | 5.418–0.960 | 0.009–0.005 | 9.074–7.389 | 1.990–0.358 | $1.37 \times 10^{-5}$–$1.87 \times 10^{-5}$ | 2202.024–752.531 | 0.115–0.023 |
| Narrative (Tale) | 3.148–0.544 | 9538.206–4171.808 | 5.823–0.816 | 0.008–0.004 | 5.652–2.372 | 1.758–0.193 | $2.1 \times 10^{-5}$–$1.61 \times 10^{-5}$ | 1121.733–411.3845 | 0.003–0.0149 |
| Narrative (Novel) | 6.197–1.796 | 25284.524–7739.246 | 4.181–0.499 | 0.020–0.010 | 20.689–10.965 | 2.489–0.379 | $1.9 \times 10^{-5}$–$1.49 \times 10^{-5}$ | 2760.667–677.608 | 0.138–0.017 |
| Research Article | 4.849–0.691 | 11758.340–6325.385 | 4.296–0.352 | 0.207–0.010 | 11.169–2.802 | 2.234–0.155 | $2.8 \times 10^{-5}$–$1.85 \times 10^{-5}$ | 1272.857–540.06 | 0.138–0.016 |

**Table 5.** Mean, and standard deviation by genre of node degree ($k$), betweenness ($b$), shortest path length ($spl$), clustering coefficient ($cc$), nearest neighbourhoods' degree ($nnd$), fractal dimension ($d_b$), complexity $c(G)$, the Area Under Box-covering ($AUB$), and the Area Under the Robustness Curve ($AURC$) of the networks built only by lemmatisation stage.

| Genre | $k_{ol}(\mu\text{–}\sigma)$ | $b_{ol}\ (\mu\text{–}\sigma)$ | $slp_{ol}\ (\mu\text{–}\sigma)$ | $cc_{ol}\ (\mu\text{–}\sigma)$ | $nnd_{ol}\ (\mu\text{–}\sigma)$ | $d_{bol}\ (\mu\text{–}\sigma)$ | $c(G)_{ol}\ (\mu\text{–}\sigma)$ | $AUB_{ol}\ (\mu\text{–}\sigma)$ | $AURC_{ol}\ (\mu\text{–}\sigma)$ |
|---|---|---|---|---|---|---|---|---|---|
| Essay | 5.377–0.94 | 2943.853–1852.075 | 2.915–0.081 | 0.413–0.093 | 280.854–232.41 | 6.041–0.775 | $7.85 \times 10^{-4}$–$3.84 \times 10^{-4}$ | 1520.381–972.673 | 0.0151–.003 |
| Narrative (Tale) | 4.756–0.534 | 1324.018–628.419 | 3.033–0.095 | 0.289–0.06 | 99.135–55.332 | 5.019–0.798 | $7.24 \times 10^{-4}$–$3.63 \times 10^{-4}$ | 693.8–316.46 | 0.0242–0.007 |
| Narrative (Novel) | 6.794–1.014 | 5035.632–1864.014 | 2.793–0.062 | 0.507–0.093 | 555.98–231.929 | 7.383–0.673 | $6.79 \times 10^{-4}$–$2.49 \times 10^{-4}$ | 2644.833–1012.555 | 0.0151–0.002 |
| Research Article | 5.777–0.466 | 2096.629–725.505 | 3.015–0.095 | 0.349–0.05 | 152.065–72.851 | 5.378–0.559 | $3.59 \times 10^{-4}$–$2.42 \times 10^{-4}$ | 1078.357–379.765 | 0.0283–0.007 |

An Analysis of Variance (ANOVA) or a Kruskal–Wallis test—an ANOVA test carried out if the normality and homoscedasticity assumptions were valid for the given measure—was performed to select the measures of complex network that are influenced by essay, tale, novel, and research article genres. The one-way ANOVA conducted on the individual influence of essay, tale, novel, and research article on $k_c$, $spl_f$, $cc_f$, $d_{bf}$, $c(G)_f$, and $AURC_f$ shows significant effects: $F(3,59) = 12.81$, $p < 0.0001$; $F(3,59) = 15.039$, $p < 0.0001$; $F(3,59) = 14.77$, $p < 0.0001$; $F(3,59) = 19.27$, $p < 0.0001$; $F(3,59) = 6.40$, $p < 0.001$; and $F(3,59) = 22.35$, $p < 0.0001$. Similarly, a Kruskal–Wallis test shows a significant difference of the literary genres on $nnd_f$, $AUB_f$, $b_f$; $H(3) = 29.44$, $p < 0.0001$; $H(3) = 27.98$, $p < 0.0001$; and $H(3) = 28.68$, $p < 0.0001$.

The one-way ANOVA conducted on the individual influence of essay, tale, novel, and research article on $spl_{nf}$, $cc_{nf}$, $d_{bnf}$, and $AURC_{nf}$ shows significant effects: $F(3,59) = 3.70$, $p = 0.016$; $F(3,59) = 6.17$, $p = 0.001$; $F(3,59) = 3.00$, $p \leq 0.037$; and $F(3,59) = 4.28$, $p = 0.008$. On the other hand, no effect on $k_{nf}$ $F(3,59) = 2.65$, $p = 0.057$; $nnd_{nf}$ $F(3,59) = 1.12$, $p = 0.347$; $b_{nf}$ $F(3,59) = 0.227$, $p = 0.877$; $c(G)_{nf}$ $H(3) = 3.99$, $p = 0.262$; and $AUB_{nf}$ $H(3) = 1.29$, $p = 0.731$ by genres were found. Although $spl_{nf}$, $cc_{nf}$, $d_{bnf}$, and $AURC_{nf}$ have a significant difference, they do not provide additional information—of those provided by the measures of full-text networks—to differentiate the genre. For example, $spl_{nf}$ is only statistically different for the novel and tale (see Table 6). However, $slp_f$ is statistically different for the novel, essay, and both the research article and tale. Thus, $spl_{nf}$, $cc_{nf}$, and $d_{bnf}$ were not included in the set of measures to build data mining models. Table 6 summarises the significant statistical difference for $spl_f$ and $spl_{nf}$.

Finally, the one-way ANOVA conducted on the individual influence of essay, tale, novel, and research article on $spl_l$ and $AURC_l$ shows significant effects: $F(3,59) = 17.62$, $p < 0.0001$; $F(3,59) = 4.28$, $p = 0.008$. Similarly, a Kruskal–Wallis test shows a significant difference of $k_l$, $cc_l$, $d_{bl}$, $c(G)_l$, $nnd_l$, $AUB_l$, and $b_l$ by genre: $H(3) = 32.98.44$, $p < 0.0001$; $H(3) = 23.38$, $p < 0.0001$; $H(3) = 30.20$, $p < 0.0001$; $H(3) = 22.03$, $p < 0.0001$; $H(3) = 29.38$, $p < 0.0001$; $H(3) = 32.40$, $p < 0.0001$, $p < 0.0001$; and $H(3) = 32.64$, $p < 0.0001$.

**Table 6.** The subsets built using the significant statistical differences between $slp_f$ and $slp_{nf}$ induced by the novel, essay, research article, and tale. The value in the intersection of each row and column is the means of each measure for a given genre.

| Genre | Subset 1 | Subset 2 | Subset 3 |
|---|---|---|---|
| Novel–$slp_f$ | 2.78 | | |
| Essay–$slp_f$ | | 2.90 | |
| Research Article–$slp_f$ | | | 3.00 |
| Tale–$slp_f$ | | | 3.02 |
| Novel–$slp_{nf}$ | 4.46 | | |
| Essay–$slp_{nf}$ | 5.10 | 5.10 | |
| Research Article–$slp_{nf}$ | 5.37 | 5.37 | |
| Tale–$slp_{nf}$ | | 5.77 | |

After these analyses, the $spl_f$, $k_f$, $nnd_f$, $cc_f$, $b_f$, $db_f$, $AURC_f$, $AUB_f$, $c(G)_f$, $spl_l$, $k_l$, $nnd_l$, $cc_l$, $b_l$, $db_l$, $AURC_l$, $AUB_l$, and $c(G)_l$ were selected to classify the genre of each literary work. This set of measures is a data mining view named $DV_1$, and $DV_1$ was compared with a data mining view named $DV_2$ that contains all the measures computed on the three types of co-occurrence networks described previously. Also, a third data mining view named $DV_3$, which contains only the measures $spl$, $k$, $nnd$, $cc$, and $b$ obtained from the three types of co-occurrence networks, was tested to show that measures such as $d_b$, $c(G)$, $AUB$, and $AURC$ contribute to capturing the features of the literary genre. Since the influences of the data mining technique and data mining view on the AROC need to be tested, a two-way ANOVA is appropriate for this purpose, providing the data is normal and homoscedastic [32,35]. However, the AROC generated by our experiments does not meet these assumptions; thus, a Scheirer–Ray–Hare test [36,37] was used instead. A Scheirer–Ray–Hare test shows there is a significant difference among the AROC of the data mining views: $H(2) = 21.496$, $p < 0.001$, the data mining techniques: $H(3) = 84.79$, $p < 0.001$, and the interaction between both: $H(6) = 30.167$, $p < 0.001$. Figure 5 summarises the effect of both data mining view and data mining technique on AROC that are detailed below.

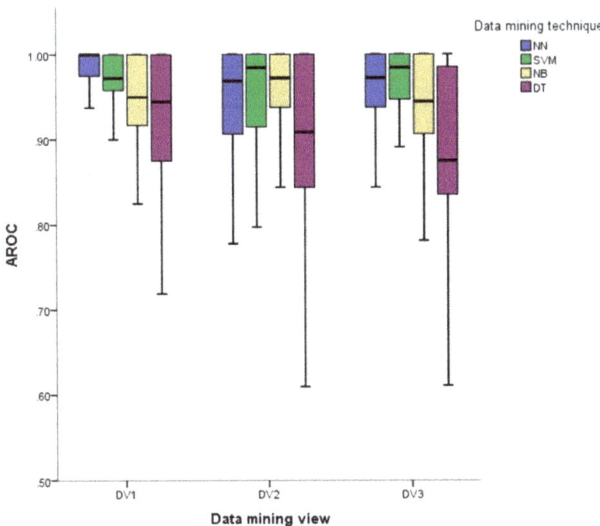


**Figure 5.** The effect of data mining view and data mining technique on Area under the Receiver Operating characteristic Curve (AROC).

A Kruskal–Wallis test shows that $DV_1$, $DV_2$, and $DV_3$ affect the median of the AROC:

$H$ (2) = 21.496, $p < 0.001$. A posthoc Mann–Whitney test using a Dunn–Sidak adjustment [38] ($\alpha = 0.0169$) shows that the median of $DV_1$ ($Mdn = 0.975$) is higher than $DV_2$ ($Mdn = 0.968$)—$U$ ($N_{Dv1} = 400$, $N_{Dv2} = 400$) = 68704, $z = -3.59$, $p < 0.001$ and $DV_3$ ($Mdn = 0.955$)—$U$ ($N_{Dv1} = 400$, $N_{Dv2} = 400$) = 66131, $z = -4.388$, $p < 0.001$. Thus, the statistical analysis carried out on the measures of three types of networks is useful to select relevant measures that increase the AROC. No statistical difference was found between $DV2$ and $DV3$, $U$ ($N_{Dv2} = 400$, $N_{Dv2} = 400$) = 78117, $z = -0.59$, $p = 0.236$. This evidence suggests that well-known measures such as: node degree, shortest path length, betweenness, clustering coefficient, and the average of nearest neighbourhoods' degree—used to build $DV_3$—applied in the previous research to identify authors' writing styles [3–8] are not enough to produce a higher AROC. On the other hand, more complex measures such as: $d_b$, $c(G)$, $AUB$, and $AURC$ improve the classification.

Similarly, the Kruskal–Wallis test shows that the medians of the AROC obtained from NN, SVM, NB, and DT affect the AROC, $H$ (3) = 84.793. A posthoc Mann–Whitney test using a Dunn–Sidak adjustment [38] ($\alpha = 0.0085$) shows that the median of both NN ($Mdn = 1.00$) and SVM ($Mdn = 0.975$) were higher than those of NB ($Mdn = 0.968$)—see the corresponding row and column of Table 7 for the result of the pair-wise test e.g., row NB and column NN show a significant difference: $U$ ($N_{NN} = 300$, $N_{NB} = 300$) = 36784.5, $z = -3.995$, $p < 0.0001$— and DT ($Mdn = 0.911$). No statistical difference between NN ($Mdn = 1.00$) and SVM ($Mdn = 0.975$) was found.

**Table 7.** Pair-wise Mann–Whitney test using a Dunn–Sidak adjustment ($\alpha = 0.0085$) among data mining techniques. The intersection of a row and a column presents the result of the test between the two data mining techniques.

| | NN | SVM | NB | DT |
|---|---|---|---|---|
| NN | — | | | |
| SVM | $U$ ($N_{NN} = 300$, $N_{SVM} = 300$) = 41859, $z = -1.55$, $p < 0.120$ | — | | |
| NB | $U$ ($N_{NN} = 300$, $N_{NB} = 300$) = 36784.5, $z = -3.995$, $p < 0.0001$ | $U$ ($N_{SVM} = 300$, $N_{NB} = 300$) = 35311, $z = -4.749$, $p < 0.0001$ | — | |
| DT | $U$ ($N_{NN} = 300$, $N_{DT} = 300$) = 29119, $z = -7.816$, $p < 0.0001$ | $U$ ($N_{NN} = 300$, $N_{DT} = 300$) = 30523, $z = -6.989$, $p < 0.0001$ | $U$ ($N_{NN} = 300$, $N_{DT} = 300$) = 35438.5, $z = -4.588$, $p < 0.0001$, | — |

Then a significant difference between NB ($Mdn = 0.968$) and DT ($Mdn = 0.911$), was found. These results suggest that the $DV_1$ and the use of NN or SVM produce statistically equal values of AROC. The accuracy of NN and SVM based on $DV_1$ are 0.93 and 0.90, respectively, based on $DV_1$.

To support the conjecture that deleting number, punctuation, and functional words do not have a significant effect on the AROC, the models of NN based on $DV_1$ and the fourth data mining view named $DV_4$, which contain the measures from the networks built using the full text ($spl_f$, $k_f$, $nnd_f$, $cc_f$, $b_f$, $db_f$, $AURC_f$, $AUB_f$, and $c(G)_f$) and those from networks built using only a lemmatisation stage ($spl_{ol}$, $k_{ol}$, $nnd_{ol}$, $cc_{ol}$, $b_{ol}$, $db_{ol}$, $AURC_{ol}$, $AUB_{ol}$, and $c(G)_{ol}$), were compared. The Mann–Whitney test shows no statistical difference—$U$ ($N_{DV1} = 100$, $N_{DV2} = 100$) = 4793, $z = -0.631$, $p = 0.528$—between the AROC of $DV_1$ ($Mdn = 1$) and $DV_4$ ($Mdn = 0.98$). The accuracy of $DV_1$ and $DV_4$ is 0.93 for both. Thus, the deletion of the number and punctuation marks is not useful to reveal stylistic attributes by genre as lemmatisation does. Furthermore, all these stages together modify the network fractality, as the evidence presented later suggests. The accuracy of the NN model based on $DV_1$, $DV_2$, $DV_3$, and $DV_4$ are 0.93, 0.90, 0.89, and 0.93, respectively.

To classify each network as fractal or non-fractal, the Akaike Information Criterion (AIC) [39] were computed for the networks based on the full text. The second network was obtained by deleting numbers, punctuation marks, and functional words. The third was created by adding a lemmatisation stage, and the fourth was attained only through a lemmatisation. The AIC is useful to classify networks as fractal and non-fractal [40]. To select the better mathematical model, first the AIC for power (denoted by subscript P) and exponential (denoted by subscript E) models—Equations (5) and (7)—were computed, then the minimum value is chosen ($AIC_{min}$). $\Delta AIC_i$ was computed by $AIC_i$ - $AIC_{min}$, where *i* is the AIC of power or exponential models. The AIC's rule of thumb is that the two models are statistically different if $\Delta AIC$ is greater than two, thus, the model with $\Delta AIC$ = 0 should be selected [41,42]. Table S2 of the supplementary material shows that the difference between $\Delta AIC_P$ and $\Delta AIC_E$ for about 87% of the full word co-occurrence network is higher than two; thus, the mathematical model for the relation *l* vs. $N_b(l)$ computed by the box-covering algorithm of these networks is the power model (see Equation (5)). Although for 13% of the networks, a model cannot be selected feasibly based on $\Delta AIC$, the power model obtained the least value. Thus, most of the full word co-occurrence networks of literary works are fractal. This result supports the fractality founded in other languages and English literature by different mathematical analyses [43–46]. Noticeably, selecting the better model based on the adjusted coefficient of determination ($R2$) is rather difficult.

Similarly, Table S3 of the supplementary material shows that the difference between $\Delta AIC_P$ and $\Delta AIC_E$ for about 89% of the word co-occurrence networks—built by deleting numbers, punctuation marks, and functional words—suggests they are fractal; 2% were classified as exponential, and 9% were undetermined (since $\Delta AIC \leq 2$). However, adding a lemmatisation stage to the previous ones dilutes the fractality (25.3% are fractal, 33.3% are exponential, and 41.3 are undetermined), see Table S4. The lemmatisation stage alone preserves the fractality of the full-text networks (87% are fractals, and 13% are undetermined); see Tables S2 and S5, which show no difference between the AROC curve of the classification of literary works according to their genre. Note that the lemmatisation stage preserves the original fractality of the networks. Thus, this supports the conjecture that lemmatisation is a kind of renormalisation of a complex network that preserves the fractality. This paves the way to compare this linguistic renormalisation with that introduced by Song, Havlin [16].

## 5. Conclusions

This research aims at showing that measures of the word co-occurrence network of literary works—by Mexican writers—classifies them according to the literary genre. The local measures—such as: node degree, the average of nearest neighbourhoods' degree, and global measures using shortest path length, betweenness, clustering coefficient, and the average of nearest neighbourhoods' degree—widely used in the previous research to identify authors' writing styles, produces acceptable values of AROC classification. However, more elaborate measures using fractal dimension, complexity, the AUB, and the AURC show an improvement of AROC. These measures capture the topology based on the minimum number of boxes to cover the network, the robustness, and the complexity measured by structural entropy and density. Precisely 87% of the full word co-occurrence networks were classified as a fractal. Thus, those findings support the conjecture that fractality occurs in the literary works of Mexican writers, as was previously reported by their English-speaking counterparts. Also, the empirical evidence suggests that the lemmatisation of literary works is a renormalisation stage that preserves the original text fractality. On the contrary, the deletion of numbers, punctuation marks, and functional works, as well as lemmatisation, dilute the fractality. The number of literary works included in this study limit the generalisation of this conjecture. Also, it would be interesting for future research directions to compare the renormalisation induced by a lemmatisation stage—linguistic renormalisation—to renormalisation of networks based on the box-covering algorithm.

**Supplementary Materials:** The following are available online at http://www.mdpi.com/1099-4300/22/8/904/s1, Table S1 Title, primary author and the gender of the literacy works, Table S2, The adjusted determination coefficient, AIC, $\Delta AIC$ and classification for the networks based the full text, Table S3 The adjusted determination coefficient $R2$, AIC, $\Delta AIC$ and classification for the networks obtained by deleting numbers, punctuation marks and functional words, Table S4 The adjusted determination coefficient $R2$, AIC, $\Delta AIC$ and classification for the networks obtained by deleting numbers, punctuation marks, functional words and a lemmatisation stage, and Table S5 The adjusted determination coefficient, AIC, $\Delta AIC$ and classification for the networks obtained only by a lemmatisation stage.

**Funding:** This research was funded by Instituto Politécnico Nacional, grant number SIP20200169.

**Acknowledgments:** An especial mention is given to Alejandro Rosas who provide the corpus of Mexican writers used for experiments.

**Conflicts of Interest:** The authors declare no conflict of interest.

## References



1. Fang, Y.; Wang, Y. Quantitative Linguistic Research of Contemporary Chinese. *J. Quant. Linguist.* **2018**, *25*, 107–121. [CrossRef]
2. Cong, J.; Liu, H. Approaching human language with complex networks. *Phys. Life Rev.* **2014**, *11*, 598–618. [CrossRef] [PubMed]
3. Amancio, D.R.; Oliveira, O.N., Jr.; Costa, L.d.F. Structure–semantics interplay in complex networks and its effects on the predictability of similarity in texts. *Phys. A Stat. Mech. Appl.* **2012**, *391*, 4406–4419. [CrossRef]
4. Akimushkin, C.; Amancio, D.R.; Oliveira, O.N. On the role of words in the network structure of texts: Application to authorship attribution. *Phys. A Stat. Mech. Appl.* **2018**, *495*, 49–58. [CrossRef]
5. Mehri, A.; Darooneh, A.H.; Shariati, A. The complex networks approach for authorship attribution of books. *Phys. A Stat. Mech. Appl.* **2012**, *391*, 2429–2437. [CrossRef]
6. Darooneh, A.H.; Shariati, A. Metrics for evaluation of the author's writing styles: Who is the best? Chaos Interdiscip. *J. Nonlinear Sci.* **2014**, *24*, 033132.
7. Machicao, J.; Corrêa Jr, E.A.; Miranda, G.H.; Amancio, D.R.; Bruno, O.M. Authorship attribution based on Life-Like Network Automata. *PLoS ONE* **2018**, *13*, e0193703. [CrossRef]
8. Stanisz, T.; Kwapień, J.; Drożdż, S. Linguistic data mining with complex networks: A stylometric-oriented approach. *Inf. Sci.* **2019**, *482*, 301–320. [CrossRef]
9. Amancio, D.R.; Nunes, M.D.; Oliveira Jr, O.N.; Pardo, T.A.; Antiqueira, L.; Costa, L.D. Using metrics from complex networks to evaluate machine translation. *Phys. A Stat. Mech. Appl.* **2011**, *390*, 131–142. [CrossRef]
10. Ferraz de Arruda, H.; Nascimento Silva, F.; Queiroz Marinho, V.; Raphael Amancio, D.; da Fontoura Costa, L. Representation of texts as complex networks: A mesoscopic approach. *J. Complex Netw.* **2017**, *6*, 125–144. [CrossRef]
11. De Arruda, H.F.; Marinho, V.Q.; Costa, L.D.; Amancio, D.R. Paragraph-based representation of texts: A complex networks approach. *Inf. Process. Manag.* **2019**, *56*, 479–494. [CrossRef]
12. Kim, J.; Wilhelm, T. What is a complex graph? *Phys. A Stat. Mech. Appl.* **2008**, *387*, 2637–2652. [CrossRef]
13. Van Steen, M. *Graph Theory and Complex Networks: An Introduction*; Cambridge University Press: Cambridge, UK, 2010.
14. Estrada, E. *The Structure of Complex Networks: Theory and Applications*; Oxford University Press: Oxford, UK, 2012.
15. Gallos, L.K.; Song, C.; Makse, H.A. A review of fractality and self-similarity in complex networks. *Phys. A Stat. Mech. Appl.* **2007**, *386*, 686–691. [CrossRef]
16. Song, C.; Havlin, S.; Makse, H.A. Origins of fractality in the growth of complex networks. *Nat. Phys.* **2006**, *2*, 275–281. [CrossRef]
17. Lei, M.; Liu, L.; Wei, D. An Improved Method for Measuring the Complexity in Complex Networks Based on Structure Entropy. *IEEE Access* **2019**, *7*, 159190–159198. [CrossRef]
18. Scott, J. Social network analysis. *Sociology* **1988**, *22*, 109–127. [CrossRef]
19. Song, C.; Gallos, L.K.; Havlin, S.; Makse, H.A. How to calculate the fractal dimension of a complex network: The box covering algorithm. *J. Stat. Mech. Theory Exp.* **2007**, *2007*, P03006. [CrossRef]
20. Holme, P.; Kim, B.J.; Yoon, C.N.; Han, S.K. Attack vulnerability of complex networks. *Phys. Rev. E Stat. Nonlinear Soft Matter Phys.* **2002**, *65 Pt 2*, 056109. [CrossRef]
21. Callaway, D.S.; Newman, M.E.; Strogatz, S.H.; Watts, D.J. Network Robustness and Fragility: Percolation on Random Graphs. *Phys. Rev. Lett.* **2000**, *85*, 5468–5471. [CrossRef]
22. Cohen, R.; Erez, K.; Ben-Avraham, D.; Havlin, S. Resilience of the Internet to Random Breakdowns. *Phys. Rev. Lett.* **2000**, *85*, 4626–4628. [CrossRef]
23. Cohen, R.; Erez, K.; Ben-Avraham, D.; Havlin, S. Breakdown of the Internet under Intentional Attack. *Phys. Rev. Lett.* **2001**, *86*, 3682–3685. [CrossRef]
24. Albert, R.; Jeong, H.; Barabási, A.-L. Error and attack tolerance of complex networks. *Nature* **2000**, *406*, 378. [CrossRef] [PubMed]

25. Gallos, L.K.; Cohen, R.; Argyrakis, P.; Bunde, A.; Havlin, S. Stability and Topology of Scale-Free Networks under Attack and Defense Strategies. *Phys. Rev. Lett.* **2005**, *94*, 188701. [CrossRef] [PubMed]
26. Gallos, L.K.; Cohen, R.; Argyrakis, P.; Bunde, A.; Havlin, S. *Fractal and Transfractal Scale-Free Networks, in Encyclopedia of Complexity and Systems Science*; Meyers, R.A., Ed.; Springer: New York, NY, USA, 2009; pp. 3924–3943.
27. Iyer, S.; Killingback, T.; Sundaram, B.; Wang, Z. Attack Robustness and Centrality of Complex Networks. *PLoS ONE* **2013**, *8*, e59613. [CrossRef]
28. Hall, M.; Frank, E.; Holmes, G.; Pfahringer, B.; Reutemann, P.; Witten, I.H. The WEKA data mining software: An update. *ACM SIGKDD Explor. Newsl.* **2009**, *11*, 10–18. [CrossRef]
29. Hutter, F.; Hoos, H.H.; Leyton-Brown, K. *Sequential Model-Based Optimisation for General Algorithm Configuration*; Springer: Berlin/Heidelberg, Germany, 2011.
30. Kotthoff, L.; Thornton, C.; Hoos, H.H.; Hutter, F.; Leyton-Brown, K. Auto-WEKA: Automatic Model Selection and Hyperparameter Optimization in WEKA. In *Automated Machine Learning: Methods, Systems, Challenges*; Hutter, F., Kotthoff, L., Vanschoren, J., Eds.; Springer International Publishing: Cham, Germany, 2019; pp. 81–95.
31. Quinlan, J.R. *C4. 5: Programs for Machine Learning*; Elsevier: Amsterdam, The Netherlands, 2014.
32. Ramirez-Arellano, A.; Bory-Reyes, J.; Hernandez-Simon, L.M. Statistical Entropy Measures in C4.5 Trees. *Int. J. Data Warehous. Min. (IJDWM)* **2018**, *14*, 1–14. [CrossRef]
33. Tharwat, A. Classification assessment methods. *Appl. Comput. Inform.* 2018. [CrossRef]
34. Hand, D.J.; Till, R.J. A Simple Generalisation of the Area Under the ROC Curve for Multiple Class Classification Problems. *Mach. Learn.* **2001**, *45*, 171–186. [CrossRef]
35. Demšar, J. Statistical comparisons of classifiers over multiple data sets. *J. Mach. Learn. Res.* **2006**, *7*, 1–30.
36. Scheirer, C.J.; Ray, W.S.; Hare, N. The Analysis of Ranked Data Derived from Completely Randomised Factorial Designs. *Biometrics* **1976**, *32*, 429–434. [CrossRef]
37. Dytham, C. *Choosing and Using Statistics: A Biologist's Guide*; Wiley: Hoboken, NJ, USA, 2011.
38. Ennos, A.R. *Statistical and Data Handling Skills in Biology*; Pearson/Prentice Hall: Upper Saddle River, NJ, USA, 2007.
39. Akaike, H. A new look at the statistical model identification. *IEEE Trans. Autom. Control* **1974**, *19*, 716–723. [CrossRef]
40. Ramirez-Arellano, A. Students learning pathways in higher blended education: An analysis of complex networks perspective. *Comput. Educ.* **2019**, *141*, 103634. [CrossRef]
41. Burnham, K.P.; Anderson, D.R. Multimodel Inference: Understanding AIC and BIC in Model Selection. *Sociol. Methods Res.* **2004**, *33*, 261–304. [CrossRef]
42. Burnham, P.K.; Anderson, D.R. *Model Selection and Multimodel Inference: A Practical Information-Theoretic Approach*; Springer: Berlin/Heidelberg, Germany, 2002; Volume 67.
43. Andres, J. On a Conjecture about the Fractal Structure of Language. *J. Quant. Linguist.* **2010**, *17*, 101–122. [CrossRef]
44. Hrebíček, L.K. Fractals in language. *J. Quant. Linguist.* **1994**, *1*, 82–86.
45. Glattre, H.R.; Glattre, E. Finding Fractal Networks in Literature. *Nonlinear Dyn. Psychol Life Sci.* **2018**, *22*, 263–282.
46. Kohler, R. Are there fractal structures in language? Units of measurement and dimensions in linguistics. *J. Quant. Linguist.* **1997**, *4*, 122–125. [CrossRef]

*Article*

# Detecting Overlapping Communities in Modularity Optimization by Reweighting Vertices

**Chen-Kun Tsung [1,*], Hann-Jang Ho [2], Chien-Yu Chen [3], Tien-Wei Chang [3] and Sing-Ling Lee [3]**

[1] Department of Computer Science and Information Engineering, National Chin-Yi University of Technology, Taichung 41170, Taiwan

[2] Department of Applied Digital Media, WuFeng University, Chiayi County 62153, Taiwan; hhj@wfu.edu.tw

[3] Department of Computer Science and Information Engineering, National Chung Cheng University, Chiayi 62102, Taiwan; kw.15@hotmail.com (C.-Y.C.); felicitylab@gmail.com (T.-W.C.); singling@ccu.edu.tw (S.-L.L.)

* Correspondence: ckt@ncut.edu.tw; Tel.: +886-4-23924505

Received: 21 June 2020; Accepted: 24 July 2020; Published: 27 July 2020

**Abstract:** On the purpose of detecting communities, many algorithms have been proposed for the disjointed community sets. The major challenge of detecting communities from the real-world problems is to determine the overlapped communities. The overlapped vertices belong to some communities, so it is difficult to be detected using the modularity maximization approach. The major problem is that the overlapping structure barely be found by maximizing the fuzzy modularity function. In this paper, we firstly introduce a node weight allocation problem to formulate the overlapping property in the community detection. We propose an extension of modularity, which is a better measure for overlapping communities based on reweighting nodes, to design the proposed algorithm. We use the genetic algorithm for solving the node weight allocation problem and detecting the overlapping communities. To fit the properties of various instances, we introduce three refinement strategies to increase the solution quality. In the experiments, the proposed method is applied on both synthetic and real networks, and the results show that the proposed solution can detect the nontrivial valuable overlapping nodes which might be ignored by other algorithms.

**Keywords:** data mining; community detection; overlapping communities; modularity

## 1. Introduction

Determining the group with some particular properties helps the analysts to capture the common properties from the members in the community. Many applications could be considered based on the community detection. For example, the precise information delivery, e.g., Google AdWords [1] increases the transaction amounts for sending the advertisement information to the right person. Therefore, detecting communities is a popular research topic [2–8].

Many results focus on the disjoin community sets that each node belongs to exactly one community [2,3]. However, in the real-world networks, many people may belong to multiple communities, so the communities may overlap with each other. For example, an engineer may belong to many projects in a company. Thus, instead of strict partitions, fuzzy partitions are more appropriate for understanding the network structures [9,10]. Fuzzy partitions allow a node belongs to multiple communities simultaneously. Considering a real-world situation, some staff work together in a building, and the manager would like to track the movement history for each staff [11]. Each one may move to various rooms, and the move purpose comes from the role of each staff. When we treat the purpose of all staff to be the communities, the staff may belong to different communities.

The modularity function proposed by Newman and Girvan [12] is the famous measurement of network partitions to measure the structure of a given network. The modularity function calculates the difference between the number of real intra-community edges and the expected number of edges to identify the qualities of the communities. The partition with larger modularity value has better community structure than those with lower modularity values. Finding the partitions with maximum modularity is a straightforward solution to the community detection. However, the modularity maximization has been proved as an NP-hard problem [13], and finding the partition with maximum modularity is difficult. Therefore, many results are proposed to calculate the near optimal solutions, such as the random walk processes [14], the structural clustering [15], and the polynomial-time approximation algorithms [16].

On the other hand, besides the computation complexity, the modularity maximization has two problems in detecting communities:

**Resolution limits** Fortunato et al. introduced that small communities cannot be detected in large networks [17,18]. Since the null model of modularity provides the global connectivity, the expected number of edges between two small groups in a large network might be very small. Eventually, the two small groups will be treated as one community. Many approaches are proposed for solving resolution limits to provide high solution qualities, such as greedy algorithms [19,20], spectral algorithms [21–23], simulating annealing algorithms [24] and mathematical programing [25].

**Overlapping community** Some nodes may belong to several communities, so simply assigning the nodes to one community is difficult. Thus, the straightforward solution is to modify the modularity for allowing the nodes belonging to multiple communities at the same time [26–30]. Figure 1 shows two benchmarks about overlapping communities. In Figure 1a, the node $v_9$ is the overlapping node, and we assign $v_9$ to community $B$ and $C$. Thus, we get three communities, and they are $\{\{v_1, v_2, v_3, v_4\}, \{v_5, v_6, v_7, v_8, v_9\}, \{v_9, v_{10}, v_{11}, v_{12}, v_{13}, v_{14}\}\}$. Moreover, $v_5$ is assigned to $A$ and $B$ in Figure 1b.

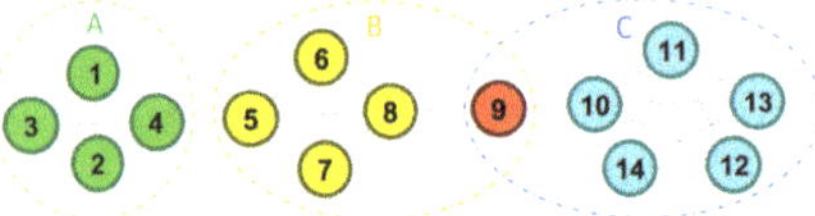

(a) *G4415*, an example with three communities

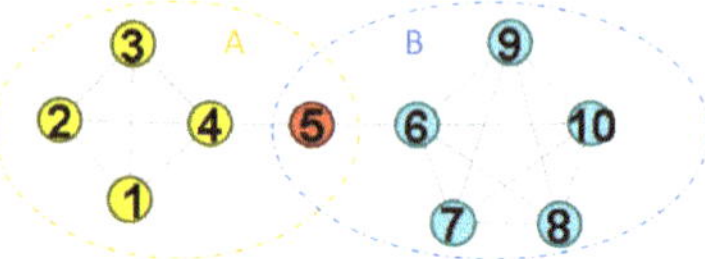

(b) *G415*, an example with two communities

**Figure 1.** The benchmark with more than two communities and two communities.

In this paper, we focus on the overlapping community detection, and propose the node weight allocation problem denoted by $NWA_{OCD}$ to formulate the community overlap. Since computing the partition with maximum modularity is NP-complete, decreasing the computation cost to seek the near optimal partitions is the popular approach in solving the overlapping community detection. The heuristic algorithms are outstanding in seeking better solutions in large search space, especially for the genetic algorithms (GAs) [2,3,8]. Therefore, some works consider GA as the core approach in their solutions. Mu et al. use a hybrid heuristic approach including GA and the simulated annealing to find out the communities [2]. Shang et al. use GA with an extra local search [3]. The heuristic algorithms perform well in seeking the solution with high quality in a large search space. However, the above results do not deal with the overlapping properties. The overlapping networks have various properties, so some approaches consider the multi-objective approach to find the balanced results [4–6,31]. The balanced results mean that most properties are considered, but the derived results may not be closed to the real-world properties. Therefore, Behera et al. check the similarity between each pair of nodes [8]. The node similarity is also considered by Ezeh et al. to the overlapping nodes and their neighbors [32]. To emphasize the community attribution of each node, Shakya et al. combine fuzzy with the GA to calculate the detail properties of the nodes [7]. Shakya et al. consider the GA to reduce the computation time without decreasing the solution quality too much and adopt the fuzzy communities to identify the overlapping nodes.

Even if some approaches provide the solutions with high modularity, the partitions may not reflect the properties of the real-world networks in some situations. We found that the solution quality could be refined by considering following issues: ignoring overlapping nodes, merging clusters, and reweighting nodes. Therefore, we consider the modularity to design the solution searcher of the approach $GA_{IMR}^{NWA}$. We firstly modify the fitness function in $GA_{IMR}^{NWA}$ to show the network properties by considering the null model, so the revised fitness function could output the partitions that are closer to the real-world behavior. Moreover, we design three refinement strategies to make the solutions to reflect the real-world properties.

In the simulation, we consider the synthetic network and popular networks that include Zachary Karate Club Network, Books about American Politics, and American College Football to evaluate the solution quality calculated by $GA_{IMR}^{NWA}$ and other approaches. The derived networks correctly reflect the real-world properties in the synthetic networks and the real-world networks. Moreover, the proposed refinement strategies are also evaluated, and the refinement strategies provide higher quality of the derived partitions in the perspective of the real-world behavior. Therefore, the simulation results show that $GA_{IMR}^{NWA}$ outputs the partitions, and the results are closed to the real-world properties.

This paper is organized as follows. The overlapping communities and the problem definition are introduced and formulated in Section 2. The proposed approach $GA_{IMR}^{NWA}$ is shown in Section 3, and the refinement strategies are also listed in this section. The simulation and comparisons are arranged in Section 4, and we show the network partitions in this section. Eventually, the conclusion and future works are stated in Section 5.

## 2. Preliminary

### 2.1. Modularity in Overlapping Communities

The community detection of a given network involves two processes. The first one is to find out the network structure and the other one is to determine the numbers of communities. Here we introduce the works proposed by Nepusz et al. [33] to explain the modularity in overlapping communities. Nepusz et al. consider a belonging coefficient matrix $U = [\alpha_{ic}]_{n \times k}$, where $n$ is the number of nodes, and $k$ is a given number of communities. Each entry $\alpha_{ic}$ shows how strongly the node $v_i$ belongs to the community $c$. The constraint of the relationship between $v_i$ and all communities is:

$$\sum_{c=1}^{k} \alpha_{ic} = 1, \forall \alpha_{ic} \in [0,1], 0 < \sum_{i}^{n} \alpha_{ic} < n. \tag{1}$$

So, the objective function is:

$$D_G(U) = \sum_{i,j=1}^{n} w_{ij}(\tilde{s}_{ij} - s_{ij})^2, \tag{2}$$

where $w_{ij}$ is the predefined weight, $s_{ij} = \sum_{c=1}^{k} \alpha_{ic}\alpha_{jc}$, and $\tilde{s}_{ij}$ is the prior similarity of $v_i$ and $v_j$. By minimizing Equation (2), the nodes with high similarity will be grouped together. So, $U$ with optimal result $D_G(U)$ is the overlapping community structure.

To determine an appropriate number of communities $k$, Nepusz et al. iteratively increase the value of $k$ from 2, and then choose the value of $k$ with the highest fuzzy modularity value calculated by Equation (3).

$$Q_{ov}^{Ne} = \frac{1}{2m} \sum_{c=1}^{k} \sum_{i,j=1}^{n} (A_{ij} - \frac{k_i k_j}{2m})\alpha_{ic}\alpha_{jc} \tag{3}$$

### 2.2. Problem Definition

The overlapping community detection problem is considered as a node weight allocation problem, denoted by $NWA_{OCD}$ for short. Given a network $G(V,E)$, a maximum number of communities $k$, and a null model weight $\gamma$. Find a modified belonging coefficient matrix $M = [\lambda_{ic}]_{n \times k}$, such that the $Q'_{ov}$ value is maximized. The objective function and constraints are:

$$\max \quad Q'_{ov} = \frac{1}{2m} \sum_{c=1}^{k} \sum_{i,j=1}^{n} (A_{ij} - \gamma \frac{k_i k_j}{2m}) \lambda_{ic} \lambda_{jc}$$

$$\text{s.t.} \quad \lambda_{ic} \in [0,1] \tag{4}$$

$$\sum_{c=1}^{|C|} \lambda_{ic}^{inc_f} = 1.$$

We consider $inc_f$ as the increasing factor. Given $inc_f > 1$, the total weight of an overlapping node over all communities is larger than one, i.e., $\sum_{c=1}^{k} \lambda_{ic} > 1$. The total weight of a non-overlapping node is still equal to one exactly, i.e., $\sum_{c=1}^{k} \lambda_{ic} = 1$.

By solving the $NWA_{OCD}$ problem, the overlapping community structure will be obtained by modifying the optimal solution. Note that if $inc_f = 1$ and $\gamma = 1$, Equation (4) is the same with Equation (5), which means the fuzzy modularity is a special case of the $NWA_{OCD}$ problem.

$$\max \quad Q_{ov} = \frac{1}{2m} \sum_{c=1}^{k} \sum_{i,j=1}^{n} (A_{ij} - \frac{k_i k_j}{2m}) \alpha_{ic} \alpha_{jc}$$

$$\text{s.t.} \quad \alpha_{ic} \in [0,1] \tag{5}$$

$$\sum_{c}^{k} \alpha_{ic} = 1.$$

Although Griechisch et al. [34] apply the fuzzy modularity to find overlapping communities, there are still some networks are unresolved. We introduce the networks with more than two communities and two communities to show this issue. The benchmark is shown in Figure 1. The values of $Q_{ov}$ for G4415 and G415 are shown in Table 1. We can see that $v_9$ belongs to $B$ in G4415 while $v_5$ belongs $A$ in G415, and they are not overlapping nodes.

The major difference between Equations (4) and (5) is the coefficient matrix. Each entry in Equation (5) is unweighted while that is weighted in Equation (4). Therefore, we need a mapping as shown in the following equations.

$$\lambda_{ic} = {}^{inc_f}\!\!\sqrt{\alpha_{ic}}$$

$$\alpha_{ic} = \lambda_{ic}^{inc_f} \tag{6}$$

**Table 1.** The values of $Q_{ov}$ in G4415 and G415.

**(a)** The $Q_{ov}$ values with different assignments of $v_9$ in G4415.

| $\alpha_{9,B}$ | $\alpha_{9,C}$ | $\alpha_{9,D}$ | $Q_{ov}$ |
|---|---|---|---|
| 1 | 0 | 0 | 0.5736 |
| 0.7 | 0.3 | 0 | 0.5709 |
| 0.3 | 0.7 | 0 | 0.5664 |
| 0 | 1 | 0 | 0.5624 |
| 0 | 0 | 1 | 0.5560 |

**(b)** The $Q_{ov}$ values with different assignments of $v_5$ in G415.

| $\alpha_{5,A}$ | $\alpha_{5,B}$ | $\alpha_{5,C}$ | $Q_{ov}$ |
|---|---|---|---|
| 1 | 0 | 0 | 0.4305 |
| 0 | 0 | 1 | 0.4151 |
| 0 | 1 | 0 | 0.4058 |

## 3. Allocate Node Weight by Genetic Algorithms

Computing the partition with maximum modularity has been proved as the NP-complete problem [13]. Even if we consider the solution with high computation performance, e.g., the cloud computing [35,36] and the parallel computing [37], to compute the partitions for maximizing the modularity, it still requires huge computation resource. Therefore, we propose a GA-based approach to get the near-optimal solution with minimum computation. The proposed algorithm $GA_{IMR}^{NWA}$ includes two steps. We first apply GA to obtain a high-quality feasible solution, and then design three refinement strategies to improve the derived solution to modify the derived partition to be closer to the real-world behavior. In the following context, we will introduce the revised GA algorithm and the refinement strategies.

### 3.1. Genetic Algorithm

The iterative process of GA as shown in Algorithm 1 includes three major processes: crossover, mutation, and selection. Before invoking the iterative process, the initial population $P$ with $indi_n$ chromosomes will be determined firstly. Each chromosome is represented by $M = [\lambda_{ic}]_{n \times k}$, as shown in Figure 2. Each entry $\lambda_{ic}$ is a weight to indicate the assignment from $v_i$ to $c$. The initial population is generated randomly, and each row of $M$ must satisfy the problem constraints. Given a maximum number of iterations $max_t$, the GA then invokes following processes.

1.  **Crossover**: we randomly select two chromosomes $C_A$ and $C_B$ form $P$, and a random column. The offspring is generated by the selected column of $C_B$ and the remaining part of $C_A$ as shown in Figure 3. The number of offsprings is determined by $indi_n$, and in other words, we will obtain $2 \times indi_n$ chromosomes after the crossover.

2.  **Mutation**: the mutation process is launched in 80% probability after finishing the crossover. Once the mutation is invoked, one $\lambda_{ic}$ of a randomly selected chromosome will be picked up within $[-0.1, 0.1]$. Eventually, the offspring will be normalized to be a feasible solution to fit the requirements in $NWA_{OCD}$.

3.  **Selection**: we consider the modularity to be the objective function, and finding the partition with maximum modularity is the purpose of GA. We use $Q'_{ov}$ to be the fitness function and calculate $Q'_{ov}$ of each solution. Moreover, all chromosomes are sorted in the descending order of $Q'_{ov}$. Computing the chromosomes with maximum $Q'_{ov}$ is the major goal of the GA, so we select top $indi_n$ individuals, and they will survive to the next generation.

---

**Algorithm 1:** Genetic algorithm for allocating node weight

---

**Data:** $max_t$: the maximum number of iteration, $indi_n$: the number of survival genes
1  $P \leftarrow$ initialization($indi_n$);
2  **for** $t = 1 : max_t$ **do**
3      $P' \leftarrow$ crossover($P$);
4      $P' \leftarrow$ mutation($P'$);
5      $P \leftarrow$ selection($P'$);

---

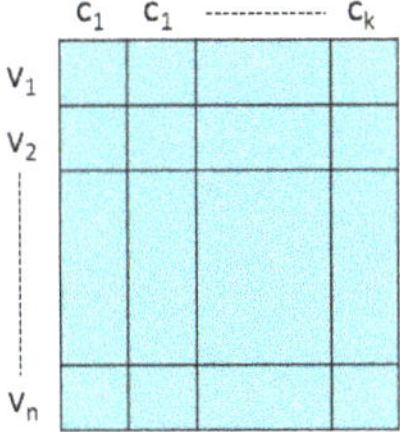

**Figure 2.** The representation of a chromosome.

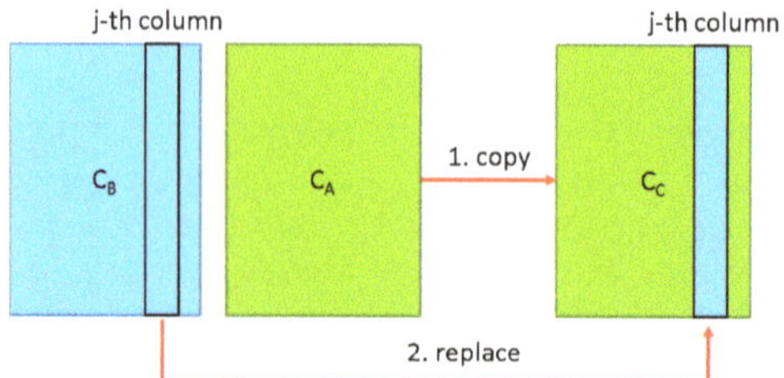


**Figure 3.** The idea of the crossover operation. Two chromosomes are switched the selected area to generate one offspring.

To keep the heavily overlapping nodes, a threshold $\alpha_T$ in terms of $\alpha$ is given. We transform $\alpha_T$ to the corresponding $\lambda$ with the threshold $\lambda_T$ by Equation (6).

*3.2. Refinement Strategies*

GA provides an elite solution from the population, but this solution may not be suitable for all instances. In the pre-analysis phase, we observed three situations derived by $GA_{IMR}^{NWA}$, and we could receive better solutions by some extra processes. The situations are (1) lightly overlapping nodes, (2) mergeable clusters, and (3) reweight nodes. We call the processes that are used to get better solutions the "refinement strategies". Therefore, we provide three refinement strategies to refine the solutions for the above situations, respectively.

**Ignore slight overlapping nodes** The overlapping degree of each $\lambda$ is important for splitting the communities. Determining the community with low value of $\lambda$ is easier than that with a higher value. We use a threshold $\lambda_T$ corresponding to Equation (6) to determine that the entry should be treated as an entry without overlaps. In addition, we also can derive $\lambda_T$ by Equation (6). When $\lambda < \lambda_T$, we set $\lambda$ as zero. When $\lambda_T$ is set as a higher value, more entries will be assigned to single community.

**Merge clusters** Some small communities should be merged by other large community. If the overlapping ratio of any two communities is larger than a given merge threshold $m_T$, they should be simply merged to a single community. Given two non-empty communities, we define $ov_{ratio} = |C_1 \cap C_2| / \min(|C_1|, |C_2|)$ to be the overlapping ratio. When $ov_{ratio}$ is larger than a given threshold, $C_1$ and $C_2$ will be merged.

**Reweight node values** To calculate the weight distribution of each overlapping node, directly converting $\lambda$ to $\alpha$ via Equation (6) results in a situation that a node belongs to multiple communities but the majority of its weight is allocated to one community. To avoid this problem, we propose the reweight strategy. The weight should be proportional to the number of edges that $v_i$ linked in $c$. Moreover, if the neighbors of $v_i$ in $c$ are more than the average number of nodes in $c$, $c$ is more important than others for $v_i$. Given a community c, $avgNighbor_c = \sum_{i,j \in V(c)} A_{ij} / |V(c)|$ represents the average number of neighbors and $\alpha_i = \sum_{c \in C(i)} \sum_{j \in V(c)} A_{ij} / avgNighbor_c$ be the normalized term. Therefore, we have the new weight is:

$$\alpha_{ic} = \frac{1}{\alpha_i} \sum_{j \in V(c)} \frac{A_{ij}}{avgNighbor_c}, \tag{7}$$

where $V(c)$ is the set of nodes belong to $c$ and $C(i)$ is the set of communities that $v_i$ belongs to. We use $\alpha_i$ for normalization, so we have $\sum_{c=1}^{k} \alpha_{ic} = 1$.

## 4. Simulations

We consider a synthetic network and three real networks including Zachary Karate Club network, Books about American Politics, and American College Football to evaluate the performance of $GA_{IMR}^{NWA}$.

The evaluation criteria involve detecting overlapping community structure, detecting meaningful communities, detecting dense overlaps, and detecting heavily overlapping nodes.

### 4.1. Synthetic Network

We consider *G210* as our synthetic network which has 210 nodes and four pre-defined communities $A$, $B$, $C$ and $D$. Each of them has 60 nodes and 10 shared by any two continuous communities, i.e., $A = \{v_1 : v_{60}\}$, $B = \{v_{51} : v_{110}\}$, $C = \{v_{101} : v_{160}\}$, and $D = \{v_{151} : v_{210}\}$. Note that $A$ and $B$ share nodes $\{v_{51}, \dots, v_{60}\}$, $B$ and $C$ share nodes $\{v_{101}, \dots, v_{110}\}$ and so on. Each pair of nodes has 3% chances to be linked to each other, and for each community they shared, an extra 55% chances for them to be linked. Thus, overlapping parts will be denser than non-overlapping parts [38].

Since the fuzzy modularity is a special case of the $NWA_{OCD}$ problem, we could use the same optimization strategy to solve the problem. The parameter settings are $inc_f = 1.5$ and 1, $\alpha_T = 0$, $m_T = -1$, $k = 6$, and $\gamma = 1$. Figure 4 shows the bitmaps of sorted adjacency matrices. The black and white points represent the entries of 1s and 0s respectively. The adjacency matrices are sorted by the following strategy:

1. Nodes are grouped by the detected community id. For the overlapping nodes, only the smallest id is counted.
2. For each $c$, all nodes are sorted in descending order of $\lambda_{ic}$. Therefore, the overlapping nodes will be in the bottom area of each community.

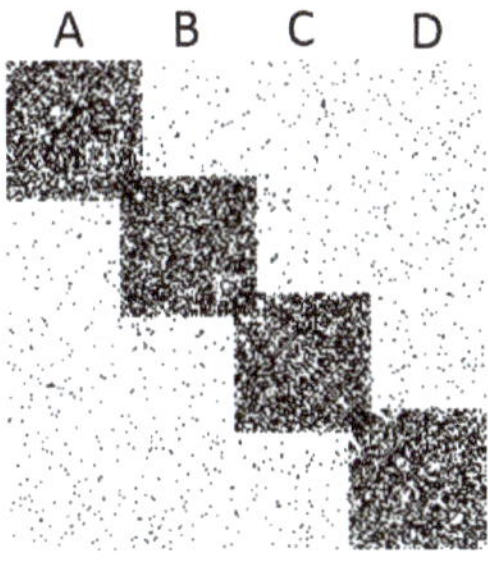

(a) Result of our method

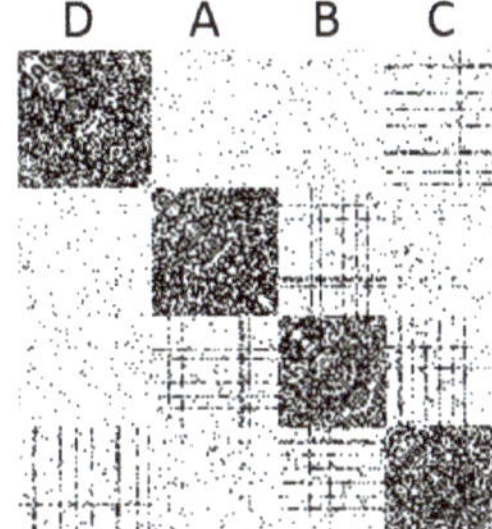


(b) Result of fuzzy modularity

**Figure 4.** The comparison between $GA_{IMR}^{NWA}$ and fuzzy modularity.

Figure 4a is the result obtained by $GA_{IMR}^{NWA}$. The dense blocks indicate four communities, and two continuous blocks have an overlapping part which is composed of overlapping nodes. In this result, all the overlapping and non-overlapping nodes are correctly identified. Figure 4b is the result of fuzzy modularity. Four communities are detected too, but no overlapping nodes are identified.

Although the maximum number of communities is six, only four communities were detected while the other two were empty communities. Since the number of communities could be captured by modularity [39], it is unnecessary to know the exactly value of number of communities in our method.

### 4.2. Zachary Karate Club Network

Zachary karate club network [40] is a popular benchmark for community detection algorithms. It has 34 nodes and 78 edges while nodes are members and edges are friendships between them. This network includes two groups due to a disagreement between the administrator and the instructor. Figure 5 is the result captured by the fuzzy modularity. In this experiment, we evaluate the results with different $inc_f$ settings, and show the importance of "ignore slight overlapping nodes" and "reweight node values". Finally, we apply our method on the case with the value $k = 2$, and halved the null model.

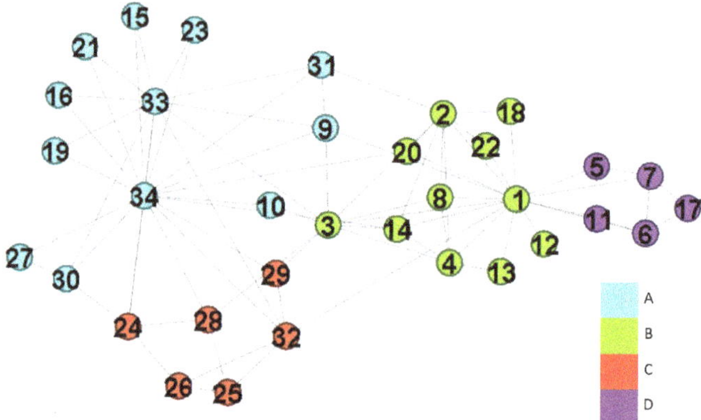



**Figure 5.** Detected communities of the karate network by fuzzy modularity.

### 4.2.1. Effects of Weight Increasing Factor

We first evaluate the communities captured by $GA_{IMR}^{NWA}$ in the networks with $inc_f = \{1, 1.2, 1.5, 1.7\}$ while $\alpha_T = 0.01$, $m_T = -1$, $k = 8$, and $\gamma = 1$. The corresponding $Q_{ov}' = \{0.419, 0.422, 0.427, 0.430\}$. We consider the fuzzy modularity with $inc_f = 1$ as our baseline since it outputs the correct solution.

Figure 6a is the result with $inc_f = 1.2$, and we get four communities and three overlapping nodes while $\lambda$ is shown in Table 2a. The network separation in Figure 6a is identical to that in Figure 5, but maximizing the modularity outputs a larger one than that we derived. When $inc_f$ is increased from 1.2 to 1.5, we get two extra overlapping nodes, and they are $v_{12}$ and $v_{34}$. When $inc_f$ is set as 1.7, the values of $\lambda$ are changes as shown in Table 2c, and others are identical to that derived by $inc_f = 1.5$. Therefore, larger settings of $inc_f$ results in more overlapping nodes.

**Table 2.** The comparison with various $inc_f$ settings.

(a) $\lambda$ values of overlapping nodes in Figure 6a with $inc_f = 1.2$

| Node | $\lambda_{iA}$ | $\lambda_{iB}$ | $\lambda_{iC}$ | $\lambda_{iD}$ |
|---|---|---|---|---|
| $v_1$ | | 0.967 | | 0.068 |
| $v_{10}$ | 0.747 | 0.362 | | |
| $v_{24}$ | 0.419 | | 0.696 | |

(b) $\lambda$ values of overlapping nodes in Figure 6b with $inc_f = 1.5$

| Node | $\lambda_{iA}$ | $\lambda_{iB}$ | $\lambda_{iC}$ | $\lambda_{iD}$ |
|---|---|---|---|---|
| $v_1$ | | 0.917 | | 0.246 |
| $v_3$ | | 0.986 | 0.075 | |
| $v_{10}$ | 0.700 | 0.556 | | |
| $v_{12}$ | | 0.993 | | 0.048 |
| $v_{24}$ | 0.553 | | 0.703 | |
| $v_{34}$ | 0.993 | | 0.048 | |

(c) $\lambda$ values of overlapping nodes in Figure 6b with $inc_f = 1.7$

| Node | $\lambda_{iA}$ | $\lambda_{iB}$ | $\lambda_{iC}$ | $\lambda_{iD}$ |
|---|---|---|---|---|
| $v_1$ | | 0.888 | | 0.369 |
| $v_3$ | | 0.987 | 0.108 | |
| $v_{10}$ | 0.694 | 0.636 | | |
| $v_{12}$ | | 0.926 | | 0.290 |
| $v_{24}$ | 0.600 | | 0.726 | |
| $v_{34}$ | 0.989 | | 0.097 | |

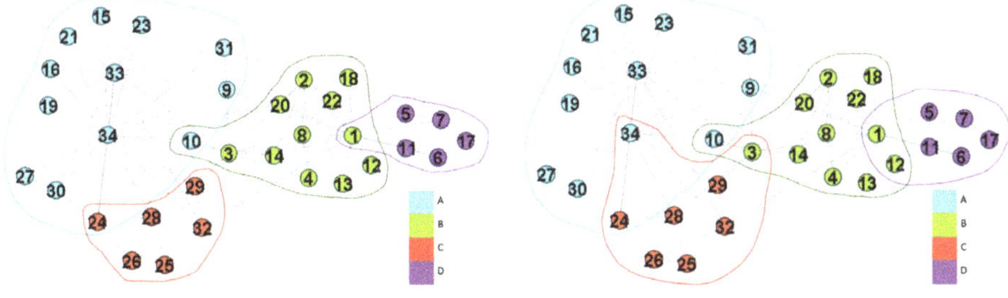

(**a**) Detected communities with $inc_f = 1.2$     (**b**) Detected communities with $inc_f = 1.5$ and 1.7

**Figure 6.** The communities detected by $GA_{IMR}^{NWA}$ under various $inc_f$ settings.

Considering that a node has only one edge connecting to an overlapping node, e.g., $v_{12}$, the isolation has the same property with that held by the overlapping node. Moreover, we found that $Q_{ov}'$ derived by $GA_{IMR}^{NWA}$ is higher than the optimal $Q$. It implies that the overlapping structure is easier to be detected as assigning higher weight to the overlapping nodes.

Here we consider an extreme case that all nodes are overlapped, i.e., $inc_f = 4$. We analyze the obtained result, and then find the "duplicate communities". Two or more communities are extremely overlapped with each other, and even some of them are just the same community.

Figure 7 shows the fuzzy partition result. Four communities are detected, but two of them denoted by dotted lines are the subsets of the rest two communities denoted by solid lines. Therefore, two sets should be merged to a correct community. After merging the communities, we derive two communities, and there is only one overlapping node $v_{10}$. However, the value of $Q_{ov}'$ is decreased from 0.526 to 0.371 simultaneously.

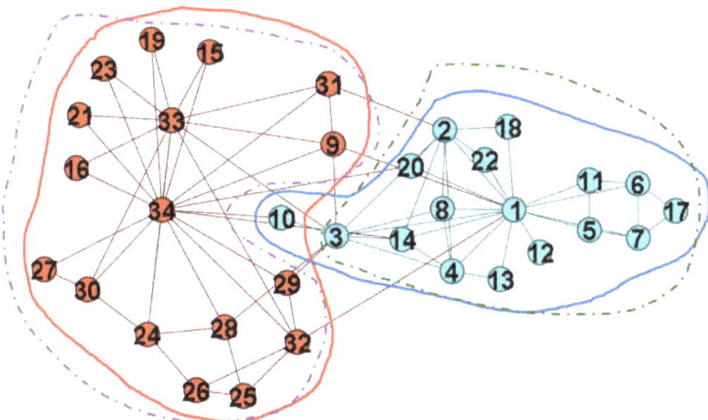

**Figure 7.** Duplicate communities result.

Even if we derive the result with maximized value of $Q_{ov}'$, the solution does not show the correct properties of the communities. We use the refinement strategies to get the solution with lower quality but more closed to the real-world properties. Therefore, the refinement strategies are useful for improving the solution quality in terms of the real-world consideration.

### 4.2.2. Effects of Ignoring Slight Overlapping Nodes

We consider the network with $inc_f = 1.5$ to evaluate the effects of the *ignore* step. The result with and without the *ignore* step are 0.427114 and 0.427117, respectively. Figure 8 and Table 3 are the detected communities and values of $\lambda$. Two overlapping nodes $v_{28}$ and $v_{30}$ are ignored. Since most of their weights were kept in a specific community, reducing the weights will not decrease $Q_{ov}'$

dramatically. Therefore, the process of ignoring slight overlapping nodes helps to keep those heavily overlapping nodes.

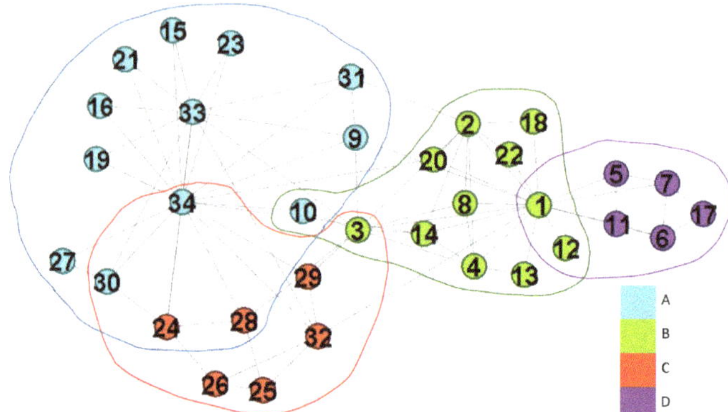

**Figure 8.** Detected communities with $inc_f = 1.5$ (before ignoring).

**Table 3.** $\lambda$ values of overlapping nodes in Figure 8 with $inc_f = 1.5$ (before ignoring).

| Node | $\lambda_{iA}$ | $\lambda_{iB}$ | $\lambda_{iC}$ | $\lambda_{iD}$ |
|------|------|------|------|------|
| $v_1$ | | 0.917 | | 0.246 |
| $v_3$ | | 0.986 | 0.075 | |
| $v_{10}$ | 0.700 | 0.556 | | |
| $v_{12}$ | | 0.993 | | 0.048 |
| $v_{24}$ | 0.553 | | 0.703 | |
| $v_{28}$ | 0.002 | | 0.999 | |
| $v_{30}$ | 0.999 | | 0.004 | |
| $v_{34}$ | 0.993 | | 0.048 | |

### 4.2.3. Effects of Reweight Strategy

To emphasize the importance of the communities, we propose a reweight strategy to assign various weights. The result with reweight strategy is identical to that shown in Figure 6b. Table 4a,b show the value of $\lambda$ without and with considering the reweight strategy, respectively. The reweight strategy reduces the gap of the number of edges for connecting the inside-community nodes and outside-community nodes. However, the structure of the main community may be changed after reweighting, because the values are inversely proportional to the average number of neighbors in the communities to that out of communities. For example, $v_{12}$ is unbalanced before reweighting, but the value of $\lambda$ of $v_{12}$ reflect the real-world behavior.

### 4.2.4. The Network with Two-Communities

We examine the network with exactly two communities to verify the property illustrated in Figure 1b can be captured by $GA_{IMR}^{NWA}$. We consider $inc_f = 1.5, \alpha_T = 0.01, m_T = -1, k = 2$, and $\gamma = 0.5$. In this case, we easily find out the overlapping nodes. The results are shown in Figure 9 and Table 5.

$GA_{IMR}^{NWA}$ derives three overlapping nodes as shown in Table 5. From Figure 9, we have $Q'_{ov} = 0.628$, and the dotted curve is the real split of the club network. $v_3$ is the main overlapping node since it has a roughly balanced weight value. In summary, the two-community problem is solved by reducing the number of expected edges.

**Table 4.** The comparison of $GA_{IMR}^{NWA}$ with reweighting and without reweighting.

| (a) Before reweighting | | | | |
|---|---|---|---|---|
| Node | $\lambda_{iA}$ | $\lambda_{iB}$ | $\lambda_{iC}$ | $\lambda_{iD}$ |
| $v_1$ | | 0.917 | | 0.246 |
| $v_3$ | | 0.986 | 0.075 | |
| $v_{10}$ | 0.700 | 0.556 | | |
| $v_{12}$ | | 0.993 | | 0.048 |
| $v_{24}$ | 0.553 | | 0.703 | |
| $v_{34}$ | 0.993 | | 0.048 | |

| (b) After reweighting | | | | |
|---|---|---|---|---|
| Node | $\alpha_{iA}$ | $\alpha_{iB}$ | $\alpha_{iC}$ | $\alpha_{iD}$ |
| $v_1$ | | 0.611 | | 0.389 |
| $v_3$ | | 0.709 | 0.291 | |
| $v_{10}$ | 0.52 | 0.48 | | |
| $v_{12}$ | | 0.440 | | 0.560 |
| $v_{24}$ | 0.468 | | 0.532 | |
| $v_{34}$ | 0.725 | | 0.275 | |

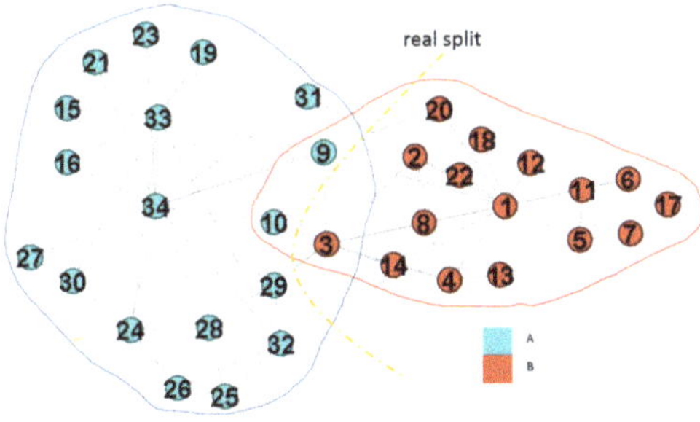


**Figure 9.** Detected communities with $k = 2$, and $\gamma = 0.5$

**Table 5.** $\lambda$ values of overlapping nodes in Figure 9.

| Node | $\lambda_{iA}$ | $\lambda_{iB}$ |
|---|---|---|
| $v_3$ | 0.493 | 0.753 |
| $v_9$ | 0.987 | 0.071 |
| $v_{10}$ | 0.984 | 0.085 |

### 4.2.5. Compare with Different Algorithms

In the above simulations, $GA_{IMR}^{NWA}$ detects two communities, and we compare the result with previous algorithms in this dataset. Shen et al. captured three overlapping communities [30], and the overlapping nodes are $v_1$, $v_3$ and $v_9$. However, $v_{12}$ is missed in the method of Shen et al. The property of the overlapping communities in $v_{12}$ is not discovered. The node $v_{12}$ has exactly one neighbor that is node $v_1$, so $v_{12}$ should have the same overlapping properties as that of $v_1$.

Chen et al. captured two overlapping communities [29], and their results are similar to ours as shown in Figure 9. Chen et al. found one overlapping node $v_{10}$. Node $v_{10}$ has two edges that one connects to the left community while the other one comments to the right community. Therefore, considering $v_{10}$ as the overlapping node is reasonable. However, the node $v_3$ has five edges where three edges connect to the left community while two connect to the right community. $v_3$ is more appropriate than $v_{10}$ to be the overlapping node.

From the above observation, the communities are split more precisely by $GA_{IMR}^{NWA}$ than the previous works. For the considerations of the split appropriateness, e.g., the number of detected communities, and the split correctness, e.g., the overlapping nodes, $GA_{IMR}^{NWA}$ provides more precise results than other approaches.

*4.3. Books about American Politics*

This network is built from the transaction data from amazon.com [41]. The network has 105 nodes and 441 edges while nodes indicate books and edges are frequent co-purchase events. The nodes are labeled by three categories including *liberal*, *neutral*, or *conservative*. Each category has 43, 13, and 49 nodes respectively. In this simulation, we consider $inc_f = 1.5$, $\alpha_T = 0.01$, $m_T = 0.5$, $k = 8$, and $\gamma = 1$. We evaluate the performance of the merge strategy. Figure 10a,b are the solutions with and without merge strategies respectively. The text on each node is the node id and the real label. The results of $Q_{ov}'$ are 0.528 and 0.533 for the results with and without merge strategy.

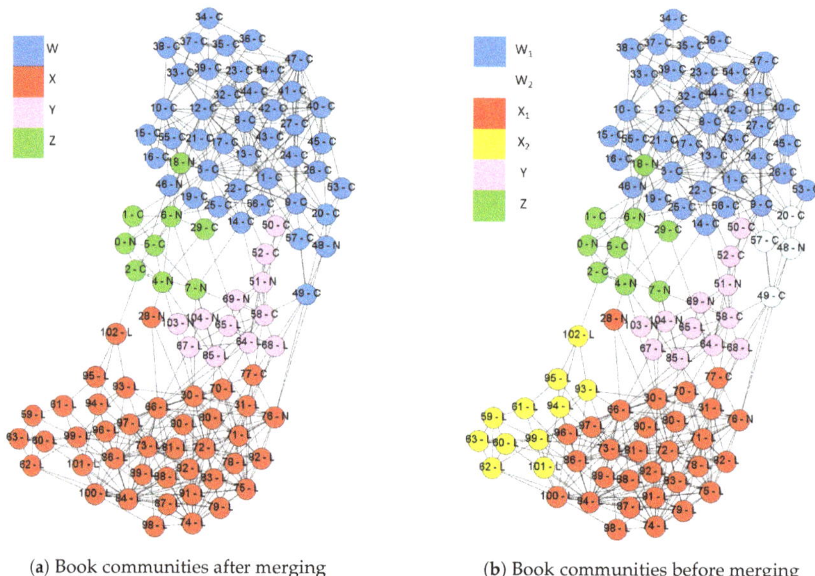

(a) Book communities after merging      (b) Book communities before merging

**Figure 10.** The book comparison between $GA_{IMR}^{NWA}$ with merging and without merging.

### 4.3.1. The Result with Merge Strategy

$GA_{IMR}^{NWA}$ with the merge strategy detects four communities denoted by $W, X, Y,$ and $Z$. Most nodes belong to two large communities $W$ and $X$, which are mainly consisted of *conservative* and *liberal* books respectively. Most *neutral* books belong to two small communities. This result is similar to that obtained by Newman [39]. Table 6 is the values of $\lambda$ for ten overlapping nodes. There are four *neutral* nodes, that is 40% of all overlapping nodes and 30% of all *neutral* nodes. The result implies that *neutral* books are often co-purchased with different books.

### 4.3.2. The Result without Merge Strategy

$GA_{IMR}^{NWA}$ without the merge strategy splits $W$ and $X$ into two parts respectively denoted by $W_1$, $W_2$, $X_1$ and $X_2$. A small community including $v_{48}$, $v_{49}$ and $v_{57}$ has been detected by the modularity maximization [25]. Therefore, we also found this community and labeled it by $W_2$.

Moreover, we also detect an extra community $X_2$. After analyzing the edge density of $X_1$ and $X_2$, they are both denser than the merged community $X$. Besides, the overlapped part is even denser as shown in Table 7. The density function definition is as follows:

$$D(c) = \frac{1}{\binom{|V(c)|}{2}} \sum_{i,j \in V(c)} \frac{A_{ij}}{2}. \qquad (8)$$

**Table 6.** $\lambda$ values of overlapping nodes in Figure 10a.

| Node | $\lambda_{iW}$ | $\lambda_{iX}$ | $\lambda_{iY}$ | $\lambda_{iZ}$ | Label |
|---|---|---|---|---|---|
| $v_3$ | 0.986 | | | 0.076 | conservative |
| $v_7$ | | | 0.254 | 0.913 | neutral |
| $v_9$ | 0.975 | | 0.11 | | conservative |
| $v_{18}$ | 0.528 | | | 0.724 | neutral |
| $v_{22}$ | 0.955 | | | 0.164 | conservative |
| $v_{25}$ | 0.922 | | | 0.236 | conservative |
| $v_{28}$ | | 0.72 | | 0.533 | neutral |
| $v_{46}$ | 0.921 | 0.238 | | | neutral |
| $v_{50}$ | 0.458 | | 0.781 | | conservative |
| $v_{85}$ | | | 0.981 | 0.093 | liberal |

**Table 7.** Density value of each part of community $X$.

| | $X$ | $X_1$ | $X_2$ | $X_1 \cap X_2$ |
|---|---|---|---|---|
| $D(c)$ | 0.20 | 0.27 | 0.34 | 0.63 |

The overlapping ratios of $(W_1, W_2)$ and $(X_1, X_2)$ are 57% and 53%, respectively. High overlapping ratios indicate that we could merge each pair of them without decreasing $Q'_{ov}$ too much. Therefore, modularity can not detect $X_2$ because of high overlapping ratio and dense overlapped part. This result shows the dense overlaps can be discovered by $GA_{IMR}^{NWA}$ correctly.

### 4.4. American College Football

This is the network of American football games between Division IA colleges in 2000 [42]. It has 115 nodes, 613 edges and 12 conferences as shown in Table 8. Nodes are teams and edges are games between the corresponding two teams while nodes are labeled by the conferences they belong to. We apply $inc_f = 1.5$, $\alpha_T = 0.01$, $m_T = -1$, $k = 15$, and $\gamma = 1$ in this simulation.

**Table 8.** Labels of conferences.

| Label | Conference | #Teams | Label | Conference | #Teams |
|---|---|---|---|---|---|
| 0 | Atlantic Coast | 9 | 6 | Mid-American | 13 |
| 1 | Big East | 8 | 7 | Mountain West | 8 |
| 2 | Big Ten | 11 | 8 | Pacific Ten | 10 |
| 3 | Big Twelve | 12 | 9 | Southeastern | 12 |
| 4 | Conference USA | 10 | 10 | Sun Belt | 7 |
| 5 | Independents | 5 | 11 | Western Athletic | 10 |

Figure 11 shows the result with $Q'_{ov} = 0.607$, true labels are on the nodes. Ten communities and 17 overlapping nodes are detected. Most conferences are well matched to the detected communities except for the conferences *Independents* (Label 5) and *Sun Belt* (Label 10). There are total seven overlapping nodes in these two conferences. From Table 9, 41% overlapping nodes and 58% nodes are in the two conferences.

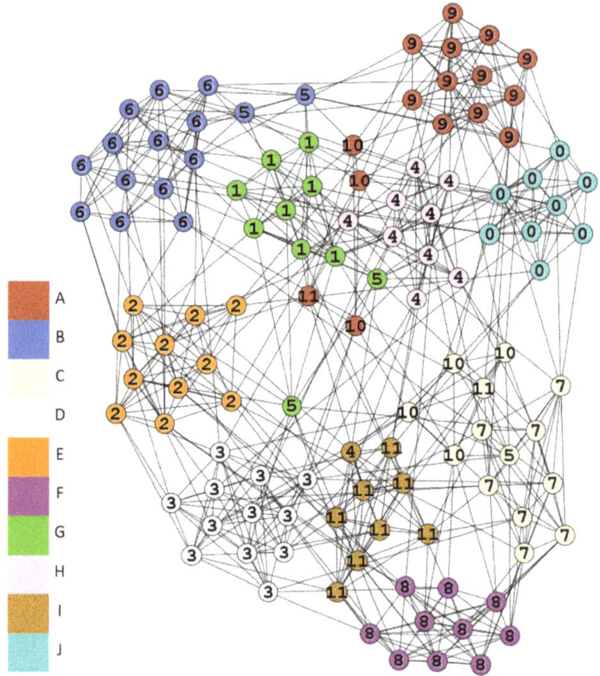

**Figure 11.** Football communities.

**Table 9.** $\lambda$ values of overlapping nodes in Figure 11.

| Node | $\lambda_{iA}$ | $\lambda_{iB}$ | $\lambda_{iC}$ | $\lambda_{iD}$ | $\lambda_{iE}$ | $\lambda_{iF}$ | $\lambda_{iG}$ | $\lambda_{iH}$ | $\lambda_{iI}$ | $\lambda_{iJ}$ | Label |
|---|---|---|---|---|---|---|---|---|---|---|---|
| $v_2$ | | | | 0.06 | 0.99 | | | | | | 2 |
| $v_8$ | | | 0.058 | | | 0.991 | | | | | 8 |
| $v_9$ | | | 0.971 | | | 0.123 | | | | | 7 |
| $v_{11}$ | | | 0.937 | 0.204 | | | | | | | 10 |
| $v_{23}$ | | | 0.981 | | | 0.094 | | | | | 7 |
| $v_{36}$ | 0.575 | 0.682 | | | | | | | | | 5 |
| $v_{44}$ | | | | | | | 0.11 | 0.975 | | | 4 |
| $v_{50}$ | | | 0.902 | | | 0.274 | | | | | 10 |
| $v_{58}$ | 0.961 | | | | | | | | 0.149 | | 11 |
| $v_{66}$ | 0.067 | | | | | | | 0.989 | | | 4 |
| $v_{67}$ | | | 0.05 | | | | | | 0.993 | | 11 |
| $v_{69}$ | | | 0.992 | | | | | | 0.054 | | 10 |
| $v_{78}$ | | | 0.05 | | | 0.992 | | | | | 8 |
| $v_{80}$ | | | | | | | 0.941 | 0.121 | | 0.126 | 5 |
| $v_{82}$ | | | | 0.065 | 0.082 | 0.097 | 0.953 | | | | 5 |
| $v_{97}$ | 0.704 | | | 0.326 | | | | 0.368 | | | 10 |
| $v_{112}$ | 0.065 | | | | | | | 0.989 | | | 4 |

The conference *Independents* has five teams, and only one game was held. This is the major reason that makes this conference undetectable. However, the teams often play with other teams in varied conferences, and this phenomenon results in the overlapping property. For example, $v_{82}$ is assigned to four communities, although it connected to community *G* with four edges. $v_{82}$ still connects to other three communities with a significant number of edges, so that is why it belongs to many communities simultaneously as shown in Figure 12. On the other hand, *Sun Belt* is in the similar situation. In this example, the heavily overlapping nodes could be detected by our method.

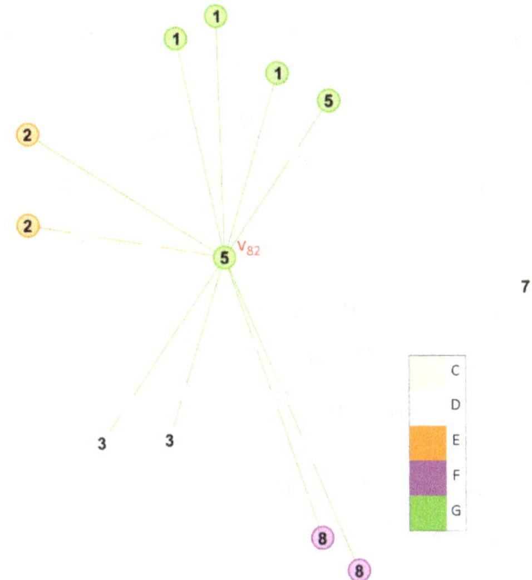

**Figure 12.** The node $v_{82}$ and its neighbors in football network.

## 4.5. Dolphin Network

The Dolphin Network is a common benchmark for evaluating the overlapping communities. Some results consider the Dolphin Network to evaluate the community quality [26,43]. We compare the proposed $GA_{IMR}^{NWA}$ with related results in this simulation. The Dolphin Network includes 62 nodes and 159 edges, and two communities are detected eventually for a long-term observation.

The distribution of $\lambda$ for overlapping nodes is listed in Table 10 while the separation with $Q' = 0.535$ is illustrated in Figure 13. According to the refinement strategy Ignore slight overlapping nodes, we get three overlapping nodes $v_{20}$, $v_{28}$, and $v_{44}$ after decreasing the setting of $\lambda_T$ from 1.0 to 0.9. The overlapping nodes are marked by the red circle with dot lines, and they are marked by the overlapping nodes based on the distribution of $\lambda$. On the other hand, we also consider $m_T = -1$ in Dolphin network as the same setting in the above simulations. The community B, C, D, and E are merged according to the refinement strategy Merge clusters. Eventually, we get two communities.

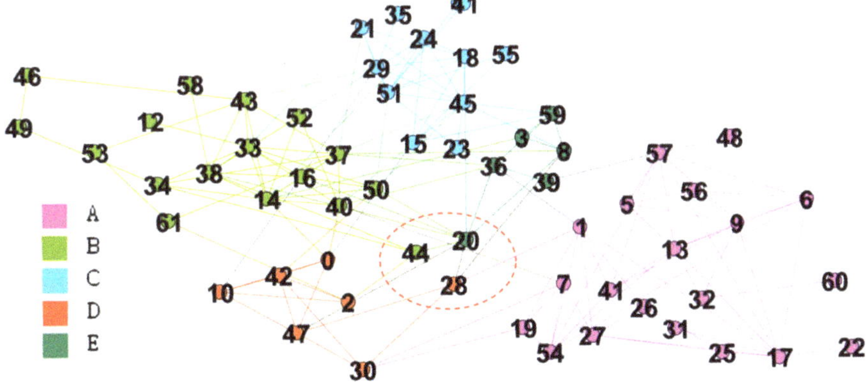

**Figure 13.** Five communities are detected by the proposed approach. There are three overlapping nodes when using $\lambda_T = 0.9$. Therefore, the community B, C, D, and E could be merged by refinement strategy Ignore slight overlapping nodes, and we find two communities eventually.

Nicosia et al. found four communities in Dolphin network [26]. The overlapping nodes are mentioned, but the authors did not list the overlapping nodes. Wang and Fleury provided detail analysis and found two communities from Dolphin network with $Q = 0.385$ [43]. The separation is acceptable, but the network structure is not so strong comparing to Figure 13. After considering the refinement strategies, the separation derived by the proposed $GA_{IMR}^{NWA}$ is similar to that provided by Wang and Fleury in [43], but the structure of our network is stronger than the network in [43]. In summary, the refinement strategies are useful in revising the network separation to be closer to the real-world behavior, and the strength of the network structure is also improved.

**Table 10.** $\lambda$ values of overlapping nodes in Figure 13.

| Node | $\lambda_{iA}$ | $\lambda_{iB}$ | $\lambda_{iC}$ | $\lambda_{iD}$ | $\lambda_{iE}$ |
|---|---|---|---|---|---|
| $v_0$ | | 0.008 | | 0.999 | |
| $v_1$ | 0.998 | | | | 0.023 |
| $v_2$ | | 0.076 | | 0.986 | |
| $v_7$ | 0.990 | | | 0.061 | |
| $v_8$ | | | | 0.022 | 0.998 |
| $v_{15}$ | | | 0.999 | 0.000 | 0.013 |
| $v_{19}$ | 0.986 | | | 0.074 | |
| $v_{20}$ | | 0.361 | | 0.364 | 0.682 |
| $v_{23}$ | | | 0.909 | | 0.261 |
| $v_{28}$ | | | | 0.823 | 0.400 |
| $v_{30}$ | 0.051 | | | 0.992 | |
| $v_{36}$ | | 0.013 | | | 0.999 |
| $v_{37}$ | | 0.990 | | | 0.062 |
| $v_{39}$ | 0.255 | | | | 0.912 |
| $v_{40}$ | | 0.998 | | | 0.021 |
| $v_{44}$ | | 0.844 | | 0.362 | 0.038 |
| $v_{45}$ | | | 0.992 | | 0.053 |
| $v_{47}$ | | | | 0.999 | 0.011 |
| $v_{50}$ | | 0.928 | 0.175 | 0.103 | |
| $v_{52}$ | | 0.999 | 0.009 | | |
| $v_{59}$ | | | 0.138 | | 0.965 |
| $v_{61}$ | | 0.925 | | 0.229 | |

## 5. Conclusion and Discussion

Given a network, the modularity is used for measuring the partition quality while the fuzzy clustering recognizes the overlapping communities. Combining above concepts together to be the fuzzy modularity is an appropriate method to formulate the structure of the given network with overlapping communities. Maximizing the modularity outputs the partition with well network structure, but computing the partition with maximum modularity requires huge computation cost. Therefore, the heuristic algorithms are outstanding in seeking high quality solution from a large search space, and we can find some research results of using heuristic algorithms for finding the partitions with maximum modularity. However, there are some special cases that we have to deal with. We find out three common situations from the partitions derived from the GA with modularity maximization and propose three solution refinement strategies to ignore overlapping nodes, merge clusters, and reweight nodes to separate the network to be closer the real-world behaviors. Moreover, we modify the fitness function of the GA to consider the null model for measuring the distance between the derived partition and the random graph. Thus, the simulation results show that the proposed $GA_{IMR}^{NWA}$ provide significant improvement comparing with previous approaches. The derived partition may not always have maximum modularity, but the community structure is more reasonable than the partitions derived by previous works. $GA_{IMR}^{NWA}$ measures the connectivity of nodes and reweight the overlapping nodes to reflect the correct properties in the given networks. Eventually, $GA_{IMR}^{NWA}$

determines the partitions appropriately, but the heavily overlapping nodes may be marked as the interior nodes by other approaches.

The overlapping nodes could be detected and provided appropriate allocation by $GA_{IMR}^{NWA}$. During the simulations, we found some extension works that will be address in the future, and they are listed as follows:

1. In our simulations, we got an interesting result as shown in Figure 14 from the karate network with $inc_f = 2$. The result consists of three communities, and they are grouped by $v_{33}$, $v_3$ and $v_1$. The community with $v_3$ that the nodes are marked by red could be consider as an overlapping set. It means that the networks not only have overlapping nodes but also overlapping groups. Thus, applying the fuzzy concept to the communities will eliminate the group with $v_3$, and they may be more closed to the real-world behavior. Since the members in the group with $v_3$ may belong to different communities based on the situations, e.g., the competitions or the events. Therefore, assigning the red nodes to any community may be inappropriate.

2. The proposed algorithm invokes GA to compute the preliminary partitions and then adopts proposed refinement strategies to correct the partitions by the secondary processes. The refinement strategies could be considered as the local search to improve the partition quality in each iteration. However, it is a tradeoff between the computation cost and the partition quality. Once the refinement strategies are modified from the external processes to the internal processes in GA, the computation cost will be increased. Moreover, the given networks may not always consist of the target properties that could be improved by the refinement strategies. Therefore, the refinement strategies could be designed as local search approaches, but the trigger of launching the local search approaches should be analyzed in the future.

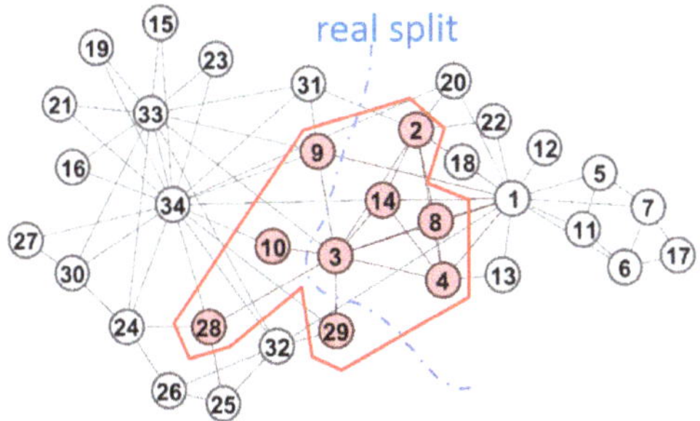


**Figure 14.** The 5th detected community of the karate network.

**Author Contributions:** Conceptualization, C.-K.T.; Data curation, C.-Y.C. and T.-W.C.; Project administration, S.-L.L.; Software, C.-Y.C. and T.-W.C.; Supervision, S.-L.L.; Writing—original draft, C.-K.T. and H.-J.H.; Writing–review & editing, C.-K.T. and H.-J.H. All authors have read and agreed to the published version of the manuscript.

**Funding:** This research was funded by Ministry of Science and Technology of the Republic of China grant number 108-2221-E-194-024- and MOST 108-2221-E-167-022-.

**Acknowledgments:** The authors would like to thank all reviewers for giving the constructive comments towards improving this paper.

**Conflicts of Interest:** The authors declare no conflict of interest.

## Abbreviations

The following abbreviations are used in this manuscript:

GA    Genetic Algorithms

## References


1. Rosso, M.A.; McClelland, M.K.; Jansen, B.J.; Fleming, S.W. Using Google AdWords in the MBA MIS course. *J. Inf. Syst. Educ.* **2019**, *20*, 6.
2. Mu, C.H.; Xie, J.; Liu, Y.; Chen, F.; Liu, Y.; Jiao, L.C. Memetic algorithm with simulated annealing strategy and tightness greedy optimization for community detection in networks. *Appl. Soft Comput.* **2015**, *34*, 485–501. [CrossRef]
3. Shang, R.; Bai, J.; Jiao, L.; Jin, C. Community detection based on modularity and an improved genetic algorithm. *Physica A* **2013**, *392*, 1215–1231. [CrossRef]
4. Bello-Orgaz, G.; Salcedo-Sanz, S.; Camacho, D. A multi-objective genetic algorithm for overlapping community detection based on edge encoding. *Inf. Sci.* **2018**, *462*, 290–314. [CrossRef]
5. Li, Z.; Liu, J. A multi-agent genetic algorithm for community detection in complex networks. *Physica A* **2016**, *449*, 336–347. [CrossRef]
6. Yuxin, Z.; Shenghong, L.; Feng, J. Overlapping community detection in complex networks using multi-objective evolutionary algorithm. *Comput. Appl. Math.* **2017**, *36*, 749–768. [CrossRef]
7. Shakya, H.K.; Singh, K.; Biswas, B. An efficient genetic algorithm for fuzzy community detection in social network. In Proceedings of the International Conference on Advanced Informatics for Computing Research, Punjab, India, 17–18 March 2017.
8. Behera, R.K.; Naik, D.; Rath, S.K.; Dharavath, R. Genetic algorithm-based community detection in large-scale social networks. *Neural Comput. Appl.* **2020**, *32*, 9649–9665. [CrossRef]
9. Binesh, N.; Rezghi, M. Fuzzy clustering in community detection based on nonnegative matrix factorization with two novel evaluation criteria. *Appl. Soft Comput.* **2018**, *69*, 689–703. [CrossRef]
10. Naderipour, M.; Zarandi, M.H.F.; Bastani, S. A type-2 fuzzy community detection model in large-scale social networks considering two-layer graphs. *Eng. Appl. Artif. Intell.* **2020**, *90*, 103206. [CrossRef]
11. Yang, C.T.; Chen, S.T.; Den, W.; Wang, Y.T.; Kristiani, E. Implementation of an intelligent indoor environmental monitoring and management system in cloud. *Future Generat. Comput. Syst.* **2019**, *96*, 731–749. [CrossRef]
12. Newman, M.E.J.; Girvan, M. Finding and evaluating community structure in networks. *Phys. Rev. E* **2004**, *69*, 026113. [CrossRef] [PubMed]
13. Brandes, U.; Delling, D.; Gaertler, M.; Goerke, R.; Hoefer, M.; Nikoloski, Z.; Wagner, D. Maximizing Modularity is hard. *arXiv* **2006**, arXiv:physics/0608255.
14. Lai, D.; Lu, H.; Nardini, C. Enhanced modularity-based community detection by random walk network preprocessing. *Phys. Rev. E* **2010**, *81*, 066118. [CrossRef] [PubMed]
15. Huang, J.; Sun, H.; Han, J.; Deng, H.; Sun, Y.; Liu, Y. SHRINK: A Structural Clustering Algorithm for Detecting Hierarchical Communities in Networks. In Proceedings of the 19th Conference on Information and Knowledge Management, Toronto, ON, Canada, 26–30 October 2010.
16. Dinh, T.N.; Thai, M.T. Community detection in scale-free networks: approximation algorithms for maximizing modularity. *IEEE J. Select. Areas Commun.* **2013**, *31*, 997–1006. [CrossRef]
17. Fortunato, S.; Barthélemy, M. Resolution limit in community detection. *Proc. Natl. Acad. Sci. USA* **2007**, *104*, 36–41. [CrossRef]
18. Arenas, A.; Fernández, A.; Gómez, S. Analysis of the structure of complex networks at different resolution levels. *New J. Phys.* **2008**, *10*, 053039. [CrossRef]
19. Newman, M.E.J. Fast algorithm for detecting community structure in networks. *Phys. Rev. E* **2004**, *69*, 066133. [CrossRef]
20. Clauset, A.; Newman, M.E.J. Moore, C. Finding community structure in very large networks. *Phys. Rev. E* **2004**, *70*, 066111. [CrossRef]
21. Newman, M.E.J. Finding community structure in networks using the eigenvectors of matrices. *Phys. Rev. E* **2006**, *74*, 036104. [CrossRef]
22. White, S.; Smyth, P. A spectral clustering approach to finding communities in graph. In Proceedings of the SIAM International Conference on Data Mining, Beach, CA, USA, 21–23 April 2005.
23. Richardson, T.; Mucha, P.J.; Porter, M.A. Spectral Tripartitioning of Networks. *Phys. Rev. E* **2009**, *80*, 0036111. [CrossRef]

24. Guimera, R.; Amaral, L.A.N. Functional cartography of complex metabolic networks. *Nature* **2005**, *433*, 895–900. [CrossRef]
25. Agarwal, G.; Kempe, D. Modularity-maximizing graph communities via mathematical programming. *EPJB* **2008**, *66*, 409–418. [CrossRef]
26. Nicosia, V.; Mangioni, G.; Carchiolo, V.; Malgeri, M. Extending the definition of modularity to directed graphs with overlapping communities. *J. Stat. Mech* **2009**, *2009*, 03024. [CrossRef]
27. Reichardt, J.; Bornholdt, S. Statistical mechanics of community detection. *Phys. Rev. E* **2006**, *74*, 016110. [CrossRef] [PubMed]
28. Liu, J. Fuzzy modularity and fuzzy community structure in networks. *Eur. Phys. J. B* **2010**, *77*, 547–557. [CrossRef]
29. Chen, D.; Shang, M.; Lv, Z.; Fu, Y. Detecting overlapping communities of weighted networks via a local algorithm. *Physica A* **2010**, *389*, 4177–4187. [CrossRef]
30. Shen, H.-W.; Cheng, X.-Q.; Guo, J.-F. Quantifying and identifying the overlapping community structure in networks. *J. Stat. Mech.* **2009**, *2009*, 07042. [CrossRef]
31. Choong, J.J.; Liu, X.; Murata, T. Optimizing Variational Graph Autoencoder for Community Detection with Dual Optimization. *Entropy* **2020**, *22*, 197. [CrossRef]
32. Ezeh, C.; Tao, R.; Zhe, L.; Yiqun, W.; Ying, Q. Multi-Type Node Detection in Network Communities. *Entropy* **2019**, *21*, 1237. [CrossRef]
33. Nepusz, T.; Petróczi, A.; Nógyessy, L.; Bazsó, F. Fuzzy communities and the concept of bridgeness in complex networks. *Phys. Rev. E* **2008**, *77*, 016107. [CrossRef]
34. Griechisch, E.; Pluhár, A. Community detection by using the extended modularity. *Acta Cybern.* **2011**, *20*, 69–85. [CrossRef]
35. Yang, C.T.; Shih, W.C.; Chen, L.T.; Kuo, C.T.; Jiang, F.C.; Leu, F.Y. Accessing medical image file with co-allocation HDFS in cloud. *Future Generat. Comput. Syst.* **2015**, *43*, 61–73. [CrossRef]
36. Yang, C.T.; Liu, J.C.; Chen, S.T.; Lu, H.W. Implementation of a big data accessing and processing platform for medical records in cloud. *J. Med. Syst.* **2017**, *41*, 149. [CrossRef] [PubMed]
37. Natarajan, S.; Vairavasundaram, S.; Ravi, L. Optimized fuzzy-based group recommendation with parallel computation. *J. Intell. Fuzzy Syst.* **2019**, *36*, 4189–4199. [CrossRef]
38. Yang, J.; Leskovec, J. Overlapping Community Detection at Scale: A Nonnegative Matrix Factorization Approach. In Proceedings of the Sixth ACM International Conference on Web Search and Data Mining, Rome, Italy, 4–8 February 2013.
39. Newman, M.E.J. Modularity and community structure in networks. *Proc. Natl. Acad. Sci. USA* **2006**, *103*, 8577–8582. [CrossRef]
40. Zachary, W.W. An information flow model for conflict and fission in small groups. *J. Anthropolog. Res.* **1977**, *33*, 452–473. [CrossRef]
41. Krebs, V. (Unpublished). Available online: http://www.orgnet.com/ (accessed on 1 February 2015).
42. Girvan, M.; Newman, M.E.J. Community structure in social and biological networks. *Proc. Natl. Acad. Sci. USA* **2002**, *99*, 7821–7826. [CrossRef]
43. Wang, Q.; Fleury, E. Uncovering overlapping community structure. In *Complex Networks*; Springer: Berlin, Germany, 2011; pp. 176–186.

*Article*

# Modelling and Recognition of Protein Contact Networks by Multiple Kernel Learning and Dissimilarity Representations

Alessio Martino [1,*], Enrico De Santis [1], Alessandro Giuliani [2] and Antonello Rizzi [1]

[1] Department of Information Engineering, Electronics and Telecommunications, University of Rome "La Sapienza", Via Eudossiana 18, 00184 Rome, Italy; enrico.desantis@uniroma1.it (E.D.S.); antonello.rizzi@uniroma1.it (A.R.)

[2] Department of Environment and Health, Istituto Superiore di Sanità, Viale Regina Elena 299, 00161 Rome, Italy; alessandro.giuliani@iss.it

[*] Correspondence: alessio.martino@uniroma1.it; Tel.: +39-06-44585745

Received: 27 June 2020; Accepted: 17 July 2020; Published: 21 July 2020

**Abstract:** Multiple kernel learning is a paradigm which employs a properly constructed chain of kernel functions able to simultaneously analyse different data or different representations of the same data. In this paper, we propose an hybrid classification system based on a linear combination of multiple kernels defined over multiple dissimilarity spaces. The core of the training procedure is the joint optimisation of kernel weights and representatives selection in the dissimilarity spaces. This equips the system with a two-fold knowledge discovery phase: by analysing the weights, it is possible to check which representations are more suitable for solving the classification problem, whereas the pivotal patterns selected as representatives can give further insights on the modelled system, possibly with the help of field-experts. The proposed classification system is tested on real proteomic data in order to predict proteins' functional role starting from their folded structure: specifically, a set of eight representations are drawn from the graph-based protein folded description. The proposed multiple kernel-based system has also been benchmarked against a clustering-based classification system also able to exploit multiple dissimilarities simultaneously. Computational results show remarkable classification capabilities and the knowledge discovery analysis is in line with current biological knowledge, suggesting the reliability of the proposed system.

**Keywords:** dissimilarity spaces; support vector machines; kernel methods; computational biology; systems biology; protein contact networks

---

## 1. Introduction

Dealing with structured data is an evergreen challenge in pattern recognition and machine learning. Indeed, many real-world systems can effectively be described by structured domains such as networks (e.g., images [1,2]) or sequences (e.g., signatures [3]). Biology is a seminal field in which many complex systems can be described by networks [4], as the biologically relevant information resides in the interaction among constituting elements: common examples include protein contact networks [5,6], metabolic networks [7] and protein–protein interaction networks [8,9].

Pattern recognition in structured domains poses additional challenges as many structured domains are non-metric in nature (namely, the pairwise dissimilarities in such domains might not satisfy the four properties of a metric: non-negativity, symmetry, identity, triangle inequality) and patterns may lack any geometrical interpretation [10].

In order to deal with such domains, five mainstream approaches can be pursued [10]:

1.  Feature generation and/or feature engineering, where numerical features are extracted ad-hoc from structured patterns (e.g., using their properties or via measurements) and can be further merged according to different strategies (e.g., in a multi-modal way [11]);
2.  Ad-hoc dissimilarities in the input space, where custom dissimilarity measures are designed in order to process structured patterns directly in the input domain without moving towards Euclidean (or metric) spaces. Common—possibly parametric—edit distances include the Levenshtein distance [12] for sequence domains and graph edit distances [13] for graphs domains;
3.  Embedding via information granulation and granular computing [3,14–25];
4.  Dissimilarity representations [26–28], where structured patterns are embedded in the Euclidean space according to their pairwise dissimilarities;
5.  Kernel methods, where the mapping between the original input space and the Euclidean space exploits positive-definite kernel functions [29–33].

This paper proposes a novel classification system based on an hybridisation of the latter two strategies: while dissimilarity representations see the (structured) patterns according to the pairwise dissimilarities, kernel methods encode pairwise similarities. Nonetheless, the class of properly-defined kernel functions is restricted: the (conditionally) positive definitiveness may not hold in case of non-metric (dis)similarities. The use of kernel methods in state-of-the-art (non-linear) classifiers such as Support Vector Machines (SVM) [34,35] is strictly related to their (conditionally) positive definitiveness due to the quadratic programming optimisation involved: indeed, non-(conditionally) positive definite kernels do not guarantee convergence to the global optimum. Although there is some research about learning from indefinite kernels (see, e.g., [36–40]), their evaluation on the top of Euclidean spaces (e.g., dissimilarity spaces) retain the (conditionally) positive definitiveness, devoting matrix regularisation or other tricks to foster positive definitiveness.

The proposed classification system is able to simultaneously explore multiple dissimilarities following a multiple kernel learning approach, where each kernel considers a different (dissimilarity) representation. The relative importance of the several kernels involved is automatically determined via genetic optimisation in order to maximise the classifier performance. Further, the very same genetic optimisation is in charge of determining a suitable subset of representative (prototypes) patterns in the dissimilarity space [27] in order to shrink the modelling complexity. Hence, the proposed system allows a two-fold a posteriori knowledge discovery phase:

1.  By analysing the kernel weights, one can determine the most suitable representation(s) for the problem at hand;
2.  The patterns elected as representatives for the dissimilarity space (hence determined as pivotal for tracking the decision boundary amongst the problem-related classes) can give some further insights for the problem at hand.

In order to validate the proposed classification system, a bioinformatics-related application is considered, namely protein function prediction. Proteins' 3D structure (both tertiary and quaternary) can effectively be modelled by a network, namely the so-called Protein Contact Network (PCN) [5]. A PCN is a minimalistic (unweighted and undirected) graph-based protein representation where nodes correspond to amino-acids and edges between two nodes exist whether the Euclidean distance between residues' $\alpha$-carbon atom coordinates is within $[4,8]$Å. The lower bound is defined in order to discard trivial connections due to closeness along the backbone (first-order neighbour contacts), whereas the upper bound is defined by considering the peptide bonds geometry (indeed, 8Å roughly correspond to two van der Waals radii between residues' $\alpha$-carbon atoms [41]). It is worth stressing that both nodes labels (i.e., the type of amino-acid) and edges labels (i.e., the distance between neighbour residues) are deliberately discarded in order to focus only on proteins' topological configuration. Despite the minimalistic representation, PCNs have been successfully used in pattern recognition problems for tasks such as solubility prediction/folding propensity [42,43] and physiological role prediction [44–46];

furthermore, their structural and dynamical properties have been extensively studied in works such as [47–50].

In order to investigate how the protein function is related to its topological structure, a subset of the entire Escherichia coli bacterium proteome, correspondent to E. coli proteins whose 3D structure is known, is considered. The problem itself is cast into a supervised pattern recognition task, where each pattern (protein) is described according to eight different representations drawn by its PCN and its respective Enzyme Commission (EC) number [51] that serves as the ground-truth class label. The EC nomenclature scheme classifies enzymes according to the chemical reaction they catalyse and a generic entry is composed by four numbers separated by periods. The first digit (1–6) indicates one of the six major enzymatic groups (EC 1: oxidoreductases; EC 2: transferases; EC 3: hydrolases; EC 4: lyases; EC 5: isomerases; EC 6: ligases) and the latter three numbers represent a progressively finer functional enzyme classification. In this work, only the first number is considered. However, proteins with no enzymatic characteristics (or proteins for which enzymatic characteristics are still unknown nowadays) are not provided with an EC number, thus an additional class of not-enzymes will be considered, identified by the categorical label 7. It is worth noting that the EC classification only loosely relates to global protein 3D configuration, given that structure is affected by many determinants other than catalysed reactions like solubility, localisation in the cell, interaction with other proteins and so forth. This makes the classification task intrinsically very difficult.

This paper is organised as follows: Section 2 overviews some theory related to kernel methods and dissimilarity spaces; Section 3 presents the proposed methodology; Section 4 shows the results obtained with the proposed approach, along with a comparison against a clustering-based classifier (also able to explore multiple dissimilarities), and we also provide some remarks on the two-fold knowledge discovery phase. Finally, Section 5 concludes the paper. The paper also features two appendices: Appendix A describes in detail the several representations used for describing PCNs, whereas Appendix B lists the proteins selected as prototypes for the dissimilarity representations.

## 2. Theoretical Background

Let $\mathcal{D} = \{x_1, \dots, x_{N_P}\}$ be the dataset at hand lying in a given input space $\mathcal{X}$. Moving the problem towards a dissimilarity space [26] consists in expressing each pattern from $\mathcal{D}$ according to the pairwise distances with respect to all other patterns, including itself. In other words, the dataset is cast into the pairwise distance matrix $\mathbf{D} \in \mathbb{R}^{N_P \times N_P}$ defined as:

$$\mathbf{D}_{i,j} = d(x_i, x_j) \qquad \forall i, j = 1, \dots, N_P , \tag{1}$$

where $d(\cdot, \cdot)$ is a suitable dissimilarity measure in $\mathcal{D}$, that is $d : \mathcal{D} \times \mathcal{D} \to \mathbb{R}$. Without loss of generality, hereinafter let us consider $\mathbf{D}$ to be symmetric: if $d(\cdot, \cdot)$ is at least symmetric, $\mathbf{D}$ is trivially symmetric; in case of asymmetric dissimilarity measures, $\mathbf{D}$ can be 'forced' to be symmetric, e.g., $\mathbf{D} := \frac{1}{2}(\mathbf{D} + \mathbf{D}^T)$. The major advantage in moving the problem from a generic input space $\mathcal{X}$ towards $\mathbb{R}^{N_P \times N_P}$ is that the latter can be equipped with algebraic structures such as the inner product or the Minkowski distance, whereas the former might not be metric altogether. As such, in the latter, standard computational intelligence and machine learning techniques can be used without alterations [10]. On the negative side, the explicit evaluation of $\mathbf{D}$ can be computationally expensive as it leads to a time and space complexity of $\mathcal{O}(N_P^2)$. To this end, in [27], a 'reduced' dissimilarity space representation is proposed, where a subset of prototype patterns $\mathcal{R} \subset \mathcal{D}$ is properly chosen and each pattern is described according to the pairwise distances with respect to the prototypes only. This leads to the definition of a 'reduced' pairwise distance matrix $\bar{\mathbf{D}} \in \mathbb{R}^{N_P \times |\mathcal{R}|}$ defined as:

$$\bar{\mathbf{D}}_{i,j} = d(x_i, x_j) \qquad \forall i = 1, \dots, N_P, \ \forall j = 1, \dots, |\mathcal{R}|. \tag{2}$$

Since usually $|\mathcal{R}| < |\mathcal{D}|$, there is no need to solve a quadratic complexity problem such as evaluating Equation (1). On the negative side, however, the selection of the subset $\mathcal{R}$ is a delicate and challenging task [10] since:

1. They must well-characterize the decision boundary between patterns in the input space;
2. The fewer, the better: the number of representatives has a major impact on the model complexity (cf. Equation (1) vs. Equation (2)).

Several heuristics have been proposed in the literature, ranging from clustering the input space to (possibly class-aware) random selection [10,27,52].

Kernel methods are usually employed whether the input space has an underlying Euclidean geometry. Indeed, the simplest kernel (namely, the linear kernel [30,53]) is the plain inner product between real-valued vectors. The kernel matrix $\mathbf{K}$ (also known as the Gram matrix) can easily be defined as:

$$\mathbf{K}_{i,j} = \langle x_i, x_j \rangle \qquad \forall i,j = 1,\ldots,N_P. \tag{3}$$

Let $K$ be a symmetric and positive semi-definite kernel function from the input space $\mathcal{X}$ towards $\mathbb{R}$, that is $K : \mathcal{X} \times \mathcal{X} \to \mathbb{R}$ such that

$$K(x_i, x_j) = K(x_j, x_i) \qquad \forall x_i, x_j \in \mathcal{X} \tag{4}$$

$$\sum_{i=1}^{N_P} \sum_{j=1}^{N_P} c_i c_j K(x_i, x_j) \geq 0 \qquad \forall c_i, c_j \in \mathbb{R}, \forall x_i, x_j \in \mathcal{X}. \tag{5}$$

As in the linear kernel case, starting from pairwise kernel evaluations, one can easily evaluate the kernel matrix as

$$\mathbf{K}_{i,j} = K(x_i, x_j) \qquad \forall i,j = 1,\ldots,N_P \tag{6}$$

and if $\mathbf{K}$ is a positive semi-definite kernel matrix, then $K$ is a positive semi-definite kernel function. One of the most intriguing kernel methods property relies on the so-called kernel trick [29,30]: kernel of the form Equations (4) and (5) are also known as Mercer's kernel as they satisfy the Mercer condition [32]. Such kernel functions can be seen as the inner product evaluation on a high-dimensional (or possibly infinite-dimensional) and usually unknown Hilbert space $\mathcal{H}$. The kernel trick is usually described by the following, seminal, equation:

$$K(x,y) = \langle \phi(x), \phi(y) \rangle_{\mathcal{H}}, \tag{7}$$

where $\phi : \mathcal{X} \to \mathcal{H}$ is the implicit (and usually unknown) mapping function. The need for using a non-linear and higher-dimensional mapping is a direct consequence of Cover's theorem [33]. Thanks to the kernel trick, one can use one of the many kernel functions available (e.g., polynomial, Gaussian, radial basis function) in order to perform such non-linear and higher-dimensional mapping without knowing and explicitly evaluating the mapping function $\phi(\cdot)$. Further, kernel methods can be used in many state-of-the-art classifiers such as (kernelised) SVM [35,54].

In multiple kernel learning, the kernel matrix $\mathbf{K}$ is defined as a properly-defined combination of a given number of $N_K$ kernels. The most intuitive combination is a linear combination of the form:

$$\mathbf{K} = \sum_{i=1}^{N_K} \beta_i \mathbf{K}^{(i)}, \tag{8}$$

where sub-kernels $\mathbf{K}^{(i)}$ are single Mercer's kernels. The weights $\beta_i$ can be learned according to different strategies and can be constrained in several ways—see, e.g., [55–61], or the survey [62]. The rationale behind using a multiple kernel learning with respect to a plain single kernel learning depends on the application: for example, if data come from different sources, one might want to explore such different sources according to several kernels or, dually, one might want to explore the same data using different kernels, where such different kernels may differ in shape and/or type. In this work, a mixture between the two approaches is pursued: same source (PCN), but different representations (see Appendix A). Further, a linear convex combination of radial basis function kernels is employed. The $i$th radial basis function kernel is defined as

$$\mathbf{K}^{(i)}_{j,k} = \exp\left\{-\gamma_i \cdot \|x_j - x_k\|^2\right\} \qquad \forall j, k = 1, \ldots, N_P \tag{9}$$

and $\gamma_i$ is its shape parameter. Further, the weights $\beta_i$ are constrained as

$$\sum_{i=1}^{N_K} \beta_i = 1 \tag{10}$$

$$\beta_i \in [0, 1] \qquad \text{for } i = 1, \ldots, N_K. \tag{11}$$

It is rather easy to demonstrate that these selections for both kernels and weights lead to the final kernel matrix (as in Equation (8)) which still is a valid Mercer's kernel, therefore it can be used on kernelised SVMs. Indeed, Cristianini and Shawe-Taylor in [31] showed that the summation of two valid kernels is still a valid kernel. Further, Horn and Johnson in [63] showed that a positive semi-definite matrix multiplied by a non-negative scalar is still a positive semi-definite matrix. Merging these two results automatically prove that kernels of the form (8) and (9) with constraints (10) and (11) are valid kernels.

## 3. Proposed Methodology

Let $\mathcal{D}$ be the dataset at hand, split into three non-overlapping subsets $\mathcal{D}_{\text{TR}}$, $\mathcal{D}_{\text{VAL}}$ and $\mathcal{D}_{\text{TS}}$ (namely training set, validation set and test set). Especially for structured data, several representations (e.g., set of descriptors) might hold for the same data, therefore let $\{\mathbf{X}^{(1)}, \ldots, \mathbf{X}^{(N_R)}\}$ be the set of $N_R$ representations, split in the same fashion (i.e., $\{\mathbf{X}^{(i)}_{\text{TR}}\}_{i=1}^{N_R}$, $\{\mathbf{X}^{(i)}_{\text{VAL}}\}_{i=1}^{N_R}$ and $\{\mathbf{X}^{(i)}_{\text{TS}}\}_{i=1}^{N_R}$). Finally, let $\{d^{(1)}(\cdot,\cdot), \ldots, d^{(N_R)}(\cdot,\cdot)\}$ be the set of dissimilarity measures suitable for working in their respective representations.

The respective training, validation and test pairwise dissimilarity matrices, as in Equation (1) can be evaluated as follows:

$$\begin{aligned}
\mathbf{D}^{(1)}_{\text{TR}} &= d^{(1)}(\mathbf{X}^{(1)}_{\text{TR}}, \mathbf{X}^{(1)}_{\text{TR}}) & \cdots & \quad \mathbf{D}^{(N_R)}_{\text{TR}} = d^{(N_R)}(\mathbf{X}^{(N_R)}_{\text{TR}}, \mathbf{X}^{(N_R)}_{\text{TR}}) \\
\mathbf{D}^{(1)}_{\text{VAL}} &= d^{(1)}(\mathbf{X}^{(1)}_{\text{VAL}}, \mathbf{X}^{(1)}_{\text{TR}}) & \cdots & \quad \mathbf{D}^{(N_R)}_{\text{VAL}} = d^{(N_R)}(\mathbf{X}^{(N_R)}_{\text{VAL}}, \mathbf{X}^{(N_R)}_{\text{TR}}) \\
\mathbf{D}^{(1)}_{\text{TS}} &= d^{(1)}(\mathbf{X}^{(1)}_{\text{TS}}, \mathbf{X}^{(1}_{\text{TR}}) & \cdots & \quad \mathbf{D}^{(N_R)}_{\text{TS}} = d^{(N_R)}(\mathbf{X}^{(N_R)}_{\text{TS}}, \mathbf{X}^{(N_R)}_{\text{TR}}).
\end{aligned} \tag{12}$$

Let $\mathbf{w} \in \{0, 1\}^{|\mathcal{D}_{\text{TR}}|}$ be a binary vector in charge of selecting columns from all matrices in Equation (12): the full pairwise dissimilarities can be sliced to their 'reduced' versions (cf. Equation (1) vs. Equation (2)), hence:

$$\begin{aligned}
\bar{\mathbf{D}}^{(1)}_{\text{TR}} &= \mathbf{D}^{(1)}_{\text{TR}}(:, \mathbf{w}) & \cdots & \quad \bar{\mathbf{D}}^{(N_R)}_{\text{TR}} = \mathbf{D}^{(N_R)}_{\text{TR}}(:, \mathbf{w}) \\
\bar{\mathbf{D}}^{(1)}_{\text{VAL}} &= \mathbf{D}^{(1)}_{\text{VAL}}(:, \mathbf{w}) & \cdots & \quad \bar{\mathbf{D}}^{(N_R)}_{\text{VAL}} = \mathbf{D}^{(N_R)}_{\text{VAL}}(:, \mathbf{w}) \\
\bar{\mathbf{D}}^{(1)}_{\text{TS}} &= \mathbf{D}^{(1)}_{\text{TS}}(:, \mathbf{w}) & \cdots & \quad \bar{\mathbf{D}}^{(N_R)}_{\text{TS}} = \mathbf{D}^{(N_R)}_{\text{TS}}(:, \mathbf{w}).
\end{aligned} \tag{13}$$

where, due to the number of subscripts and superscripts in Eq. (13), for ease of notation, we used a MATLAB®-like notation for indexing matrices.

In other words, $\mathbf{w}$ acts as a feature (prototype) selector. Given this newly obtained dataset, it is possible to train a kernelised $\nu$-SVM [64] whose multiple kernel has the form Equation (8) where each one has the form Equation (9), thus:

$$\mathbf{K} = \sum_{i=1}^{N_R} \beta_i \cdot \exp\left\{-\gamma_i \cdot \|\bar{\mathbf{D}}_{TR}^{(i)} \ominus \bar{\mathbf{D}}_{TR}^{(i)}\|^2\right\}, \tag{14}$$

where $\ominus$ denotes the pairwise difference. Hence, each dissimilarity representation is subject to a proper non-linear kernel ($N_K \equiv N_R$).

A genetic algorithm [65] acts as a wrapper method in order to automatically tune in a fully data-driven fashion the several free parameters introduced in this problem. The choice behind a genetic algorithm stems from them being widely famous in the context of derivative-free optimisation, embarrassingly easy to parallelise and for the sake of consistency with competing techniques (see Section 4.4). For our problem, the genetic code has the form:

$$\begin{bmatrix} \nu & \beta & \gamma & \mathbf{w} \end{bmatrix}, \tag{15}$$

where $\nu \in (0, 1]$ is the SVM regularisation term, $\beta = [\beta]_{i=1}^{N_R}$ contains the kernel weights, $\gamma = [\gamma_i]_{i=1}^{N_R}$ contains the kernel shapes and $\mathbf{w}$ properly selects prototypes in the dissimilarity space, as described above.

For the sake of argument, it is worth remarking that there have been several attempts to use evolutionary strategies in order to tune multiple kernel machines: for example in [66] a genetic algorithm has been used in order to tune the kernel shapes (namely, $\gamma$), whereas in [67] both the kernel shapes and the kernel weights have been tuned by means of a $(\mu + \lambda)$ evolution strategy [68]. Conversely, the idea of using a genetic algorithm for prototypes selection in the dissimilarity space has been inherited from a previous work [44].

The fitness function to be maximised is the informedness $J$ (also known as Youden's index [69]) defined as:

$$J = \text{specificity} + \text{sensitivity} - 1, \tag{16}$$

which is, by definition, bounded in range $[-1, 1]$ (the closer to 1, the better). For the sake of comparison with other performance measures (e.g., accuracy, F-score and the like) which are, by definition, bounded in $[0, 1]$, the fitness function sees a scaled version of the informedness [23–25], hence:

$$f_1 \equiv \bar{J} = \frac{J - (-1)}{1 - (-1)} = \frac{J + 1}{2} \in [0, 1]. \tag{17}$$

The rationale behind using the informedness rather than other most common performance measures (mainly accuracy and F-score) is that the informedness is well suited for unbalanced classes without being biased towards the most frequent class (the same is not true for accuracy) and whilst considering also true negative predictions (the same is not true for F-score) [70].

By assuming that the full dissimilarity matrices are pre-evaluated beforehand, the objective function evaluation is performed for each individual from the current generation as follows:

1. The individual receives the $N_R$ full dissimilarity matrices between training data samples, i.e., $\mathbf{D}_{TR}^{(1)}, \ldots, \mathbf{D}_{TR}^{(N_R)}$ as in Equation (12);
2. According to the $\mathbf{w}$ portion of its genetic code (see Equation (15)), a subset of prototypes is selected, leading to the 'reduced' dissimilarity matrices between training data, i.e., $\bar{\mathbf{D}}_{TR}^{(1)}, \ldots, \bar{\mathbf{D}}_{TR}^{(N_R)}$ as in Equation (13);
3. Considering the $\beta$ and $\gamma$ values in its genetic code, the (multiple) kernel matrix is evaluated by using Equation (14);
4. A $\nu$-SVM is trained using the regularisation term $\nu$ from the genetic code and the kernel matrix from step #3;

5. The individual receives the $N_R$ full dissimilarity matrices between training and validation data, each of which is computed by considering all possible $\langle x, y \rangle$-pairs where $x$ belongs to the validation set and $y$ belongs to the training set, i.e., $\mathbf{D}_{\text{VAL}}^{(1)}, \ldots, \mathbf{D}_{\text{VAL}}^{(N_R)}$ as in Equation (12);

6. The 'reduced' dissimilarity matrices are projected thanks to $\mathbf{w}$, i.e., $\bar{\mathbf{D}}_{\text{VAL}}^{(1)}, \ldots, \bar{\mathbf{D}}_{\text{VAL}}^{(N_R)}$ as in Equation (13);

7. The (multiple) kernel matrix between training and validation data is evaluated thanks to $\beta$ and $\gamma$, alike Equation (14);

8. The (multiple) kernel matrix from step #7 is fed to the SVM trained on step #4 and the predicted classes on the validation set are returned;

9. The fitness function is evaluated.

At the end of the evolution, the best individual (i.e., the one with best performances on the validation set) is retained and its final performances are evaluated on the test set.

Finally, it is worth remarking the rationale behind the proposed, structured, genetic code since a genetic code of the form Equation (15) allows, in a two-fold manner, a deeper a posteriori knowledge discovery phase. Indeed, using upfront good classification results (for the sake of reliability), by looking at $\beta$, it is possible to check which kernels (representations) are considered as the most important (higher weights) for the learning machine in order to solve the problem at hand. Similarly, by looking at $\mathbf{w}$, it is possible to check which training set patterns have been selected as representatives and ask why those patterns have been selected instead of others, leading to a pattern-wise check (possibly with help by field-experts). Especially the latter a posteriori check might be troublesome if a huge number of representatives is selected. In order to alleviate this problem (if present), it is possible to re-state the fitness function (formerly (17)) by considering a convex linear combination between the performance index and the feature selector sparsity, hence:

$$f_2 = \omega \left(1 - \bar{J}\right) + (1 - \omega) \frac{|\{i : \mathbf{w}_i = 1\}|}{|\mathbf{w}|}, \tag{18}$$

where $\omega \in [0, 1]$ in a user-defined parameter which tunes the convex linear combination by weighting the rightmost term (sparsity) against the leftmost term (performance). It is worth noting that whilst fitness (17) should be maximised, (18) should be minimised.

## 4. Tests and Results

### 4.1. Data Collection and Pre-Processing

The data retrieval processing can be summarised as follows. Using the Python BioServices library [71]:

1. The entire protein list for Escherichia coli str. K12 has been retrieved from UniProt [72];

2. This list has been cross-checked with Protein Data Bank [73] in order to discard unresolved proteins (i.e., proteins whose 3D structure is not available).

Then, using the BioPython library [74]:

1. .pdb files have been downloaded for all resolved proteins;

2. information such as the EC number and the measurement resolution (if present) have been parsed from the .pdb file header;

3. proteins having multiple EC numbers have been discarded.

Finally, using the BioPandas library [75]:

1. $\alpha$-carbon atoms 3D coordinates have been parsed from each .pdb file;

2. In case of multiple equivalent models within the same .pdb file, only the first model is retained;

3.  Similarly, for atoms having alternate coordinate locations, only the first location is retained.

After this retrieval stage, a total number of 6685 proteins has been successfully collected. Some statistics on the measurement resolutions and the number of nodes are sketched in Figure 1a,b, respectively.

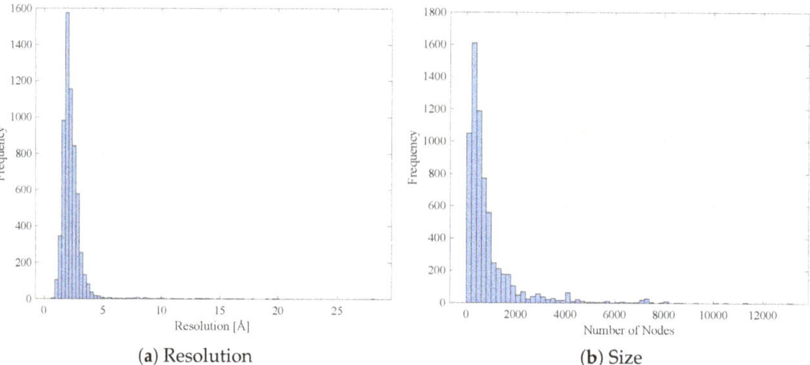

(a) Resolution         (b) Size

**Figure 1.** Distributions within the original 6685 proteins set.

In order to keep only good quality structures (with reliable atomic coordinates), all proteins with missing resolution in their respective .pdb files and proteins whose resolution is greater than 3Å have been discarded. Further, proteins having more than 1500 nodes have been discarded as well. These filtering procedures dropped the number of available proteins from 6685 to 4957. The class labels (EC number) distribution is summarised in Table 1.

**Table 1.** Classes distribution within the filtered 4957 proteins set.

|            |       |       |       |       |       |       |             | Total |
| ---------- | ----- | ----- | ----- | ----- | ----- | ----- | ----------- | ----- |
| **Class**      | EC1   | EC2   | EC3   | EC4   | EC5   | EC6   | not-enzymes |       |
| **Count**      | 540   | 1017  | 919   | 329   | 182   | 244   | 1726        | 4957  |
| **Percentage** | 10.89 | 20.52 | 18.54 | 6.64  | 3.67  | 4.92  | 34.82       | 100%  |

For each of the 4957 available proteins, its respective eight representations (see Appendix A) have been evaluated using the following tools:

- The NetworkX library [76] (Python) for evaluating centrality measures ($\mathbf{X}^{(2)}$) and the Vietoris–Rips complex ($\mathbf{X}^{(1)}$);
- The Numpy and Scipy libraries [77,78] (Python) for several algebraic computations, mainly spectral decompositions for energy, Laplacian energy, heat trace, heat content invariants ($\mathbf{X}^{(3)}$, $\mathbf{X}^{(5)}$, $\mathbf{X}^{(6)}$, $\mathbf{X}^{(8)}$) and the homology group rank ($\mathbf{X}^{(1)}$);
- The Rnetcarto (https://cran.r-project.org/package=rnetcarto) library (R) for network cartography ($\mathbf{X}^{(4)}$).

As in previous works [45,46] the 7-class classification problem is cast into seven binary classification problems in one-against-all fashion, hence the $i$th classifier sees the $i$th class as positive and all other classes as negative. The eight representations $\mathbf{X}^{(1)}, \ldots, \mathbf{X}^{(8)}$ are split into training, validation and test set in a stratified manner in order to preserve labels' distribution across splits. Thus, each of the seven classifiers sees a different training-validation-test split due to the one-against-all labels recoding. The genetic optimisation and classification stage has been performed in MATLAB® R2018a using the built-in genetic algorithm and LibSVM [79] for $\nu$-SVMs.

### 4.2. Computational Results with Fitness Function $f_1$

The first test suite sees $f_1$ (17) as the fitness function, hence the system aims at the maximisation of the (normalised) informedness.

The genetic algorithm has been configured to host 100 individuals for a maximum of 100 generations and each individual's genetic code (upper/lower bounds and constraints, if any) is summarised in Table 2. At each generation, the elitism is set to the top 10% individuals; the crossover operates in a scattered fashion; the selection operator follows the roulette wheel heuristic and the mutation adds to each real-valued gene ($v, \beta, \gamma$) a random number extracted from a zero-mean Gaussian distribution whose variance shrinks as generations go by, whereas it acts in a flip-the-bit fashion for boolean-valued genes (**w**).

**Table 2.** Genetic algorithm parameters description.

| Parameter | Bounds | Contraints |
|---|---|---|
| $v$ | $(0,1]$ by definition | |
| $\beta$ | $\beta_i \in [0,1], \forall i = 1, \ldots, N_R$ | $\sum_{i=1}^{N_R} \beta_i = 0$ |
| $\gamma$ | $\gamma \in (0,100], \forall i = 1, \ldots, N_R$ | |
| **w** | $w_i \in \{0,1\}, \forall i = 1, \ldots, |\mathcal{D}_{TR}|$ | |

Table 3 shows the performances obtained by the proposed Multiple Kernels over Multiple Dissimilarities (MKMD, for short) approach using the fitness function $f_1$. Due to randomness in genetic optimisation, five runs have been performed for each classifier and the average results are shown. Figures of merit include:

- Accuracy $= \dfrac{TP + TN}{TP + FP + TN + FN}$;
- Precision $= \dfrac{TP}{TP + FP}$;
- Recall (Sensitivity) $= \dfrac{TP}{TP + FN}$;
- (Normalised) Informedness as in Equation (17);
- Area Under the Curve (AUC), namely the area under the Receiver Operating Characteristic (ROC) curve [80];

where $TP$, $TN$, $FP$ and $FN$ indicate true positives, true negatives, false positives and false negatives, respectively.

**Table 3.** Test Set Performances with Fitness Function $f_1$.

| Class | Performances | | | | | Complexity |
|---|---|---|---|---|---|---|
| | Accuracy | Precision | Recall | Informedness [†] | AUC | Sparsity |
| 1 (EC1) | 0.95 | 0.87 | 0.68 | 0.83 | 0.92 | 49.43 |
| 2 (EC2) | 0.91 | 0.88 | 0.66 | 0.82 | 0.90 | 49.62 |
| 3 (EC3) | 0.90 | 0.84 | 0.58 | 0.78 | 0.88 | 49.48 |
| 4 (EC4) | 0.97 | 0.90 | 0.56 | 0.78 | 0.88 | 49.42 |
| 5 (EC5) | 0.98 | 0.83 | 0.44 | 0.72 | 0.78 | 50.78 |
| 6 (EC6) | 0.99 | 0.94 | 0.76 | 0.88 | 0.95 | 49.28 |
| 7 (not-enzymes) | 0.82 | 0.77 | 0.70 | 0.79 | 0.89 | 50.52 |

[†] Normalised.

Similarly, Figure 2 shows the ROC curves for all classifiers by considering their respective run with greatest AUC.

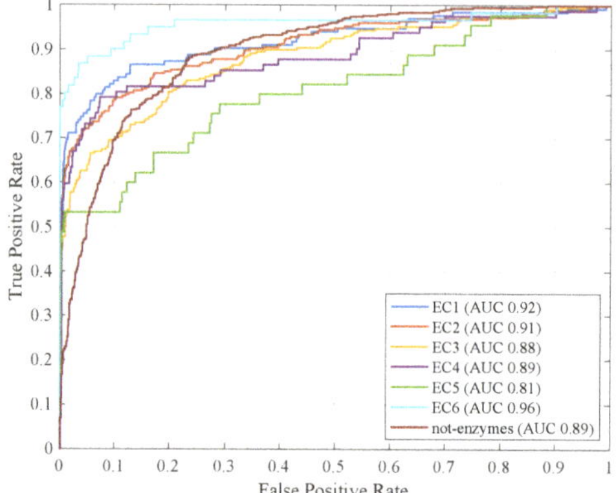

**Figure 2.** ROC curves with fitness function $f_1$. In brackets, the respective AUC values.

### 4.3. Computational Results with Fitness Function $f_2$

These experiments see the fitness function $f_2$ (Equation (18)) in lieu of $f_1$ (Equation (17)), where the weighting parameter $\omega$ is set to 0.5 in order to give the same importance to performances and sparsity. In order to ensure a fair comparison with the previous analysis, the same training-validation-test splits have been used for all seven classifiers, along with the same genetic algorithm setup (genetic code, number of individuals and generations, genetic operators). Table 4 shows the average performances obtained by the seven classifiers across five genetic algorithm runs. As in the previous case, Figure 3 shows the ROC curves for all classifiers by considering their respective run with greatest AUC.

**Table 4.** Test set performances with fitness function $f_2$ and $\omega = 0.5$.

| Class | Performances | | | | | Complexity |
|---|---|---|---|---|---|---|
| | Accuracy | Precision | Recall | Informedness [†] | AUC | Sparsity |
| 1 (EC1) | 0.95 | 0.86 | 0.69 | 0.84 | 0.92 | 33.08 |
| 2 (EC2) | 0.91 | 0.88 | 0.67 | 0.82 | 0.90 | 32.48 |
| 3 (EC3) | 0.90 | 0.83 | 0.57 | 0.77 | 0.87 | 29.94 |
| 4 (EC4) | 0.97 | 0.88 | 0.54 | 0.77 | 0.88 | 33.89 |
| 5 (EC5) | 0.98 | 0.85 | 0.45 | 0.73 | 0.79 | 35.54 |
| 6 (EC6) | 0.98 | 0.91 | 0.76 | 0.88 | 0.95 | 35.38 |
| 7 (not-enzymes) | 0.82 | 0.77 | 0.69 | 0.79 | 0.88 | 33.37 |

[†] Normalised.

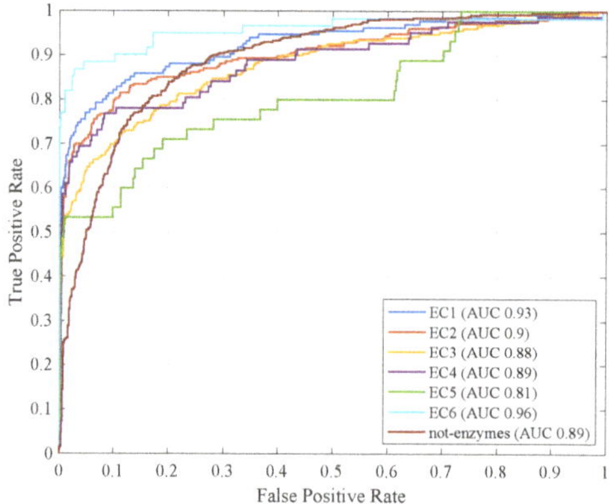

**Figure 3.** ROC curves with fitness function $f_2$ and $\omega = 0.5$. In brackets, the respective AUC values.

### 4.4. Benchmarking against a Clustering-Based One-Class Classifier

In order to properly benchmark the proposed MKMD system, a One-Class Classification System (hereinafter OCC or OCC_System) capable of exploiting multiple dissimilarities is used. This classification system has been initially proposed in [81] and later used for modelling complex systems such as smart grids [81–83] and protein networks [44].

The main idea in order to build a model through the One-Class Classifier is to use a clustering-evolutionary hybrid technique [81,82]. The main assumption is that similar protein types have similar chances of generating a specific class, reflecting the cluster model. Therefore, the core of the recognition system is a custom-based dissimilarity measure computed as a weighted Euclidean distance, that is:

$$d(\breve{\vec{x}}_1, \breve{\vec{x}}_2; \vec{W}) = \sqrt{(\breve{\vec{x}}_1 \ominus \breve{\vec{x}}_2)^T \vec{W}^T \vec{W}(\breve{\vec{x}}_1 \ominus \breve{\vec{x}}_2)}, \tag{19}$$

where $\breve{\vec{x}}_1, \breve{\vec{x}}_2$ are two generic patterns and $\vec{W}$ is a diagonal matrix whose elements are generated through a suitable vector of weights $\vec{w}$. The dissimilarity measure is component-wise, therefore the $\ominus$ symbol represents a generic dissimilarity measure, tailored on each pattern subspace, that has to be specified depending on the semantic of data at hand.

In this study, patterns are represented by dissimilarity vectors extracted from each sub-dissimilarity matrix, one for each feature adopted to describe the protein (see Section 2). In other words, patterns pertain to a suitable dissimilarity space.

The decision region of each cluster $C_i$ is constructed around the medoid $c_i$ bounded by the average radius $\delta(C_i)$ plus a threshold $\sigma$, considered together with the dissimilarity weights $\vec{w} = diag(\vec{W})$ as free parameters. Given a test pattern $\breve{\vec{x}}$ the decision rule consists in evaluating whether it falls inside or outside the overall target decision region, by checking whether it falls inside the closest cluster. The learning procedure consists in clustering the training set $\mathcal{D}_{\text{TR}}$ composed by target patterns, adopting a standard genetic algorithm in charge of evolving a family of cluster-based classifiers considering the weights $\vec{w}$ and the thresholds of the decision regions as search space, guided by a proper objective function. The latter is evaluated on the validation set $\mathcal{D}_{\text{VAL}}$, taking into account a linear combination of the accuracy of the classification (that we seek to maximise) and the extension of the thresholds (that should be minimised). Note that in building the classification model we use only target patterns, while non-target ones are used in the cross-validation phase, hence the adopted learning paradigm is the One-Class classification one [84,85]. Moreover, in order to outperform the well-known limitations of the initialization of the standard $k$-means algorithm, the OCC_System initializes more than one instance of the clustering algorithm with random starting representatives, namely medoids, since the OCC_System is capable of dealing with arbitrarily structured data [86–88]. At test stage (or during validation) a voting procedure for each cluster model is performed. This technique allows building a more robust proteins model.

Figure 4 shows the schematic representing the core subsystems of the proposed OCC_System, such as the ones performing the clustering procedure and the genetic algorithm. Moreover, it is shown the Test subsystem, where given a generic test pattern and given a learned model, it is possible to associate a score value (soft-decision) besides the Boolean decision. Hence, we equip each cluster $C_i$ with a suitable membership function, denoted in the following as $\mu_{C_i}(\cdot)$. In practice, we generate a fuzzy set [89] over $C_i$. The membership function allows quantifying the uncertainty (expressed by the membership degree in $[0,1]$) of a decision about the recognition of a test pattern. Membership values close to either 0 or 1 denote "certain" and hence reliable decisions. When the membership degree assigned to a test pattern is close to 0.5, there is no clear distinction about the fact that such a test pattern is really a target pattern or not (regardless of the correctness of the Boolean decision).

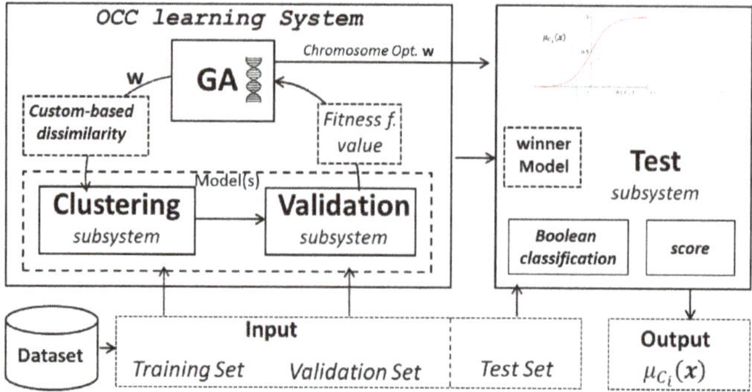

**Figure 4.** Schematic of the classification system able to learn a classification model for each positive class. The model provides the crisp decision as well as a score (a real number) encoding the decision reliability.

For this purpose, we adopt a parametric sigmoid model for $\mu_{C_i}(\cdot)$, which is defined as follows:

$$\mu_{C_i}(x) = \frac{1}{1 + \exp\{(d(c_i, x) - b_i)/a_i\}}, \tag{20}$$

where $a_i, b_i \geq 0$ are two parameters specific to $C_i$, and $d(\cdot, \cdot)$ is the dissimilarity measure (19). Notably, $a_i$ is used to control the steepness of the sigmoid (the lower the value, the faster the rate of change), and $b_i$ is used to translate the function in the input domain. If a cluster (that models a typical protein found in the training set) is very compact, then it describes a very specific scenario. Therefore, no significant variations should be accepted to consider test patterns as members of this cluster. Similarly, if a cluster is characterised by a wide extent, then we might be more tolerant in the evaluation of the membership. Accordingly, the parameter $a_i$ is set equal to $\delta(C_i)$. On the other hand, we define $b_i = \delta(C_i) + \sigma_i/2$. This allows us to position the part of the sigmoid that changes faster right in-between the area of the decision region determined by the dissimilarity values falling in $[B(C_i) - \sigma_i, B(C_i)]$, where in turn $B(C_i) = \delta(C_i) + \sigma_i$ is the boundary of the decision region related to the $i$th cluster.

Finally, the soft decision function, $s(\cdot)$, is defined as

$$s(\bar{x}) = \mu_{C^*}(\bar{x}), \tag{21}$$

where $C^*$ is the cluster where the test (target) pattern falls.

With the aim of making a synthesis, we remark that the OCC_System works in two phases:

1. Learning a cluster model of proteins through a suitable dataset divided into two disjoint sets, namely training and validation set;
2. Using the learned model in order to recognise or classify unseen proteins drawn from the test set, assigning to each pattern a probability value.

The OCC parameters defining the model are optimised by means of a genetic algorithm guided by a suitable objective function that takes into account the classification accuracy. For the sake of comparison, the same genetic operators (selection, mutation, crossover, elitism) as per the MKMD system and have been considered (see Section 4.2). As concerns the complexity of the model, measured as the cardinality of the partition $k$, we choose a suitable value $k = 120$.

Table 5 shows the comparison between the OCC_System and the MKMD approach. In order to ensure a fair comparison, since the OCC_System does not perform representatives selection in the dissimilarity space, in the MKMD genetic code (cf. Equation (15)), the weights vector $\mathbf{w}$ has been removed and all weights have been considered unitary (i.e., no representative selection). Similarly, Figure 5b and Figure 5a show the ROC curves for OCC and MKMD, respectively.

From Table 5 is evident that MKML outperforms OCC in terms of accuracy, informedness and AUC (see also the ROC curves in Figure 5b and Figure 5a), but a clear winner does not exist as regards precision and recall. As regards the structural complexity, OCC is bounded by the number of clusters $k$, whereas MKMD is bounded by the number of support vectors as returned by the training phase [24]. Indeed, the computational burden required to classify new test data is given by:

- The pairwise distances between the test data and the $k$ clusters centres (for OCC);
- The dot product between the test data and the support vectors (for MKMD).

Specifically, for OCC, a suitable number of 120 clusters has been defined for all classes, whereas the training phase for MKMD returned an average of 1300 support vectors ($\sim$52% of the training data) for class 1, 1881 support vectors ($\sim$76%) for class 2, 1745 support vectors ($\sim$70%) for class 3, 1213 support vectors ($\sim$49%) for class 4, 767 support vectors ($\sim$31%) for class 5, 864 support vectors ($\sim$35%) for class 6 and 1945 support vectors ($\sim$78%) for class 7. In conclusion, whilst MKMD outperforms OCC in terms of performances, the latter outperforms the former in terms of structural complexity.

**Table 5.** Test set performances with the one-class classifier.

| Class | Classifier | Performances | | | | |
|---|---|---|---|---|---|---|
| | | Accuracy | Precision | Recall | Informedness [†] | AUC |
| 1 (EC1) | OCC | 0.92 | **0.97** | 0.35 | 0.67 | 0.85 |
| | MKMD | **0.95** | 0.88 | **0.67** | **0.83** | **0.91** |
| 2 (EC2) | OCC | 0.83 | 0.87 | 0.45 | 0.69 | 0.76 |
| | MKMD | **0.91** | **0.89** | **0.66** | **0.82** | **0.91** |
| 3 (EC3) | OCC | 0.83 | **0.86** | 0.49 | 0.70 | 0.77 |
| | MKMD | **0.90** | 0.84 | **0.57** | **0.77** | **0.88** |
| 4 (EC4) | OCC | 0.68 | 0.60 | **0.78** | 0.61 | 0.72 |
| | MKMD | **0.97** | **0.89** | 0.53 | **0.76** | **0.87** |
| 5 (EC5) | OCC | 0.85 | 0.75 | 0.37 | 0.62 | 0.69 |
| | MKMD | **0.98** | **0.82** | **0.44** | **0.72** | **0.78** |
| 6 (EC6) | OCC | 0.97 | 0.96 | 0.57 | 0.78 | 0.88 |
| | MKMD | **0.99** | **0.92** | **0.77** | **0.88** | **0.95** |
| 7 (not-enzymes) | OCC | 0.68 | 0.60 | **0.78** | 0.61 | 0.72 |
| | MKMD | **0.82** | **0.78** | 0.68 | **0.79** | **0.88** |

[†] Normalised.

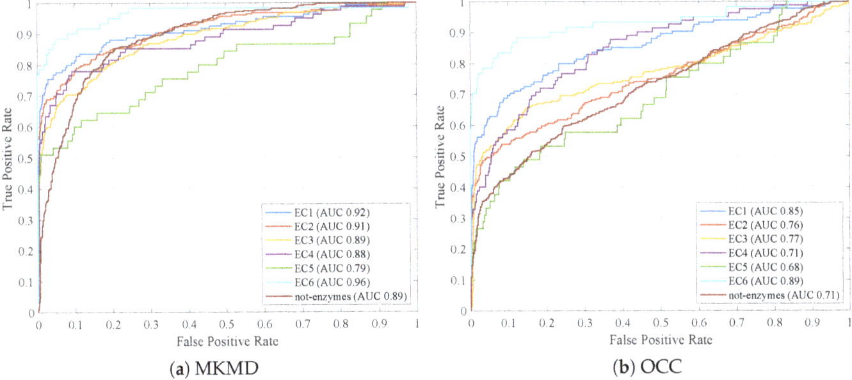


(a) MKMD

(b) OCC

**Figure 5.** ROC curves comparison (best run for all classes). In brackets, the respective AUC values.

## 4.5. Comparing against Previous Works

In Table 6 are reported the performances (in terms of AUC only, for the sake of shorthand) between the proposed MKMD approach with fitness function $f_1$ (Table 3), with fitness function $f_2$ (Table 4) and with no representatives selection in the embedding space (Table 5) against our previous studies for solving the same classification problem. For the sake of completeness, the results obtained by OCC (Table 5) are also included.

**Table 6.** Comparison (in terms of AUC) between the proposed MKMD approach and previous studies.

| Approach | EC1 | EC2 | EC3 | EC4 | EC5 | EC6 | Not-Enzymes |
|---|---|---|---|---|---|---|---|
| DME + Logistic Regression [44] | – | – | – | – | – | – | 0.62 |
| DME + SVM [44] | – | – | – | – | – | – | 0.64 |
| DME + Naïve Bayes [44] | – | – | – | – | – | – | 0.62 |
| DME + Decision Tree [44] | – | – | – | – | – | – | 0.60 |
| DME + Neural Network [44] | – | – | – | – | – | – | 0.63 |
| OCC [44] | – | – | – | – | – | – | 0.63 |
| Feature Generation via Betti Numbers + SVM [46] | 0.79 | 0.75 | 0.73 | 0.73 | 0.46 | 0.77 | 0.77 |
| Feature Generation via Spectral Density + SVM [45] | 0.85 | 0.82 | 0.85 | 0.81 | 0.59 | 0.81 | 0.82 |
| MKMD with $f_1$ (Table 3) | 0.92 | 0.90 | 0.88 | 0.88 | 0.78 | 0.95 | 0.89 |
| MKMD with $f_2$ (Table 4) | 0.92 | 0.90 | 0.87 | 0.88 | 0.79 | 0.95 | 0.88 |
| MKMD with no representative selection (Table 5) | 0.91 | 0.91 | 0.88 | 0.87 | 0.78 | 0.95 | 0.88 |
| OCC (Table 5) | 0.85 | 0.76 | 0.77 | 0.72 | 0.69 | 0.88 | 0.72 |

In [44], two experiments have been performed: the first relied on the Dissimilarity Matrix Embedding (DME) by considering different protein representations (similar to the ones considered in this work) and the second one relied on OCC being able to explore those different representations simultaneously (alike this work). There are three main differences between this work and [44]: first, the set of representations is different; second, we only managed to solve the binary classification problem between enzymes and not-enzymes; third, the set of considered proteins is different. In fact, in [44], we performed an additional filtering stage in order to select (for the same UniProt ID) only the PDB entry with best resolution: we found that this heavily limits the number of protein samples available, possibly reducing the learning capabilities.

In [45,46] we used the sampled spectral density of the protein contact networks (more information can be found in Appendix A.8) and the Betti numbers (more information can be found in Appendix A.1), respectively: the results in Table 6 feature the same proteins set used in this work. Indeed, thanks to the observation above, experiments have been repeated with an augmented number of protein samples [90,91].

Results in Table 6 highlight that:

1. Avoiding to filter out PDB structures by considering only the best resolution for a given UniProt ID (as carried out also in this work) helps in improving classification models: indeed, performances from [44] are amongst the lowest ones;
2. The proposed MKMD approach, regardless of the fitness function and/or representative selection, outperforms all competitors for all EC classes (including not-enzymes).

## 4.6. On the Knowledge Discovery Phase

Apart from the good generalisation capabilities, it is worth remarking that an interesting aspect of the proposed multiple kernel approach is the two-fold knowledge discovery phase:

1. By analysing the kernel weights $\beta$, it is possible to determine the most important representations for the problem at hand;

2. By analysing **w**, namely the binary vector in charge of selecting prototypes from the dissimilarity space, it is possible to determine and analyse the patterns (proteins, in this case) elected as prototypes.

Let us start our discussion from the latter point. From a chemical viewpoint, proteins are linear hetero-polymers in the form of non-periodic sequences of 20 different monomers (amino-acids residues). While artificial polymers (periodic) are very large extended molecules forming a matrix, the majority of proteins fold as self-contained water-soluble structures. Thus, we can consider the particular linear arrangement of amino-acid residues as a sort of 'recipe' for making a water-soluble polymer with a well-defined three-dimensional architecture [92]. "Well-defined three-dimensional structure" should not be intended as a 'fixed architecture': many proteins appear as partially or even totally disordered when analysed with spectroscopic methods. This apparent disorder corresponds to an efficient organisation as for protein physiological role giving to the molecule the possibility to adapt to rapidly changing microenvironment conditions [93].

This implies the two main drivers of amino-acid residues 3D arrangement (from where the particular properties of relative contact networks derive) are:

1. To efficiently accomplish the task of being water soluble while maintaining a stable structure (or dynamics);
2. To allow for an efficient spreading of the signal across amino-acid residues contact network so to sense relevant microenvironment changes and to reshape accordingly—allosteric effect, see [94].

Currently, we have only a coarse-grain knowledge of such complex tasks, and biochemists are still very far to be able to reproduce this behaviour by synthetic constructs.

The ability to catalyse a specific class of chemical reactions (the property the EC classification is based upon), while being crucial for the biological role of protein molecules is, from the point of view of topological and geometrical proteins structure, only a very minor modulation of their global shape [92]. Notwithstanding that, the thorough analysis of representative proteins (thus pivotal for discrimination) can give us some general hints, not only confined to the specific classification task, but extending to all the 'hard' classification problems based upon very tiny details of the statistical units.

Looking at the representative proteins (hence, endowed with meaningful discriminative power) in Tables A1–A7 (Appendix B) we immediately note that the pivotal proteins come from all the analysed EC categories and not only from the specific class to be discriminated. This is expected by the absence of a simple form-function relation, hence they can be considered as an 'emergent property' of the discrimination task. The presence of molecules of different classes crucial for a specific category modelling and thus the image in light of a peculiar strategy adopted by the system is analogue to the use of 'paired samples' in statistical investigation [95,96]. When in presence of only minor details discriminating statistical units pertaining to different categories, the only possibility to discriminate is to adopt a paired samples strategy in which elements of a category is paired with a very similar example of another category so to rely on their differences (on a sample-by-sample basis) instead of looking for a general 'class-specific' properties. This is the case of proteins whose general shape is only partially determined by the chemical reaction they catalyse: looking at the 3D structures of relevant proteins, we can easily verify they pertain to three basic patterns (Figure 6):

1. Cyclic pattern with an approximately spherical symmetry (Figure 6a);
2. A globular pattern with 'duplication': protein can be considered as two identical half-structures (Figure 6b);
3. Elongated non-cyclic pattern, typical of membrane-bound proteins (Figure 6c).

Even if the three above-mentioned patterns have slightly different relative frequencies in the EC classes (e.g., pattern 3 is more frequent in non-enzymatic proteins), they are present in all the analysed classes so allowing for the 'between-categories' sample-by-sample pairing mentioned above.

This peculiar situation is in line with current biochemical knowledge (minimal effect exerted by catalysed reaction on global structure) and it is a relevant proof-of-concept of both the reliability of the classification solution and of the power of the proposed approach. On the other hand, it is very hard to de-convolve the discriminating structural nuances from the obtained solution that, as it is, only confirms the presence of 'tiny and still unknown' structural details linked to the catalytic activity of the studied molecules.

As regards the former point, Figure 7 shows the average weights vector $\beta$ across the aforementioned five runs for $\omega = 0.5$, showing that the MKMD approach considers for almost all classes centrality measures ($\mathbf{X}_2$) and the protein size ($\mathbf{X}_7$) as the most relevant representations, followed by the Betti numbers sequence ($\mathbf{X}_1$), heat content invariants ($\mathbf{X}_5$) and heat kernel trace ($\mathbf{X}_6$).

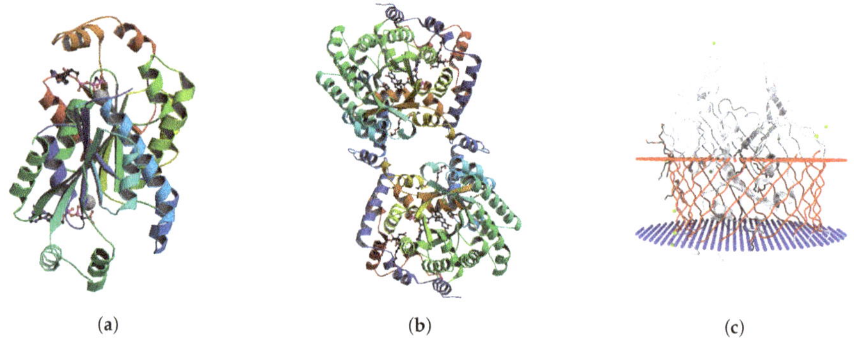

(a)      (b)      (c)

**Figure 6.** Three basic patterns in protein 3D structures. (**a**) Transferase—PDB ID 1KOF, (**b**) Proline dehydrogenase (oxidoreductase)—PDB ID 3E2R, (**c**) Transport Protein (Non-Enzyme)—PDB ID 3RGM.

| | $\mathbf{X}_1$ | $\mathbf{X}_2$ | $\mathbf{X}_3$ | $\mathbf{X}_4$ | $\mathbf{X}_5$ | $\mathbf{X}_6$ | $\mathbf{X}_7$ | $\mathbf{X}_8$ |
|---|---|---|---|---|---|---|---|---|
| EC 1 | 0.1270 | 0.2791 | 0.0465 | 0.0546 | 0.1287 | 0.1225 | 0.1883 | 0.0534 |
| EC 2 | 0.1771 | 0.3478 | 0.0742 | 0.0500 | 0.0242 | 0.0780 | 0.1695 | 0.0792 |
| EC 3 | 0.0847 | 0.2849 | 0.0646 | 0.0241 | 0.0809 | 0.1576 | 0.2338 | 0.0695 |
| EC 4 | 0.0525 | 0.3758 | 0.0820 | 0.0364 | 0.1270 | 0.1164 | 0.1710 | 0.0390 |
| EC 5 | 0.2438 | 0.2239 | 0.0853 | 0.0939 | 0.0468 | 0.1361 | 0.1025 | 0.0677 |
| EC 6 | 0.1223 | 0.2962 | 0.0327 | 0.0301 | 0.1464 | 0.0517 | 0.2790 | 0.0415 |
| not-enzymes | 0.2179 | 0.2027 | 0.1033 | 0.0916 | 0.0226 | 0.0604 | 0.1937 | 0.1077 |

**Figure 7.** Average kernel weights vectors $\beta$.

It is worth noting that enzymes have a more pronounced allosteric effect with respect to non-enzymatic structures. This is a consequence of the need to modulate chemical kinetics according to microenvironment conditions—allostery is the modulating effect of a modification happening in a site different from catalytic site on the efficiency of the reaction [97]. Allostery implies an efficient transport of the signal along protein structure and it was discovered to be efficiently interpreted in terms of PCN descriptors [98] thus, the observed kernel weights fit well with the current biochemical knowledge.

## 5. Conclusions

In this paper, we proposed a classification system able to explore simultaneously multiple representations following an hybridisation between multiple kernel learning and dissimilarity spaces, hence exploiting the discriminative power of kernel methods and the customisability of dissimilarity spaces.

Specifically, several representations are treated using their respective dissimilarity representations and combined in a multiple kernel fashion, where each kernel function considers a specific dissimilarity representation. A genetic algorithm (although any derivative-free evolutive metaheuristic can be placed instead) is able to simultaneously select suitable representatives in the dissimilarity space and tune the kernel weights, allowing a two-fold a posteriori knowledge discovery phase regarding the most suitable representations (higher kernel weights) and the patterns elected as prototypes in the dissimilarity space.

The proposed MKMD system has been applied for solving a real-world problem, namely protein function prediction, with satisfactory results, greatly outperforming our previous works in which graph-based descriptors extracted from PCNs have been tested for solving the very same problem. Further, the proposed system has been benchmarked against a One-Class Classifier, also able to simultaneously explore multiple dissimilarities: whilst the former outperforms the latter in terms of accuracy, AUC and informedness, a clear winner between the two methods does not exist in terms of precision and recall.

As far as the two-fold knowledge discovery phase for the proposed application is concerned, results both in terms of selected representatives in the dissimilarity space and weights automatically assigned to different representations are in line with current biological knowledge, showing the reliability of the proposed system.

Furthermore, due to its flexibility, the proposed system can be applied to any input domain (not necessarily graphs), provided that several representations can be extracted by the structured data at hand and that suitable dissimilarity measures can be defined for such heterogeneous representations.

**Author Contributions:** Conceptualization, A.M.; Data curation, A.M. and A.G.; Formal analysis, A.M., A.G. and E.D.S.; Investigation, A.M., A.G., E.D.S. and A.R.; Methodology, A.M.; Resources, A.M. and A.G.; Software, A.M. and E.D.S; Supervision, A.G. and A.R.; Validation, A.M. and A.G.; Writing–original draft, A.M., A.G. and E.D.S.; Writing—review & editing, A.M., A.G., E.D.S. and A.R. All authors have read and agreed to the published version of the manuscript.

**Funding:** This work has been partially supported by the Sapienza Research Calls project "PARADISE - PARAllel and DIStributed Evolutionary agent-based systems for machine learning and big data mining", 2018.

**Conflicts of Interest:** The authors declare no conflict of interest.

## Abbreviations

The following abbreviations are used in this manuscript:

| | |
|---|---|
| AUC | Area Under the Curve |
| DME | Dissimilarity Matrix Embedding |
| MKMD | Multiple Kernels over Multiple Dissimilarities |
| OCC | One-Class Classification (also OCC_System) |
| PCN | Protein Contact Networks |
| PDB | Protein Data Bank |
| ROC | Receiver Operating Characteristic |
| SVM | Support Vector Machine |

## Appendix A. Selected Representations

The set of eight representations $\mathbf{X}^{(1)}, \ldots, \mathbf{X}^{(8)}$ used to characterise PCNs are described in the following eight subsections.

### Appendix A.1. Betti Numbers

Topological Data Analysis [99,100] is a novel data analysis approach useful whenever data can be described by topological structures (networks) as it consists in a set of techniques in order to extract information from data (starting from topological information) by means of dimensionality reduction, manifold estimation and persistent homology in order to study how components lying in a multi-dimensional space are connected (e.g., in terms of loops and multi-dimensional surfaces). One can start either from so-called point clouds, where objects are described by their coordinates in a multi-dimensional space equipped with notion of distance, or by explicitly providing the pairwise distance matrix between objects. Hereinafter, the former case is considered.

The most intuitive scenario in order to study how components lying in a multi-dimensional space are connected is (trivially) by studying the connectivity itself. To this end, it is worth defining simplices as (multi-dimensional) topological objects which can be extracted from a given topological space $\mathcal{X}$: points, lines, triangles and tetrahedrons are (for example) 0-dimensional, 1-dimensional, 2-dimensional, 3-dimensional simplices and, obviously, higher-order analogues exist. Simplices can be seen as descriptors of the space under analysis, thus worthy of attention when studying $\mathcal{X}$. Starting from simplices, it is possible to define simplicial complexes as properly-constructed collection of simplices able to capture the multi-scale organisation (or multi-way relations) in complex networks [101–103]. The two seminal examples of simplicial complexes are the Čech complex and the Vietoris–Rips complex [99,100,104,105], however due to its intuitiveness and lighter computational complexity, one in practice uses the latter. The Vietoris–Rips complex can be built according to the following rule: initially, all 0-dimensional simplices belong to the complex, then a given set of $k$ points forms a $(k-1)$-dimensional simplicial complex to be included in the Vietoris–Rips complex if the pairwise distances are all less than or equal to a user-defined threshold $\epsilon$.

The homology of a simplicial complex can be described by its Betti numbers. Formally, the $i$th Betti number is the rank of the $i$th homology group in the simplicial complex. Informally, the $i$th Betti number corresponds to the number of $i$-dimensional 'holes' in a topological surface. In this work, 3-dimensional graphs are considered and the first three Betti numbers have the following interpretations: the 0th Betti number is the number of connected components, the 1st Betti number is the number of 1-dimensional (circular) holes, the 2nd Betti number is the number of 2-dimensional holes (cavities). The Betti numbers vanish after the spatial dimension.

From the above Vietoris–Rips complex definition, it is clear that the choice of $\epsilon$ is critical as it somewhat defines the resolution of the simplicial complex. In many cases, one builds a sequence of Vietoris–Rips complexes as $\epsilon$ varies in order to study how 'holes' appear and disappear as the resolution changes and then selects a desired value $\epsilon^*$ by studying the 'holes' lifetime in order to obtain a useful homology summary: in algebraic topology, this concept is known as persistence [106].

Instead of having a 'topological summary', following a previous work [46], the rationale is to keep proper track of the number of holes as $\epsilon$ changes. To this end, the range $\epsilon \in [4, 8]$ with sampling step 1 is considered, according to the PCN connectivity range. Hence, the first representation $\mathbf{X}^{(1)}$ sees each protein as a 15-length integer-valued vector obtained by the concatenation of $\mathbf{b}_4, \mathbf{b}_5, \mathbf{b}_6, \mathbf{b}_7, \mathbf{b}_8$, where $\mathbf{b}_i$ is (in turn) a 3-dimensional vector containing the first three Betti numbers for $\epsilon = i$. Technically speaking, for a given $\epsilon$, the Vietoris–Rips complex can be evaluated in two steps [107]:

1. Build the Vietoris–Rips neighbourhood graph $\mathcal{G}_{VR}(\mathcal{V}, \mathcal{E})$: an undirected graph where edges between two nodes, say $v_i, v_j \in \mathcal{V}$, are scored if $d(v_i, v_j) \leq \epsilon$;
2. The set of maximal cliques in $\mathcal{G}_{VR}$ form the Vietoris–Rips complex.

Let $\partial_k : \mathcal{S}_k \to \mathcal{S}_{k-1}$ be the boundary operator, an incidence-like matrix which maps $\mathcal{S}_k$ (i.e., the set of simplices of order $k$) with the set of simplices of order $k - 1$. The $k$-order homology group is defined as [108]:

$$\mathfrak{H}_k = \ker\{\partial_k\} / \mathrm{im}\{\partial_{k+1}\}, \tag{A1}$$

where $\ker\{\cdot\}$ and $\mathrm{im}\{\cdot\}$ denote the kernel and image operators. The rank of $\mathfrak{H}_k$, namely the $k^{\text{th}}$ Betti number is then defined as [102]:

$$b^{(k)} = \mathrm{rank}\{\ker\{\partial_k\}\} - \mathrm{rank}\{\mathrm{im}\{\partial_{k+1}\}\} \tag{A2}$$

or, thanks to the Rank–Nullity theorem [109]:

$$b^{(k)} = (\dim\{\partial_k\} - \mathrm{rank}\{\mathrm{im}\{\partial_k\}\}) - \mathrm{rank}\{\mathrm{im}\{\partial_{k+1}\}\}, \tag{A3}$$

where the rank of the image corresponds to the plain matrix rank in linear algebra.

*Appendix A.2. Centrality Measures*

In graph theory and network analysis, centrality measures indicate the node/edge importance with respect to a given criterion. Let $\mathcal{G} = (\mathcal{V}, \mathcal{E})$ be a graph and let $\mathcal{V}$ and $\mathcal{E}$ be the set of nodes and edges, respectively. The following centrality measures are considered:

- The degree centrality [110] $DC(v_i)$ for node $v_i \in \mathcal{V}$, defined as the percentage of nodes connected to it:

$$DC(i) = \frac{1}{|\mathcal{V}| - 1} \sum_j \mathbf{A}_{i,j}, \tag{A4}$$

where $\mathbf{A}$ is the adjacency matrix, defined as in Equation (A22). The normalisation coefficient $\frac{1}{|\mathcal{V}|-1}$ takes into account the maximum attainable degree in a simple graph, thus making the degree centrality in Equation (A4) independent from the number of nodes in the graph;
- The eigenvector centrality [110] highly rank nodes whether they are connected to other high-rank nodes. Formally, the eigenvector centrality $\mathbf{e}_i$ for node $v_i \in V$ is given by:

$$\mathbf{e}_i = \frac{1}{\lambda} \sum_j \mathbf{A}_{j,i} \mathbf{e}_j, \tag{A5}$$

where $\lambda \neq 0$ is a scalar constant. Equation (A5) can be re-written in matrix form as:

$$\lambda \mathbf{e} = \mathbf{e}\mathbf{A}. \tag{A6}$$

Hence, the eigenvector centrality vector $\mathbf{e}$ is the left-hand eigenvector of the adjacency matrix $\mathbf{A}$ associated with the eigenvalue $\lambda$. According to the Perron–Frobenius theorem, by choosing $\lambda$ as the largest (in absolute value) eigenvalue of $\mathbf{A}$, the solution $\mathbf{e}$ is unique and all its entries are positive;
- The PageRank centrality [110] $\mathbf{p}_i$ for node $v_i \in \mathcal{V}$ is given by:

$$\mathbf{p}_i = \alpha \sum_j \frac{\mathbf{A}_{j,i}}{D(v_j)} \mathbf{p}_j + \frac{1 - \alpha}{|\mathcal{V}|}, \tag{A7}$$

where $\alpha$ is a scalar constant (usually $\alpha = 0.85$) and $D(v_j)$ is the degree of node $v_j$. It is worth remarking the difference between degree and degree centrality: the degree is the number of nodes connected to a given node (namely Equation (A4) without the normalisation term), whereas the degree centrality includes the normalisation term. As in the eigenvector centrality case, Equation (A7) can be re-written in matrix form as:

$$\mathbf{p} = \alpha \mathbf{p} \mathbf{D}^{-1} \mathbf{A} + \beta, \tag{A8}$$

where $\mathbf{D}^{-1}$ is a diagonal matrix whose $i$th element equals $1/D(v_i)$ and $\beta$ is a vector whose elements are all equal to $\frac{1-\alpha}{|\mathcal{V}|}$;
- The Katz centrality [110,111] $\mathbf{k}_i$ for node $v_i \in \mathcal{V}$ is given by:

$$\mathbf{k}_i = \alpha \sum_j \mathbf{A}_{i,j} \mathbf{k}_j + \beta, \tag{A9}$$

where $\beta$ controls the initial centrality (first neighbourhood weights) and $\alpha < 1/\lambda_{\max}$ attenuates the importance with respect to higher-order neighbours (in turn, $\lambda_{\max}$ is the largest eigenvalue of $\mathbf{A}$). It is worth noting that if $\alpha = 1/\lambda_{\max}$ and $\beta = 0$, the Katz centrality equals the eigenvector centrality;

- The closeness centrality [110] $CC(v_i)$ for node $v_i \in V$ is the inverse sum of shortest path distances between node $v_i \in V$ and all other $n-1$ reachable nodes. Formally:

$$CC(v_i) = \frac{n-1}{|V|-1} \frac{n-1}{\sum_{j=1}^{n-1} \delta(v_i, v_j)}, \tag{A10}$$

where $\delta(\cdot, \cdot)$ indicates the shortest path distance. The normalisation factor takes into account the graph size in order to allow comparison between nodes of graphs having different sizes, also in case of multiple connected components [112]. Indeed, $n$ can be seen as the number of nodes in the connected component in which $v_i$ lies. In case of one connected component, the scale factor $(n-1)/(|V|-1)$ can be neglected since $n = |V|$;

- The betweenness centrality [110] $BC(v_i)$ quantifies how many times a given node $v_i \in V$ acts as a bridge along the shortest paths between any two nodes:

$$BC(v_i) = \sum_{v_i \neq v_j \neq v_k} \frac{s^{(v_i)}(v_j, v_k)}{s(v_j, v_k)}, \tag{A11}$$

where $s(v_j, v_k)$ is the number of shortest paths from $v_i$ to $v_j$ and $s^{(v_i)}(v_j, v_k)$ is the number of shortest paths from $v_i$ to $v_j$ passing through $v_i$. As in the closeness centrality case, it is often customary to normalise the betweenness centrality in order to avoid dependency from the number of nodes, thus:

$$BC(v_i) := \frac{2 \cdot BC(v_i)}{(|V|-1) \cdot (|V|-2)}; \tag{A12}$$

- The edge betweenness centrality [113] $EBC(e_i)$ is the edge counterpart of the "standard" (node) betweenness centrality as it quantifies how many times a given edge $e_i \in \mathcal{E}$ acts as a bridge along the shortest paths between two nodes:

$$EBC(e_i) = \sum_{v_i, v_j \in V} \frac{s^{(e_i)}(v_i, v_j)}{s(v_i, v_j)}, \tag{A13}$$

where $s^{(e_i)}(v_i, v_j)$ is the number of shortest paths between nodes $v_i$ and $v_j$ passing through edge $e_i$ and $s(v_i, v_j)$ is the total number of shortest paths between nodes $v_i$ and $v_j$. As in the "standard" betweenness centrality, the edge betweenness centrality can be normalised as follows:

$$EBC(e_i) := \frac{2 \cdot EBC(e_i)}{(|V|-1) \cdot |V|}; \tag{A14}$$

- The load centrality [113,114] $LC(v_i)$ for node $v_i \in V$ is the percentage of the total number of shortest paths passing through $v_i$;
- The edge load centrality $ELC(e_i)$ for edge $e_i \in \mathcal{E}$ is the edge-related counterpart of the load centrality (like betweenness vs. edge betweenness): it is defined as the percentage of the total number of shortest paths crossing edge $e_i$;
- The subgraph centrality [115] $SC(v_i)$ for node $v_i \in V$ is the sum of (weighted) closed walks (i.e., connected subgraphs) starting and ending at $v_i$ (the longer the walk, the lower the weight). It can be evaluated thanks to the spectral decomposition of the adjacency matrix, which reads as $\mathbf{A} =$

$\mathbf{B}\Lambda^{(\mathbf{A})}\mathbf{B}^T$ where $\Lambda^{(\mathbf{A})} = \mathrm{diag}\left\{\lambda_1^{(\mathbf{A})}, \ldots, \lambda_{|\mathcal{V}|}^{(\mathbf{A})}\right\}$ is a diagonal matrix containing the eigenvalues in increasing order and $\mathbf{B}$ contains the corresponding unitary-length eigenvectors, thus:

$$SC(v_i) = \sum_{j=1}^{|\mathcal{V}|} e^{\lambda_j^{(\mathbf{A})}} \left(\mathbf{b}_j(v_i)\right)^2, \tag{A15}$$

where $\lambda_j$ and $\mathbf{b}_j$ are the eigenvalue and eigenvector associated to node $v_j \in \mathcal{V}$ and $\mathbf{b}_j(v_i)$ indicates the value related to $v_i$ in the $j^{\text{th}}$ eigenvector;

- The Estrada Index [116] $EI(\mathcal{G})$ of a graph $\mathcal{G}$ quantifies the compactness (or 'folding', since the Estrada Index was indeed originally proposed in order to study molecular 3D compactness) of a graph starting from the spectral decomposition of the adjacency matrix (as in the subgraph centrality):

$$EI(\mathcal{G}) = \sum_{j=1}^{|\mathcal{V}|} e^{\lambda_j^{(\mathbf{A})}}; \tag{A16}$$

- The harmonic centrality [117] $HC(v_i)$ is the sum of inverse shortest paths distances from a given node $v_i \in \mathcal{V}$ to all other nodes:

$$HC(v_i) = \sum_{\substack{j=1 \\ j \neq i}}^{|\mathcal{V}|} \frac{1}{\delta(v_i, v_j)}; \tag{A17}$$

- The global reaching centrality [118] $GRC(\mathcal{G})$ of a graph $\mathcal{G}$ is the average (over all nodes) of the difference between the maximum local reaching centrality and each node's local reaching centrality. Formally:

$$GRC(\mathcal{G}) = \frac{\sum_{i=1}^{|\mathcal{V}|} \left(LRC_{\max} - LRC(v_i)\right)}{|\mathcal{V}| - 1}, \tag{A18}$$

where $LRC(v_i)$ is the local reaching centrality of node $v_i \in \mathcal{V}$ and $LRC_{\max}$ is the maximum local reaching centrality amongst all nodes. In turn, the local reaching centrality for a given node $v_i$ is defined as the percentage of nodes reachable from $v_i$;

- The average clustering coefficient [119] $ACC(\mathcal{G})$ of a graph $\mathcal{G}$ is given by:

$$ACC(\mathcal{G}) = \frac{1}{|\mathcal{V}|} \sum_{i=1}^{|\mathcal{V}|} cc(v_i), \tag{A19}$$

where $cc(v_i)$ is the clustering coefficient for node $v_i$, defined as:

$$cc(v_i) = \frac{2 \cdot tri(v_i)}{DC(v_i) \cdot (DC(v_i) - 1)}, \tag{A20}$$

where, in turn, $tri(v_i)$ is the number of triangles passing through node $v_i$ and $D(v_i)$ is its degree;

- The average neighbour degree [120] $AND(v_i)$ of node $v_i \in \mathcal{V}$ is given by:

$$AND(v_i) = \frac{1}{|\mathcal{N}(v_i)|} \sum_{v_j \in \mathcal{N}(v_i)} D(v_j), \tag{A21}$$

where $\mathcal{N}(v_i)$ is the set of neighbours of node $v_i$.

Apart from $ACC$, $EI$ and $GRC$, which are global characteristics (i.e., related to the whole graph), the others are local characteristics (i.e., related to each node or edge). As such, it is impossible to compare graphs having different sizes (number of nodes and/or edges) by considering their local centralities. The second representation $\mathbf{X}^{(2)}$ sees each protein as a 27-length real-valued vector containing $\bar{DC}$, $\tilde{DC}$, $\bar{\mathbf{e}}$, $\tilde{\mathbf{e}}$, $\bar{\mathbf{p}}$, $\tilde{\mathbf{p}}$, $\bar{\mathbf{k}}$, $\tilde{\mathbf{k}}$, $\bar{CC}$, $\tilde{CC}$, $\bar{BC}$, $\tilde{BC}$, $\bar{EBC}$, $\tilde{EBC}$, $\bar{LC}$, $\tilde{LC}$, $\bar{ELC}$, $\tilde{ELC}$, $\bar{SC}$, $\tilde{SC}$, $EI$,

$\bar{HC}$, $\tilde{HC}$, $GRC$, $ACC$, $A\bar{N}D$, $A\tilde{N}D$ (where bar and tilde indicate the average and standard deviation centrality across nodes/edges).

*Appendix A.3. Energy and Laplacian Energy*

Let $\mathcal{G} = (\mathcal{V}, \mathcal{E})$ be a graph and let $\mathcal{V}$ and $\mathcal{E}$ be the set of nodes and edges, respectively. Since in this work unweighted and undirected graphs are considered, the adjacency matrix $\mathbf{A}$ is a binary $|\mathcal{V}| \times |\mathcal{V}|$ matrix defined as:

$$\mathbf{A}_{i,j} = \begin{cases} 1 & \text{if } (v_i, v_j) \in \mathcal{E} \\ 0 & \text{otherwise.} \end{cases} \tag{A22}$$

From $\mathbf{A}$, it is possible to define the diagonal $|\mathcal{V}| \times |\mathcal{V}|$ degree matrix as:

$$\mathbf{D}_{i,j} = \begin{cases} D(i) & \text{if } i = j \\ 0 & \text{otherwise,} \end{cases} \tag{A23}$$

where $D(i)$ is the degree of the $i$th node. In turn, from $\mathbf{A}$ and $\mathbf{D}$, it is possible to define the Laplacian matrix as:

$$\mathbf{L} = \mathbf{D} - \mathbf{A}, \tag{A24}$$

The spectrum and Laplacian spectrum of $\mathcal{G}$ are defined as the set of eigenvalues from $\mathbf{A}$ and $\mathbf{L}$, respectively [121]:

$$\lambda^{(\mathbf{A})} = \left\{ \lambda_1^{(\mathbf{A})}, \ldots, \lambda_{|\mathcal{V}|}^{(\mathbf{A})} \right\}, \tag{A25}$$

$$\lambda^{(\mathbf{L})} = \left\{ \lambda_1^{(\mathbf{L})}, \ldots, \lambda_{|\mathcal{V}|}^{(\mathbf{L})} \right\}. \tag{A26}$$

From Equations (A25) and (A26), it is possible to define the graph energy $E$ and the Laplacian energy $LE$ as

$$E = \sum_{i=1}^{|\mathcal{V}|} \left| \lambda_i^{(\mathbf{A})} \right|, \tag{A27}$$

$$LE = \sum_{i=1}^{|\mathcal{V}|} \left| \lambda_i^{(\mathbf{L})} - \frac{2|\mathcal{E}|}{|\mathcal{V}|} \right|. \tag{A28}$$

The third representation $\mathbf{X}^{(3)}$ sees each protein as a 2-length real-valued vector containing $E$ and $LE$.

*Appendix A.4. Nodes Functional Cartography*

Guimerà and Amaral in their seminal work [122] proposed a methodology in order to extract functional modules from a graph by maximising its modularity using simulated annealing [123]. Their definition of modularity takes into account both within-module degree and between-module degree with the idea that a good graph partition (i.e., high modularity) must have many within-module links and few between-module links.

Each node is then assigned with two scores: the $z$-score and the participation coefficient $P$. The former measures how well-connected a given node is with respect to other nodes in its own module. The latter quantifies how many connections a given nodes has with respect to nodes belonging to different modules.

The $z - P$ plane has been heuristically divided into seven regions and each node can be classified into one of seven functional roles by considering its $z$-score and its participation coefficient $P$. Nodes having $z < 2.5$ are non-hubs, whereas nodes having $z \geq 2.5$ are hubs. In turn, non-hub nodes can be divided in: ultra-peripherals (if $P \leq 0.05$), peripherals (if $P \in (0.05, 0.62]$), non-hub connectors (if

$P \in (0.62, 0.8]$) and non-hub kinless (if $P > 0.8$). Finally, hub nodes can be divided in: provincial hubs (if $P \leq 0.3$), connector hubs (if $P \in (0.3, 0.75]$) and kinless hubs (if $P > 0.75$).

The fourth representation $\mathbf{X}^{(4)}$ sees each protein as an 8-length real-valued vector containing the modularity (as returned by the simulated annealing) and the percentage of nodes belonging to each functional role.

*Appendix A.5. Heat Content Invariant*

From the graph Laplacian and degree matrices (Equations (A24) and (A23), respectively), the normalised Laplacian matrix can be evaluated as:

$$\tilde{\mathbf{L}} = \mathbf{D}^{-\frac{1}{2}} \mathbf{L} \mathbf{D}^{-\frac{1}{2}}, \tag{A29}$$

The spectral decomposition of $\tilde{\mathbf{L}}$ reads as:

$$\tilde{\mathbf{L}} = \mathbf{V} \mathbf{\Lambda} \mathbf{V}^T, \tag{A30}$$

where $\mathbf{\Lambda} = \mathrm{diag}\left\{ \lambda_1^{(\tilde{\mathbf{L}})}, \ldots, \lambda_{|\mathcal{V}|}^{(\tilde{\mathbf{L}})} \right\}$ is a diagonal matrix containing the eigenvalues in increasing order and $\mathbf{V}$ contains the corresponding unitary-length eigenvectors.

The heat equation associated to $\tilde{\mathbf{L}}$ is given by [124,125]:

$$\frac{\partial \mathbf{H}(t)}{\partial t} = -\tilde{\mathbf{L}} \mathbf{H}(t), \tag{A31}$$

where $\mathbf{H}(t)$ is the $|\mathcal{V}| \times |\mathcal{V}|$ heat kernel matrix at time $t$. The heat content $HC(t)$ of $\mathbf{H}(t)$ is given by:

$$
\begin{aligned}
HC(t) &= \sum_{v_i \in \mathcal{V}} \sum_{v_j \in \mathcal{V}} \mathbf{H}_{i,j}(t) \\
&= \sum_{v_i \in \mathcal{V}} \sum_{v_j \in \mathcal{V}} \sum_{k=1}^{|\mathcal{V}|} \exp\left\{ -\lambda_k^{(\tilde{\mathbf{L}})} t \right\} \mathbf{v}_k(v_i) \mathbf{v}_k(v_j),
\end{aligned} \tag{A32}
$$

where $\mathbf{v}_k(v_i)$ is the value related to node $v_i$ in the $k$th eigenvector.

The MacLaurin series for the negative exponential reads as:

$$\exp\left\{ -\lambda_k^{(\tilde{\mathbf{L}})} t \right\} = \sum_{m=0}^{\infty} \frac{\left( -\lambda_k^{(\tilde{\mathbf{L}})} t \right)^m t^m}{m!} \tag{A33}$$

and substituting Equation (A33) in Equation (A32) yields:

$$HC(t) = \sum_{v_i \in \mathcal{V}} \sum_{v_j \in \mathcal{V}} \sum_{k=1}^{|\mathcal{V}|} \sum_{m=0}^{\infty} \frac{\left( -\lambda_k^{(\tilde{\mathbf{L}})} t \right)^m t^m}{m!} \mathbf{v}_k(v_i) \mathbf{v}_k(v_j) \tag{A34}$$

By re-writing Equation (A32) in terms of power series as:

$$HC(t) = \sum_{m=0}^{\infty} q_m t^m, \tag{A35}$$

where the set of coefficients $q_m$ are the so-called heat content invariants and can be evaluated in closed-form as:

$$q_m = \sum_{i=1}^{|\mathcal{V}|} \left( \left( \sum_{v \in \mathcal{V}} \mathbf{v}_i(v) \right)^2 \right) \frac{\left( -\lambda_i^{(\tilde{\mathbf{L}})} \right)^m}{m!}. \tag{A36}$$

The fifth representation $\mathbf{X}^{(5)}$ sees each protein as a 4-length real-valued vector containing the first four coefficients from Equation (A36); that is $q_1, q_2, q_3, q_4$.

*Appendix A.6. Heat Kernel Trace*

Recalling the heat equation from Equation (A31) and the spectral decomposition of the normalised Laplacian matrix from Equation (A30), the solution to the former (already in Equation (A32)) reads as:

$$
\mathbf{H}(t) = \exp\left\{-t\tilde{\mathbf{L}}\right\} = \mathbf{V}\exp\left\{-t\Lambda\right\}\mathbf{V}^T
$$
$$
= \sum_{i=1}^{|\mathcal{V}|} \exp\left\{-\lambda_i^{(\tilde{\mathbf{L}})}t\right\}\mathbf{v}_i\mathbf{v}_i^T. \tag{A37}
$$

The heat kernel trace is evaluated by taking the trace of $\mathbf{H}(t)$:

$$
HT(t) = \mathrm{Tr}\left\{\mathbf{H}(t)\right\} = \sum_{i=1}^{|\mathcal{V}|} \exp\left\{-\lambda_i^{(\tilde{\mathbf{L}})}t\right\}. \tag{A38}
$$

The sixth representation $\mathbf{X}^{(6)}$ sees each protein as a 10-length real-valued vector containing the heat kernel trace for $t = 1, 2, \ldots, 10$. These values for $t$ have been chosen by visual inspection: indeed, for $t > 10$ the heat kernel trace decay makes proteins undistinguishable one another.

*Appendix A.7. Size*

The seventh representation $\mathbf{X}^{(7)}$ sees each protein as a 4-length real-valued vector containing the number of nodes, the number of edges, the number of protein chains and the radius of gyration. Whilst the first two items are rather straightforward, the latter two items deserve some further comments. Proteins are composed by one or more amino-acids chains (linear polymers), thus the number of chains may impact on the overall protein size. Finally, the radius of gyration [126] is a measure of how-compact is the overall folded protein structure with respect to its centre of mass.

*Appendix A.8. Normalised Laplacian Spectral Density*

Recalling the spectral decomposition of the normalised Laplacian matrix from Equation (A30), let $\lambda^{(\tilde{\mathbf{L}})} = \left\{\lambda_1^{(\tilde{\mathbf{L}})}, \ldots, \lambda_{|\mathcal{V}|}^{(\tilde{\mathbf{L}})}\right\}$ be the normalised Laplacian spectrum (namely, the set of eigenvalues from $\tilde{\mathbf{L}}$). One of the interesting properties of the normalised Laplacian matrix is that its spectrum lies in range $[0, 2]$, regardless of the underlying graph [127]. The size of the spectrum, however, equals the number of nodes and therefore one cannot easily compare graphs having different sizes just by considering their respective spectra. In order to overcome this problem, following previous works [45,50], it is possible to estimate the (normalised Laplacian) spectral density using a kernel density estimator (also known as Parzen window [128]) equipped with the Gaussian kernel. The spectral density thus has the form:

$$
p(x) = \frac{1}{|\mathcal{V}|}\sum_{i=1}^{|\mathcal{V}|}\frac{1}{\sqrt{2\pi\sigma^2}}\cdot\exp\left\{\frac{-\left(x - \lambda_i^{(\tilde{\mathbf{L}})}\right)^2}{2\sigma^2}\right\}, \tag{A39}
$$

where $\sigma$ is the kernel bandwidth which determines the estimate resolution. Following [50], the Scott's rule [129] has been used in order to determine the proper bandwidth value, hence:

$$
\sigma = \frac{3.5\cdot\mathrm{std}\left\{\lambda^{(\tilde{\mathbf{L}})}\right\}}{|\lambda^{(\tilde{\mathbf{L}})}|^{1/3}}. \tag{A40}
$$

In this manner, the bandwidth scales in a graph-wise fashion by considering each graph's spectrum size (denominator) and its standard deviation (numerator). Let $\mathcal{G}_1$ and $\mathcal{G}_2$ be two graphs, their distance can be evaluated by considering the $\ell_2$ norm between their respective spectral densities $p_1(x)$ and $p_2(x)$:

$$d(\mathcal{G}_1, \mathcal{G}_2) = \int_0^2 (p_1(x) - p_2(x))^2 dx. \tag{A41}$$

The same operation can be carried in the discrete domain by extracting $n$ samples from $p(x)$ (equal for all graphs) and the latter collapses into the standard Euclidean distance.

The eighth representation $\mathbf{X}^{(8)}$ sees each protein as an 100-length real-valued vector containing $n = 100$ samples uniformly drawn from their respective normalised Laplacian spectral densities.

### Appendix B. Selected Prototypes

In the following, the sets of proteins elected as prototypes for each of the seven classification problems are shown. In order to shrink the output size, our a posteriori analysis has been carried only on proteins which have been selected in all of the five runs of the genetic algorithm (in order to remove 'spurious' representatives due to randomness in the optimisation procedure).

**Table A1.** Selected proteins in order to discriminate EC 1 (oxidoreductases) vs. all the rest.

| PDB ID | Notes/Description |
|---|---|
| 1KOF | Transferase |
| 1XFG | Transferase |
| 3E2R | Oxydoreductase |
| 4TS9 | Transferase |
| 1ZDM | Signalling Protein |
| 1MPG | Hydrolase |
| 1QQQ | Transferase |

**Table A2.** Selected proteins in order to discriminate EC 2 (transferases) vs. all the rest.

| PDB ID | Notes/Description |
|---|---|
| 3EDC | LAC repressor (signalling protein) |
| 1DKL | Hydrolase |
| 1JKJ | Ligase |
| 2DBI | Unknown function |
| 3UCS | Chaperone |
| 1LX7 | Transferase |
| 2GAR | Transferase |
| 3ILI | Transferase |
| 1S08 | Transferase |
| 4IXM | Hydrolase |
| 4XTJ | Isomerase |
| 1KW1 | Lyase |
| 1BDH | Transcription factor (DNA-binding) |
| 4PC3 | Elongation factor (RNA-binding) |
| 5G1L | Isomerase |

**Table A3.** Selected proteins in order to discriminate EC 3 (hydrolases) vs. all the rest.

| PDB ID | Notes/Description |
|---|---|
| 4RZS | Transcription factor (signalling protein) |
| 1ZDM | Signalling protein |
| 3I7R | Lyase |
| 1HW5 | Signalling protein |
| 1SO5 | Lyase |

**Table A4.** Selected proteins in order to discriminate EC 4 (lyases) vs. all the rest.

| PDB ID | Notes/Description |
| --- | --- |
| 2BWX | Hydrolase |
| 3UWM | Oxydoreductase |
| 2H71 | Electron transport |
| 1D7A | Lyase |
| 4DAP | DNA-binding |
| 1SPV | Structural genomics, unknown function |
| 1EXD | Ligase + RNA-binding |
| 1X83 | Isomerase |
| 3ILJ | Transferase |
| 2D4U | Signalling protein |
| 1JNW | Oxydoreductase |
| 1TRE | Oxydoreductase |
| 1ZPT | Oxydoreductase |
| 3LGU | Hydrolase |
| 1IB6 | Oxydoreductase |
| 3C0U | Structural genomics, unknown function |
| 5GT2 | Oxydoreductase |
| 2RN2 | Hydrolase |
| 4L4Z | Transcription regulator |
| 3CMR | Hydrolase |
| 1NQF | Transport protein |
| 1GPQ | Hydrolase |
| 4ODM | Isomerase + chaperone |
| 2NPG | Transport protein |
| 2UAG | Ligase |
| 1OVG | Transferase |
| 3AVU | Transferase |
| 1RBV | Hydrolase |
| 5AB1 | Cell adhesion |
| 1TMM | Transferase |
| 4NIY | Hydrolase |
| 4WR3 | Isomerase |

**Table A5.** Selected proteins in order to discriminate EC 5 (isomerases) vs. all the rest.

| PDB ID | Notes/Description |
| --- | --- |
| 4ITX | Lyase |
| 2BWW | Hydrolase |
| 5IU6 | Transferase |
| 1ODD | Gene regulatory |
| 5G5G | Oxydoreductase |
| 1G7X | Transferase |
| 2E0Y | Transferase |
| 2SCU | Ligase |
| 1HO4 | Hydrolase |
| 3RGM | Transport Protein |
| 1OAC | Oxydoreductase |
| 5MUC | Oxydoreductase |
| 3OGD | Hydrolase + DNA binding |
| 4K34 | Membrane protein |
| 1Q0L | Oxydoreductase |
| 1G58 | Isomerase |
| 5M3B | Transport protein |
| 2WOH | Oxydoreductase |
| 2PJP | Translation regulation (RNA-binding) |

**Table A6.** Selected proteins in order to discriminate EC 6 (ligases) vs. all the rest.

| PDB ID | Notes/Description |
| --- | --- |
| 2OLQ | Lyase |
| 1JDI | Isomerase |
| 4NIG | Oxydoreductase + DNA-binding |
| 5T03 | Transferase |
| 5FNN | Oxydoreductase |
| 2Z9D | Oxydoreductase |
| 2V3Z | Hydrolase |
| 4ARI | Ligase + RNA-binding |
| 3LBS | Transport protein |
| 4QGS | Oxydoreductase |
| 5B7F | Oxydoreductase |
| 2ABH | Transferase |

**Table A7.** Selected proteins in order to discriminate not-enzymes vs. all the rest.

| PDB ID | Notes/Description |
| --- | --- |
| 1SPA | Transferase |
| 2YH9 | Membrane protein |
| 1NQF | Transport protein |
| 1LDI | Transport protein |
| 1TIK | Hydrolase |
| 1MWI | Hydrolase + DNA-binding |
| 1GEW | Transferase |
| 5CKH | Hydrolase |
| 3ABQ | Lyase |
| 3B6M | Oxydoreductase |

## References


1. Bianchi, F.M.; Scardapane, S.; Livi, L.; Uncini, A.; Rizzi, A. An interpretable graph-based image classifier. In Proceedings of the 2014 International Joint Conference on Neural Networks (IJCNN), Beijing, China, 6–11 July 2014; pp. 2339–2346. [CrossRef]
2. Bianchi, F.M.; Scardapane, S.; Rizzi, A.; Uncini, A.; Sadeghian, A. Granular Computing Techniques for Classification and Semantic Characterization of Structured Data. *Cogn. Comput.* **2016**, *8*, 442–461. [CrossRef]
3. Del Vescovo, G.; Rizzi, A. Online Handwriting Recognition by the Symbolic Histograms Approach. In Proceedings of the 2007 IEEE International Conference on Granular Computing (GRC 2007), San Jose, CA, USA, 2–4 November 2007; pp. 686. [CrossRef]
4. Giuliani, A.; Filippi, S.; Bertolaso, M. Why network approach can promote a new way of thinking in biology. *Front. Genet.* **2014**, *5*, 83. [CrossRef]
5. Di Paola, L.; De Ruvo, M.; Paci, P.; Santoni, D.; Giuliani, A. Protein contact networks: an emerging paradigm in chemistry. *Chem. Rev.* **2012**, *113*, 1598–1613.[CrossRef]
6. Krishnan, A.; Zbilut, J.P.; Tomita, M.; Giuliani, A. Proteins as networks: usefulness of graph theory in protein science. *Curr. Protein Pept. Sci.* **2008**, *9*, 28–38. [CrossRef]
7. Jeong, H.; Tombor, B.; Albert, R.; Oltvai, Z.N.; Barabási, A.L. The large-scale organization of metabolic networks. *Nature* **2000**, *407*, 651. [CrossRef]
8. Di Paola, L.; Giuliani, A. *Protein–Protein Interactions: The Structural Foundation of Life Complexity*; American Cancer Society: Atlanta, GA, USA, 2017; pp. 1–12. [CrossRef]
9. Wuchty, S. Scale-Free Behavior in Protein Domain Networks. *Mol. Biol. Evol.* **2001**, *18*, 1694–1702. [CrossRef] [PubMed]
10. Martino, A.; Giuliani, A.; Rizzi, A. Granular Computing Techniques for Bioinformatics Pattern Recognition Problems in Non-metric Spaces. In *Computational Intelligence for Pattern Recognition*; Pedrycz, W., Chen, S.M., Eds.; Springer International Publishing: Cham, Switzerland, 2018; pp. 53–81.

11. Ieracitano, C.; Mammone, N.; Hussain, A.; Morabito, F.C. A novel multi-modal machine learning based approach for automatic classification of EEG recordings in dementia. *Neural Netw.* **2020**, *123*, 176–190. [CrossRef]

12. Cinti, A.; Bianchi, F.M.; Martino, A.; Rizzi, A. A Novel Algorithm for Online Inexact String Matching and its FPGA Implementation. *Cogn. Comput.* **2019**. [CrossRef]

13. Bunke, H. On a relation between graph edit distance and maximum common subgraph. *Pattern Recognit. Lett.* **1997**, *18*, 689–694. [CrossRef]

14. Bargiela, A.; Pedrycz, W. *Granular Computing: An Introduction*; Kluwer Academic Publishers: Boston, MA, USA, 2003.

15. Pedrycz, W. Granular computing: an introduction. In Proceedings of the 9th IFSA World Congress and 20th NAFIPS International Conference, Vancouver, BC, Canada, 25–28 July 2001; Volume 3, pp. 1349–1354. [CrossRef]

16. Bargiela, A.; Pedrycz, W. Granular Computing. In *Handbook on Computational Intelligence*; World Scientific Publishers: Singapore, 2016; Chapter 2, pp. 43–66. [CrossRef]

17. Singh, P.K. Similar Vague Concepts Selection Using Their Euclidean Distance at Different Granulation. *Cogn. Comput.* **2018**, *10*, 228–241. [CrossRef]

18. Lin, T.Y.; Yao, Y.Y.; Zadeh, L.A. *Data Mining, Rough Sets and Granular Computing*; Springer: Berlin/Heidelberg, Germany, 2013; Volume 95.

19. Bianchi, F.M.; Livi, L.; Rizzi, A.; Sadeghian, A. A Granular Computing approach to the design of optimized graph classification systems. *Soft Comput.* **2014**, *18*, 393–412. [CrossRef]

20. Rizzi, A.; Del Vescovo, G.; Livi, L.; Frattale Mascioli, F.M. A new Granular Computing approach for sequences representation and classification. In Proceedings of the 2012 International Joint Conference on Neural Networks (IJCNN), Brisbane, Australia, 10–15 June 2012; pp. 1–8. [CrossRef]

21. Del Vescovo, G.; Rizzi, A. Automatic Classification of Graphs by Symbolic Histograms. In Proceedings of the 2007 IEEE International Conference on Granular Computing (GRC 2007), San Jose, CA, USA, 2–4 November 2007; p. 410. [CrossRef]

22. Baldini, L.; Martino, A.; Rizzi, A. Stochastic Information Granules Extraction for Graph Embedding and Classification. In Proceedings of the 11th International Joint Conference on Computational Intelligence—Volume 1: NCTA, (IJCCI 2019), Vienna, Austria, 17–19 September 2019; pp. 391–402. [CrossRef]

23. Martino, A.; Giuliani, A.; Todde, V.; Bizzarri, M.; Rizzi, A. Metabolic networks classification and knowledge discovery by information granulation. *Comput. Biol. Chem.* **2020**, *84*, 107187. [CrossRef] [PubMed]

24. Martino, A.; Frattale Mascioli, F.M.; Rizzi, A. On the Optimization of Embedding Spaces via Information Granulation for Pattern Recognition. In Proceedings of the 2020 International Joint Conference on Neural Networks (IJCNN), Glasgow, UK, 19–24 July 2020.

25. Martino, A.; Giuliani, A.; Rizzi, A. (Hyper)Graph Embedding and Classification via Simplicial Complexes. *Algorithms* **2019**, *12*. [CrossRef]

26. Pękalska, E.; Duin, R.P. *The Dissimilarity Representation for Pattern Recognition: Foundations and Applications*; World Scientific: London, UK, 2005.

27. Pękalska, E.; Duin, R.P.; Paclík, P. Prototype selection for dissimilarity-based classifiers. *Pattern Recognit.* **2006**, *39*, 189–208. [CrossRef]

28. Duin, R.P.; Pękalska, E. The dissimilarity space: Bridging structural and statistical pattern recognition. *Pattern Recognit. Lett.* **2012**, *33*, 826–832. [CrossRef]

29. Shawe-Taylor, J.; Cristianini, N. *Kernel Methods for Pattern Analysis*; Cambridge University Press: Cambridge, UK, 2004.

30. Schölkopf, B.; Smola, A.J. *Learning with Kernels: Support Vector Machines, Regularization, Optimization, and Beyond*; MIT Press: Cambridge, MA, USA, 2002.

31. Cristianini, N.; Shawe-Taylor, J. *An Introduction to Support Vector Machines and Other Kernel-Based Learning Methods*; Cambridge University Press: Cambridge, UK, 2000.

32. Mercer, J. Functions of positive and negative type, and their connection with the theory of integral equations. *R. Soc.* **1909**, *209*, 415–446.

33. Cover, T.M. Geometrical and statistical properties of systems of linear inequalities with applications in pattern recognition. *IEEE Trans. Electron. Comput.* **1965**, *EC-14*, 326–334. [CrossRef]

34. Boser, B.E.; Guyon, I.; Vapnik, V. A training algorithm for optimal margin classifiers. In Proceedings of the Fifth Annual Workshop on Computational Learning Theory, Pittsburgh, PA, USA, 27–29 July 1992; pp. 144–152. [CrossRef]

35. Cortes, C.; Vapnik, V. Support-vector networks. *Mach. Learn.* **1995**, *20*, 273–297. [CrossRef]

36. Haasdonk, B. Feature space interpretation of SVMs with indefinite kernels. *IEEE Trans. Pattern Anal. Mach. Intell.* **2005**, *27*, 482–492. [CrossRef]

37. Laub, J.; Müller, K.R. Feature Discovery in Non-Metric Pairwise Data. *J. Mach. Learn. Res.* **2004**, *5*, 801–818.

38. Ong, C.S.; Mary, X.; Canu, S.; Smola, A.J. Learning with Non-positive Kernels. In Proceedings of the Twenty-first International Conference on Machine Learning, Banff, AL, Canada, 4–8 July 2004. [CrossRef]

39. Chen, Y.; Gupta, M.R.; Recht, B. Learning kernels from indefinite similarities. In Proceedings of the 26th Annual International Conference on Machine Learning, Montreal, QC, Canada, 14–18 June 2009; pp. 145–152. [CrossRef]

40. Chen, Y.; Garcia, E.K.; Gupta, M.R.; Rahimi, A.; Cazzanti, L. Similarity-based classification: Concepts and algorithms. *J. Mach. Learn. Res.* **2009**, *10*, 747–776.

41. Pauling, L.; Corey, R.B.; Branson, H.R. The structure of proteins: two hydrogen-bonded helical configurations of the polypeptide chain. *Proc. Natl. Acad. Sci. USA* **1951**, *37*, 205–211. [CrossRef] [PubMed]

42. Livi, L.; Giuliani, A.; Sadeghian, A. Characterization of graphs for protein structure modeling and recognition of solubility. *Curr. Bioinform.* **2016**, *11*, 106–114. [CrossRef]

43. Livi, L.; Giuliani, A.; Rizzi, A. Toward a multilevel representation of protein molecules: Comparative approaches to the aggregation/folding propensity problem. *Inf. Sci.* **2016**, *326*, 134–145. [CrossRef]

44. De Santis, E.; Martino, A.; Rizzi, A.; Frattale Mascioli, F.M. Dissimilarity Space Representations and Automatic Feature Selection for Protein Function Prediction. In Proceedings of the 2018 International Joint Conference on Neural Networks (IJCNN), Rio de Janeiro, Brazil, 8–13 July 2018; pp. 1–8. [CrossRef]

45. Martino, A.; Maiorino, E.; Giuliani, A.; Giampieri, M.; Rizzi, A. Supervised Approaches for Function Prediction of Proteins Contact Networks from Topological Structure Information. In *Image Analysis, Proceedings of the 20th Scandinavian Conference, SCIA 2017, Tromsø, Norway, 12–14 June 2017*; Sharma, P., Bianchi, F.M., Eds.; Springer International Publishing: Cham, Switzerland, 2017; pp. 285–296.

46. Martino, A.; Rizzi, A.; Frattale Mascioli, F.M. Supervised Approaches for Protein Function Prediction by Topological Data Analysis. In Proceedings of the 2018 International Joint Conference on Neural Networks (IJCNN), Rio de Janeiro, Brazil, 8–13 July 2018; pp. 1–8. [CrossRef]

47. Livi, L.; Maiorino, E.; Giuliani, A.; Rizzi, A.; Sadeghian, A. A generative model for protein contact networks. *J. Biomol. Struct. Dyn.* **2016**, *34*, 1441–1454. [CrossRef]

48. Livi, L.; Maiorino, E.; Pinna, A.; Sadeghian, A.; Rizzi, A.; Giuliani, A. Analysis of heat kernel highlights the strongly modular and heat-preserving structure of proteins. *Phys. A Stat. Mech. Its Appl.* **2016**, *441*, 199–214. [CrossRef]

49. Maiorino, E.; Livi, L.; Giuliani, A.; Sadeghian, A.; Rizzi, A. Multifractal characterization of protein contact networks. *Phys. A Stat. Mech. Its Appl.* **2015**, *428*, 302–313. [CrossRef]

50. Maiorino, E.; Rizzi, A.; Sadeghian, A.; Giuliani, A. Spectral reconstruction of protein contact networks. *Phys. A Stat. Mech. Its Appl.* **2017**, *471*, 804–817. [CrossRef]

51. Webb, E.C. *Enzyme Nomenclature 1992. Recommendations of the Nomenclature Committee of the International Union of Biochemistry and Molecular Biology on the Nomenclature and Classification of Enzymes*, 6th ed.; Academic Press: Cambridge, MA, USA, 1992.

52. Livi, L.; Rizzi, A.; Sadeghian, A. Optimized dissimilarity space embedding for labeled graphs. *Inf. Sci.* **2014**, *266*, 47–64. [CrossRef]

53. Bishop, C.M. *Pattern Recognition and Machine Learning*; Springer: New York, NY, USA, 2006.

54. Vapnik, V. *Statistical Learning Theory*; Wiley: New York, NY, USA, 1998.
55. Sonnenburg, S.; Rätsch, G.; Schäfer, C.; Schölkopf, B. Large scale multiple kernel learning. *J. Mach. Learn. Res.* **2006**, *7*, 1531–1565.
56. Lewis, D.P.; Jebara, T.; Noble, W.S. Nonstationary kernel combination. In Proceedings of the 23rd International Conference on Machine Learning, Pittsburgh, PA, USA, 25–29 June 2006; pp. 553–560. [CrossRef]
57. Lanckriet, G.R.; Cristianini, N.; Bartlett, P.; Ghaoui, L.E.; Jordan, M.I. Learning the kernel matrix with semidefinite programming. *J. Mach. Learn. Res.* **2004**, *5*, 27–72.
58. Gönen, M.; Alpaydin, E. Localized multiple kernel learning. In Proceedings of the 25th International Conference on Machine Learning, Helsinki, Finland, 5–9 July 2008; pp. 352–359. [CrossRef]
59. Cortes, C.; Mohri, M.; Rostamizadeh, A. Learning non-linear combinations of kernels. In *Advances in Neural Information Processing Systems 22, Proceedings of the 23rd Annual Conference on Neural Information Processing Systems 2009, Vancouver, BC, Canada, 7–10 December 2009*; Curran Associates Inc.: Nice, France, 2009; pp. 396–404.
60. Bach, F.R.; Lanckriet, G.R.; Jordan, M.I. Multiple kernel learning, conic duality, and the SMO algorithm. In Proceedings of the Twenty-first International Conference on Machine Learning, Banff, AL, Canada, 4–8 July 2004; p. 6. [CrossRef]
61. Hu, M.; Chen, Y.; Kwok, J.T.Y. Building sparse multiple-kernel SVM classifiers. *IEEE Trans. Neural Netw.* **2009**, *20*, 827–839. [CrossRef]
62. Gönen, M.; Alpaydın, E. Multiple kernel learning algorithms. *J. Mach. Learn. Res.* **2011**, *12*, 2211–2268.
63. Horn, R.A.; Johnson, C.R. *Matrix Analysis*; Cambridge University Press: Cambridge, UK, 1985.
64. Schölkopf, B.; Smola, A.J.; Williamson, R.C.; Bartlett, P.L. New support vector algorithms. *Neural Comput.* **2000**, *12*, 1207–1245. [CrossRef] [PubMed]
65. Goldberg, D.E. *Genetic Algorithms in Search, Optimization and Machine Learning*, 1st ed.; Addison-Wesley Longman Publishing Co., Inc.: Boston, MA, USA, 1989.
66. Rojas, S.A.; Fernandez-Reyes, D. Adapting multiple kernel parameters for support vector machines using genetic algorithms. In Proceedings of the 2005 IEEE Congress on Evolutionary Computation, Scotland, UK, 2–5 September 2005; Volume 1, pp. 626–631. [CrossRef]
67. Phienthrakul, T.; Kijsirikul, B. Evolving Hyperparameters of Support Vector Machines Based on Multi-Scale RBF Kernels. In Proceedings of the International Conference on Intelligent Information Processing, Adelaide, Australia, 20–23 September 2006; pp. 269–278. [CrossRef]
68. Beyer, H.G.; Schwefel, H.P. Evolution strategies—A comprehensive introduction. *Nat. Comput.* **2002**, *1*, 3–52. [CrossRef]
69. Youden, W.J. Index for rating diagnostic tests. *Cancer* **1950**, *3*, 32–35. [CrossRef]
70. Powers, D.M.W. Evaluation: From precision, recall and f-measure to roc., informedness, markedness & correlation. *J. Mach. Learn. Technol.* **2011**, *2*, 37–63.
71. Cokelaer, T.; Pultz, D.; Harder, L.M.; Serra-Musach, J.; Saez-Rodriguez, J. BioServices: A common Python package to access biological Web Services programmatically. *Bioinformatics* **2013**, *29*, 3241–3242. [CrossRef]
72. The UniProt Consortium. UniProt: The universal protein knowledgebase. *Nucleic Acids Res.* **2017**, *45*, D158–D169. [CrossRef]
73. Berman, H.M.; Westbrook, J.; Feng, Z.; Gilliland, G.; Bhat, T.N.; Weissig, H.; Shindyalov, I.N.; Bourne, P.E. The Protein Data Bank. *Nucleic Acids Res.* **2000**, *28*, 235–242. [CrossRef]
74. Cock, P.J.A.; Antao, T.; Chang, J.T.; Chapman, B.A.; Cox, C.J.; Dalke, A.; Friedberg, I.; Hamelryck, T.; Kauff, F.; Wilczynski, B.; et al. Biopython: freely available Python tools for computational molecular biology and bioinformatics. *Bioinformatics* **2009**, *25*, 1422–1423.. [CrossRef]
75. Raschka, S. BioPandas: Working with molecular structures in pandas DataFrames. *J. Open Source Softw.* **2017**, *2*. [CrossRef]
76. Hagberg, A.; Swart, P.; Schult, D. Exploring network structure, dynamics, and function using NetworkX. In Proceedings of the 7th Python in Science Conference (SciPy), Pasadena, CA, USA, 19–24 August 2008; pp. 11–15.
77. Virtanen, P.; Gommers, R.; Oliphant, T.E.; Haberland, M.; Reddy, T.; Cournapeau, D.; Burovski, E.; Peterson, P.; Weckesser, W.; Bright, J.; et al. SciPy 1.0: fundamental algorithms for scientific computing in Python. *Nat. Methods* **2020**, *17*, 261–272. [CrossRef] [PubMed]

78. Oliphant, T.E. Python for scientific computing. *Comput. Sci. Eng.* **2007**, *9*. [CrossRef]

79. Chang, C.C.; Lin, C.J. LIBSVM: A library for support vector machines. *ACM Trans. Intell. Syst. Technol.* **2011**, *2*, 27:1–27:27. [CrossRef]

80. Fawcett, T. An Introduction to ROC Analysis. *Pattern Recognit. Lett.* **2006**, *27*, 861–874. [CrossRef]

81. De Santis, E.; Livi, L.; Sadeghian, A.; Rizzi, A. Modeling and recognition of smart grid faults by a combined approach of dissimilarity learning and one-class classification. *Neurocomputing* **2015**, *170*, 368–383. [CrossRef]

82. De Santis, E.; Rizzi, A.; Sadeghian, A. A cluster-based dissimilarity learning approach for localized fault classification in Smart Grids. *Swarm Evol. Comput.* **2018**, *39*, 267–278. . [CrossRef]

83. De Santis, E.; Paschero, M.; Rizzi, A.; Frattale Mascioli, F.M. Evolutionary Optimization of an Affine Model for Vulnerability Characterization in Smart Grids. In Proceedings of the 2018 International Joint Conference on Neural Networks (IJCNN), Rio de Janeiro, Brazil, 8–13 July 2018; pp. 1–8. [CrossRef]

84. Khan, S.S.; Madden, M.G. A Survey of Recent Trends in One Class Classification. In *Artificial Intelligence and Cognitive Science*; Coyle, L., Freyne, J., Eds.; Springer: Berlin/Heidelberg, Germany, 2010; Volume 6206, pp. 188–197. [CrossRef]

85. Pimentel, M.A.F.; Clifton, D.A.; Clifton, L.; Tarassenko, L. A review of novelty detection. *Signal Process.* **2014**, *99*, 215–249. [CrossRef]

86. Martino, A.; Rizzi, A.; Frattale Mascioli, F.M. Efficient Approaches for Solving the Large-Scale k-medoids Problem. In Proceedings of the 9th International Joint Conference on Computational Intelligence—Volume 1, Madeira, Portugal, 1–3 November2017; pp. 338–347. [CrossRef]

87. Martino, A.; Rizzi, A.; Frattale Mascioli, F.M. Distance Matrix Pre-Caching and Distributed Computation of Internal Validation Indices in k-medoids Clustering. In Proceedings of the 2018 International Joint Conference on Neural Networks (IJCNN), Rio de Janeiro, Brazil, 8–13 July 2018; pp. 1–8. [CrossRef]

88. Martino, A.; Rizzi, A.; Frattale Mascioli, F.M. Efficient Approaches for Solving the Large-Scale k-Medoids Problem: Towards Structured Data. In *Computational Intelligence, Proceedings of the 9th International Joint Conference, IJCCI 2017 Funchal-Madeira, Portugal, 1–3 November 2017*; Sabourin, C., Merelo, J.J., Madani, K., Warwick, K., Eds.; Springer International Publishing: Cham, Switzerland, 2019; pp. 199–219. [CrossRef]

89. Mendel, J.M. Fuzzy logic systems for engineering: A tutorial. *Proc. IEEE* **1995**, *83*, 345–377. [CrossRef]

90. Martino, A.; De Santis, E.; Baldini, L.; Rizzi, A. Calibration Techniques for Binary Classification Problems: A Comparative Analysis. In Proceedings of the 11th International Joint Conference on Computational Intelligence—Volume 1, Vienna, Austria, 17–19 September 2019; pp. 487–495. [CrossRef]

91. Martino, A. Pattern Recognition Techniques for Modelling Complex Systems in Non-Metric Domains. Ph.D. Thesis, University of Rome "La Sapienza", Rome, Italy, 2020.

92. Branden, C.I.; Tooze, J. *Introduction to Protein Structure*; Garland Publishing Inc.: New York, NY, USA, 1991.

93. Giuliani, A.; Benigni, R.; Zbilut, J.P.; Webber, C.L.; Sirabella, P.; Colosimo, A. Nonlinear Signal Analysis Methods in the Elucidation of Protein Sequence-Structure Relationships. *Chem. Rev.* **2002**, *102*, 1471–1492. [CrossRef]

94. Di Paola, L.; Giuliani, A. Protein contact network topology: a natural language for allostery. *Curr. Opin. Struct. Biol.* **2015**, *31*, 43–48. [CrossRef] [PubMed]

95. Devore, J.L.; Peck, R. *Statistics: The Exploration and Analysis of Data*, 4th ed.; Brooks/Cole: Pacific Grove, CA, USA, 2001.

96. Bartz, A.E. *Basic Statistical Concepts*; Macmillan Pub Co.: New York, NY, USA, 1988.

97. Guarnera, E.; Berezovsky, I.N. Allosteric sites: remote control in regulation of protein activity. *Curr. Opin. Struct. Biol.* **2016**, *37*, 1–8. [CrossRef]

98. Negre, C.F.A.; Morzan, U.N.; Hendrickson, H.P.; Pal, R.; Lisi, G.P.; Loria, J.P.; Rivalta, I.; Ho, J.; Batista, V.S. Eigenvector centrality for characterization of protein allosteric pathways. Proc. Natl. Acad. Sci. USA **2018**. *115*, E12201–E12208. [CrossRef]

99. Carlsson, G. Topology and data. *Bull. Am. Math. Soc.* **2009**, *46*, 255–308. [CrossRef]

100. Wasserman, L. Topological Data Analysis. *Annu. Rev. Stat. Its Appl.* **2018**, *5*, 501–532. [CrossRef]

101. Estrada, E.; Rodriguez-Velazquez, J.A. Complex networks as hypergraphs. *arXiv* **2005**, arXiv:physics/0505137.

102. Horak, D.; Maletić, S.; Rajković, M. Persistent homology of complex networks. *J. Stat. Mech. Theory Exp.* **2009**, *2009*, p03034. [CrossRef]

103. Barbarossa, S.; Sardellitti, S. Topological Signal Processing over Simplicial Complexes. *IEEE Trans. Signal Process.* **2020**. [CrossRef]

104. Ghrist, R.W. *Elementary Applied Topology*; Createspace: Seattle, WA, USA, 2014.

105. Hausmann, J.C. On the Vietoris-Rips complexes and a cohomology theory for metric spaces. *Ann. Math. Stud.* **1995**, *138*, 175–188.

106. Zomorodian, A.; Carlsson, G. Computing persistent homology. *Discret. Comput. Geom.* **2005**, *33*, 249–274. [CrossRef]

107. Zomorodian, A. Fast construction of the Vietoris-Rips complex. *Comput. Graph.* **2010**, *34*, 263–271. [CrossRef]

108. Munkres, J.R. *Elements of Algebraic Topology*; Addison-Wesley: Cambridge, MA, USA, 1984.

109. Artin, M. *Algebra*; Prentice Hall: Englewood Cliffs, NJ, USA, 1991.

110. Newman, M.E.J. *Networks: An Introduction*; Oxford University Press: New York, NY, USA, 2010.

111. Katz, L. A new status index derived from sociometric analysis. *Psychometrika* **1953**, *18*, 39–43. [CrossRef]

112. Wasserman, S.; Faust, K. *Social Network Analysis: Methods and Applications*; Cambridge University Press: Cambridge, UK, 1994.

113. Brandes, U. On variants of shortest-path betweenness centrality and their generic computation. *Soc. Netw.* **2008**, *30*, 136–145. [CrossRef]

114. Goh, K.I.; Kahng, B.; Kim, D. Universal behavior of load distribution in scale-free networks. *Phys. Rev. Lett.* **2001**, *87*, 278701. [CrossRef]

115. Estrada, E.; Rodriguez-Velazquez, J.A. Subgraph centrality in complex networks. *Phys. Rev. E* **2005**, *71*, 056103. [CrossRef] [PubMed]

116. Estrada, E. Characterization of 3D molecular structure. *Chem. Phys. Lett.* **2000**, *319*, 713–718. [CrossRef]

117. Boldi, P.; Vigna, S. Axioms for centrality. *Internet Math.* **2014**, *10*, 222–262. [CrossRef]

118. Mones, E.; Vicsek, L.; Vicsek, T. Hierarchy measure for complex networks. *PLoS ONE* **2012**, *7*, e33799. [CrossRef] [PubMed]

119. Saramäki, J.; Kivelä, M.; Onnela, J.P.; Kaski, K.; Kertesz, J. Generalizations of the clustering coefficient to weighted complex networks. *Phys. Rev. E* **2007**, *75*, 027105. [CrossRef] [PubMed]

120. Barrat, A.; Barthélemy, M.; Pastor-Satorras, R.; Vespignani, A. The architecture of complex weighted networks. *Proc. Natl. Acad. Sci. USA* **2004**, *101*, 3747–3752. [CrossRef]

121. Gutman, I.; Zhou, B. Laplacian energy of a graph. *Linear Algebra Appl.* **2006**, *414*, 29–37. [CrossRef]

122. Guimera, R.; Amaral, L.A.N. Functional cartography of complex metabolic networks. *Nature* **2005**, *433*, 895. [CrossRef]

123. Kirkpatrick, S.; Gelatt, C.D.; Vecchi, M.P. Optimization by simulated annealing. *Science* **1983**, *220*, 671–680. [CrossRef] [PubMed]

124. Xiao, B.; Hancock, E.R.; Wilson, R.C. Graph characteristics from the heat kernel trace. *Pattern Recognit.* **2009**, *42*, 2589–2606. [CrossRef]

125. Xiao, B.; Hancock, E.R. Graph clustering using heat content invariants. In Proceedings of the Iberian Conference on Pattern Recognition and Image Analysis, Estoril, Portugal, 7–9 June 2005; pp. 123–130. [CrossRef]

126. Lobanov, M.Y.; Bogatyreva, N.; Galzitskaya, O. Radius of gyration as an indicator of protein structure compactness. *Mol. Biol.* **2008**, *42*, 623–628. [CrossRef]

127. Butler, S. Algebraic aspects of the normalized Laplacian. In *Recent Trends in Combinatorics*; Springer: Berlin/Heidelberg, Germany, 2016; pp. 295–315. [CrossRef]

128. Parzen, E. On estimation of a probability density function and mode. *Ann. Math. Stat.* **1962**, *33*, 1065–1076. [CrossRef]

129. Scott, D.W. On optimal and data-based histograms. *Biometrika* **1979**, *66*, 605–610. [CrossRef]

*Article*

# Cross-Domain Recommendation Based on Sentiment Analysis and Latent Feature Mapping

Yongpeng Wang [1], Hong Yu [1,*], Guoyin Wang [1] and Yongfang Xie [2]

[1] Chongqing Key Laboratory of Computational Intelligence, Chongqing University of Posts and Telecommunications, Chongqing 400065, China; wangyp946@foxmail.com (Y.W.); wanggy@cqupt.edu.cn (G.W.)

[2] School of Information Science and Engineering, Central South University, Changsha 410083, China; yfxie@csu.edu.cn

* Correspondence: yuhong@cqupt.edu.cn; Tel.: +86-23-6246-1432

Received: 25 February 2020; Accepted: 17 April 2020; Published: 20 April 2020

**Abstract:** Cross-domain recommendation is a promising solution in recommendation systems by using relatively rich information from the source domain to improve the recommendation accuracy of the target domain. Most of the existing methods consider the rating information of users in different domains, the label information of users and items and the review information of users on items. However, they do not effectively use the latent sentiment information to find the accurate mapping of latent features in reviews between domains. User reviews usually include user's subjective views, which can reflect the user's preferences and sentiment tendencies to various attributes of the items. Therefore, in order to solve the cold-start problem in the recommendation process, this paper proposes a cross-domain recommendation algorithm (CDR-SAFM) based on sentiment analysis and latent feature mapping by combining the sentiment information implicit in user reviews in different domains. Different from previous sentiment research, this paper divides sentiment into three categories based on three-way decision ideas—namely, positive, negative and neutral—by conducting sentiment analysis on user review information. Furthermore, the Latent Dirichlet Allocation (LDA) is used to model the user's semantic orientation to generate the latent sentiment review features. Moreover, the Multilayer Perceptron (MLP) is used to obtain the cross domain non-linear mapping function to transfer the user's sentiment review features. Finally, this paper proves the effectiveness of the proposed CDR-SAFM framework by comparing it with existing recommendation algorithms in a cross-domain scenario on the Amazon dataset.

**Keywords:** cross-domain recommendation; sentiment analysis; latent sentiment review feature; non-linear mapping

---

## 1. Introduction

A recommendation system helps a user discover the information he/she wants such as products and content from the massive information produced by the Internet. Recommendation systems are primarily used in commercial applications. A recommendation system helps users find valuable information as the interested information can be recommended to users. This is a win-win situation for both consumers and manufacturers. A good recommendation system can not only accurately detect the user's behavior, but also help users find the potential information they are interested in.

There are a lots of achievements in recommendation systems, which try to enhance the accuracy, diversity and novelty of recommendation. For example, a collaborative filtering-based recommendation algorithm [1] is one of the most popular and widely used algorithms and it can be divided into two categories—user-based recommendations and item-based recommendations. Among model-based collaborative filtering methods, matrix factorization [2] technology is popular because it is extremely scalable and easy to implement. The accuracy of the matrix factorization recommendation to a great extent depends on the rating matrix. However, in real life, with the rapid growth of users and items, the rating matrix is very sparse, which has a great impact on the recommendation of new users and new items. Thus, the cold-start and data sparsity problems in the recommendation process arise.

Recently, more and more researchers [3] have researched cross-domain recommendation by introducing the concepts of source domain and target domain in order to solve the problem of data sparsity and cold-start in the single-domain recommendation process. The purpose of cross-domain based recommendation is to use the richer information in multiple domains than in a single-domain and to transfer the knowledge between different domains effectively based on the idea of transfer learning. One of the key assumptions that cross-domain recommendation can work is that there exist consistency or correlation between users' interest preferences or item features between domains. This hypothesis is also supported by some research work. The cross-domain recommendation utilizes the consistency or correlation between domains, such as the intersection of users and items, the similarity between user interests, the similarity between item features, and the relationship between latent factors, and so forth, to make up for the problem of insufficient information in the target domain.

However, the existing cross-domain recommendation methods are only based on the sharing and transfer of knowledge in the text information such as rating, tag or review, and ignore the latent sentiment information in the review. User reviews usually include user's subjective views, which can reflect the user's preferences and sentiment tendencies to various attributes of the item. Fully mining and using the implied sentiment information is helpful to solve the cold-start problems and data sparsity in the process of cross-domain recommendation. The existing cross-domain recommendation algorithms using user reviews do not make full use of the sentiment information in these reviews. They mixed positive sentiment, neutral sentiment and negative sentiment together to realize knowledge transfer, which will weaken or even lose some sentiment information of users, especially negative sentiment. Therefore, it has a great significance to make cross-domain recommendations by combining the user's sentimental features implicit in the review information.

To address the problem of cold-start in the process of recommendation, we propose a cross-domain recommendation algorithm based on sentiment analysis and latent feature mapping (shorted by CDR-SAFM ) in this paper, by combining with the implicit sentiment information in user reviews. First, this paper divides the sentiment of user review information into three categories based on the theory of three-way decisions [4,5], namely positive, negative and neutral. Then, the Latent Dirichlet Allocation method is used to model users' semantic orientation to generate users' latent sentiment review features. Finally, the Multi-Layer Perceptron method is used to obtain the cross-domain non-linear mapping function to transfer the user's sentiment review features. The main contributions are concluded as follows:

- A novel algorithm for cross-domain recommendation named CDR-SAFM is proposed for cold-start users in target domain. It employs sentiment analysis and latent feature mapping and it can transfer latent sentiment review feature from source domain to target domain and make recommendation for cold-start users in target domain.
- Basing on the idea of three-way decisions, we take into account neutral sentiment to generate the latent sentiment review feature from both the source and target domains, which can affect ratings in the two domains.
- The LDA models is used to generate user latent review features. When generating features, we consider the sentiment information from reviews, and generate the user sentiment review features in different domains.
- The Multi-layer Perceptron is employed to accurately map the latent sentiment review feature from the source domain to the target domain, which improves recommendation accuracy.

The rest of the paper is organized as follows. Section 2 introduces the related work. In Section 3, the preliminary work is reviewed. A detailed description of our algorithm is stated in Section 4.

Subsequently, in Section 5, we discuss experimental settings and the comparative results. Finally, we conclude the paper in Section 6.

## 2. Related Work

Aiming at the cold-start problem caused by too sparse rating matrices in different domains, some scholars [6–8] tried to use the common users in the two domains as bridges, using the data in the auxiliary domain to solve the cold start problem in the target domain by feature mapping. Pan and Yang [6] learned a transformation matrix based on the feature representation of common users in the two domains, and realized the mapping of features between different domains. The transformation matrix implements a linear mapping, and the mapping relationship of features in different domains may be non-linear. Then, Xin et al. [7] modeled a non-linear feature mapping function through a multi-layer perceptron, and obtained a better mapping effect than the transformation matrix. Wei et al. [8] implemented the recommendation of e-commerce website products to social networking sites through common users and the recommendations for cold-start users are made through the mapping of user features.

Moreover, in order to make full use of the hidden user and item relationships between domains, some works [9–11] proposed combining them with transfer learning. For example, Jiang et al. [9] connected different domains with each other through social networks, forming a hybrid graph with a social network-centric star structure, and used a random walk algorithm to predict the user and item relationship. In addition, some scholars [12–14] try to analyze the behavior of users in multiple social web platforms. The semantic relationships of items in each domain are also used for knowledge transfer. Yang et al. [15] introduced the tag system into the cross-domain recommendation, and successfully implemented the cold-start problem of recommending, that is, to recommend movies to the new user based on the blog posts on Weibo. The basic idea of the work is to use the semantic relationship between tags on user blog posts and movie tags as a bridge to associate users with movies, and then it predicts user preferences based on graph models. Shi et al. [16] proposed a cross-domain recommendation algorithm for collaborative filtering with fused labels. The model first uses the rich label information in the labeling system that the user has labeled the item to calculate the user-user similarity matrix and the item-item similarity matrix. Then it uses the information as a smoothing term to improve the probability matrix decomposition model PMF [17]. The trained user and item feature vectors can also satisfy the similarity relationship between users and items on the basis of minimizing the error between the predicted rating and the actual rating. Kumar et al. [18] used the Latent Direchlet Allocation (LDA) topic model [19] to model the user's tagging information to build a user feature topic sharing space shared by different domains and then, based on this space, to find users with similar preferences in different domains and implement cross-domain recommendation.

Furthermore, Song et al. [20] believed that, compared with the rating information, the user review information cannot only express the user's preferences for the item, but also cover other user interest preferences. Therefore, they proposed a joint tensor decomposition model based on review information for cross-domain recommendation. The model is trained by using the AIRS method rating information proposed in Reference [21], analyzing user reviews from multiple different angles, and obtaining the user's rating and degree of interest at each angle. This is achieved by sharing feature vectors in the source and target domain Knowledge transfer. Hu et al. [22] aimed at the problem of data sparseness and integrated auxiliary information such as product reviews and news headlines to form a hybrid filtering method transferring knowledge from other source domains, such as improving movie recommendations with knowledge in the book domain, and thus forming a transfer of learning methods.

## 3. Preliminary Work

### 3.1. Sentiment Analysis

Text sentiment analysis refers to the process of analyzing, processing, and extracting subjective text with emotion using natural language processing and text mining technology [23]. The sentiment analysis task can be divided into chapter level, sentence level, word or phrase level according to its analysis granularity; according to its processing text category, it can be divided into sentiment analysis based on product reviews and sentiment analysis based on news reviews; according to its research tasks types can be divided into sub-problems such as sentiment classification, sentiment retrieval and sentiment extraction.

Sentiment classification refers to the identification of subjective text in a given text, whether it is positive or negative, which is the most researched in the field of sentiment analysis. There are usually a lot of subjective texts and objective texts in network texts. Objectivity text is an objective description of things, without emotion color and emotional tendency, and subjective text is the author's views or ideas on various things, with emotional tendencies such as the author's likes and dislikes. The object of sentiment classification is subjective text with emotional tendency, so emotion classification must first be subjective and objective classification of text. The subjective and objective classification of texts is mainly based on the recognition of sentiment words. Using different text feature representation methods and classifiers for classification, subjective and objective classification of web texts in advance can improve the speed and accuracy of sentiment classification. Looking at the current research work on subjective text sentiment analysis, the main research ideas are divided into semantic-based sentiment dictionary methods and machine learning-based methods.

In the semantic-based sentiment dictionary method, the construction of sentiment dictionary is the premise and basis of sentiment classification. At present, it can be classified into four categories: general sentiment words, degree adverbs, negative words, and domain words. The construction method of emotional dictionaries is to use existing electronic dictionary extensions to generate emotional dictionaries. English is mainly based on the expansion of the English dictionary WordNet to form SentiWordNet lexicon. Hu and Liu [24] have manually established the seed adjective vocabulary and used the synonymous relationship between words in WordNet to determine the emotion tendency of emotion words, and use this to judge the emotional polarity of the point of view.

The tendency calculation of semantic-based sentiment dictionaries is different from the machine learning algorithms that require a large number of training datasets. It mainly analyzes the special structure and sentiment tendency words of text sentences by using sentiment dictionary and sentence lexicon, and uses weight algorithm instead of traditional manual discrimination or simply statistical method for sentiment classification. Emotional words with different sentiment intensity are assigned different weights, and then weighted summation is performed.

Finally, the threshold is determined to judge the tendency of the text. In general, the weighted calculation result is positive to indicate positive tendency; the result is negative to indicate negative tendency, and the score is zero to indicate no tendency. Compared with the classification algorithm based on machine learning, the sentiment dictionary-based method is a coarse-grained tendency classification method, but because it does not rely on a well-labeled training set, implementation is relatively simple. It can effectively and quickly classify sentiment for web texts in universal fields.

Sentiment analysis has been widely used in recommendation systems. Calculating the sentiment orientation of user reviews has been studied by some researchers. Diao [25] built a language model component in the JMARS model they proposed to capture hidden points in reviews. Zhang [26] performed phrase-level sentiment analysis on user reviews to extract clear product features and user opinions to generate interpretable recommendation results. Li [27] proposed a SUIT model for sentiment analysis using both text themes and user items. In this article, we apply sentiment analysis to cross-domain recommendation tasks, focusing on finding latent sentiment review features of users and mapping them from the source domain to the target domain.

### 3.2. Topic Model

In natural language processing, the Latent Dirichlet Allocation (LDA) [19] is a powerful and practical tool for analyzing large text documents. Latent Dirichlet Allocation is also a common method to solve the cold start problem. LDA can automatically cluster words into topics and discover relationships between documents from a dataset. It assumes that the authors of the resource have multiple themes; based on their vocabulary, the author chooses specific vocabulary to describe their topic. Formally, resources are distributed on topics and topics are distributed on vocabularies. Vocabulary consists of different words in the corpus.

As shown in Figure 1, the symbolic representation of standard LDA is shown. $\varphi$ refers to the representation of topic in vocabulary, and $\theta$ refers to the distribution of resources across topics. Variables $\alpha$ and $\beta$ are hyperparameters of the model. Parameter $\alpha$ controls the distribution of resources on topics, and parameter $\beta$ controls the distribution of topics in vocabulary. Variable $z$ represents subject assignment, while variable $w$ is the observed word. $R$ is the number of resources in the corpus, and $N$ is the number of words in the resources. Parameter $K$ indicates the number of topics suitable for corpus. $K$ is allocated during initialization. Among the variables, only $w$ is the observation variable and the rest is the latent variable.

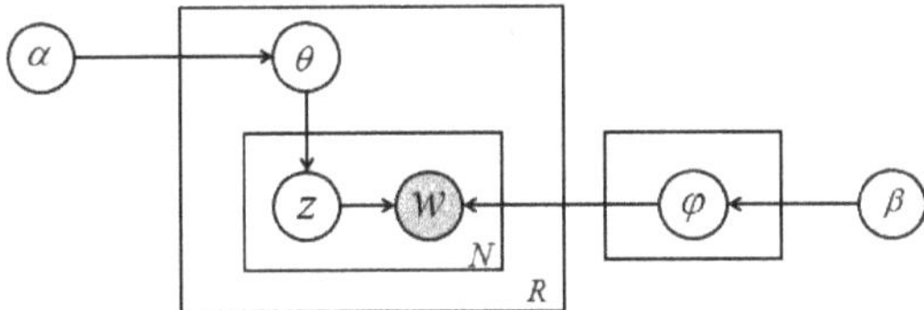

**Figure 1.** Standard Latent Dirichlet Allocation (LDA).

The process of LDA starts from the sampling of topic $z$. Based on topic $z$, the word $w$ is obtained by sampling $\varphi$ with a polynomial, which is described as follows (1)–(5):

$$\varphi(w_i|k,\beta) = \frac{n(w_i,k) + \beta}{\sum_{w \in V} n(w,k) + (\beta - 1)}. \tag{1}$$

$$\theta(k|d,\alpha) = \frac{n(r,k) + \alpha}{\sum_{k \in K} n(r,k) + (\alpha - 1)}. \tag{2}$$

Variable $k$ represents the $k$ topic of model sampling, and $\varphi(w_i|k,\beta)$ calculates the probability that word $w_i$ is the $k$ topic in the dictionary. $n(w_i,k)$ indicates the number of occurrences of word $w_i$ assigned to topic $k$. $\theta(k|d,\alpha)$ calculates the probability of document $d$ for topic $k$. $n(r,k)$ is the number of times resource $r$ is assigned to topic $k$. The joint distribution of the model is as follows:

$$p(\theta,z,w|\alpha,\beta) = p(\theta|\alpha) \prod_{n=1}^{N} p(z_n|\theta)p(w_n|z_n,\beta). \tag{3}$$

$$p(\theta,z|w,\alpha,\beta) = p(\theta,z,w|\alpha,\beta)p(w|\alpha,\beta). \tag{4}$$

Equation (4) is used to infer markers, but it is difficult to calculate the real posterior distribution. Gibbs sampling is used to estimate the posterior distribution in order to deal with the difficulty. Gibbs sampling starts by randomly assigning a word to a topic. In subsequent iterations, it assigns a word to a topic based on the following equation:

$$p(z_i = k|z_{-i},w,\theta) = \frac{n(w_i,k)_{-i} + \beta}{\sum_{w \in V} n(w,k) + (\beta - 1)} \times \frac{n(r,k)_{-i} + \alpha}{\sum_{k \in K} n(r,k) + (\alpha - 1)}. \tag{5}$$

where $n(w_i, k)_{-i}$ represents the number of times a word $w_i$ appears in topic $k$ without including the currently assigned task.

## 4. The CDR-SAFM Algorithm

### 4.1. Notations

We suppose that there are two domains sharing the same user. Users who appear in one domain can appear in another domain. In this sense, the two domains share the same user. Without losing generality, one domain is called the source domain. The other is called the target domain.

Define the recommended objects in the target domain as items. Let $U = \{u_1, u_2, ..., u_{|U|}\}$ represent the common users of source domain and target domain, that is overlapping users. Let $J_S = \{i_1, i_2, ..., i_{|J_S|}\}$ and $J_T = \{l_1, l_2, ..., l_{|J_T|}\}$ be the item sets from the source and target domains respectively. The user review dataset is represented as $SR_U = \{r_{u_1}, r_{u_2}, ..., r_{u_{|U|}}\}$ in source domain and $TR_U = \{r_{u_1}, r_{u_2}, ..., r_{u_{|U|}}\}$ in target domain, where $r_{u_i}$ is all of reviews of user $u_i$ in the corresponding domain. Similarly, we let $TR_I = \{r_{i_1}, r_{i_2}, ..., r_{i_{|J_T|}}\}$ denote the item review dataset in target, where $r_{i_j}$ is all of reviews which item $i_j$ acquired in target domain. $R^s$ and $R^t$ be two rating matrices from the source and target domains respectively, where $R^s_{ij}$ is the rating that user $u_i$ gives to item $i_j$ in the source domain and $R^t_{ij}$ is the corresponding rating in the target domain.

### 4.2. Problem Formulation

Given the review information $SR_U$ and $TR_U$ of two domains, and overlapping user sets $U$ across domains, we aim to analyze the sentiment information of the source domain and the target domain, use the common user as a bridge to realize the knowledge transfer from the source domain to the target domain, and solve the problem of rating prediction of cold-start users in the target domain. For this purpose, we propose a cross-domain recommendation algorithm, CDR-SAFM, based on sentiment analysis and latent feature mapping. This framework contains three major steps, that is, latent sentiment review features modeling, latent sentiment review features mapping and cross-domain recommendation, as illustrated in Figure 2.

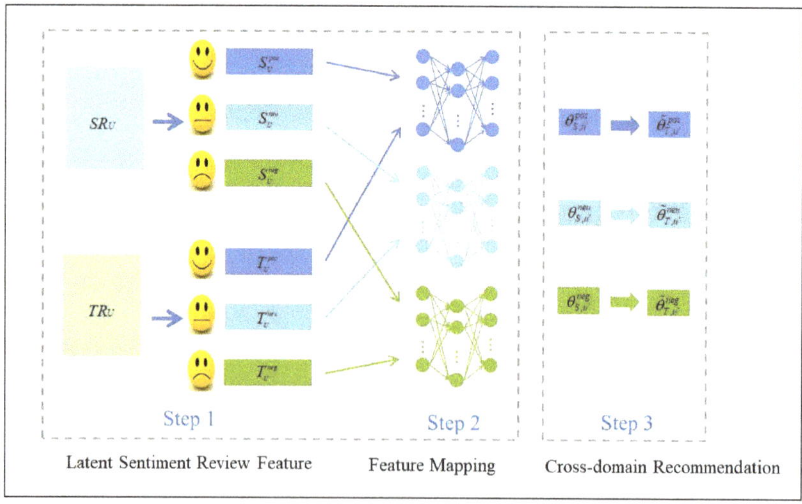


**Figure 2.** Illustrative diagram of the CDR-SAFM algorithm.

At the first step, we aim to analyze the emotional orientation of user reviews in two domains through sentiment analysis methods, so that the original dataset in the two domains is divided into three parts: positive reviews, neutral reviews, and negative reviews. Reviews in the source domain are divided into $SR_U^{pos}$, $SR_U^{neu}$ and $SR_U^{neg}$, and reviews in the target domain are divided into $TR_U^{pos}$, $TR_U^{neu}$ and $TR_U^{neg}$. Then we aim to find the representation of latent sentiment review features by LDA, including positive sentiment review features, neutral sentiment review features, and negative sentiment review features. The latent sentiment review feature assumes the association between a user's reviews on an item, and the user's reviews on an item are actually the result of the combination of the user and the sentiment review features. That is to say, users' reviews always contain sentiment features. In order to reduce the influence of previous sentiment classification methods on the overall algorithm results, we classify users' review information by users' rating of the item, because a user's review emotion polarity of an item can be reflected in the rating information. For example, if a user likes an item, the user's final reviews will be more positive, and the user's rating will be higher.

In the second step, we aim to obtain a mapping function for modeling cross-domain relationship of sentiment review features. We assume that there is a latent mapping relationship between the sentiment review features from source domain and target domain, and then capture this relationship by mapping function. To avoid the mutual interference among the features of positive, neutral and negative sentiment review during the process of knowledge transfer, we use mapping function to model the cross-domain relationship of different sentiments respectively. In order to avoid the lack of different sentiment review features of users in the mapping process, we train the mapping function of different emotions by preprocessing the data and using the common users in the two domains whose sentiment review features are not missing.

Finally, we recommend a cold-start user in the target domain. Using this method, we can get the corresponding latent sentiment review feature in the target domain and use these features to affect the final recommendation results. Different mapping results of sentiment review features of cold-start users have different influence on users' ratings. Therefore, we set different weights for different results of sentiment review features to get the emotional ratings of cold-start users. The complete CDR-SAFM algorithm is presented in Algorithm 1.

---

**Algorithm 1** The CDR-SAFM algorithm.

---

**Require:**

 Source domain $SR_U$, target domain $TR_U$;

 Common User set $U$;

**Ensure:**

 Make recommendation for cold-start users in the target domain;

 **Sentiment Analysis**
1:  Learn $\{SR_U^{pos}, SR_U^{neu}, SR_U^{neg}\}$ from $SR_U$;
2:  Learn $\{TR_U^{pos}, TR_U^{neu}, TR_U^{neg}\}$ from $TR_U$;

 **Latent sentiment review feature**
3:  Learn $\{S_U^{pos}, S_U^{neu}, S_U^{neg}\}$ from $\{SR_U^{pos}, SR_U^{neu};, SR_U^{neg}\}$;
4:  Learn $\{T_U^{pos}, T_U^{neu}, T_U^{neg}\}$ from $\{TR_U^{pos}, TR_U^{neu};, TR_U^{neg}\}$;

 **Latent Sentiment Review Feature Mapping**
5:  Learn the mapping function $f_{pos}(\cdot)$, $f_{neu}(\cdot)$ and $f_{neg}(\cdot)$ by users across domain;

 **Cross-domain Recommendation**
6:  Get affine features $U_{new}^{T,pos}$, $U_{new}^{T,neu}$ and $U_{new}^{T,neg}$ of target users;
7:  Make recommendation for target users.

---

### 4.3. Latent Sentiment Review Feature Modeling

In this section, we aim to analyze the sentiment information hidden in the user's review information and extract the user's review features under different sentiments from the source domain and the target domain, as illustrated in Figure 3.

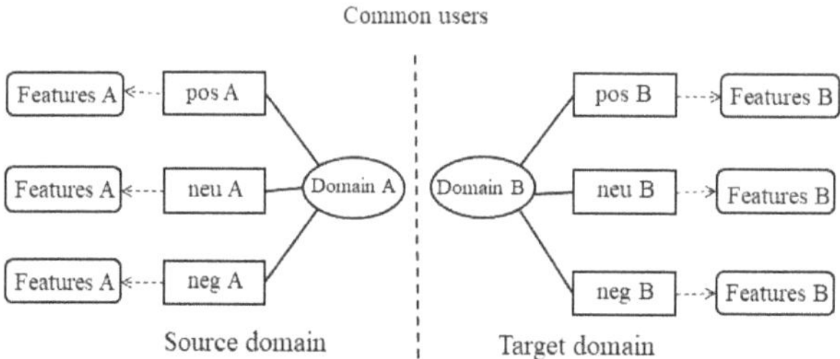


**Figure 3.** Sentiment analysis and feature extraction.

### 4.3.1. Sentiment Analysis

According to general psychology, sentiment has an important influence on one's behavior and choices. Sentiment analysis plays an important role in information retrieval. It clarifies people's thoughts and feelings about something or someone in a certain situation. This kind of high level information can be used in many applications, such as customer review analysis, business and government intelligence, personalized recommendation and so on. User reviews on online platforms show similar sentiment expressions, which is generated by similar psychological stimulation. Therefore, it is valuable to combine the latent sentiment information in reviews with cross domain recommendation.

Sentiment analysis (SA) is a process of analyzing, processing, summarizing and reasoning subjective characters with emotional color. Among them, sentiment analysis can also be divided into emotional orientation analysis, emotional level analysis, subjective and objective analysis, and so forth. The purpose of emotional orientation analysis is to judge the positive, negative and neutral meaning of the text. In most application scenarios, there are only two types. For example, the two words "love" and "disgust" belong to different emotional orientations.

However, in the past work based on sentiment analysis, most of the sentiment analysis problems are expressed as—given a set of review $R$, a sentiment classification algorithm can divide each sentence of a review $r \in R$ into two categories—positive $R^{pos}$ and negative $R^{neg}$. In real life, some texts can not be directly classified into positive sentiment or negative sentiment. Therefore, in this paper, based on the three-way decision ideas, we divide the source domain review data information into positive $SR_U^{pos}$, neutral $SR_U^{neu}$ and negative $SR_U^{neg}$, and the target domain review data into $TR_U^{pos}$, $TR_U^{neu}$ and $TR_U^{neg}$. Because the sentiment analysis algorithm is not the focus of this paper, it is suggested that readers refer to the relevant literature. To achieve this goal, we use a method based on statistics to do sentiment analysis on the review dataset.

We extract "ratings" and "reviews" from the dataset for analysis. "Ratings" represents the user rating of the item in the review, and the rating range is 1–5. "reviews" is the text information of the user review. We randomly selected 1000 reviews each of 1–5, and performed basic de-punctuation, lowercase, and stop word processing on the text to count the word frequency to make a word cloud diagram. Based on the analysis of word cloud diagram with different ratings, most reviews rated as 1, 2 are negative sentiment reviews, and reviews rated as 4, 5 are mostly positive sentiment reviews. However, it is difficult to define the division of reviews with a score of 3, which also proves the idea of three-way decision ideas, so we regard them as a separate category. Therefore, we can divide the dataset into positive sentiment reviews with a rating of 4 or 5, neutral sentiment reviews with a rating of 3 and negative sentiment reviews with a rating of 1 or 2.

### 4.3.2. Latent Sentiment Review Feature

Obtaining the latent sentiment feature of users is the key for improving the performance of recommendation algorithms. Review-based recommendation algorithms tend to extract latent feature by topic models. In this paper, we use an LDA topic model to extract latent feature. Each latent feature (topic) extracted by LDA is associated with a set of keywords. Thus, we can get interpretable recommendation results by matching the topic distribution of users and items.

Take the analysis of user features under positive sentiment in source domain as an example (other similar). We use the review information of all users as a corpus for LDA model training, and use all the reviews of each user as a document. Finally, we find the word distribution of the topics and the topic distribution of the document under the positive sentiment of the source domain. The topic distribution of the document is the user's latent sentiment review feature, which represented as $S_U^{pos} = \{\theta_{S,u_1}^{pos}, \theta_{S,u_2}^{pos}, ..., \theta_{S,u_{|U|}}^{pos}\}$. Similarly, we can calculate the latent sentiment review feature of users in the source domain and target domain under the positive, negative and neutral sentiment. The sentiment review features in the source domain are $S_U^{pos} = \{\theta_{S,u_1}^{pos}, \theta_{S,u_2}^{pos}, ..., \theta_{S,u_{|U|}}^{pos}\}$, $S_U^{neu} = \{\theta_{S,u_1}^{neu}, \theta_{S,u_2}^{neu}, ..., \theta_{S,u_{|U|}}^{neu}\}$ and $S_U^{neg} = \{\theta_{S,u_1}^{neg}, \theta_{S,u_2}^{neg}, ..., \theta_{S,u_{|U|}}^{neg}\}$, where $\theta_{S,u_i}^{pos}$ represent positive review features of user $u_i$ in the source domain. Similarly, $\theta_{S,u_i}^{neu}$ and $\theta_{S,u_i}^{neg}$ represent the neutral review features and the negative review features, respectively. The sentiment review features in the target domain are $T_U^{pos} = \{\theta_{T,u_1}^{pos}, \theta_{T,u_2}^{pos}, ..., \theta_{T,u_{|U|}}^{pos}\}$, $T_U^{neu} = \{\theta_{T,u_1}^{neu}, \theta_{T,u_2}^{neu}, ..., \theta_{T,u_{|U|}}^{neu}\}$ and $T_U^{neg} = \{\theta_{T,u_1}^{neg}, \theta_{T,u_2}^{neg}, ..., \theta_{T,u_{|U|}}^{neg}\}$, where $\theta_{T,u_i}^{pos}$ represent positive review features of user $u_i$ in the target domain. $\theta_{T,u_i}^{neu}$ and $\theta_{T,u_i}^{neg}$ represent the neutral review features and the negative review features, respectively. In addition, $T_I^{pos} = \{\theta_{T,l_1}^{pos}, \theta_{T,l_2}^{pos}, ..., \theta_{T,l_{|I|}}^{pos}\}$, $T_I^{neu} = \{\theta_{T,l_1}^{neu}, \theta_{T,l_2}^{neu}, ..., \theta_{T,l_{|I|}}^{neu}\}$ and $T_I^{neg} = \{\theta_{T,l_1}^{neg}, \theta_{T,l_2}^{neg}, ..., \theta_{T,l_{|I|}}^{neg}\}$ respectively represent the distribution of sentiment review topics of all items in the target domain, and $\theta_{T,l_i}^{pos}$ represents the distribution of positive review topics of items $l_i$ in the target domain.

### 4.4. Sentiment Review Feature Mapping

How to transfer users' sentiment information effectively is an important problem to be solved in this paper. Existing cross-domain recommendation algorithms that use user reviews do not make full use of the sentiment information in these reviews. They achieve knowledge transfer by mixing positive, neutral, and negative sentiment, which will weaken or even lose some of the user's sentiment Information, especially negative sentiment. For example, users may be very concerned about the plot of a novel and make positive reviews on the plots of some novels in the field of e-books (source domain), while making negative reviews on the plots of other novels. If we transfer the knowledge obtained from user reviews from the source domain to the target domain without distinguishing the emotional orientation of these reviews, the underlying sentiment factors in the reviews will be mixed together and transferred to the target domain as user features. In movies (target domain), movies with poor plots will match user features transferred to the target domain. But users may not like the movie.

To connect the source domain and the target domain, we assume that we can get the relationship between domains through a mapping function. The mapping architecture is shown in Figure 4.

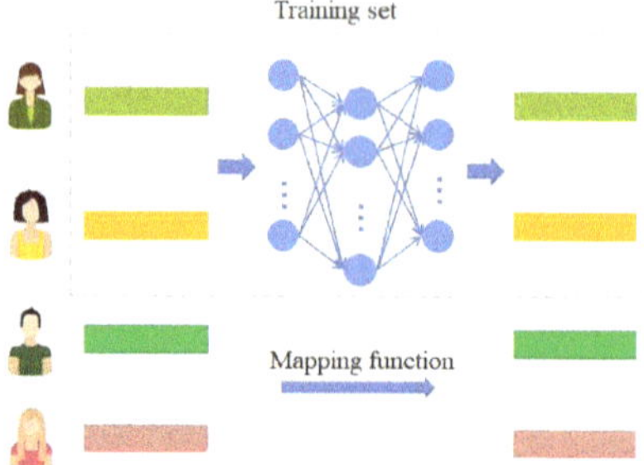


**Figure 4.** Mapping architecture.

To obtain the mapping function, we can formalize the learning process into a supervised regression problem. In particular, we minimize the mapping loss to obtain the mapping function.

$$\min_{\theta} \sum_{ui \in U} L\left(f_U(U_i^s; \theta), U_i^t\right), \tag{6}$$

where $L(\cdot, \cdot)$ is the loss function that defines the corresponding vector in the source domain and the target domain. Because the input and output of the mapping function are multi-dimensional numerical vectors, we choose the square error as the loss function.

Multilayer Perceptron (MLP) is a nonlinear transformation, which is more flexible than a linear mapping function. Thus, we choose mapping based on multi-layer perceptron to realize cross domain connection between different domains. It makes the user's sentiment review feature in the source domain play a complementary and perfect role in the target domain recommendation.

MLP can be optimized using back propagation. The optimization problem can be formalized as

$$\min_{\theta} \sum_{u_i \in U} L\left(f_{mlp}(U_i^s; \theta), U_i^t\right), \tag{7}$$

where $f_{mlp}(\cdot; \theta)$ is MLP mapping function, $\theta$ is its parameter set, which are the weight matrices and bias terms between layers. In this paper, we use the method of random gradient descent to study its parameters and get MLP mapping function. By traversing the training set, refresh the parameters of MLP with any user $u_i$ across the source and target domains. The gradient of parameters is calculated by back propagation algorithm. Until the model converges, the iterative process stops.

In this paper, through sentiment analysis and feature extraction of the source domain and target domain datasets, we can get the user sentiment review features $S_U^{pos}$, $S_U^{neu}$ and $S_U^{neg}$ in the source domain, and the user sentiment review features $T_U^{pos}$, $T_U^{neu}$ and $T_U^{neg}$ in the target domain. In order to connect the source domain and the target domain, we train the feature mapping relationship $f_{MLP}^{pos}$ of positive sentiment in different domains through $S_U^{pos}$ and $T_U^{pos}$ (the negative and neutral feature mapping relationships are similar, respectively $f_{MLP}^{neu}$ and $f_{MLP}^{neg}$).

Let $S_U^{pos} = \{\theta_1^S, \theta_2^S, ..., \theta_N^S\}$ denote s set of sentiment review feature in the source domain, $T_U^{pos} = \{\theta_1^T, \theta_2^T, ..., \theta_N^T\}$ denote s set of sentiment review feature in the target domain. $N$ is the number of common users. We formalized mapping problem as: given $N$ training samples $(\theta_i^S, \theta_i^T)$, $\theta_i^S, \theta_i^T \in R^M$ $(i = 1, 2, ..., N)$, we aim to learn a mapping function so that we can get the sentiment review features in the target domain through the sentiment review features in the source domain.

That is to say, $S_U^{pos}$ and $T_U^{pos}$ are used to train the mapping function of source domain and target domain in positive sentiment review feature. Similarly, $S_U^{neu}$ and $T_U^{neu}$ are used to train the mapping function of neutral sentiment review feature, and $S_U^{neg}$ and $T_U^{neg}$ are used to train the mapping function of negative sentiment review features. Specifically, given positive sentiment review features $S_{u'}^{pos}$ and $T_{u'}^{pos}$ in two domains of user $u'$, we use a mapping function $f(\cdot; \theta)$ to capture the cross domain relationship $T_{u'}^{pos} = f_{pos}(S_{u'}^{pos}; \theta)$, where $\theta$ is the parameter of the mapping function. Similarly, we can get $T_{u'}^{neu} = f_{neu}(S_{u'}^{neu}; \theta)$ and $T_{u'}^{neg} = f_{neg}(S_{u'}^{neg}; \theta)$.

## 4.5. Cross-Domain Recommendation

Given a cold-start user, we cannot recommend it directly because of its sparse data. We first can get the user's sentiment review feature in the source domain and get the sentiment review feature in the target domain through the learned sentiment feature mapping function. Then by combining the existing user benchmark recommendation results and the influence of sentiment information on user recommendations, we can get the final recommendation results. The specific formal description is as follows:

Given a cold-start user $u'$ in the target domain, we classify all the user's review information into $S_{u'}^{pos}$, $S_{u'}^{neu}$ and $S_{u'}^{neg}$ in the source domain. We can obtain the sentiment review features $\theta_{S,u'}^{pos}$, $\theta_{S,u'}^{neu}$ and $\theta_{S,u'}^{neg}$ by Latent Dirichlet Allocation. Then we aim to get sentiment review features $\tilde{\theta}_{T,u'}^{pos}$, $\tilde{\theta}_{T,u'}^{neu}$ and $\tilde{\theta}_{T,u'}^{neg}$ of user $u'$ in the target domain is defined as:

$$\tilde{\theta}_{T,u'}^{pos} = f_{MLP}^{pos}(\theta_{S,u'}^{pos}; \theta pos), \tag{8}$$

$$\tilde{\theta}_{T,u'}^{neu} = f_{MLP}^{neu}(\theta_{S,u'}^{neu}; \theta neu), \tag{9}$$

$$\tilde{\theta}_{T,u'}^{neg} = f_{MLP}^{neg}(\theta_{S,u'}^{neg}; \theta neg). \tag{10}$$

Calculating the topic similarity under the same sentiment in different domains is defined as:

$$SIM^{pos} = \{sim_{i,j}^{pos}\}, i, j = 1, 2, ..., M, \tag{11}$$

$$SIM^{neu} = \{sim_{i,j}^{neu}\}, i, j = 1, 2, ..., M, \tag{12}$$

$$SIM^{neg} = \{sim_{i,j}^{neg}\}, i, j = 1, 2, ..., M. \tag{13}$$

Therefore, we can get the predicted sentiment rating of cold-start users. That is, the impact of sentiment information in user reviews on the final recommendation is defined as:

$$e(u') = w_1 \times \tilde{\theta}_{T,u'}^{pos} \cdot SIM^{pos} \cdot T_I^{posT} - w_2 \times \tilde{\theta}_{T,u'}^{neg} \cdot SIM^{neg} \cdot T_I^{negT} + w_3 \times \tilde{\theta}_{T,u'}^{neu} \cdot SIM^{neu} \cdot T_I^{neuT}, \tag{14}$$

where the parameters $w_1$, $w_2$ and $w_3$ represent different weights. Finally, combined with the user's benchmark rating and prediction sentiment rating, we can get the final prediction rating of cold-start user $u'$ on the item $l_i$, which can be expressed as:

$$pre(u') = R\_base + e(u'), \tag{15}$$

where $R\_base = b_T + b_{u'} + b_{l_j}$, $b_T$ represents the average rating of all items in the target domain. The parameter $b_{u'}$ is the user rating bias in the source domain and $b_{l_j}$ is the item rating bias in the target domain.

## 5. Experiments

In this section, we test the CDR-SAFM algorithm proposed in this paper with a real-world dataset. Firstly, the experimental dataset is introduced, and the possible rating of cold-start users in the target domain is predicted by using review data from two different domains—Electronics, and Movies and TV. Then we compare the proposed model with the common methods in the recommendation system. Finally, we randomly select different percentages of cold-start users and analyze the experimental results.

*5.1. Experimental Settings*

**Datasets.** We employ Amazon cross-domain dataset [28] in our experiment. This dataset contains product reviews and star ratings with 5-star scale from Amazon, including 142.8 million reviews spanning May 1996–July 2014. In our experiment, we select the top two domains with the most widely used in previous studies to employ in our cross-domain experiment: Electronics and Movies & TV. The global statistics of two domains used in our experiment are shown in Table 1.

**Table 1.** Statistics of the Amazon dataset.

| Datasets | Electronics | Movies & TV |
|---|---|---|
| Num. of users | 192,403 | 123,960 |
| Num. of Items | 63,001 | 50,052 |
| Num. of reviews | 8,898,041 | 1,697,533 |

**Experimental Settings.** In the experiment, we set Electronics as the source domain and Movies & TV as the target domain. After data preprocessing, the number of common users in the two domains is 2406. The items in these two domains are very different, forming a cross domain user sharing scenario. To evaluate the validity and efficiency of the proposed algorithm on cross-domain recommendation task, we randomly remove all the rating information of a fraction of entities in the target domain and take them as cross-domain cold-start entities for making recommendation. For the sake of stringency of the experiments, we set different fraction for cold-start entities, namely, 20%, 50% and 70%. In addition, since different sets of cold-start entities may affect the final recommendation results, we repeatedly sample users for 10 times to generate different sets. Dimension K of latent sentiment review features is set as 50 and 100. For the MLP mapping function, we choose the structure of the MLP as one-hidden layer, the dimension of input and output of the MLP is set as K, whilst the number of nodes in the hidden layer is set as 2K. The weight and bias parameters of the MLP is initialized according to the rule in Reference [29]. Finally, a tan-sigmoid function is employed as the activation function. In order to obtain the mapping function of MLP, we use stochastic gradient descent to learn the parameters. Through the loop on the training set, the parameters of the MLP are updated. This back-propagation algorithm is used to calculate the gradient of the parameters, and this process continues until the model converges.

**Models for Comparison.** We compare the CDR-SAFM algorithm with the following baseline models and algorithms for validating the performance.

- **AVE**: It predicts ratings by the following equation: $r_{ui} = b_t + b_u + b_i$ where $b_t$ is the overall average ratings of all items in the target domain, $b_u$ denotes the user rating bias in the source domain and $b_i$ represents item bias in the target domain.
- **MF**: This is the single-domain matrix factorization algorithm proposed in [2].
- **MF_MLP**: This is a cross-domain recommendation algorithm based on MF and MLP, which is proposed by [30]. In our experiment, for MF_MLP, the structure of the MLP is set as one-hidden layer, and the number of nodes in the hidden layer is set as 2M.

**Evaluation Metric.** We adopt the Root Mean Square Error (RMSE) to evaluate the prediction performance. It is defined as:

$$RMSE = \sqrt{\sum_{y_{u_i} \in T'} \frac{(y_{u_i} - \tilde{y}_{u_i})^2}{|T'|}}, \tag{16}$$

where $T'$ is the set of test ratings, $y_{u_i}$ denotes an observed rating in the test set, and $\tilde{y}_{u_i}$ represents the predictive value, $|T'|$ is the number of test ratings.

### 5.2. Experimental Results

The experimental results in terms of RMSE on the Amazon dataset are presented in Table 2, where K represents the number of topics. For the sake of stringency of the experiments, we set different fractions for cold-start entities, namely, 20%, 50% and 70%. It can be seen that the performance of our proposed method is superior to other comparative methods.

**Table 2.** Recommendation performance in terms of RMSE on the Amazon dataset.

| K: the Number of Topics | Algorithms | 20% | 50% | 70% |
|---|---|---|---|---|
| 50 | AVE | 1.4998 | 1.5020 | 1.5163 |
| 50 | MF | 1.6300 | 1.7071 | 1.7682 |
| 50 | MF_MLP | 1.4996 | 1.5021 | 1.5113 |
| 50 | CDR-SAFM | 1.4650 | 1.4912 | 1.5073 |
| 100 | AVE | 1.4998 | 1.5020 | 1.5163 |
| 100 | MF | 1.7494 | 1.8300 | 1.8950 |
| 100 | MF_MLP | 1.5338 | 1.5726 | 1.5929 |
| 100 | CDR-SAFM | 1.4829 | 1.4940 | 1.5100 |

As the proportion of cold-start users increases, the recommended performance decreases. The decline of single-domain recommendation method MF is the most obvious, while the performance of cross-domain recommendation method is relatively good, which also proves the effectiveness of knowledge transfer in cross-domain recommendation. The result of MF_MLP is better than MF, which also shows that the mapping function based on MLP is feasible in knowledge transfer. Compared with MF_MLP, the CDR-SAFM method proposed in this paper has been improved in terms of RMSE. When the number of topics K = 50, the mean square errors of our method are reduced by 0.0346, 0.0109 and 0.004 respectively. When the number of topics K = 100, the mean square errors are reduced by 0.0509, 0.0786 and 0.0829, respectively. These results demonstrate that the CDR-SAFM is more suitable for making recommendations to cold-start users compared to other cross-domain baseline methods, and also proves the effectiveness of our method in the cross-domain recommendation scenarios.

In the topic model, the number of topics is very important. The number of topics directly affects the results of the experiment. However, the number of topics is not directly proportional to the results. As the number of topics increases, more computing costs will be required and there will be a risk of overfitting. In our method, we select the number of topics K as 50 and 100, and analyze the results under different number of topics. As shown in Figure 5, we tested RMSE with the number of topics K = 50 and K = 100 on the Electronics and Movies & TV datasets, respectively. We can see that the comparison results of AVG and MF_MLP are most obvious with the change of topic number. When the topic number K = 100, the difference between the two methods is more obvious. At the same time, we can see that when the number of topics K = 100, the CDR-SAFM method is relatively stable for different proportions of cold-start users in term of RMSE, which also proves the rationality of our method.

A cross-domain recommendation algorithm is an effective recommendation method. Its purpose is to transfer the knowledge in the source domain to the target domain, so as to improve the quality of recommendations and alleviate the cold start problem in the recommendation system. However, existing works on cross-domain algorithms mostly consider ratings, tags and text information such as reviews, but cannot use the sentiments implicated in the reviews efficiently. In this paper, we propose a cross-domain recommendation algorithm (CDR-SAFM) based on sentiment analysis and latent feature mapping by combining the sentiment information implicit in user reviews in different domains. The results of comparative experiments show that our method is effective. At the same time, it also proves that it is feasible for us to consider the implicit sentiment information in the reviews into the cross-domain recommendation method. However, there are still many deficiencies in our work. In this paper, we only validate our method on one dataset, and need to validate our method on more datasets in different fields. At the same time, there is a lot of space for data preprocessing and method improvement, which is also the problem we need to solve in future work.

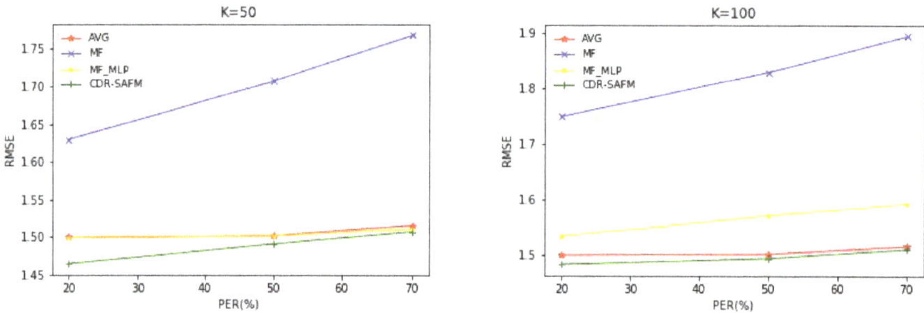

**Figure 5.** RMSE of the number of different topics.

## 6. Conclusions

In this paper, we address the cold-start problem in the recommendation process. We proposed a cross-domain recommendation algorithm based on sentiment analysis and latent feature mapping by combining the sentiment information implicit in user reviews in different domains. We first employed the latent feature model to project users in both source and target domains into two different feature spaces. Then, we learned an appropriate non-linear mapping function to capture the coordinate relationship across the two domains. To avoid mutual interference between different sentiment features during the process of knowledge transfer, we have learned three different types of sentiment mapping function, respectively based on three-way decision ideas, including positive, neutral and negative mapping functions. For a cold-start user in the target domain, we made recommendations by mapping a user's sentiment features from the source domain to the target domain. Experimental results from a cross-domain recommendation scenario on the Amazon dataset demonstrate that the proposed framework can improve the quality of cross-domain recommendation. There are possible minimum biases in the rating process in relation to some factors as interests of raters. This work does not make a contribution to the bias. Developing an efficient method to preprocess the rating data is part of our planned future work.

**Author Contributions:** Conceptualization, Y.W.; Data curation, Y.W.; Funding acquisition, H.Y. and G.W.; Investigation, Y.W.; Methodology, Y.W. and H.Y.; Project administration, H.Y., G.W. and Y.X.; Resources, H.Y.; Supervision, H.Y. and G.W.; Validation, Y.W.; Visualization, Y.W.; Writing–original draft, Y.W.; Writing–review and editing, H.Y. All authors have read and agreed to the published version of the manuscript.

**Funding:** This work was supported in part by the National Natural Science Foundation of China under Grant Nos. 61751312, 61533020 & 61876027 and the key T&A program of Chongqing under grant No. cstc2019jscx-mbdxX0048.

**Conflicts of Interest:** The authors declare no conflict of interest.

## References

1. Herlocker, J.L.; Konstan, J.A.; Borchers, A. An algorithmic framework for performing collaborative filtering. In Proceedings of the 22nd Annual International ACM SIGIR Conference on Research and Development in Information Retrieval, New York, NY, USA, 15–19 August 1999; pp. 230–237.
2. Koren, Y.; Bell, R.; Volinsky, C. Matrix factorization techniques for recommender systems. *IEEE Comput.* **2009**, *42*, 30–37. [CrossRef]
3. Cremonesi, P.; Tripodi, A.; Turrin, R. Cross-domain recommender systems. In Proceedings of the 11th International Conference on Data Mining Workshops, Vancouver, BC, Canada, 11 December 2011; pp. 496–503.

4. Yao, Y.Y. Three-way decisions and cognitive computing. *Cogn. Comput.* **2016**, *8*, 543–554. [CrossRef]

5. Yu, H.; Wang, X.; Wang, G. An active three-way clustering method via low-rank matrices for multi-view data. *Inf. Sci.* **2020**, *507*, 823–839. [CrossRef]

6. Pan, W.; Yang, Q. Transfer learning in heterogeneous collaborative filtering domains. *Artif. Intell.*, **2013**, *197*, 39–55. [CrossRef]

7. Xin, X.; Liu, Z.; Lin C.Y. Cross-domain collaborative filtering with review text. In Proceedings of the 24th International Joint Conference on Artificial Intelligence, Buenos Aires, Argentina, 25 July–1 August 2015; pp. 1827–1834.

8. Wei, C.; Hsu, W.; Lee, M.L. A unified framework for recommendations based on quaternary semantic analysis. In Proceedings of the 34th International ACM SIGIR Conference on Research and Development in Information Retrieval, Beijing, China, 24–28 July 2011; pp. 1023–1032.

9. Jiang, M.; Cui, P.; Chen, X. Social recommendation with cross-domain transferable knowledge. *IEEE Trans. Knowl. Data Eng.*, **2015**, *27*, 3084–3097. [CrossRef]

10. Nakatsuji, M.; Fujiwara, Y.; Tanaka, A. Recommendations Over Domain Specific User Graphs. In Proceedings of the 19th European Conference on Artificial Intelligence, Lisbon, Portugal, 16–20 August 2010; pp. 607–612.

11. Tiroshi, A.; Berkovsky, S.; Kaafar, M.A. Cross social networks interests predictions based on graph features. In Proceedings of the 7th ACM Conference on Recommender Systems, Hong Kong, China, 12–16 October 2013; pp. 319–322.

12. Gong, Q.; Chen, Y.; Hu, J.; Cao, Q.; Hui, P.; Wang, X. Understanding cross-site linking in online social networks. *ACM Trans. Web.* **2018**, *12*, 1–29. [CrossRef]

13. Meo, P.D.; Ferrara, E.; Abel, F.; Aroyo, L.; Houben, G.J. Analyzing user behavior across social sharing environments. *ACM Trans. Intell. Syst. Technol.* **2014**, *5*, 1–31. [CrossRef]

14. Li, Y.; Su, Z.; Yang, J.; Gao, C. Exploiting similarities of user friendship networks across social networks for user identification. *Inf. Sci.* **2020**, *506*, 78–98. [CrossRef]

15. Yang, D.; He, J.; Qin, H. A graph-based recommendation across heterogeneous domains. In Proceedings of the 24th ACM International Conference on Information and Knowledge Management, Melbourne, Australia, 19–23 October 2015; pp. 463–472.

16. Shi, Y.; Larson, M.; Hanjalic, A. Tags as bridges between domains: Improving recommendation with tag-induced cross-domain collaborative filtering. In Proceedings of the 19th International Conference on User Modeling, Adaptation, and Personalization, Girona, Spain, 11–15 July 2011; pp. 305–316.

17. Mnih, A.; Salakhutdinov, R.R. Probabilistic matrix factorization. In Proceedings of the 22nd Annual Conference on Neural Information Processing Systems, Vancouver, BC, Canada, 8–10 December 2008; pp. 1257–1264.

18. Kumar, A.; Kumar, N.; Hussain, M. Semantic clustering-based cross-domain recommendation. In Proceedings of the 5th Edition Conference Ieee Symposium on Computational Intelligence and Data Mining, Orlando, FL, USA, 9–12 December 2014; pp. 137–141.

19. Blei, D.M.; Ng, A.Y.; Jordan, M.I. Latent dirichlet allocation. *J. Mach. Learn. Res.*, **2003**, *3*, 993–1022.

20. Song, T.; Peng, Z.; Wang, S. Review-based cross-domain recommendation through joint tensor factorization. In Proceedings of the 22nd International Conference on Database Systems for Advanced Applications, Suzhou, China, 27–30 March 2017; pp. 525–540.

21. Li, H.; Lin, R.; Hong, R. Generative models for mining latent aspects and their ratings from short reviews. In Proceedings of the 15th IEEE International Conference on Data Mining, Atlantic City, NJ, USA, 14–17 November 2015; pp. 241–250.

22. Hu, G.; Zhang, Y.; Yang, Q. Transfer Meets Hybrid: A Synthetic Approach for Cross-Domain Collaborative Filtering with Text. In Proceedings of the 28th International World Wide Web Conferences, San Francisco, CA, USA, 13–17 May 2019; pp. 2822–2829.

23. Pang, B.; Lee, L. Opinion mining and sentiment analysis. *Found. Trends Inf. Retr.* **2008**, *2*, 1–135. [CrossRef]

24. Hu, M.; Liu, B. Mining and summarizing customer reviews. In Proceedings of the 10th ACM SIGKDD International Conference on Knowledge Discovery and Data Mining, Seattle, WA, USA, 22–25 August 2004; pp. 168–177.

25. Diao, Q.; Qiu, M.; Wu, C.Y. Jointly modeling aspects, ratings and sentiments for movie recommendation (JMARS). In Proceedings of the 20th ACM SIGKDD international conference on Knowledge discovery and data mining, New York, NY, USA, 24–27 August 2014; pp. 193–202.

26. Zhang, Y.; Lai, G.; Zhang, M. Explicit factor models for explainable recommendation based on phrase-level sentiment analysis. In Proceedings of the 37th international ACM SIGIR conference on Research & development in information retrieval, Gold Coast, Queensland, Australia, 6–11 July 2014; pp. 83–92.
27. Li, F.; Wang, S.; Liu, S. Suit: A supervised user-item based topic model for sentiment analysis. In Proceedings of the 28th AAAI Conference on Artificial Intelligence, Québec City, QC, Canada, 27–31 July 2014; pp. 1636–1642.
28. He R, McAuley J. Ups and downs: Modeling the visual evolution of fashion trends with one-class collaborative filtering. In Proceedings of the 25th International World Wide Web Conference, Montreal, QC, Canada, 11–15 April 2016; pp. 507–517.
29. Glorot, X.; Bengio, Y. Understanding the difficulty of training deep feedforward neural networks. In Proceedings of the 13th International Conference on Artificial Intelligence and Statistics, Athens, Greece, 28–29 May 2010; pp. 249–256.
30. Man, T.; Shen, H.; Jin, X. Cross-Domain Recommendation: An Embedding and Mapping Approach. In Proceedings of the 26th International Joint Conference on Artificial Intelligence, Melbourne, Australia, 19–25 August 2017; pp. 2464–2470.

*Article*

# Complex Contagion Features without Social Reinforcement in a Model of Social Information Flow

Tyson Pond [1], Saranzaya Magsarjav [2], Tobin South [2], Lewis Mitchell [2] and James P. Bagrow [1,*]

[1] Department of Mathematics & Statistics, University of Vermont, Burlington, VT 05405, USA;
tyson.pond@uvm.edu
[2] School of Mathematical Sciences, The University of Adelaide, Adelaide, SA 5005, Australia;
saranzaya.magsarjav@adelaide.edu.au (S.M.); tobin.south@adelaide.edu.au (T.S.);
lewis.mitchell@adelaide.edu.au (L.M.)
* Correspondence: james.bagrow@uvm.edu

Received: 1 February 2020; Accepted: 24 February 2020; Published: 26 February 2020

**Abstract:** Contagion models are a primary lens through which we understand the spread of information over social networks. However, simple contagion models cannot reproduce the complex features observed in real-world data, leading to research on more complicated complex contagion models. A noted feature of complex contagion is social reinforcement that individuals require multiple exposures to information before they begin to spread it themselves. Here we show that the quoter model, a model of the social flow of written information over a network, displays features of complex contagion, including the weakness of long ties and that increased density inhibits rather than promotes information flow. Interestingly, the quoter model exhibits these features despite having no explicit social reinforcement mechanism, unlike complex contagion models. Our results highlight the need to complement contagion models with an information-theoretic view of information spreading to better understand how network properties affect information flow and what are the most necessary ingredients when modeling social behavior.

**Keywords:** online social networks; social media; information spreading; information diffusion; cross-entropy

---

## 1. Introduction

Social networks mediated through online platforms are an increasingly important way in which individuals send and receive information, and their influence is now felt in economics, politics, and the workplace [1–6]. These platforms provide rich opportunities for researchers to collect and study real-world data related to human behavior and the spread of information. In concert with these datasets, considerable research has worked towards better statistical and information-theoretic tools to quantify information flow [7–9] and towards more accurate mathematical models to understand and even predict information flow [10–12].

A common approach to measuring information flow over a network is to idealize information as a collection of 'packets,' and then track the spread of those packets throughout the network. This approach is especially common when studying social media where keywords such as hashtags or URLs are easily tracked. More complex phenomena, such as the adoption of behaviors can also be monitored and used as a proxy for information flow [13]. Treating information flow in this way brings to mind the spread of infections and the use of epidemiologically inspired models is popular. In this context, the social "diffusion" of information is often characterized as either a simple contagion or a complex contagion [14]. Simple contagions are those where each exposure can independently lead to an infection. Complex contagions, in contrast, introduce a social reinforcement mechanism where multiple exposures are needed before the contagion can spread.

However, despite its simplicity and popularity, there can be drawbacks to treating information as the contagion of discrete packets. Within social media, for example, there is a wealth of written information being posted by users that is ignored when focusing only on particular keywords. Likewise, considerable information could be exchanged between individuals without leading to an observable adoption of behavior. Therefore, we argue in this work that a more nuanced approach grounded in information theory can give a better view of information flow in online social networks while more fully using the available data.

The goal of this work is to study how network properties can affect information flow when taking an information-theoretic view on information flow, and how this information-theoretic view compares to contagion. We study the quoter model [12], a simple model for individuals generating text data within social media and apply information-theoretic estimators to the model text. Using both network models and real-world network data, we compare the behavior of information flow in this model with traditional simple and complex contagion, to see the similarities and differences we may observe through these contrasting viewpoints. Interestingly, we find that the quoter model exhibits several phenomena characteristic of complex contagion, despite lacking an explicit social reinforcement mechanism, the key feature of complex contagion.

The rest of this work is organized as follows. In Section 2 we describe information-theoretic estimators of information flow and mathematical models of information flow and contagion. In Section 3 we describe the materials and methods used in this study, including simulation details, measures of information flow, the network properties we investigate, and the network data we use. Section 4 presents our results comparing contagion models with the information-theoretically motivated quoter model and exploring how various network properties affect information flow in the quoter model. We conclude with a discussion in Section 5.

## 2. Background

### 2.1. Measuring Information Flow

Suppose an individual within a social network generates a stream of text representing posts shared online on Twitter, for example. The entropy rate $h$ of this text captures the information present within it. It can be challenging to estimate $h$ for natural language data as information is present in the ordering of the words, not just the relative frequencies of words [15]. To help address this challenge, Kontoyianni et al. [16] proved that the estimator

$$\hat{h} = \frac{T \log_2 T}{\sum_{t=1}^{T} \Lambda_t},$$

(1)

converges to the true entropy rate $h$ of a text, where $T$ is the length of the sequence of words and $\Lambda_t$ is the *match length* of the prefix at position $t$: it is the length of the shortest substring (of words) starting at $t$ that has not previously appeared in the text. This estimator has been used to study human dynamics including mobility patterns and social media predictability [11,17].

Equation (1) generalizes to an estimator of the **cross-entropy** $h_\times$ between two texts $A$ and $B$ [11,18]:

$$\hat{h}_\times(A \mid B) = \frac{T_A \log_2 T_B}{\sum_{t=1}^{T_A} \Lambda_t(A \mid B)},$$

(2)

where $T_A$ and $T_B$ are the lengths of the two texts, and $\Lambda_t(A|B)$ is the length of the shortest substring $[A_t, A_{t+1}, \ldots, A_{t+\Lambda_t(A|B)+1}]$ starting at position $t$ of text $A$ not previously seen in text $B$. Previously, in this case, refers to all the words of $B$ written prior to the time when the $t$th word of $A$ was written. Specifically, compute $\Lambda_t(A|B)$ by searching for each substring $[A_t]$, $[A_t, A_{t+1}]$, ... within $B_{:t} \equiv [B_j \mid \text{time}(B_j) < \text{time}(A_t)]$, the ordered sequence of words in $B$ that appear before the time of the $t$-th word in $A$, until the first substring $[A_t, \ldots, A_{t+\Lambda_t(A|B)+1}]$ that is not seen in $B_{:t}$. By matching the future text of $A$ (words posted at times $\geq \text{time}(A_t)$) against the past text of $B$ (words posted at times $< \text{time}(A_t)$) at every $t$, only $B$'s past predictive information about $A$'s future is estimated and *temporal precedence* is satisfied. The cross-entropy can be applied directly to the texts of a pair of individuals by choosing $B$ to be the text stream of one individual and $A$ the text stream of the other, and Equation (2) can be used to measure the information flow between those individuals by asking how much predictive information about one text is contained within the other. This can be a quite powerful and effective measure of information flow, as it satisfies temporal precedence of the text streams and it uses all of the available (text) data for the pair of users [7,11,12,16,18].

We focus on the cross-entropy estimated using Equation (2) as a pairwise measure of information flow, but generalizations can capture information flow from multiple social ties towards a single individual [11,12]. Doing so allows for measures of more complex information flow such as analogs of transfer entropy or causation entropy [7,8,19]. The best extensions of information flow estimators beyond pairwise measures remains an active and fruitful area of research (see also our discussion in Section 5).

Closely associated with the cross-entropy is the predictability $\Pi$. Predictability, given by Fano's Inequality [20], provides a bound on how accurately an *ideal* predictive method can perform when working with data of a given entropy: $\Pi$ is the probability the most accurate possible method will correctly predict the subsequent word with the given information's uncertainty (i.e., the cross-entropy).

$$h(\Pi) + (1 - \Pi)\log(z - 1) \geq h_\times \tag{3}$$

where $h(\Pi) = -\Pi\log(\Pi) - (1 - \Pi)\log(1 - \Pi)$ and $z$ is the cardinality of the sample space; in our problem, this is the vocabulary size or number of unique words for the quoter model (Section 3.1). The predictability is then given by finding numerically the largest $\Pi$ that satisfies Equation (3). Equation (3) demonstrates that $h_\times$ and $\Pi$ are functionally equivalent (and inversely related, with higher $h_\times$ corresponding to lower $\Pi$ and vice versa) as $z$ is a constant for the model we study here (see also discussion in Section 5). Higher values of $\Pi$ (lower $h_\times$) correspond to higher amounts of information flow.

### 2.2. Quoter Model

To study the effects of network properties on information flow, we use the recently proposed quoter model [12]. The quoter model represents an idealized model of social conversations, meant to capture some of the processes by which individuals in an online social network post text while also being analytically tractable. Nodes in a network generate text streams both by sampling from a given vocabulary distribution and by copying ("quoting") short sub-sequences of text from their neighbors. This model provides a parameter $q$, the quote probability that tunes the degree of information flow. (Full details of the model and how we simulate it are given in Section 3.1.) After simulating the quoter model for a given number of time steps (Section 3.1), a text stream has been generated by each node in the network, and we can estimate the cross-entropies between these texts to study the social flow of written information. See Bagrow and Mitchell [12] for full details on the quoter model.

### 2.3. Other Models of Information Flow

Contagion approaches are often used to model information flow [14]. A classic simple contagion approach to information flow is compartment models, taken from models of epidemics. Two simple compartment models are Susceptible-Infected (SI) and Susceptible-Infected-Recovered (SIR) models. On a network, a small number of nodes are initially "infected" while the remaining nodes are susceptible. The contagion then spreads from those infected nodes with a constant transmission rate per link so that each node in the "S" compartment has a constant probability to move to the "I" compartment with any given exposure. For SIR models, an additional "R" compartment is used to model a recovery process where infected nodes cease spreading the contagion while also becoming immune to reinfection. Many variants on these models exist.

Complex contagion phenomena are typically captured with threshold models [21,22]. Here nodes are again labeled as susceptible or infected, but the probability for a node $i$ to become "infected" is a function of the number of neighbors of that node already infected. If too few neighbors are infected there is zero probability that $i$ will be infected. Yet if a sufficient fraction of $i$'s neighbors become infected, then $i$ has a non-zero probability of becoming infected. This *social reinforcement* mechanism is intended to capture the cognitive mechanisms underlying opinion change, knowledge acquisition, and other facets of how individuals respond to and adopt information and ideas [23,24].

Complex contagion leads to several phenomena that differ from simple contagion. For one, there is an interesting *cascade window* where network density leads to a non-monotonic relationship with the spread of the contagion. Often denser networks lead to less spread, unlike simple contagion where a contagion will spread more easily as denser networks afford more opportunities (links) for spreading. Another feature of complex contagion is the complicated role of clustering where clustering can appear to either promote or inhibit contagion [25–28]. Complex contagion also exhibits a "weakness of long ties" effect, where long ties impede the flow of contagion [29], in contrast with the seminal "strength of weak ties" result [30] that implies long-range ties have an out-sized role in promoting information flow. The goal of our work here is to study the information-theoretic view of information flow we adopt here with the quoter model and compare to the effects of complex contagion that is commonly used as a *non*-information-theoretic view to study information flow.

### 3. Materials and Methods

In this study, we use the quoter model on networks to elucidate the role of network structure on information flow. Here we describe the procedures to simulate the quoter model, measure information flow between nodes in networks, we describe the network features we study in relation to information flow, and we provide the details on the network models (random graphs) and real-world network datasets we study.

### 3.1. The Quoter Model

We use the following process to simulate the quoter model on a given network. The quoter model requires a directed graph $G = (V, E)$ (where $N = |V|$ is the number of nodes and $M = |E|$ is the number of edges) and, in the most general case, quote probabilities $q_{uv}$ on each directed edge (we say node $v$ (ego) may quote $u$ (alter) if the edge $u \to v$ exists and has $q_{uv} > 0$). We simplify this for our simulations: when an ego generates new text, with probability $q$ (bidirectional quoting) we pick an alter (predecessor) uniformly at random to quote from; otherwise, with probability $1 - q$ the ego generates new content. If an ego quotes an alter (probability $q$), copy a random segment of the alter's past text and append this onto the ego's growing text stream. We take the "quote length" (number of words) being copied to be Poisson-distributed (with mean $\lambda$) for all users; Otherwise, if not quoting (probability $1 - q$), generate new content by sampling with replacement from a vocabulary distribution $W(w)$ and appending those samples onto the ego's growing text stream, where the number of samples is again Poisson-distributed with mean $\lambda$. We assume a common, fixed vocabulary distribution $W(w)$ that follows a Zipf law of word use, as in prior studies and motivated by real-world language usage patterns [12]. Specifically, a Zipf law defines the probability of using word $w$ to be a power law based on the rank $r_w$ of $w$: $W(w) = H_{z,\alpha}^{-1} r_w^{-\alpha}$, where $z$ is the vocabulary size and $H_{z,\alpha} = \sum_{r=1}^{z} r^{-\alpha}$. Here we take $z = 1000$ as in [12] and, unless otherwise stated, focus on the exponent $\alpha = 1.5$, a value typical of social media data. We focus in this work on $q = 1/2$ and $\lambda = 3$ but we explore the robustness of our results to other parameter choices in Appendix A. This process repeats for $T = 1000N$ time steps so that each user has generated approximately $1000\lambda = 3000$ words when complete. This number of time steps was chosen to ensure the entropy estimator would converge (see [16,18] for convergence proofs).

While very short amounts of text will make the estimated entropy too uncertain to be reliable, this length of text is in line with the empirical convergence of $h_\times$ reported in real data [11].

### 3.2. Measuring Information Flow over the Network

After generating text streams for all nodes in $G$ by iterating the quoter model, the cross-entropy estimator (Equation (2)) is then applied as needed to compute $h_\times$. We compute the cross-entropy over all edges, $\{h_\times\} = \{h_\times(u \mid v) \mid (u,v) \in E\}$, and report the mean $\langle h_\times \rangle$ and variance $\text{Var}(h_\times)$ of these values. (We examine the distribution of $h_\times$ in Appendix B to show that $\langle h_\times \rangle$ and $\text{Var}(h_\times)$ are reasonable summaries of the distribution of $h_\times$.) Likewise, the predictability $\Pi$, given by Fano's Inequality [20], is a functionally equivalent measure of information flow (as we assume the same vocabulary sizes for nodes in the quoter model). We focus on link-based cross-entropies although the cross-entropy estimator can be applied to non-neighboring nodes. Indeed, when studying the role of community structure in modular networks (see Section 3.4), we also consider cross-entropies between nodes in different modules, to assess information flow between and within said modules.

### 3.3. Simulating Contagion Models

To compare and contrast information flow in the quoter model, we also simulate traditional models of information flow, specifically simple and complex contagion. For simple contagion we simulate a stochastic SIR model on different networks (1000-node Erdős-Rényi and Barabási-Albert networks, as well as a sample of real-world networks) using [31]. For the simulations here we set the transmission rate 20 and recovery rate 1. We initialize with a random 5% of the nodes infected, and run 10 outbreaks on 100 realizations of the network for each choice of average degree $\langle k \rangle$. For complex contagion we use exactly the same parameters, except we introduce a threshold function for transmission as in [22], where the transmission rate is set to zero if the proportion of infected neighbors is below some threshold $\phi$ (and we set $\phi = 0.18$ following [22]). For all simple and complex contagion simulations we measure the peak outbreak size, noting that larger outbreak sizes conventionally correspond to greater information flow.

### 3.4. Assessing the Impact of Structure on Dynamics

In this work we use several network models (random graphs) tailored to control for various network properties such as density, clustering, and modular structure. Here we describe the models and properties we study in relation to information flow in the quoter model.

#### Density and Average Degree

To explore how network density relates to information flow, we create Erdős-Rényi and Barabási-Albert networks of $N$ nodes with varying average degree, $\langle k \rangle$, allowing us to the tune their densities. For the Erdős-Rényi networks we add edges independently with probability $p = \langle k \rangle / (N - 1)$. For the Barabási-Albert model we start with $m = \langle k \rangle / 2$ nodes with no edges and add nodes which each form $m$ links with previous nodes according to preferential attachment. Here we measure how cross-entropies varies with the densities of the networks using their average degree $\langle k \rangle$ and edge density $M / \binom{N}{2}$ where $M$ is the total number of edges in the network. To complement the Erdős-Rényi and Barabási-Albert results, we also compare the densities of real networks with their average cross-entropy.

#### Degree Heterogeneity

To assess the role of degree heterogeneity on information flow, we study the simplest random graph model with tunable degree heterogeneity, termed "dichotomous networks" in [32]. Dichotomous networks are generated via the configuration model. They have only two types of nodes—those with degree $k_1$ and those with degree $k_2$. We assume there are $N/2$ nodes of each degree and fix $k_1 + k_2$

so that the average degree is fixed. The mean and variance of the degree distribution, respectively, are given by $\mu = \frac{1}{2}(k_1 + k_2)$ and $\sigma^2 = (k_1 - k_2)^2/4$. We are interested in how the cross-entropy varies with $k_1/k_2$. When $k_1/k_2 = 1$ the network reduces to a random $k$-regular graph ($\sigma^2 = 0$), while $\sigma^2 \to \infty$ as $k_1/k_2 \to 0$.

Clustering

Clustering or triadic closure, the tendency towards forming triangles, is a key feature of social networks. We studied clustering using a network model with tunable numbers of triangles and with a randomization procedure that can lower the number of triangles in an existing network. We quantify a network's clustering using *transitivity* $T(G)$, the fraction of possible triangles in the network which actually exist: $T(G) = 3N_{\text{triangles}}/N_{\text{triads}}$, where $N_{\text{triangles}}$ counts the number of triangles in the network and $N_{\text{triads}}$ is the number of triads or paths of length 2.

We constructed "small-world" networks using the Watts–Strogatz (WS) model [33] to tune their clustering. We generated a one-dimensional periodic lattice of $N$ nodes with $k$ nearest-neighbor connections, and randomly rewired lattice edges with a rewiring probability $p$. Varying the rewiring probability $p$ allows us to tune the network diameter and clustering.

While the Watts–Strogatz model lets us generate networks with different clustering values, a generic challenge when assessing the impact of clustering (and other network properties) on dynamics is generating networks with tunable clustering, but for which other structural properties, such as density or diameter, can be controlled for. To study the relationship between transitivity and information flow, we apply the established degree-preserving stochastic rewiring or "x-swap" method [34–36], in which we repeatedly choose two links at random and two randomly selected endpoints of those links are swapped as long as the number of links does not change by swapping and the network does not become disconnected. These swaps lower transitivity while fixing the number of links and degrees of all nodes in the network. We performed $5M$ swaps for each real network. Examining information flow on the randomized network compared with information flow on the original network can then illustrate what effect, if any, transitivity had on information flow.

Community Structure and Modularity

Community structure is another inherent property of social networks. It is commonly quantified using modularity [37]:

$$Q = \frac{1}{2M} \sum_{i,j} \left( a_{ij} - \frac{k_i k_j}{2M} \right) \delta(c_i, c_j),$$

where $M$ is the total number of links, the sum runs over all pairs of nodes in the network, $\mathbf{A} = [a_{ij}]$ is the adjacency matrix of the network, $k_i$ is the degree of node $i$, $\delta$ is the Kronecker delta, and $c_i$ denotes the community containing $i$. The community structure encoded in the $\{c_i\}$ can be found using a community detection algorithm or it may be planted within a network model. To investigate community structure within a network model, we examined instances of the stochastic block model (SBM) [38,39] with $N$ nodes and two planted blocks, or groups of nodes, denoted $A$ and $B$, of equal size $m = N/2$. Here there are two connection probabilities: $p_0$ (the within-block connection probability) and $p_1$ (the between-block connection probability) governing the probability for a link to form between nodes in the same block and in different blocks, respectively. The expected modularity in this two-block stochastic block model is

$$Q = \frac{1}{2} \left( \frac{p_0 - p_0 m + p_1 m}{p_0 - p_0 m - p_1 m} \right).$$

Our main quantities of interest are the average cross-entropy on within-block edges, $\langle h_\times(\text{within}) \rangle$, the average cross-entropy on between-block edges $\langle h_\times(\text{between}) \rangle$ and their difference, $\Delta h_\times \equiv \langle h_\times(\text{between}) \rangle - \langle h_\times(\text{within}) \rangle$. These quantities describe to what extent information flows within and between communities.

We also computed modularity for real networks using the Louvain method [40]. The Louvain method is a hierarchical community detection algorithm that finds a partition of nodes that maximizes modularity $Q$. As commonly done, we initialize each node in its own community.

Multiple Vocabulary Distributions

A recent study [41] showed that heterogeneity in the dynamical parameters can be as important as structural heterogeneity. Communities offer an obvious way to implement such heterogeneity: We also investigate a two-block SBM where we distinguish the two groups $A$ and $B$ by giving them different Zipf exponents $\alpha_A, \alpha_B$, respectively, for their vocabulary distributions.

### 3.5. Network Datasets

To supplement the above graph models, we also studied contagion and quoter model dynamics on real-world networks. We developed a corpus of 10 social networks spanning a range of sizes and densities that were used as the basis for simulation. See Appendix C for details on network sources and processing. Table 1 shows several descriptive statistics for the networks we analyzed.

**Table 1.** Descriptive statistics for real-world networks used in this study. ASPL: Average Shortest Path Length. Modularity computed using the Louvain method [40].

| Network | $|V|$ | $|E|$ | $\langle k \rangle$ | Density | Transitivity | ASPL | Modularity | Assortativity |
|---|---|---|---|---|---|---|---|---|
| Sampson's monastery | 18 | 71 | 7.9 | 0.464 | 0.53 | 1.54 | 0.29 | −0.07 |
| Freeman's EIES | 34 | 415 | 24.4 | 0.740 | 0.82 | 1.26 | 0.07 | −0.15 |
| Kapferer tailor | 39 | 158 | 8.1 | 0.213 | 0.39 | 2.04 | 0.32 | −0.18 |
| Hollywood music | 39 | 219 | 11.2 | 0.296 | 0.56 | 1.86 | 0.20 | −0.08 |
| Golden Age | 55 | 564 | 20.5 | 0.380 | 0.53 | 1.64 | 0.45 | −0.13 |
| Dolphins | 62 | 159 | 5.1 | 0.084 | 0.31 | 3.36 | 0.52 | −0.04 |
| Terrorist | 62 | 152 | 4.9 | 0.080 | 0.36 | 2.95 | 0.52 | −0.08 |
| Les Miserables | 77 | 254 | 6.6 | 0.087 | 0.50 | 2.64 | 0.56 | −0.17 |
| CKM physicians | 110 | 193 | 3.5 | 0.032 | 0.16 | 4.24 | 0.61 | 0.11 |
| Email Spain | 1133 | 5452 | 9.6 | 0.009 | 0.17 | 3.61 | 0.57 | 0.08 |

## 4. Results

Here we compare information flow in the quoter model with traditional simple and complex contagion (Section 4.1), then investigate how degree heterogeneity (Section 4.1), clustering (Section 4.2) and network modularity (Section 4.3) affect information flow. We also study how heterogeneity in the parameters affects information flow compared to the effects of network structure (Section 4.4).

### 4.1. Information Flow and Models of Contagion

A distinguishing feature of simple and complex contagion is that denser networks lead to higher spreading for simple contagion and lower spreading (mostly) for complex contagion. We illustrate this difference using simulations in Figure 1A,B. For the simple and complex contagion models we use the average peak size of the outbreak as our measure of information flow in the network, whereas for the quoter model we use the average predictability over links. The decrease in spreading in complex contagion is due to its social reinforcement mechanism: it is more difficult for a contagion to spread when egos have many alters as more alters must adopt the contagion before the ego does. Yet we see in Figure 1C that the quoter model, which lacks an explicit social reinforcement mechanism, also exhibits lower information flow at higher density. Here we measure information flow using predictability on links (Section 3.2), which is functionally equivalent (Section 2.1) in our simulations to the cross-entropy $h_\times$ (Figure 1C inset). Please note that while the curve for $h_\times$ looks visually similar to that of simple contagion's average peak size, it is measuring the opposite effect: higher $h_\times$ corresponds to lower information flow. These results also hold on our corpus of real-world networks (Figure 2).

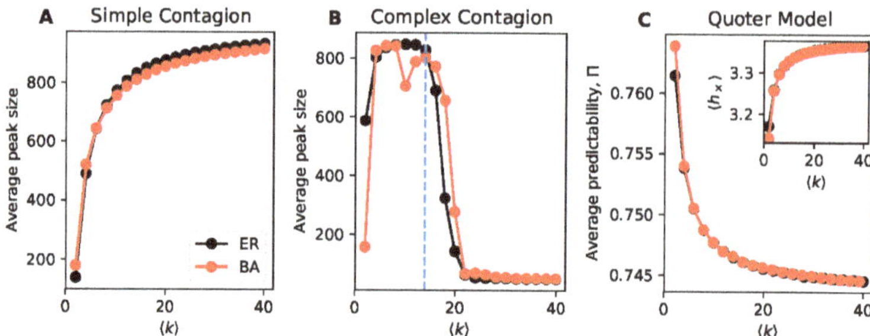

**Figure 1.** Denser networks are associated with higher information flow for simple contagion but lower information flow for both complex contagion and the quoter model. Here density is measured by average degree $\langle k \rangle$ for Erdős-Rényi (ER) & Barabási-Albert (BA) model networks. (**A**) Simple contagion. (**B**) Complex contagion (**C**) Quoter model. (Panel C, inset) Average cross-entropy on links; higher cross-entropies correspond to lower predictabilities and lower information flow, unlike for contagions where higher average peak sizes correspond to higher information flow. Networks consisted of $N = 1000$ nodes and each point constitutes 200 simulations; parameters for simulating information flow in these models are described in Section 3.

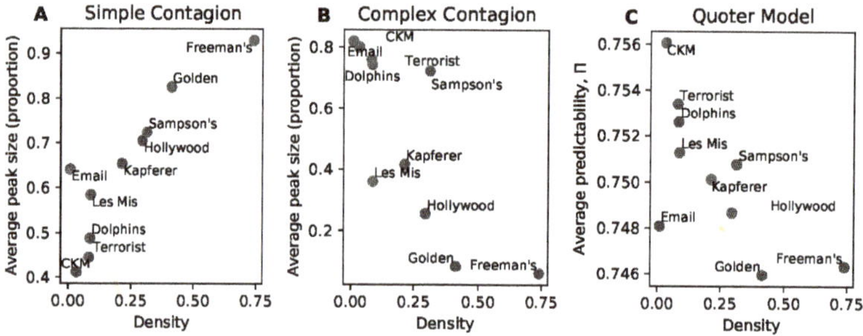

**Figure 2.** Information flow on real-world networks. (**A**) Simple contagion. (**B**) Complex contagion. (**C**) Quoter model. Here information flow measures (average peak size, average text predictability) are compared to network density $M/\binom{N}{2}$. The association between information flow and density, either positive (simple contagion) or negative (complex contagion, quoter model), is significant (Wald test on non-zero regression slope, $p < 0.05$). Each point constitutes 300 simulations.

Somewhat surprisingly, in Figure 1C we see that Erdős-Rényi (ER) and Barabási-Albert (BA) networks are qualitatively indistinguishable in terms of information flow, despite the preponderance of hubs in the latter that we expect would play an out-sized role in information flow. To better understand this observation, we investigated the variance of $h_\times$ over links in Figure 3A. We see that the cross-entropy varies more from link to link in the BA networks than for ER networks, indicating that hubs do not move the average information flow but do create fluctuations in the flow, especially for sparser networks.

To further explore the role of network structure heterogeneity, we investigate dichotomous networks (Section 3.4). Here half the nodes have degree $k_1$ and the other half have degree $k_2$. Varying the degree ratio $k_1/k_2$ allows us to tune the degree variance within this simplified network model. In Figure 3B we see that the total number of nodes and average degree change the average information flow while the degree heterogeneity ($k_1/k_2$) has little effect. Yet degree heterogeneity does affect the variance of information flow (Figure 3C). These simpler dichotomous networks show the same effects as observed previously in BA networks.

The simplified bimodal degree distribution of dichotomous networks also lets us explore the effects of ego and alter degrees by computing conditional expectations of $h_\times$ conditioned on degree.

We see from the grouping of curves in Figure 3D that the degree of the ego (the node being predicted) but not the alter (the node predicting) plays a role in the information flow: degree-$k_1$ egos have more information flow than degree-$k_2$ egos regardless of the degree of the alter.

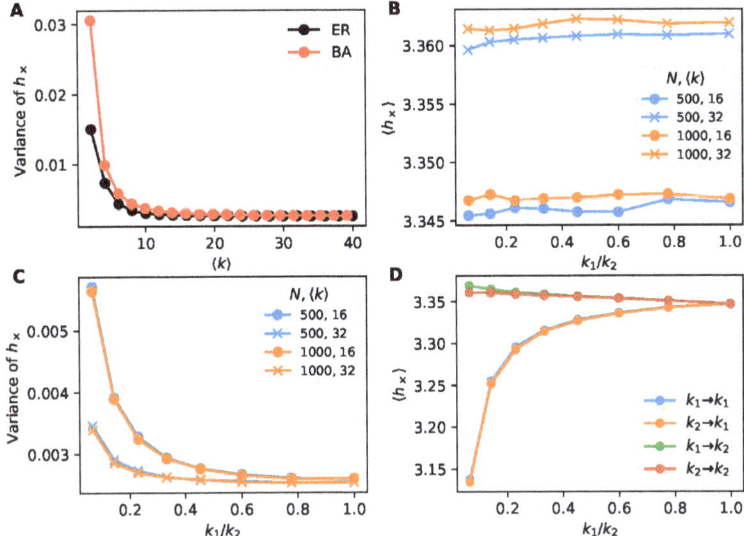

**Figure 3.** Exploring the variance of information flow. (**A**) Variance of cross-entropy is higher at low densities for BA than ER networks despite the average $h_\times$ being similar (Figure 1C). (**B–D**) Information flow on dichotomous networks (random networks where all nodes have degree $k_1$ or degree $k_2$, allowing tunable degree heterogeneity) of size $N \in \{500, 1000\}$ with $\langle k \rangle \in \{16, 32\}$. Each point constitutes 500 trials. (**B**) Average cross-entropy versus $k_1 / k_2$. Degree heterogeneity does not affect average cross-entropy, supporting Figure 1C. Network size has a smaller effect on $h_\times$ compared to the average degree. (**C**) Variance of cross-entropy versus $k_1 / k_2$. Higher degree heterogeneity (lower $k_1 / k_2$) leads to higher variation in $h_\times$ over links, indicating the existence of highly predictive nodes and nodes that contribute little predictive information within heterogeneous networks. (**D**) Dichotomous networks of size $N = 1000$ and $\langle k \rangle = 16$. Average cross-entropy over links conditioned on degrees of endpoints (predicting ego from alter). Only the degree of the ego matters, approximately, not the degree of the alter.

## 4.2. Interplay of Clustering and Information Flow

Next, we study how clustering (transitivity) affects information flow. Clustering plays a complicated role in both simple and complex contagion [25,27] and we report interesting, if mixed, results in Figure 4 with the quoter model's information flow.

First, in Figure 4A we study information flow for small-world networks that are randomly rewired to remove clustering [33]. Regardless of network size or average degree, information flow decreases (higher $h_\times$ in top panel of Figure 4A) as clustering decreases (Figure 4A bottom panel). Please note that rewiring also changes the diameter of the small-world network, but we see that the main increase in $h_\times$ occurs when clustering begins to drop. In small-world networks, clustering tends to promote information flow.

Next, in Figure 4B we investigate transitivity in the corpus of real-world networks. For each network, we compute information flow on the original network and on a replicate of the network that is randomized by the "x-swap" method. The x-swap method lowers transitivity for all networks but for half of the networks it also lowers $h_\times$, contradicting the previous results on small-world networks by indicating that transitivity *inhibits* information. However, it is challenging to draw a sharp conclusion from this x-swap procedure as it also affects other network properties simultaneously. We illustrate this in Figure 4C where we compare four network properties in the original and x-swapped networks. X-swapping affects transitivity but also average shortest path length (ASPL), modularity and assortativity (degree correlations). This means the changes in information flow seen in Figure 4B may be due to changes in a combination of these (and possibly other) network properties. Unfortunately, it remains an open research problem how best to systematically control for network properties to uncover their effects on dynamics.

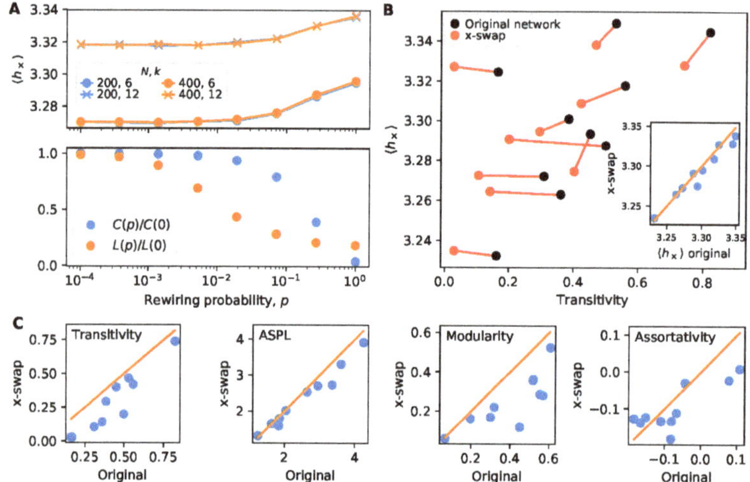

**Figure 4.** Mixed effects of clustering on information flow. (**A**) Information flow on small-world networks of size $N \in \{200, 400\}$ and average degree $k \in \{6, 12\}$. As network rewiring increases (and clustering decreases) $h_\times$ increases. This suggests that clustered networks promote information flow. Rewiring a small-world network changes the diameter ($L$) as well the clustering (panel A, bottom); however, $h_\times$ begins to increase primarily when the clustering begins to drop, not when diameter begins to drop. Each point constitutes 300 trials. (**B**) Average cross-entropy versus transitivity for real-world networks. By randomizing networks using the standard "x-swap" method (Section 3.4), we can lower the transitivity and investigate how $h_\times$ changes. Some networks show little change in $h_\times$ on randomized networks compared with the original networks, while others show a slight decrease in $h_\times$. This is especially visible in the inset comparing $h_\times$ directly. Each point constitutes 300 simulations. (**C**) Several network properties before and after the x-swap method. While the x-swap method lowers transitivity, it also alters other important network properties, making it challenging to isolate the role of clustering from other properties.

### 4.3. Community Structure and the Weakness of Long Ties

The effects of long-range links on information flow have been investigated for some time, from the seminal "strength of weak ties" [30] and the contrasting "weakness of long ties" in complex contagion [29]. Here we investigate long ties in the context of community structure: In networks with densely connected groups of nodes, long ties act to bridge nodes in different groups. How does information flow differ between groups compared to flow within groups?

Using the stochastic block model (Section 3.4) with two groups of equal size as a model for networks with dense modules, we study in Figure 5 information flow between and within groups. The two-group SBM is parameterized by two connection probabilities, the probability for a link within each group ($p_0$) and the probability for a link between the two groups ($p_1$). In Figure 5A we see that information flow decreases as $p_0$ increases and the network becomes denser. Likewise, the difference in information flow $\Delta h_\times$ increases due to between-block links containing less predictive information

(Figure 5B). This supports the well-known "weakness of long ties" feature of complex contagion. For larger values of $p_1$, when there are more links connecting the groups making them less distinct, this difference decreases. The collapse of curves in Figure 5C indicates $\Delta h_\times$ is entirely predicated on the network modularity $Q$.

Interestingly, we also remark that $\Delta h_\times$ is always positive—even when $p_0 < p_1$ (equivalently, $Q < 0$). We would expect more information flow between groups than within when within this "anti-community" regime of the SBM, when there are more links between groups than within groups, yet we observe a weak effect otherwise.

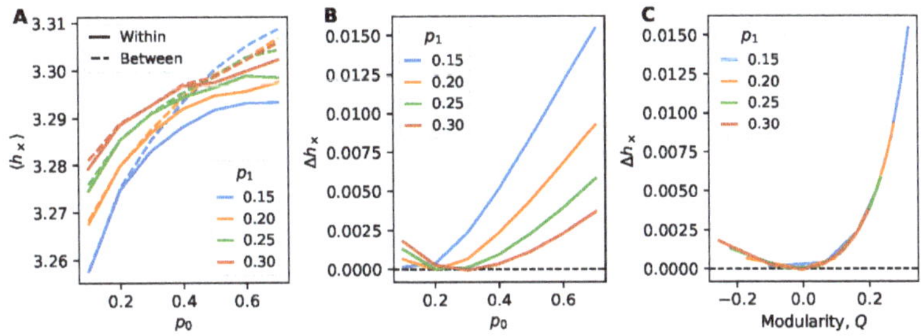

**Figure 5.** Information flow within the stochastic block model (SBM) of $N = 100$ (two blocks of size $N = 50$). Each point constitutes 10k trials. (**A**) Average cross-entropy on within-block edges and between-block edges as a function of the within-block connection probability $p_0$ for different between-block connection probabilities $p_1$. (**B**, **C**) Examining the cross-entropy difference $\Delta h_\times \equiv \langle h_\times(\text{between}) \rangle - \langle h_\times(\text{within}) \rangle$ across (**B**) connection probabilities and (**C**) modularity $Q$. Examining $\Delta h_\times$ as a function of modularity $Q$ shows a clear collapse across values of SBM probabilities. Interestingly, anti-community structure ($Q < 0$) still leads to positive $\Delta h_\times$, indicating that information flow is still more prevalent within blocks.

### 4.4. The Role of Dynamic Heterogeneity

In our results so far, we have treated nodes as identical within the quoter model and focused only on their topological differences within the network. Yet recent studies have underlined the importance of comparing dynamic heterogeneity with structural heterogeneity [41]. Here we taken an exploratory step in this direction by considering a generalization of the quoter model where nodes have different vocabulary distributions.

We explored how information flow changes in the stochastic block model when the nodes in the two blocks have different vocabulary distributions. This is intended to model a difference in the nodes between the two groups, capturing in the quoter model a social homophily in how egos write. Specifically, we assume they have the same vocabularies and follow Zipf distributions, but the exponent of the Zipf distribution is different: nodes in block A have exponent $\alpha_A$ and nodes in block B have exponent $\alpha_B$. A larger $\alpha$ (steeper distribution) corresponds to a less diverse vocabulary, and could capture a group of nodes that is more consistent and repetitive in their dialog. In contrast, a lower $\alpha$ (shallower distribution) may describe a group of nodes that uses more diverse words.

Figure 6 shows how information flow changes when the two blocks have different vocabulary distributions (Figure 6A,C) compared with the same distribution (Figure 6B). For illustration, we show the Zipfian vocabulary distributions for the two groups as insets in Figure 6. We observe a much larger trend in how cross-entropy changes with modularity when the exponents are not equal compared to when they are equal. This underscores how structural features (the degree of modularity) greatly magnifies the effects of intrinsic dynamic heterogeneity (different vocabulary distributions). While modularity plays a role even when the two groups have identical vocabulary distributions (Figure 5), this difference is challenging to detect in Figure 6B when viewed on the scale of groups with different vocabulary distributions (Figure 6A,C).

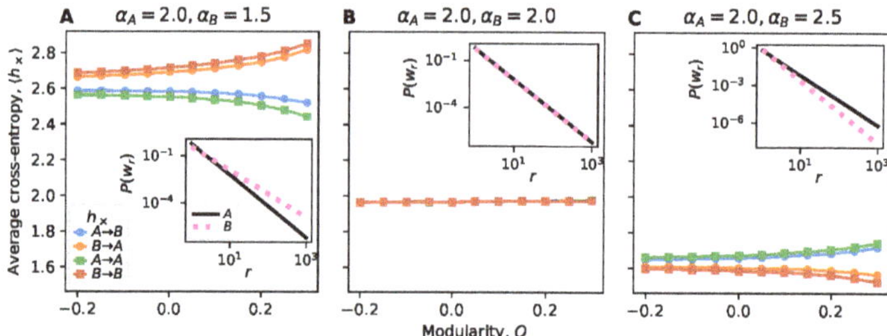

**Figure 6.** Effects of dynamic heterogeneity on information flow in the stochastic block model. Nodes in group $A$ have Zipfian vocabulary distribution with exponent $\alpha_A$ while nodes in $B$ have exponent $\alpha_B$. The between-block connection probability is fixed ($p_1 = 0.15$) and the within-block connection probability $p_0$ is varied to generate a range of modularities. Since the structure is symmetric (subgraphs $A$ and $B$ have the same size and expected density), we only show the result of fixing $\alpha_A = 2$ and varying $\alpha_B$. Each point constitutes 150 trials. (**A**) The vocabulary distribution of group $A$ has a lower Shannon entropy than of $B$, and this difference is visible from examining links $A \rightarrow A$ and $B \rightarrow B$. When examining links $A \rightarrow B$ and $B \rightarrow A$, the cross-entropy is mainly dependent on the vocabulary distribution of the alter. As modularity increases, differences between the predictabilities of various nodes are exaggerated. (**B**) In homogeneous communities, the cross-entropy does not vary with modularity at such a scale. (**C**) The vocabulary distribution of group $A$ has a higher Shannon entropy than of $B$. Similar mirror results are seen as in panel A.

## 5. Discussion

In this work, we have studied how the social flow of written information can be affected by network properties such as the density of links, preponderance of triangles, and modular or community structure. We focused on the quoter model, a toy model for a network of individuals to communicate by generating text sequences and applied information-theoretic estimators of the information flow to these texts. We compared results of information flow in the quoter model with traditional simple and complex contagion models.

A particularly intriguing facet of the interplay between quoter model dynamics and network topology is how the quoter model exhibits both the density-driven inhibition of information flow and the weakness of long ties that are signatures of complex contagion, despite lacking an explicit mechanism of social reinforcement. Social reinforcement, the idea that individuals adopt a piece of information only after receiving repeat exposure from social ties, is considered one of the characteristics that distinguishes complex contagion from epidemic spreading. Social reinforcement mechanisms better model how people perceive and react to information. Yet we found here that social reinforcement is not strictly necessary when modeling a more nuanced view of information flow. In particular, considering text streams (as generated by the quoter model) and predictive measures of information flow (as quantified using cross-entropy estimators) allows us to capture how information can be "drowned out" by the increased "cross-talk" that occurs in denser networks, showing how increased density can inhibit information flow. Further pursuing this line of investigation may give more insight into information flow and even human behavior within social networks.

We also found a mixed combination of results relating clustering to information flow. For small-world (Watts–Strogatz) networks, increasing the clustering leads to a significant increase in information flow (decrease in cross-entropy). At the same time, however, experiments on real-world networks showed the opposite effect: randomizing networks to lower transitivity while preserving connectedness and the degree distribution leads to a *decrease* in information flow. However, this well-established randomization procedure does not control for other network properties such as modularity or average shortest path length, so it remains an open question if the interplay of multiple effects may resolve the discrepancy between these results.

Another interesting result related information flow to community structure, with the modularity $Q$ used to measure the strength of the modular divide. When $Q > 0$, meaning there were fewer links between modules than expected, we found in Figure 5 an increase in cross-entropy between modules compared with the cross-entropy between nodes that share a module, as expected by the "weakness of long ties". However, we found the same increase in cross-entropy when $Q < 0$, where there were more links between modules than expected. We would initially expect this regime of "anti-community" structure to have more information flow between modules as there exist more links to facilitate this flow. One possible reason for this anti-community result is that nodes in the same group, while having fewer direct links to one another, may have many links to common nodes in the other group, leading to more similar inputs to their texts. This nonlocal interplay of information flow and network structure is an intriguing avenue for future work.

There are some important limitations to discuss regarding this work. We only considered undirected, unweighted networks. In the context of social networks, this implies all relationships are reciprocal and equal in strength. Future work should extend to directed, weighted networks. Furthermore, a more exhaustive study of the robustness of results to parameter choices is necessary (we take a first step towards this in Appendix A). Vocabulary size is another parameter worth exploring; here we assume it is constant across all nodes. Likewise, cross-entropy (Equation (2)) is a somewhat simplistic information-theoretic measure of information flow, and it is important to consider more advanced measures. Measures such as transfer or causation entropy can offer more insight, quantifying non-redundant information and allowing us to identify indirect influences [7,8]. However, in the context of time-ordered social text data, it is challenging to estimate conditional entropies, making it non-obvious how to implement such measures [12]. Finally, while we observed several features that are signatures of complex contagion, not all features of complex contagion are exhibited by the quoter model. For example, there is an optimal modularity that maximizes spreading of complex contagions within the stochastic block model: if $Q$ is either too small or too large then the contagion will not spread [42]. We were unable to observe a corresponding feature within the quoter model. This warrants further investigation, in particular to understand if this is due to how the quoter model differs from complex contagion models, or if it is due to the information-theoretic measure of information, or a combination of the two.

In general, contagion models are a successful way to study information flow in social networks, but to gain more insight it is necessary to adopt more nuanced views of information flow. We argue here that information theory can provide a pathway towards these insights, especially when combined with models such as the quoter model that capture features of human behavior while also modeling key aspects of the data being generated by social network platforms.

**Author Contributions:** Conceptualization, T.P., L.M. and J.P.B.; Funding acquisition, L.M. and J.P.B.; Investigation, T.P., S.M. and L.M.; Methodology, T.P., T.S. and J.P.B.; Project administration, J.P.B.; Software, T.P., T.S. and L.M.; Supervision, L.M. and J.P.B.; Validation, T.P., S.M. and L.M.; Visualization, T.P.; Writing–original draft, T.P. and J.P.B.; Writing–review & editing, T.P. and J.P.B. All authors have read and agreed to the published version of the manuscript.

**Funding:** This material is based upon work supported by the National Science Foundation under Grant No. IIS-1447634.

**Conflicts of Interest:** The authors declare no conflict of interest.

## Abbreviations

The following abbreviations are used in this manuscript:

| | |
|---|---|
| ASPL | Average Shortest Path Length |
| BA | Barabási-Albert |
| ER | Erdős-Rényi |
| SBM | Stochastic Block Model |
| SI | Susceptible-Infected |
| SIR | Susceptible-Infected-Recovered |
| WS | Watts–Strogatz |

## Appendix A. Further Exploring Quoter Model Parameters

To support our results, here we explore other choices of quoter model parameters ($q$ and $\lambda$). The simulations are done on smaller networks to make it less computationally expensive to do a wide sweep of the parameter space. We first simulate the quoter model on ER, BA, and small-world networks for $q \in \{0.1, 0.5, 0.9\}$ and vary $\langle k \rangle$ or the rewiring probability, $p$, to support results from Section 4.1 and Section 4.2. We then simulate the ER, BA, and small-world experiments again for various combinations of the quote probability $q$ and mean quote length $\lambda$. We evaluate the robustness of results for ER networks as follows. For each combination of $(q, \lambda)$, we calculate the difference $\langle h_\times \rangle_{k=20} - \langle h_\times \rangle_{k=6}$, whereby $\langle h_\times \rangle_{k=20}$ we mean the average cross-entropy on ER networks of average degree $k = 20$. The quantity will be positive if density inhibits information flow. This allows us to assess the how the magnitude of our results vary with $(q, \lambda)$, although it does not confirm a monotonic trend holds. We repeat these calculations with the BA networks and extend them to the small-world networks by replacing $\langle k \rangle$ with $p \in \{0, 1\}$. In general, we find in Figures A1 and A2 that our results are qualitatively robust to parameter choices, with the exception of very small values of $q$, as we expect.

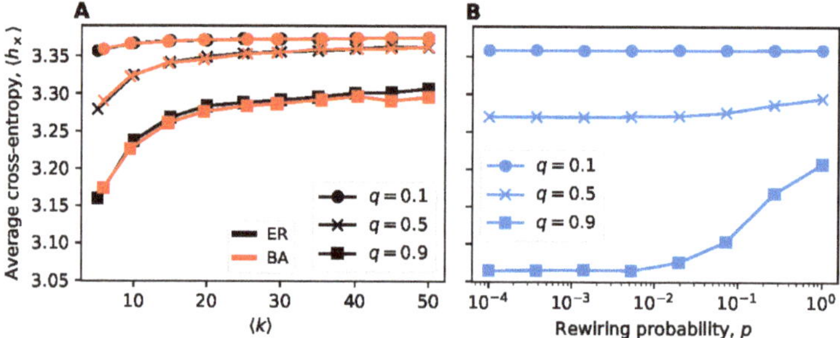

**Figure A1.** Trends in information flow in ER, BA, and small-world networks for $q \in \{0.1, 0.5, 0.9\}$. Except for very low quote probabilities, we see qualitatively similar trends. (**A**) ER & BA networks of size $N = 100$ with varying average degree. Each point constitutes 200 simulations. (**B**) Small-world networks of size $N = 200$ with $k = 6$ with varying rewiring probability. Each point constitutes 500 simulations.

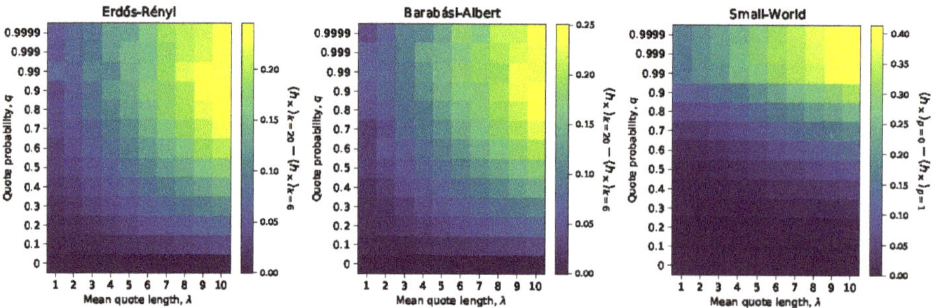

**Figure A2.** Effects of quoter model parameter choices on observed trends. Information flow is lower for denser ER and BA networks across a range of $q$ and $\lambda$ with the effect being more pronounced at higher values of $q$ and $\lambda$. Likewise, for small-world networks, more clustering (lower $p$) exhibits higher $h_\times$ than less clustering (higher $p$), with the effect being most pronounced at $q > 0.5$ regardless of $\lambda$. Here, ER & BA networks had $N = 100$ and small-world networks had $N = 200$ and $k = 6$. Each cell constitutes 100 simulations.

## Appendix B. Summarizing $h_\times$

In this work, we summarized $h_\times$ by the mean $\langle h_\times \rangle$ and variance $\mathrm{Var}(h_\times)$. In Figure A3, we see that this choice was appropriate: examining the distributions of $h_\times$ for various networks shows that they are approximately normal. We also find the mean and median $h_\times$ to be approximately equal.

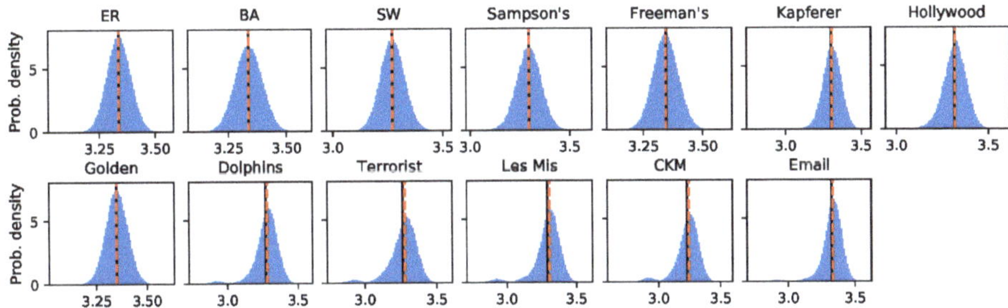

**Figure A3.** The distributions of $h_\times$ for quoter model simulations on various networks. Examining the distributions supports using $\langle h_\times \rangle$ and $\mathrm{Var}(h_\times)$ as summary statistics, although some real networks show a small bimodality (an excess of $h_\times < 3$ bits). We also remark that the mean and median are approximately equal (solid line shows $\langle h_\times \rangle$, dashed line shows median $h_\times$) for all networks. ER & BA networks have $N = 1000$ nodes with $\langle k \rangle = 12$, and 200 simulations as in Figure 1. Small-world networks have $N = 200$ nodes with $k = 6$ and $p = 10^{-4}$, and 500 simulations as in Figure 4A. Real-world networks are from 300 simulations as in Figures 2 and 4B,C. Quoter model parameters are given in Section 3.1.

## Appendix C. Network Corpus

All networks studied here can be found through the Index of Complex Networks (ICON) [43]. We converted any directed or weighted networks to undirected (bidirectional) and unweighted. Details for each of the ten networks:

1. Les Miserables co-appearances [44] [Undirected, Weighted].
2. Hollywood film music [45] [Undirected, Weighted]. This is a bipartite network; we converted it to a one-mode projection (nodes are composers and two composers are linked if they worked with the same producer).
3. Freeman's EIES dataset [46] [Directed, Weighted]. We used the "personal relationships (time 1)" network.
4. Sampson's monastery [47] [Directed, Weighted]. We used the Pajek dataset. The weight of a directed link represents how an individual rates the other. The rating can be positive (1,2,3 = top 3 ranked) or negative (-1,-2,-3 = worst 3 ranked). We chose to only keep links which were positive.
5. Golden Age of Hollywood [48] [Directed, Weighted]. We used the aggregated network over 1909-2009.
6. 9-11 terrorist network [49] [Undirected, Unweighted].
7. CKM physicians social network [50] (1966) [Directed, Unweighted]. We used "CKM physicians Freeman" networks hosted by Linton Freeman, and chose the "friend" network (i.e., the third adjacency matrix). We took only the giant component.
8. Kapferer tailor shop [51] (1972) [Undirected, Unweighted]. We used the "Kapferer tailor shop 1" Pajek dataset (kapfts1.dat).
9. Dolphin social network [52] (1994-2001) [Undirected, Unweighted].
10. Email network (Uni. R-V, Spain, 2003) [53] [Directed, Unweighted]. We used the "email-uni-rv-spain-arenas" network.

## References


1. Lazer, D.; Pentland, A.; Adamic, L.; Aral, S.; Barabasi, A.L.; Brewer, D.; Christakis, N.; Contractor, N.; Fowler, J.; Gutmann, M.; et al. SOCIAL SCIENCE: Computational Social Science. *Science* **2009**, *323*, 721–723. [CrossRef]
2. Tumasjan, A.; Sprenger, T.O.; Sandner, P.G.; Welpe, I.M. Predicting elections with twitter: What 140 characters reveal about political sentiment. In Proceedings of the Fourth International AAAI Conference on Weblogs and Social Media, Washington, DC, USA, 23–26 May 2010; pp. 178–185.
3. Conover, M.D.; Ferrara, E.; Menczer, F.; Flammini, A. The Digital Evolution of Occupy Wall Street. *PLoS ONE* **2013**, *8*, e64679. [CrossRef]
4. Castells, M. *Networks of Outrage and Hope: Social Movements in the Internet Age*; John Wiley & Sons: Hoboken, NJ, USA, 2015.
5. de Montjoye, Y.A.; Radaelli, L.; Singh, V.; Pentland, A. Unique in the shopping mall: On the reidentifiability of credit card metadata. *Science* **2015**, *347*, 536–539. [CrossRef] [PubMed]
6. Garcia, D. Leaking privacy and shadow profiles in online social networks. *Sci. Adv.* **2017**, *3*, e1701172. [CrossRef] [PubMed]
7. Schreiber, T. Measuring Information Transfer. *Phys. Rev. Lett.* **2000**, *85*, 461–464. [CrossRef] [PubMed]
8. Sun, J.; Bollt, E.M. Causation entropy identifies indirect influences, dominance of neighbors and anticipatory couplings. *Phys. D* **2014**, *267*, 49–57. [CrossRef]
9. Borge-Holthoefer, J.; Perra, N.; Gonçalves, B.; González-Bailón, S.; Arenas, A.; Moreno, Y.; Vespignani, A. The dynamics of information-driven coordination phenomena: A transfer entropy analysis. *Sci. Adv.* **2016**, *2*, e1501158. [CrossRef]
10. Wang, D.; Wen, Z.; Tong, H.; Lin, C.Y.; Song, C.; Barabási, A.L. Information spreading in context. In Proceedings of the 20th international conference on World wide web (WWW 2011), Hyderabad, India, 28 March–1 April 2011; pp. 735–744.
11. Bagrow, J.P.; Liu, X.; Mitchell, L. Information flow reveals prediction limits in online social activity. *Nat. Hum. Behav.* **2019**, *3*, 122–128. [CrossRef]
12. Bagrow, J.P.; Mitchell, L. The quoter model: A paradigmatic model of the social flow of written information. *Chaos* **2018**, *28*, 075304. [CrossRef]
13. Centola, D. The Spread of Behavior in an Online Social Network Experiment. *Science* **2010**, *329*, 1194–1197. [CrossRef]
14. Borge-Holthoefer, J.; Banos, R.; Gonzalez-Bailon, S.; Moreno, Y. Cascading behaviour in complex socio-technical networks. *J. Complex Netw.* **2013**, *1*, 3–24. [CrossRef]
15. Shannon, C. Prediction and Entropy of Printed English. *Bell Labs Tech. J.* **1951**, *30*, 50–64. [CrossRef]
16. Kontoyiannis, I.; Algoet, P.; Suhov, Y.; Wyner, A. Nonparametric entropy estimation for stationary processes and random fields, with applications to English text. *IEEE Trans. Inf. Theory* **1998**, *44*, 1319–1327. [CrossRef]
17. Song, C.; Qu, Z.; Blumm, N.; Barabasi, A.L. Limits of Predictability in Human Mobility. *Science* **2010**, *327*, 1018–1021. [CrossRef] [PubMed]
18. Ziv, J.; Merhav, N. A Measure of Relative Entropy between Individual Sequences with Application to Universal Classification. In Proceedings of the IEEE International Symposium on Information Theory, San Antonio, TX, USA, 17–22 January 1993; p. 352.
19. Sun, J.; Taylor, D.; Bollt, E.M. Causal Network Inference by Optimal Causation Entropy. *SIAM J. Appl. Dyn. Syst.* **2015**, *14*, 73–106. [CrossRef]
20. Cover, T.M.; Thomas, J.A. *Elements of Information Theory*; John Wiley & Sons, Inc.: Hoboken, NJ, USA, 1991.
21. Granovetter, M. Threshold Models of Collective Behavior. *Am. J. Sociol.* **1978**, *83*, 1420–1443. [CrossRef]
22. Watts, D. A simple model of global cascades on random networks. *Proc. Natl. Acad. Sci. USA* **2002**, *99*, 5766–5771. [CrossRef]
23. Centola, D.; Eguíluz, V.M.; Macy, M.W. Cascade dynamics of complex propagation. *Phys. A* **2007**, *374*, 449–456. [CrossRef]
24. Ugander, J.; Backstrom, L.; Marlow, C.; Kleinberg, J. Structural diversity in social contagion. *Proc. Natl. Acad. Sci. USA* **2012**, *109*, 5962–5966. [CrossRef]

25. Miller, J.C. Percolation and epidemics in random clustered networks. *Phys. Rev. E* **2009**, *80*, 020901. [CrossRef]

26. Pastor-Satorras, R.; Castellano, C.; Van Mieghem, P.; Vespignani, A. Epidemic processes in complex networks. *Rev. Mod. Phys.* **2015**, *87*, 925–979. [CrossRef]

27. O'Sullivan, D.J.; O'Keeffe, G.J.; Fennell, P.G.; Gleeson, J.P. Mathematical modeling of complex contagion on clustered networks. *Front. Phys.* **2015**, *3*, 71. [CrossRef]

28. Gray, C.; Mitchell, L.; Roughan, M. Super-blockers and the effect of network structure on information cascades. In Proceedings of the Companion Proceedings of the The Web Conference 2018, Lyon, France, 23–27 April 2018; pp. 1435–1441.

29. Centola, D.; Macy, M. Complex Contagions and the Weakness of Long Ties. *Am. J. Sociol.* **2007**, *113*, 702–734. [CrossRef]

30. Granovetter, M.S. The Strength of Weak Ties. In *Social Networks*; Elsevier: New York, NY, USA, 1977; pp. 347–367.

31. Miller, J.; Ting, T. EoN (Epidemics on Networks): A fast, flexible Python package for simulation, analytic approximation, and analysis of epidemics on networks. *J. Open Source Softw.* **2019**, *4*, 1731. [CrossRef]

32. Lambiotte, R. How does degree heterogeneity affect an order-disorder transition? *Europhys. Lett.* **2007**, *78*, 68002. [CrossRef]

33. Watts, D.J.; Strogatz, S.H. Collective dynamics of 'small-world' networks. *Nature* **1998**, *393*, 440–442. [CrossRef]

34. Singh, P.; Sreenivasan, S.; Szymanski, B.; Korniss, G. Threshold-limited spreading in social networks with multiple initiators. *Sci. Rep.* **2013**, *3*, 2330. [CrossRef]

35. Milo, R.; Kashtan, N.; Itzkovitz, S.; Newman, M.E.; Alon, U. On the uniform generation of random graphs with prescribed degree sequences. *arXiv* **2003**, arXiv:cond-mat/0312028.

36. Blitzstein, J.; Diaconis, P. A Sequential Importance Sampling Algorithm for Generating Random Graphs with Prescribed Degrees. *Internet Math.* **2011**, *6*, 489–522. [CrossRef]

37. Newman, M.; Girvan, M. Finding and evaluating community structure in networks. *Phys. Rev. E* **2004**, *69*, 026113. [CrossRef]

38. Danon, L.; Díaz-Guilera, A.; Duch, J.; Arenas, A. Comparing community structure identification. *J. Stat. Mech.* **2005**, *2005*, P09008. [CrossRef]

39. Karrer, B.; Newman, M. Stochastic blockmodels and community structure in networks. *Phys. Rev. E* **2011**, *83*, 016107. [CrossRef] [PubMed]

40. Blondel, V.D.; Guillaume, J.L.; Lambiotte, R.; Lefebvre, E. Fast unfolding of communities in large networks. *J. Stat. Mech.* **2008**, *2008*, P10008. [CrossRef]

41. de Arruda, G.F.; Petri, G.; Rodrigues, F.A.; Moreno, Y. Impact of the distribution of recovery rates on disease spreading in complex networks. *Phys. Rev. Res.* **2020**, *2*, 013046. [CrossRef]

42. Nematzadeh, A.; Ferrara, E.; Flammini, A.; Ahn, Y.Y. Erratum: Optimal Network Modularity for Information Diffusion. *Phys. Rev. Lett.* **2014**, *113*, 088701. [CrossRef] [PubMed]

43. Clauset, A.; Tucker, E.; Sainz, M. The Colorado index of complex networks. *Retrieved July* **2016**, *20*, 2018. Available online: https://icon.colorado.edu (accessed on 25 February 2020).

44. Knuth, D.E. *Stanford GraphBase: A platform for Combinatorial Computing*; Addison-Wesley: Boston, MA, USA, 1993.

45. Faulkner, R.R. *Music on Demand: Composers and Careers in the Hollywood Film Industry*; Transaction Books: New Brunswick, NJ, USA, 1983.

46. Freeman, S.C.; Freeman, L.C. *The Networkers Network: A Study of the Impact of a New Communications Medium on Sociometric Structure*; University of California: Irvine, CA, USA, 1979.

47. Sampson, S.F. A novitiate in a Period of Change: An Experimental and Case Study of Social Relationships. Ph.D. Thesis, Cornell University, Ithaca, NY, USA, 1968.

48. Taylor, D.; Myers, S.A.; Clauset, A.; Porter, M.A.; Mucha, P.J. Eigenvector-Based Centrality Measures for Temporal Networks. *Multiscale Model. Simul.* **2017**, *15*, 537–574. [CrossRef]

49. Krebs, V. Uncloaking Terrorist Networks. *First Monday* **2002**, *7*, 43–52. [CrossRef]

50. Burt, R.S. Social Contagion and Innovation: Cohesion versus Structural Equivalence. *Am. J. Sociol.* **1987**, *92*, 1287–1335. [CrossRef]

51. Kapferer, B. *Strategy and Transaction in an African Factory: African Workers and Indian Management in a Zambian Town*; Manchester University Press: Manchester, UK, 1972.

52.  Lusseau, D.; Schneider, K.; Boisseau, O.J.; Haase, P.; Slooten, E.; Dawson, S.M. The bottlenose dolphin community of Doubtful Sound features a large proportion of long-lasting associations. *Behav. Ecol. Sociobiol.* **2003**, *54*, 396–405. [CrossRef]
53.  Guimerà, R.; Danon, L.; Díaz-Guilera, A.; Giralt, F.; Arenas, A. Self-similar community structure in a network of human interactions. *Phys. Rev. E* **2003**, *68*, 065103. [CrossRef] [PubMed]

*Article*

# Optimizing Variational Graph Autoencoder for Community Detection with Dual Optimization

Jun Jin Choong [1,*], Xin Liu [2] and Tsuyoshi Murata [1]

[1] Department of Computer Science, Tokyo Institute of Technology, Tokyo 152-8552, Japan; murata@c.titech.ac.jp

[2] National Institute of Advanced Industrial Science and Technology, Tokyo 135-0064, Japan; xin.liu@aist.go.jp

* Correspondence: junjin.choong@net.c.titech.ac.jp

Received: 31 December 2019; Accepted: 4 February 2020; Published: 7 February 2020

**Abstract:** Variational Graph Autoencoder (VGAE) has recently gained traction for learning representations on graphs. Its inception has allowed models to achieve state-of-the-art performance for challenging tasks such as link prediction, rating prediction, and node clustering. However, a fundamental flaw exists in Variational Autoencoder (VAE)-based approaches. Specifically, merely minimizing the loss of VAE increases the deviation from its primary objective. Focusing on Variational Graph Autoencoder for Community Detection (VGAECD) we found that optimizing the loss using the stochastic gradient descent often leads to sub-optimal community structure especially when initialized poorly. We address this shortcoming by introducing a dual optimization procedure. This procedure aims to guide the optimization process and encourage learning of the primary objective. Additionally, we linearize the encoder to reduce the number of learning parameters. The outcome is a robust algorithm that outperforms its predecessor.

**Keywords:** variational inference; graph neural network; variational autoencoder; network embedding

---

## 1. Introduction

Networks (graphs) with nodes (vertices) and edges (links) are a considerable simplification of complex patterns observed in real life, thus permitting studies of complex systems. For instance, the study of social interactions between individuals can be represented in the form of social networks [1]. Researchers who published together can be related in a collaboration network [2]. Movies and their respective critics can be presented as a bipartite graph with the edge-weight indicating a user-movie rating [3] which further allow applications like recommender systems [4]. The flexibility of networks and its vast literature on graph theory makes network science [5,6] extremely appealing to researchers.

An area of interest with significant importance is community detection, also known as graph clustering [7], i.e., identifying groups of densely connected nodes. Traditionally, researchers have measured communities in terms of partition quality, known as modularity [8]. A recovered community structure with high modularity implies good partitioning. To this date, community detection algorithms have evolved from traditional algorithms to the usage of complex learning algorithms like graph representation learning [9,10]. In graph representation learning, one can enforce nodes within the same community to share similar representations. These representations are learned by aggregating features from neighboring nodes. In addition, graph representation learning is extremely appealing because it provides a generalized application for downstream tasks such as link prediction [11], classification [12] and clustering [13]. By exploiting existing literature on representation learning, these tasks can be solved simply by reusing existing machine learning techniques.

Among many types of graph representation learning algorithms, Graph Neural Network (GNN) has recently gained significant popularity. Inspired by Deep Learning methodologies, GNN is designed to follow a similar learning approach, but with graphs as its primary application. For instance, in graphs, convolutional layers are replaced with graph convolutional layers [14]. The outcome is a translation of Deep Learning techniques from computer vision readily applied to graph data. Likewise, GNN inherits similar disadvantages from deep learning algorithms, which is widely known to be a black-box learning algorithm. To overcome this problem, machine learning researchers have explored explainable artificial intelligence (XAI) algorithms. Causal inference [15] and Bayesian Deep Learning [16,17] are some examples of attempts to unravel the mysteries behind machine learning algorithms by presenting uncertainties and causal reasons.

From a different paradigm, generative models are equally appealing for introducing explainability. Stochastic Blockmodel (SBM) [18,19] is a popular approach to model networks. By proposing an assortative configuration on the stochastic matrix, one can generate networks that exhibit community structures. Leveraging on the reparameterization trick, Variational Autoencoder (VAE) [20] improves explainability by introducing uncertainty to an autoencoder. Recently, Kipf and Welling [21] proposed Variational Graph Autoencoder (VGAE), which results in research variants such as VGAECD [22] and ARVGA [23].

Albeit powerful, VGAE-based algorithm suffers from an optimization problem. When trained, it has tendencies to deviate from its primary objective in favor of the reconstruction of the input network and eventually lead to a posterior collapse [24]. In this work, we focus our attention on a variant of VGAE, namely Variational Graph Autoencoder for Community Detection (VGAECD) [22]. We found that optimizing the loss using stochastic gradient descent often leads to sub-optimal community structure when the model is initialized poorly. Figure 1 demonstrates an example of the quality of detected communities (measured by NMI) with respect to the loss during training in a synthetic network generated by the LFR benchmark with $\mu = 0.60$. We can observe that the training loss is consistently decreasing as expected. However, the NMI suddenly drops approximately after 80 epochs and gradually begins its re-ascent. This tendency has also been observed in other unsupervised deep learning algorithms [25–27]. To circumvent this problem, one can train the unsupervised algorithm with a meta-learner [25,26]. More specifically, one can introduce a guide that prevents the algorithm from going astray. This new formulation comes with an advantage and disadvantage. Specifically, the weakness comes at the cost of more computation complexity.

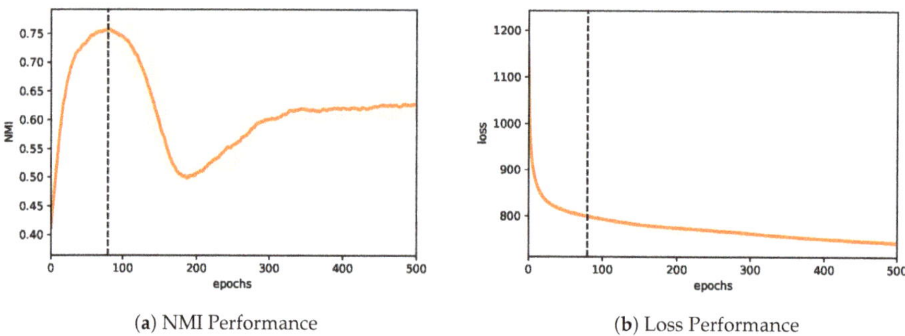

(a) NMI Performance            (b) Loss Performance

**Figure 1.** Left: The deviation problem exhibited when training VGAECD. The NMI drops approximately after 80 epochs and gradually begins its re-ascent. In most cases, it deteriorates in favor of its secondary objective of minimizing the reconstruction loss. Right: The performance of loss continues to drop regardless of its NMI performance.

Additionally, generalization becomes difficult because of the high coupling between the learner and the task of interest (task-dependent). Instead, in our work, we chose to leverage a variational solution. More specifically, we maximize the lower bound introduced in the Variational Autoencoder such that no additional modification is required on the original loss function. Instead, the optimization procedure follows a Neural Expectation-Maximization (NEM) algorithm [28], which guarantees that communities do not collapse. This is possible because NEM can be formulated in terms of maximizing a variational lower bound [29]. Maximizing this lower bound ensures that every new update is an improvement over the previous step. Furthermore, it has a theoretical guarantee for convergence up to local optima. We term our improved version, Variational Graph Autoencoder for Community Detection - Optimized (VGAECD-OPT).

To summarize, we improve VGAECD and propose a robust algorithm (VGAECD-OPT) for community detection. Our contributions are as follows,

- Demonstrate the efficacy of linearization on VGAECD's encoder in community detection task [30,31].
- Propose a dual optimization approach to alleviate the deviation of objective functions (community detection vs. network reconstruction)

The proposed algorithm, like VGAECD, inherits properties unique to generative models such as the possibility to generate a synthetic network with community structure. Such models can be useful to application areas such as high-performance computing [32,33]. On the other hand, the model itself can be used as a network anonymizer by inducing *artificial* links or nodes on an existing social network [34].

## 2. Related Work

From a probabilistic modeling perspective, community detection can be divided into two classes, namely discriminative and generative models. The former is a class of algorithms that attempt to maximize the community structure recovery while the latter considers the process of generating a network that exhibits community structure with high fidelity. In this section, we briefly explain recent work on these algorithms.

### 2.1. Discriminative Models

Traditionally, community structures were identified via connectivity patterns such as density within a community [35,36]. In practice, these patterns can be measured by a quality metric such as modularity [8] and conductance [37]. The Louvain method [38], is a greedy algorithm that maximizes the modularity objective function. Although popular, modularity maximization is known to exhibit a resolution limit [39] and degeneracies [40]. On the other hand, propagation algorithms such as label propagation [41] are popular for detecting communities in networks at scale. Other approaches such as WalkTrap [42] and Infomap [43] balances scalability with computational performance [44]. Representation learning methods such as GraRep [45] and CFOND [46] cast community detection as a matrix factorization problem. Models like GA-Net [36] employ a traditional genetic algorithm while maximizing a *community score* defined on the maximization of the dense internal sub-matrices. Moscato et al. [47] formulates community detection as a game model, employing Game Theory approaches to maximize the community assignment.

Recent successes [48,49] in deep learning rekindled interest in unsupervised learning models such as autoencoders for networks. In particular, GraphEncoder proposed by Tian et al. [50] showed that optimizing the objective function of the autoencoder is similar to finding a solution for Spectral Clustering [50]. Leveraging on deep learning's non-linearity and recent advances in Convolutional Neural Networks, Defferrard et al. [51] proposed Graph Neural Network (GNN), with Kipf and Welling [14] further simplifying to Graph Convolutional Neural Network (GCN).

On the other hand, DeepWalk [12] and node2vec [11] are popular algorithms for graph embedding. To generate a co-occurrence context, random walks are used in conjunction with negative sampling for large scale datasets. More recently, this line of algorithms can be generalized to a class of matrix factorization algorithms [52,53]. Albeit powerful, the model has many hyperparameters to tune, which can be time-consuming.

*2.2. Generative Models*

Generative models can be classified into two types: algorithmic and statistical models. In the former case, graphs are generated under assumptions of prior knowledge. For instance, Kronecker Graphs [54] considers the generation of graphs via a Kronecker product. The Block Two-Level Erdős-Rényi (BTER) model [55] considers a greedy approach for matching clustering coefficient and degree distribution. Simpler models include benchmark graph models such as Girvan-Newman [35], Lancichinetti–Fortunato–Radicchi (LFR) Graph [56] and mLFR [57]. The latter considers a class of algorithms from the lens of probabilistic graphical modeling. Given a network as the input, its goal is to maximize the likelihood of the latent variables which generate the same input network. For example, the Stochastic Blockmodel (SBM) [18] considers a stochastic matrix **B** as the probability of connectivity under the assumption of stochastic equivalence (i.e., nodes within the same community shares the same connectivity pattern). Karrer and Newman [19] further extends this work to community detection by introducing a degree correction procedure to the algorithm. Extensions to SBM includes the Mixed Membership SBM (MMSBM) [58] for identifying mixed community participation and bipartite SBM (biSBM) [59] for finding communities in bipartite networks. Today, SBM is well explored, and its limitations have been widely studied [60,61]. However, SBM is not a network representation learning model. In other words, SBM's learning paradigm differs from representation learning, which is the goal of this work.

From the lens of representation learning, autoencoders are considered the closest cousin to generative models. With an encoder and decoder framework, it is no surprise that one considers autoencoders as a generative model. In reality, the autoencoder lacks sampling capability, which is the core of a generative model. To alleviate this problem, recent literature considers alternative models such as Generative Adversarial Networks (GAN) or Variational Autoencoder (VAE), which introduces an approximate posterior. For graphs, Kipf and Welling [21] introduced a variant of VAE for link prediction tasks in graphs and Pan et al. [23] recently introduced Adversarially Regularized Graph Autoencoder (ARGA) using GAN.

## 3. Problem Definition

Formally, a network with $N$ nodes can be defined as $\mathcal{G} = (\mathcal{V}, \mathcal{E})$, where $\mathcal{V} = \{v_i, \ldots, v_N\}$ denotes the set of nodes and $\mathcal{E} = \{e_{ij}\}$ is the set of edges. Incidentally, each node may be described by some features $\mathbf{X} = \{\mathbf{x}_1, \ldots, \mathbf{x}_N\}$ where $\mathbf{x}_i \in \mathbb{R}^D$ defines a vector of real-values associated with node $v_i$ with $D$-dimension. Vectorizing the notations, $\mathbf{A} = \{\mathbf{a}_1, \ldots, \mathbf{a}_N\} \in \mathbb{R}^{N \times N}$ is the adjacency matrix of $\mathcal{G}$. In this work, we consider the undirected and unweighted network $\mathcal{G}$, such that $A_{ij} = 1$ if $e_{ij} \in \mathcal{E}$ otherwise 0. Given the network $\mathcal{G}$, we aim to partition the nodes in $\mathcal{G}$ into $K$ disjoint groups $\{c_1, c_2, \ldots, c_K\}$ such that nodes grouped within the same communities share a similar connectivity pattern. We define the connectivity pattern by the node's edge probability $p$. Specifically, $p_{in}$ is the probability of connecting between nodes of the same community and $p_{out}$ is the probability of connecting nodes between other communities. Consequently, the community structure is defined as,

$$p_{in} > p_{out}. \tag{1}$$

We refer to this as the *modern view* community structure definition as suggested in [62]. This notion of community structure is a generalization to a probabilistic perspective.

Additionally, we further constrain our problem definition to the view of a generative model. Given a generative model, $p(\theta \mid \mathbf{X}, \mathbf{A})$ infers the model parameters $\theta$ from the observed network $\mathcal{G}$. Concretely, we are interested to maximize,

$$\arg\max_{\theta} p(\mathbf{A} \mid \theta). \tag{2}$$

Under this optimization condition, a similar graph, $\mathcal{G}'$ with adjacency matrix $\mathbf{A}'$ can be generated from the same set of parameters such that $p(\mathbf{A}' \mid \theta)$ defines the reconstruction probability of the original adjacency matrix $\mathbf{A}$. According to Bayesian principles, one can say that the model is a good model when $\mathbf{A}' \approx \mathbf{A}$ and satisfies the condition of having community structures as defined in Equation (1).

## 4. Model Description

### 4.1. Variational Graph Autoencoder for Community Detection

Kipf and Welling [21] introduced Variational Graph Autoencoder (VGAE) by replacing the encoder of Graph Autoencoder (GAE) [21,50] with a Graph Convolutional Network [14] and an inner product decoder. Formally, VGAE's decoder can be defined as

$$p(\mathbf{A} \mid \mathbf{Z}) = \prod_{i=1}^{N} \prod_{j=1}^{N} p(A_{ij} = 1 \mid \mathbf{z}_i \mathbf{z}_j) = \tau(\mathbf{z}_i^{\top} \mathbf{z}_j), \tag{3}$$

where $\mathbf{Z} \in \mathbb{R}^{N \times F}$ is the latent representation with $N$ nodes and $F$ is the size of the latent representation, given as a hyperparameter. Additionally, we denote the latent representation of node $v_i$ as $\mathbf{z}_i$ such that $\{\mathbf{z}_i, ... \mathbf{z}_N\} \in \mathbf{Z}$. The decoding process then follows a sampling process from the variational distribution $q(\cdot)$. Specifically, the model samples from a Gaussian prior distribution, $\mathcal{N}(\cdot \mid \mu_i, \sigma_i^2 \mathbf{I})$ with mean $\mu$, variance $\sigma^2$ and the identity matrix $\mathbf{I}$. Samples are then mapped through a non-linear function denoted by $\tau(\cdot)$. Most commonly, the non-linear function can be a logistic sigmoid function, $\tau(t) = 1/(1 + e^{-t})$) or a ReLU function, $\text{ReLU}(t) = \max(0, t)$. The encoder is then defined as

$$q(\mathbf{Z} \mid \mathbf{X}, \mathbf{A}) = \prod_{i=1}^{N} q(\mathbf{z}_i \mid \mathbf{X}, \mathbf{A}) \tag{4}$$

$$q(\mathbf{z}_i \mid \mathbf{X}, \mathbf{A}) = \mathcal{N}(\mathbf{z}_i \mid \mu_i, \sigma_i^2 \mathbf{I}).$$

The encoder $q(\cdot)$ is a variational distribution that approximates the true distribution $p(\cdot)$ [20,29]. By mathematical convenience, $q(\cdot)$ is usually a member of the exponential family. The mean $\mu$ and standard deviation $\sigma$ are obtained through *amortization* using a two-layer GCN defined as,

$$\text{GCN}(\mathbf{X}, \mathbf{A}) = \hat{\mathbf{A}} \tau(\hat{\mathbf{A}} \mathbf{X} \mathbf{W}_0) \mathbf{W}_1, \tag{5}$$

where $\hat{\mathbf{A}}$ is obtained through a *renormalization trick* [14], $\hat{\mathbf{A}} = \mathbf{D}^{-\frac{1}{2}} \mathbf{A} \mathbf{D}^{-\frac{1}{2}}$ and $\{\mathbf{W}_0, \mathbf{W}_1\}$ are the trainable weight filters for each GCN layer. To train VGAE, we optimize the evidence lower bound (ELBO) $\mathcal{L}(\cdot)$,

$$\log p_\theta(\mathbf{X}) \geq \mathcal{L}(\theta, \phi; \mathbf{X}) = \mathbb{E}_{q_\phi(\mathbf{Z}|\mathbf{X},\mathbf{A})} [\log p_\theta(\mathbf{A} \mid \mathbf{Z})] \\ - \mathcal{D}_{KL}[q_\phi(\mathbf{Z} \mid \mathbf{X}) \parallel p_\theta(\mathbf{Z})]. \tag{6}$$

$\mathcal{D}_{KL}[q_\phi(\cdot) \parallel p_\theta(\cdot)]$ defines the Kullback–Leibler (KL) divergence between $q_\phi(\cdot)$ and $p_\theta(\cdot)$. The lower bound can be maximized with respect to the variational parameters $(\theta, \phi) = \mathbf{W}_i$ via stochastic gradient descent. Here, the prior is defined as $p_\theta(\mathbf{Z}) = \prod_{i=1}^{N} \mathcal{N}(\mathbf{z}_i \mid 0, \mathbf{I})$, the isotropic Gaussian. Since this requires sampling from a Gaussian white noise, backpropagation from a stochastic variable is not trivial. Equation (6) via stochastic gradient descent. However, by applying a *reparameterization trick* [20], gradients can backpropagate to deterministic variables and stochastic variables can be effectively ignored.

Following prior work, Variational Graph Autoencoder for Community Detection (VGAECD) [22] generalizes the generation process of VGAE by introducing a mixture of Gaussians in the generation process (decoder). The generation process can be generalized to a mixture of Gaussians by introducing a community assignment parameter $c$. Specifically, we would like to calculate the joint probability distribution of $p(\mathbf{a}, \mathbf{z}, c)$ such that

$$
\begin{aligned}
p(\mathbf{a}, \mathbf{z}, c) &= p(\mathbf{a} \mid \mathbf{z}) p(\mathbf{z} \mid c) p(c) \\
p(c) &= Cat(\cdot \mid \boldsymbol{\gamma}) \\
p(\mathbf{z} \mid c) &= \mathcal{N}(\cdot \mid \boldsymbol{\mu}_c, \sigma_c^2 \mathbf{I}) \\
p(\mathbf{a} \mid \mathbf{z}) &= \phi(\mathbf{z}_i^T \mathbf{z}; \mathcal{N}(\cdot \mid \boldsymbol{\mu}_a, \sigma_a^2 \mathbf{I})).
\end{aligned}
\tag{7}
$$

For brevity, we drop the explicit subscript notation $\mathbf{z} = \mathbf{z}_i$ and $\mathbf{a} = \mathbf{a}_i$. In Equation (7), we obtain $p(c)$ from the categorical distribution parameterized by $\gamma$ with $K$ communities. The parameter $\gamma$ encodes our prior belief and is commonly initialized with a non-informative priors such as a uniform probability distribution. The reconstruction probability, $p(\mathbf{a} \mid \mathbf{z})$ is the inner product between latent representations $\mathbf{z}$ parameterized by embeddings sampled from the Gaussian distribution. Effectively, two nodes $v_i$ and $v_j$ are more likely to have an edge $e_{ij}$ when their latent representations are closer to one another.

*4.2. Linearization of the Encoder*

VGAECD uses Graph Convolution layer (GCN) for its encoder to approximate parameters $\mu$ and $\sigma$. Albeit powerful, GCN is more computationally expensive due to its non-linearity and increase in training parameters required. Wu et al. [30] recently proposed a simplification of GCN by removing the non-linear component, $\tau(\cdot)$, effectively linearizing GCN.

$$
\text{SGC}(\mathbf{X}, \mathbf{A}) = \hat{\mathbf{A}} \ldots \hat{\mathbf{A}} \hat{\mathbf{A}} \mathbf{X} \mathbf{W}^{(1)} \mathbf{W}^{(2)} \ldots \mathbf{W}^{(L)}.
\tag{8}
$$

Equation (8) describes SGC layer formally. In Equation (8), the non-linear function $\tau$ is removed and features from $L$-hop neighbors are accumulated. Equation (8) further simplifies to

$$
\text{SGC}(\mathbf{X}, \mathbf{A}) = \hat{\mathbf{A}}^L \mathbf{X} \mathbf{W},
\tag{9}
$$

with $\mathbf{W} = \mathbf{W}^{(1)} \mathbf{W}^{(2)} \ldots \mathbf{W}^{(L)}$. Similar to the *renormalization trick*, A two-layer $L = 2$, SGC, $\hat{\mathbf{A}}^L$ can be pre-computed before training. Extending Wu et al. [30]'s work, Salha et al. [31] demonstrated performance improvement upon linearization of the encoder on GAE and VGAE in link prediction and clustering tasks. From the perspective of graph signal processing, NT and Maehara [63] considered GCN & SGC to be equally powerful since both encoders resemble low-pass filters. Under the aforementioned motivation, we experimentally show that the linearization of the encoder reduces training time for convergence and time complexity. We further discuss about the implications of changing the encoder in Section 6.

*4.3. Dual Optimization*

In Section 4.1, we briefly discuss about the weakness of VGAECD. Formally, the objective function of VGAECD can be formulated into two losses,

$$
\mathcal{L} = \mathcal{L}_{\text{recon}} + \mathcal{L}_{\text{comm}},
\tag{10}
$$

such that the reconstruction loss $\mathcal{L}_{\text{recon}}$ and the community's quality loss, $\mathcal{L}_{\text{comm}}$ is minimized. It follows that optimizing the loss is not trivial. Given enough capacity, an autoencoder trained with stochastic gradient descent would favor optimizing its reconstruction loss ($\mathcal{L}_{\text{recon}}$), eventually leading to a posterior collapse [64]. Furthermore, as studied in [25–27], unsupervised deep learning algorithms tend to deviate from their main objective. They converge slowly especially when no guidance is given. In a similar fashion, we depict this problem exhibited by VGAECD in Figure 1. To rectify this issue, we propose a dual optimization algorithm based on Neural Expectation-Maximization (NEM) [28].

Unlike the Expectation-Maximization algorithm [65], Neural Expectation-Maximization (NEM) can be trained with gradient descent. As a result, VGAECD can be trained end-to-end. From Equation (7), the objective function can be defined as,

$$\log p(\mathbf{a}) \geq \mathcal{L}_{\text{ELBO}}(\mathbf{a}) = \mathbb{E}_{q(\mathbf{z},c|\mathbf{a})} \left[ \log \frac{p(\mathbf{a},\mathbf{z},c)}{q(\mathbf{z},c \mid \mathbf{a})} \right]. \tag{11}$$

Reformulating Equation (11), we obtain

$$\mathcal{L}_{\text{ELBO}}(\mathbf{a}) = \underbrace{\mathbb{E}_{q(\mathbf{z},c|\mathbf{a})}[\log p(\mathbf{a} \mid \mathbf{z})]}_{\text{reconstruction loss}} - \underbrace{\mathcal{D}_{KL}[q(\mathbf{z},c \mid \mathbf{a}) \parallel p(\mathbf{z},c)]}_{\text{community loss}}. \tag{12}$$

In Equation (12), by using a dual optimization process, the reconstruction loss is first optimized, followed by the community loss. This process is then repeated until convergence. Similar to [21,22], the reconstruction loss is minimized using binary cross-entropy and optimized using Adam [66]. The community loss is then minimized using an Expectation-Maximization (EM) algorithm [65] which guarantees a local optimum. Given $\psi_{i,k} = f_\phi(\mu_k)_i$ parameterized by $\phi$ and $\theta = \mu_1, \dots, \mu_K$, the loss of our variational distribution follows,

$$\mathcal{L}_{\text{comm}}\left(\theta, \theta^{\text{old}}\right) = \sum_c p\left(c \mid a, \psi^{\text{old}}\right) \log p(a,c \mid \psi). \tag{13}$$

To optimize Equation (13), we use NEM as the optimization algorithm. First, we compute the expectation, obtaining $\gamma$, the soft assignment of each node $v_i$,

$$\gamma_{i,k} := p\left(c_{i,k} = 1 \mid \mathbf{z}_i, \psi_i^{\text{old}}\right). \tag{14}$$

Next, the maximization step follows,

$$\theta^{\text{new}} = \theta^{\text{old}} + \eta \frac{\partial \mathcal{L}}{\partial \theta} \tag{15}$$

where

$$\frac{\partial \mathcal{L}}{\partial \mu_k} \propto \sum_{i=1}^{N} \gamma_{i,k}\left(\psi_{i,k} - \mathbf{a}_i\right) \frac{\partial \psi_{i,k}}{\partial \mu_k}. \tag{16}$$

In Equation (15), $\eta$ is the learning rate hyperparameter,. This process can be repeated $R$-times or until convergence. In practice, we found that $R \approx 5$ and $\eta = 0.01$ would suffice to achieve convergence. The complete algorithm is described in Algorithm 1. For our decoding function $f_\phi(\cdot)$, we use a single layer Multilayer Perceptron (MLP) in this work.

---

**Algorithm 1** VGAECD-OPT

---

**Input:** Features $\mathbf{X}$, Adjacency Matrix $\mathbf{A}$, no. of comm. $K$, filter size $\mathcal{D}$, number of epochs $L$, NEM steps $R$.

**Output:** Community Assignment Probability $\gamma$ and Reconstructed Adjacency matrix $\tilde{\mathbf{A}}$

1: $\pi \sim \mathcal{U}(0,1)$
2: **for** $l = 1, ..., L$ **do**
3:     **for** $i = 1, ..., N$ **do**
4:         $\mu_i = \text{SGC}_\mu(\mathbf{x}_i, \mathbf{a}_i)$
5:         $\sigma_i = \text{SGC}_\sigma(\mathbf{x}_i, \mathbf{a}_i)$
6:         Sample $\mathbf{z}_i \sim \mathcal{N}(\mu_x|_i, \text{diag}(\sigma_x^2|_i))$
7:         Obtain $\tilde{\mathbf{a}}_i = \sigma(\mathbf{z}_i^\top \mathbf{z}_j)$
8:         Compute loss, $\mathcal{L}_{\text{ELBO}}$                      ▷ From Equation (12)
9:           and backpropagate gradients.
10:     **end for**
11:     **for** $r = 1, ..., R$ **do**
12:         Compute E-Step: $\gamma, \{\mu_c, \sigma_c\}$           ▷ From Equation (14)
13:         Compute M-Step: $\mu_c, \sigma_c, \{\gamma\}$          ▷ From Equation (16)
14:         Compute loss, $\mathcal{L}_{\text{comm}}$
15:           and backpropagate gradients.
16:     **end for**
17: **end for**
18: Extract community assignment $\arg\max_k \gamma$
19: Return $\tilde{\mathbf{A}} = \{\tilde{\mathbf{a}}_1, ..., \tilde{\mathbf{a}}_N\}$

---

To explain this intuition, we refer to the theoretical formulation of VAE [66] and the bits-back argument [67]. Intuitively, while training, gradient signals would favor $\mathcal{L}_{\text{recon}}$ over minimizing $\mathcal{L}_{\text{comm}}$ when the model has high capacity. Consequently, each centroid $\mu_k$ is neglected; the centroid's is randomly positioned in the embedding's manifold and remain unoptimized. In an extreme case, the model would choose to collapse the posterior; resulting in a single cluster. To discourage this behavior, a dual optimization process allows gradient signals to be backpropagated to $\mu_k$ and $\sigma_k$. More specifically, turning $\mathcal{L}_{\text{comm}}$ to a variational EM optimization problem guarantees that the centroid's embedding has a higher presence of encoding useful information. Moreover, this formulation retains the main characteristic of VAE without requiring auxiliary loss functions commonly found in other literature [24]. The complete algorithm can be found in Algorithm 1 with illustration shown in Figures 2 and 3.

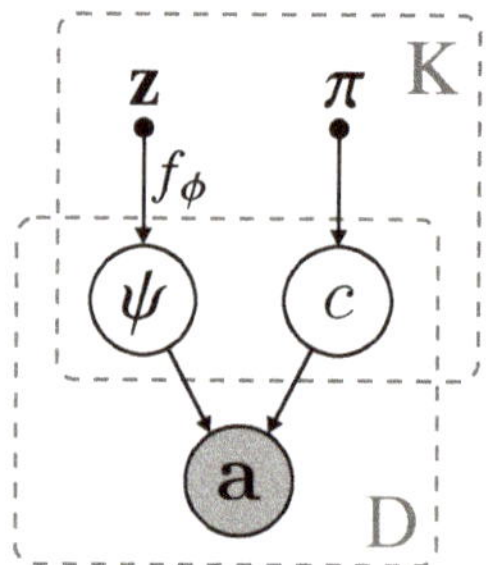


**Figure 2.** The probabilistic graphical model of VGAECD-OPT. The variable $\mathbf{z}$ is acquired from sampling of the variational distribution $p(\mathbf{z} \mid c)$, $\pi$ is the non-informative prior initialized uniformly. $f_\phi$ is the decoding function to obtain logits $\psi$. $K$ is the number of clusters and $D$ is the total number of data samples (i.e., $|\mathcal{V}|$).

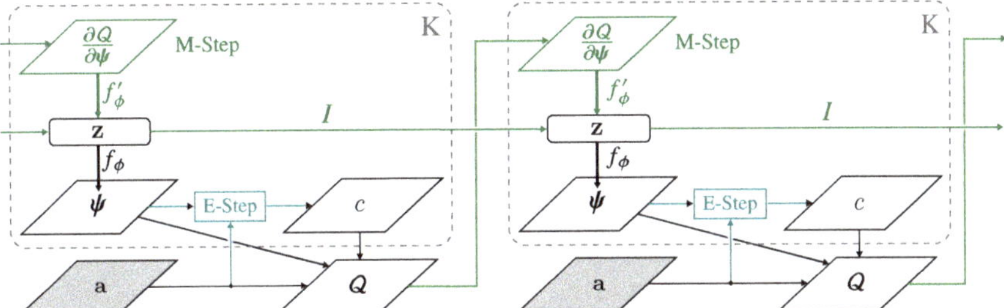

**Figure 3.** VGAECD optimized under Neural Expectation-Maximization algorithm (NEM). In the first iteration, the community assignment probability $c$ is first computed (Expectation) followed by the Maximization step. We obtain probability $\psi$ from the decoding function $f_\phi(\cdot)$ with embeddings $\mathbf{z}$.

## 5. Experiments

In this section, we evaluate the optimized proposed algorithm (VGAECD-OPT). Similar to [22], we first evaluate on two benchmark graphs followed by real-world datasets. We note that all datasets have associated ground truths. In later subsections, we list our experiment settings and evaluation metric. We leave the discussion of our experimental findings to Section 6.

### 5.1. Synthetic Datasets

Two synthetic benchmarks are used in our evaluation. We consider the two most common benchmarks used for benchmarking community detection algorithms. Specifically, synthetic networks with community structures were generated with Girvan-Newman (GN) [7,35] benchmark and the LFR [56] benchmark. The result is a set of generated graphs with ground-truth labels (true partitions) used for evaluation purposes.

The GN benchmark is a variant of the planted $l$-partition model. Given a fixed number of communities $c = 4$, and fixed number of nodes $n = 128$, the GN benchmark graph generator generates a graph with $M$ number of edges with an average degree $k = 16$. A mixture variable $z_{out}$ is manipulated from $\{1, ..., 8\}$, effectively controlling the connectivity pattern between intra-community $p_{in}$ and inter-community $p_{out}$ probabilities.

The LFR benchmark is an extension of the GN benchmark. It is considered to be more realistic than the GN benchmark while accounting for network heterogeneity and follows a power law distribution for the degree and community size distributions. The result is a generated network with variable communities of different sizes. To ensure consistency, default parameters were used [56]. These parameters are, number of nodes ($N = 1000$), average degree ($k = 15$), minimum ($c_{min} = 30$) and maximum ($c_{max} = 50$) number of nodes per community. The generation follows the *scale-free* parameters settings of exponents $\tau_1 = -2$ and $\tau_2 = -1$ respectively. On average, between 20 to 30 communities are generated.

### 5.2. Real-World Datasets

To evaluate performance of VGAECD-OPT, real-world datasets were used. These datasets are divided into two categories; networks with and without features. All datasets have ground-truth labels associated with them. The datasets are listed as follows:

- **Karate**: A social network that represents friendship among 34 members of a karate club at a US University observed by Zachary [1]. Community assignment corresponds to the clubs that members went to after the club split.
- **PolBlogs**: A network of political blogs assembled by Adamic and Glance [68]. The nodes are blogs, and web links between them are represented by its edge. These blogs have known political leanings and were labeled by hand by Adamic and Glance [68].

- **Cora**: A citation network with 2708 nodes and 5429 edges. Each node corresponds to a document and the edges are citation links [69,70]. Class labels correspond to each paper's topic curated by Cora's site portal [70] and were compiled by Sen et al. [69].
- **PubMed**: A network consisting of 19,717 scientific publications from PubMed database pertaining to diabetes classified into one of three classes ("Diabetes Mellitus, Experimental", "Diabetes Mellitus Type 1", "Diabetes Mellitus Type 2"). The citation network consists of 44,338 links. Each publication in the dataset is described by a TF-IDF weighted word vector from a dictionary, which consists of 500 unique words.

For starters, we perform experiments on datasets following Karrer and Newman [19]. These networks (Karate and PolBlogs) are featureless and only contain structural information. The Karate network is a commonly studied real-world network for community detection. Similar to [19], we consider the largest connected component and its undirected form for the PolBlogs dataset. Next, we used two networks containing features (Cora and PubMed) [10,14]. Table 1 summarizes the list of datasets and their respective properties.

**Table 1.** Empirical network datasets

| Dataset | Type | Nodes | Edges | Communities ($K$) | Features |
|---------|------|-------|-------|-------------------|----------|
| Karate | Social | 34 | 78 | 2 | *N/A* |
| PolBlogs | Blogs | 1222 | 16,717 | 2 | *N/A* |
| Cora | Citation | 2708 | 5429 | 7 | 1433 |
| PubMed | Citation | 19,717 | 44,338 | 3 | 500 |

*5.3. Experimental Settings*

For a baseline comparison, we chose to compare with VGAE and VGAECD. For generative models, we chose SBM [18], SBM (D.C) [19], VGAE and VGAECD as baseline comparisons. SBM and SBM (D.C) requires a specific optimization algorithm. In this case, it is optimized with Variational Expectation-Maximization (VEM) for the best performance. The encoder of VGAE and VGAECD consists of a 2-layer GCN ($L = 2$) with configuration settings of (32-16), (32-16), (32-8), and (32-8) $D$-dimension for Karate, PolBlogs, Cora and PubMed respectively. Since VGAECD-OPT consists of only a single layer **W**, for a fair comparison, we use (16), (16), (8), and (8) for Karate, PolBlogs, Cora, and PubMed which are the deepest layer's dimension in VGAECD. Additionally, we set the number of hops the same in Equation (9), i.e., $L = 2$ and the fixed number of epochs to 200 [21].

For generative models, we chose SBM [18], SBM (D.C) [19], VGAE and VGAECD as baseline methods. SBM and SBM (D.C) employ different optimization strategies. For runtime feasibility, we have chosen to use a Markov Chain Monte Carlo (MCMC) sampling strategy. The encoder of VGAE and VGAECD consists of a 2-layer GCN with configuration settings of (32-16), (32-16), (32-8), and (32-8) $D$-dimension for Karate, PolBlogs, Cora and PubMed respectively. Since VGAECD-OPT consists of only a single layer **W**, for a fair comparison, we use (16), (16), (8), and (8) for Karate, PolBlogs, Cora, and PubMed which are the deepest layer's dimension in VGAECD.

All experiments are conducted on a Linux Machine with Intel i9-7900X CPU @ 3.30GHz, 64GB @ 2666 MHz DDR3 memory and Nvidia GeForce GTX 1080Ti (12GB GPU memory) ×2 GPU with `PyTorch` framework. By default, all compatible algorithms were performed on GPU; otherwise, they are experimented using CPU computation.

*5.4. Evaluation Metric*

For evaluation purposes, we chose standard baseline metrics from [71]. These metrics are divided into two types. Specifically, the first three metrics have known ground truth, and the last three do not. In this case, the previous three metrics are useful to determine the quality of communities recovered.

- Accuracy measures the number of correctly classified clusters given the ground-truth. Formally, given two sets of community labels, i.e., $C$ is the ground-truth and $C'$ is the detected community labels, the accuracy can be calculated by,

$$ACC(C') = \frac{\sum_{i=1}^{|C|} \delta(c_i, c_i')}{|C|} \times 100\%.$$

  $c_i \in C, c_i' \in C'$, where $\delta(\cdot)$ denotes the Kronecker delta, $\delta(c_i, c_i') = 1$ when both labels matches and $|\cdot|$ denotes the cardinality of a set. For clustering tasks, accuracy is usually not emphasized as labels are known to oscillate between clusters.

- NMI and VI are based on information theory. NMI measures the 'similarity' between two community covers, while VI measures their 'dissimilarity' in terms of uncertainty. Correspondingly, a higher NMI indicates a better match between both covers while VI indicates the opposite. Formally [72],

$$\text{NMI}(C, C') = \frac{2I(C, C')}{(H(C) + H(C'))}$$

  and

$$\text{VI}(C, C') = H(C) + H(C') - 2I(C, C'),$$

  where $H(\cdot)$ is the entropy function, and $I(C, C') = H(C) + H(C') - H(C, C')$ is the mutual information function.

- Modularity (Q) [73] measures the quality of a particular community structure when compared to a null (random) model. Intuitively, intra-community links are expected to be stronger than inter-community links. Specifically,

$$Q = \frac{1}{2m} \sum_{ij} \left( A_{ij} - \frac{k_i k_j}{2m} \right) \delta(c_i, c_j),$$

  where $A_{ij} - k_i k_j / 2m$ measures the actual edge connectivity versus the expectation at random and $\delta(c_i, c_j)$ defines the Kronecker delta, where $\delta(c_i, c_j) = 1$ when both node $i$ and $j$ belongs to the same community, and 0 otherwise. The modularity score $Q \in [-1, 1]$ approaches 1 when partitioning is close to optimum.

- Conductance (CON) [37,71] measures the separability of a community across the fraction of outgoing local volume of links in the community, which is defined as,

$$\text{CON}(C) = \frac{\sum_{i \in C, j \in C'} A_{ij}}{\min(a(C), a(C'))},$$

  where the nominator defines the total number of edges within community $C$ and $a(C) = \sum_{i \in C} (j \in V)$ defines the volume of set $C \subseteq V$. A better local separability of community is achieved when the overall conductance value is the smallest.

- Triangle Participation Ratio (TPR) [71] measures the fraction of triads within the community $C$.

$$\text{TPR}(C) = |\{v_i \in C, \{(v_j, v_k) : v_j, v_k \in C,$$
$$(v_i, v_j), (v_j, v_k), (v_i, v_k) \in E\} \neq \varnothing\}| / |C|,$$

  where $E$ denotes the total number of edges in the graph $G$. A larger TPR value indicates a denser community structure.

## 6. Results and Discussion

In this section, we compare our proposed model (VGAECD-OPT) with several baseline methods. Among the generative models, SBM is an unsupervised model that does not use representation learning. For methods such as VGAE, DeepWalk, and node2vec, the latent representation is first learned then a clustering algorithm such as $k$-means is used. The * symbol denotes methods that were confined to structural information only.

We begin by discussing the stability performance of VGAECD-OPT in contrast to VGAECD in Section 6.1. In Section 6.2 we discuss the performance on several synthetic datasets. Section 6.3 provides in-depth discussion on VGAECD-OPT's performance and Section 6.4 compares the runtime and time complexity of VGAECD-OPT against baseline methods. Finally, we end this section with a discussion on the limitations of the VGAE framework in Section 6.5.

### 6.1. Stability Performance

As discussed in Section 4.3 and illustrated in Figure 1, we show that optimizing VGAECD with standard SGD will eventually result in a deviation. To show that VGAECD-OPT does not exhibit such property, we illustrate VGAECD-OPT's performance curve in Figure 4. Since VGAECD-OPT optimizes $\mathcal{L}_{\text{recon}}$ and $\mathcal{L}_{\text{comm}}$ loss in two different steps, VGAECD-OPT is more stable under the same training settings as VGAECD. On the contrary, our proposed algorithm achieves much higher NMI performance when initialized (epoch 0). At epoch 80, we achieve similar performance as VGAECD (Figure 1). After epoch 80, our algorithm continues to ascend without any decline in performance. Although we note that the rate of convergence for our loss is much slower in comparison to VGAECD, it does not hinder our main objective of achieving better community structure recovery.

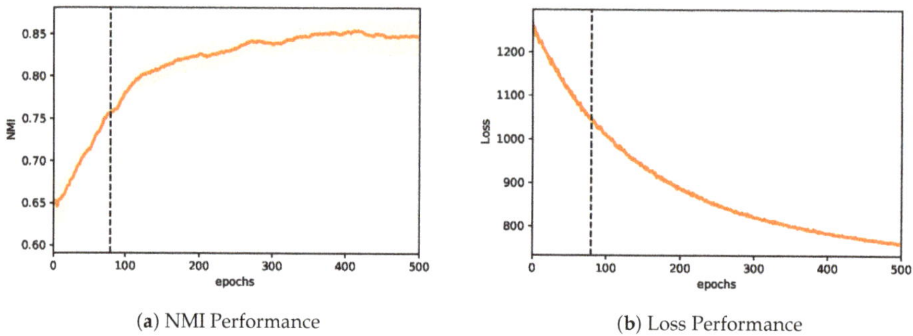

(a) NMI Performance      (b) Loss Performance

**Figure 4.** The proposed algorithm, VGAECD-OPT with Dual Optimization. In contrast to VGAECD, performance deviation is alleviated.

### 6.2. Performance on Synthetic Datasets

For starters, VGAECD-OPT is evaluated against existing discriminative and generative methods on two synthetic datasets. For methods that were confined to structural information only (i.e., absence of features), we denote them with an asterisk (*) sign. The results performed on synthetic datasets are shown in Figure 5. As shown, VGAECD-OPT can recover community structures even in difficult settings (i.e., $z_{\text{out}} \geq 6$ and $\mu \geq 0.50$) meanwhile other algorithms exhibit difficulties. This proves that with linearization and dual optimization, VGAECD-OPT is much more stable than VGAECD. Its resiliency to posterior collapse has been mitigated, subsequently increasing the probability of community structure recovery.

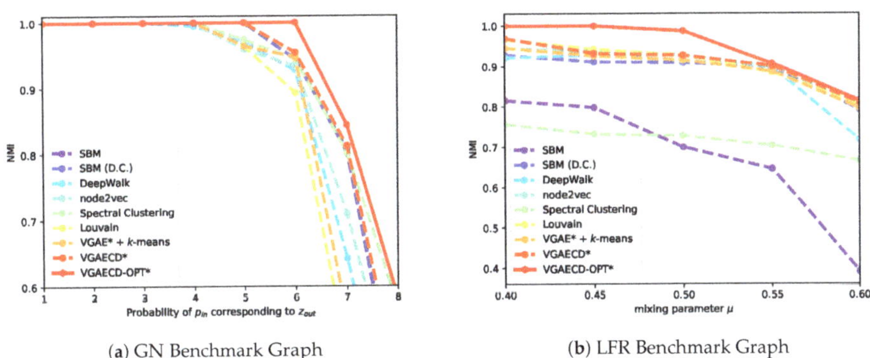

(**a**) GN Benchmark Graph        (**b**) LFR Benchmark Graph

**Figure 5.** Comparative performance of VGAECD-OPT against VGAECD on Synthetic Networks

### 6.3. Performance on Real-World Datasets

We now demonstrate the effectiveness of linearization and dual optimization approach towards real-world datasets. Tables 2 and 3 demonstrates the performance of VGAECD-OPT on datasets without the presence of features. The arrows ($\uparrow$ and $\downarrow$) indicate the direction towards better performance. For example, NMI ($\uparrow$) indicates that the higher the value, the better the performance. Values marked in bold denote best-performing results. Additionally, we also note that the performance measure on ACC is subject to label oscillation. For instance, in a binary community detection task, communities are measured by an overlap between two covers (ground-truth and detected communities), but in classification tasks, exact label assignment assignments are accounted for (i.e., labeling a cat as a dog is a false positive).

As for competing baselines, we show that discriminative models such as Louvain's algorithm, DeepWalk, and node2vec have competitive baseline performance. In general, Louvain's method performs very well with modularity score. This is because Louvain's method is by design an algorithm that maximizes modularity. However, this does not translate to true performance, as shown metrics with ground-truth measures (i.e., NMI, VI, and ACC). Node2vec and DeepWalk remain competitive in all datasets but performs poorly in Cora and PubMed datasets. These datasets contain features that determine the outcome of the algorithms' performance. Among discriminative methods, Spectral Clustering has the highest variance in terms of performance. It performs extremely poor in PolBlogs like its generative model counterpart, the SBM. We reason that the algorithm is affected by hubs with high degrees like PolBlogs dataset [19]. As a result, both algorithms pick these hubs as single node communities resulting in poorer performance.

On the other hand, generative models such as SBM performed poorly, mainly when datasets such as Cora and PubMed are used. Since SBM does not support features, it is difficult for SBM to thrive, especially when these datasets are feature-driven [63]. Indeed, with VGAE-based approaches, the performance increases significantly. Most importantly, we note that VGAECD-OPT achieves the best performance among other variants.

In Karate dataset, it can be observed that VGAECD-SGC* and VGAECD-OPT* both performed poorer than VGAECD* in terms of NMI, VI, and ACC. Further analysis shows that the non-linearity of VGAECD was a contributing factor to its higher performance. However, as we demonstrate in Tables 3–5, the presence of non-linearity was mostly negligible. Instead, the NMI performance improved when VGAECD adopts an SGC encoder. For baseline purposes, we introduced VGAECD-SGC, which comes with a linearized encoder but an absent dual optimizer. Coupled with dual optimization, VGAECD-OPT consistently shows improvements in contrast to VGAECD. We describe this with Equation (13) such that $\theta > \theta^{\text{old}}$ ensures each new community membership proposal follows proportionally to the loss.

**Table 2.** Experimental results on karate dataset.

| | NMI (↑) | VI (↓) | ACC (↑) | Q (↑) | CON (↓) | TPR (↑) |
|---|---|---|---|---|---|---|
| Spectral Clustering | 0.7323 | 0.8742 | 0.6765 | 0.3599 | 0.1313 | 0.9403 |
| Louvain | 0.4900 | 1.5205 | 0.3235 | **0.4188** | 0.2879 | 0.7333 |
| DeepWalk | 0.7198 | 0.8812 | 0.9353 | 0.3582 | 0.1337 | 0.9353 |
| node2vec | 0.8372 | 0.8050 | 0.9706 | 0.1639 | 0.4239 | 0.4549 |
| Stochastic Blockmodel | 0.0105 | 1.1032 | 0.4412 | −0.2084 | 0.7154 | 0.4034 |
| Stochastic Blockmodel (D.C) | 0.8372 | 0.8050 | 0.9706 | 0.3718 | **0.1282** | **0.9412** |
| VGAE* + $k$-means | 0.6486 | 0.8189 | 0.9647 | 0.3669 | 0.1295 | 0.9407 |
| VGAECD* | **1.0000** | **0.6931** | **1.0000** | 0.3582 | 0.1412 | **0.9412** |
| VGAECD-SGC* | 0.8372 | 0.8050 | 0.9706 | 0.3714 | **0.1282** | 0.9409 |
| VGAECD-OPT* | 0.8372 | 0.8050 | 0.9706 | 0.3742 | **0.1282** | 0.9409 |

**Table 3.** Experimental results on PolBlogs dataset.

| | NMI (↑) | VI (↓) | ACC (↑) | Q (↑) | CON (↓) | TPR (↑) |
|---|---|---|---|---|---|---|
| Spectral Clustering | 0.0014 | 1.1152 | 0.4828 | −0.0578 | 0.5585 | 0.7221 |
| Louvain | 0.6446 | 1.0839 | 0.9149 | 0.2987 | 0.8130 | 0.1922 |
| DeepWalk | 0.7367 | 1.0839 | 0.9543 | 0.0980 | 0.3873 | 0.6870 |
| node2vec | 0.7545 | 0.8613 | 0.9586 | 0.1011 | 0.3827 | 0.6863 |
| Stochastic Blockmodel | 0.0002 | 1.2957 | 0.4905 | −0.0235 | 0.5329 | 0.5657 |
| Stochastic Blockmodel (D.C) | 0.7145 | 0.8890 | 0.9496 | **0.4256** | **0.0730** | 0.8101 |
| VGAE* + $k$-means | 0.7361 | 0.8750 | 0.9552 | 0.4238 | 0.0752 | 0.8089 |
| VGAECD* | 0.7583 | 0.8583 | **0.9601** | 0.4112 | 0.0880 | 0.7913 |
| VGAECD-SGC* | 0.7235 | 0.8808 | 0.9492 | 0.4248 | 0.0735 | **0.8142** |
| VGAECD-OPT* | **0.7620** | **0.8558** | **0.9601** | 0.4252 | 0.0734 | 0.8086 |

**Table 4.** Experimental results on Cora dataset.

| | NMI (↑) | VI (↓) | ACC (↑) | Q (↑) | CON (↓) | TPR (↑) |
|---|---|---|---|---|---|---|
| Spectral Clustering | 0.2623 | **2.4183** | 0.1770 | 0.0011 | 0.8527 | 0.0577 |
| Louvain | 0.4336 | 4.0978 | 0.0081 | **0.8142** | 0.0326 | 0.2821 |
| DeepWalk | 0.3796 | 2.7300 | 0.1626 | 0.6595 | **0.0396** | 0.4949 |
| node2vec | 0.3533 | 2.9947 | 0.1359 | 0.6813 | 0.1078 | 0.4902 |
| Stochastic Blockmodel | 0.0917 | 3.5108 | 0.1639 | 0.4068 | 0.4280 | 0.3376 |
| Stochastic Blockmodel (D.C.) | 0.1679 | 3.4547 | 0.1176 | 0.6809 | 0.1736 | 0.5112 |
| VGAE* + $k$-means | 0.2384 | 3.3151 | 0.1033 | 0.6911 | 0.1615 | 0.4906 |
| VGAE + $k$-means | 0.3173 | 3.1277 | 0.1589 | 0.6981 | 0.1517 | 0.5031 |
| VGAECD* | 0.2822 | 3.1606 | 0.1532 | 0.6674 | 0.1808 | 0.5076 |
| VGAECD | 0.5072 | 2.7787 | 0.1101 | 0.7029 | 0.1371 | 0.4987 |
| VGAECD-SGC* | 0.3003 | 3.1734 | 0.1418 | 0.6116 | 0.2125 | 0.4479 |
| VGAECD-SGC | 0.5170 | 2.7707 | 0.2610 | 0.7138 | 0.1345 | 0.5053 |
| VGAECD-OPT* | 0.3735 | 2.4200 | 0.2717 | 0.4930 | 0.1792 | 0.4921 |
| VGAECD-OPT | **0.5437** | 2.6877 | **0.3190** | 0.7213 | 0.1227 | **0.5324** |

**Table 5.** Experimental results on PubMed dataset.

|  | NMI (↑) | VI (↓) | ACC (↑) | Q (↑) | CON (↓) | TPR (↑) |
|---|---|---|---|---|---|---|
| Spectral Clustering | 0.1829 | **1.4802** | 0.3405 | 0.4327 | **0.0249** | 0.1850 |
| Louvain | 0.1983 | 3.6667 | 0.0954 | **0.7726** | 0.1388 | 0.1592 |
| DeepWalk | 0.2946 | 1.7865 | 0.3101 | 0.5766 | 0.0499 | 0.2461 |
| node2vec | 0.1197 | 1.9849 | 0.2228 | 0.3501 | 0.3170 | 0.2269 |
| Stochastic Blockmodel | 0.0004 | 1.9340 | 0.3080 | −0.1620 | 0.1038 | 0.1965 |
| Stochastic Blockmodel (D.C.) | 0.1325 | 2.0035 | 0.3118 | 0.5622 | 0.8121 | 0.2441 |
| VGAE* + $k$-means | 0.2041 | 1.8096 | 0.3724 | 0.5273 | 0.1320 | 0.2898 |
| VGAE + $k$-means | 0.1981 | 1.8114 | 0.2751 | 0.5297 | 0.1283 | 0.2900 |
| VGAECD* | 0.1642 | 1.8320 | 0.1956 | 0.4966 | 0.1252 | 0.2692 |
| VGAECD | 0.3252 | 1.7056 | **0.4216** | 0.6878 | 0.1636 | **0.4827** |
| VGAECD-SGC* | 0.2350 | 1.8630 | 0.4155 | 0.5501 | 0.1163 | 0.2524 |
| VGAECD-SGC | 0.2948 | 1.7960 | 0.2396 | 0.5413 | 0.1044 | 0.2463 |
| VGAECD-OPT* | 0.2505 | 1.8517 | 0.3223 | 0.5853 | 0.0800 | 0.2519 |
| VGAECD-OPT | **0.3552** | 1.7082 | 0.3223 | 0.5378 | 0.0830 | 0.2446 |

### 6.4. Time Complexity Analysis

We now discuss the time complexity and runtime of the proposed algorithm. We divide our runtime analysis into four parts. In the first part, we will discuss the convergence rate of our proposed method. The second part analyzes the runtime performance of all methods on real-world datasets. The third part explores the scalability performance of our proposed method on synthetic networks. Finally, we present our analysis of the time complexity of all methods.

To measure convergence rate, we introduced an early stopping criterion (The early stopping criterion serves the purpose of measuring convergence rate only. In practice, algorithms run with a fixed number of epochs); when a specific NMI threshold is achieved, we terminate the algorithm. This allows fairer comparison since VGAECD-SGC and VGAECD-OPT converges faster than VGAECD. We present this result in Table 6. We show a marginal improvement in speed when the encoder has been replaced with a linear encoder. Coupled with a dual optimization process, we can obtain a faster convergence rate, resulting in a fewer number of training iterations. For Karate dataset, VGAECD-OPT is ahead by 1 s, whereas in the Cora dataset, it achieved a speedup of almost 2×. In contrast to VGAECD-SGC, the proposed algorithm is faster on Karate and Cora datasets but is slower on PolBlogs on average run time. We find this to be insignificant since the standard deviation is more unstable.

**Table 6.** Convergence rate of VGAECD-OPT vs. VGAECD.

|  | Karate | PolBlogs | Cora | PubMed |
|---|---|---|---|---|
| VGAECD | 3.3297 ± 0.0336 | 8.6538 ± 0.2808 | 6.6419 ± 0.1886 | 82.2131 ± 0.1321 |
| VGAECD-SGC | 2.8960 ± 0.0320 | 4.7735 ± 0.0372 | 3.9832 ± 0.0209 | 68.2313 ± 0.0332 |
| VGAECD-OPT | 2.1015 ± 0.0100 | 5.0768 ± 0.0120 | 3.6996 ± 0.0275 | 67.8840 ± 0.0313 |

For complete runtimes analysis, we allow all VGAE variants to complete 200 epochs without early stopping. We present these results in Table 7. Overall, the fastest algorithm is Louvain's method, which has been shown to run near-linear time in a very sparse network [7]. In the worst case, it performs with time complexity of $\mathcal{O}(N \log N)$ as presented in Table 8. On the other hand, our proposed algorithm performs better than state-of-the-art representation learning methods such as DeepWalk and node2vec on all real-world datasets (Karate, PolBlogs, Cora, and PubMed) despite being a generative model.

**Table 7.** Runtime comparison between VGAECD-OPT and baseline methods in (s)econds.

|  | Karate | PolBlogs | Cora | Pubmed |
|---|---|---|---|---|
| Spectral Clustering | $0.0111 \pm 0.0004$ | $0.0981 \pm 0.0129$ | $0.1932 \pm 0.0247$ | $14.835 \pm 0.1107$ |
| Louvain | $0.0020 \pm 0.0003$ | $0.2765 \pm 0.0204$ | $0.2571 \pm 0.0201$ | $3.1068 \pm 0.0021$ |
| DeepWalk | $0.2805 \pm 0.0204$ | $29.3969 \pm 1.7295$ | $60.2633 \pm 3.3005$ | $446.1594 \pm 1.5393$ |
| node2vec | $4.1691 \pm 0.0071$ | $73.8038 \pm 0.2477$ | $59.8279 \pm 0.0681$ | $451.6884 \pm 0.1085$ |
| Stochastic Blockmodel | $0.2126 \pm 0.0030$ | $0.2831 \pm 0.0078$ | $7.4576 \pm 4.7685$ | $6.3896 \pm 3.9298$ |
| Stochastic Blockmodel (D.C.) | $0.1452 \pm 0.0336$ | $0.2344 \pm 0.0796$ | $3.2463 \pm 1.7783$ | $3.3545 \pm 2.7707$ |
| VGAE* + $k$-means | $3.2319 \pm 0.1204$ | $18.6163 \pm 0.3803$ | $6.5510 \pm 0.2043$ | $93.4253 \pm 0.2476$ |
| VGAECD* | $3.3363 \pm 0.0539$ | $21.3191 \pm 0.2571$ | $7.4428 \pm 0.1177$ | $93.5190 \pm 0.3785$ |
| VGAECD-SGC* | $3.3503 \pm 0.0418$ | $19.0820 \pm 0.0386$ | $4.7377 \pm 0.1175$ | $89.8966 \pm 0.0844$ |
| VGAECD-OPT* | $2.4467 \pm 0.0238$ | $20.2052 \pm 0.0649$ | $7.4037 \pm 0.0342$ | $92.1212 \pm 0.1192$ |

**Table 8.** Time complexity.

| Method | Complexity |
|---|---|
| Spectral Clustering | $\mathcal{O}(N^3)$ |
| Louvain | $\mathcal{O}(N \log N)$ |
| DeepWalk | $\mathcal{O}(\gamma NTW(D + D \log N))$ |
| node2vec | $\mathcal{O}(\gamma NTW(D + D \log N))$ |
| Stochastic Blockmodel | $\mathcal{O}(N^2 K)$ |
| Stochastic Blockmodel (D.C.) | $\mathcal{O}(N^2 K)$ |
| VGAE + $k$-means | $\mathcal{O}(NXD^2) + \mathcal{O}(NK)$ |
| VGAECD | $\mathcal{O}(NXD^2) + \mathcal{O}(N^2)$ |
| VGAECD-SGC | $\mathcal{O}(NXD) + \mathcal{O}(N^2)$ |
| VGAECD-OPT | $\mathcal{O}(NXD) + \mathcal{O}(NK) + \mathcal{O}(N^2)$ |

To demonstrate runtime scalability, Figure 6 shows the algorithm's expected runtime as the number of nodes and community increases. Each network is generated from an LFR benchmark with standard parameters (see Section 5.3), but with a variable number of nodes and communities. The resulting network is summarized in Table 9. Due to the nature of our proposed method being a generative model, the runtime performance approximately polynomial in runtime. Although SBM is a generative model, it employs different optimization strategies. For instance, the original implementation by Karrer and Newman [19] struggles beyond 5000 number of nodes. To overcome this, we used a Markov Chain Monte Carlo sampling strategy to obtain reasonable runtime results. When nodes and communities are fewer than 10,000 and 10, respectively, the performance of our method is comparable to DeepWalk and node2vec. This is because both discriminative methods do not account for communities and $k$-means is used instead, resulting in faster runtimes.

**Table 9.** Networks used.

| Number of Nodes | Edges | Communities |
|---|---|---|
| 5000 | 74,278 | 5 |
| 10,000 | 148,427 | 10 |
| 20,000 | 295,857 | 20 |
| 40,000 | 599,396 | 40 |
| 80,000 | 1,189,991 | 80 |

We now analyze the time complexity of VGAECD-OPT. From Algorithm 1, the encoder has a time complexity of $\mathcal{O}(2N^2 XD^l)$ where $N$ is the number of nodes, $D$ is the size of the trainable graph filter, $l$ is the number of linear layers, and $X$ is the dimension of each node features. Since the number of filters is constant with respect to the number of layers, we have $l = 1$. The constant 2 accounts for the computation of $\mu$ and $\sigma$. If we assume that the adjacency matrix is sparse, we can have an encoder with a complexity of $\mathcal{O}(NXD)$. With NEM, we introduce two additional steps, which has a time complexity of $\mathcal{O}(2NK)$ for performing the Expectation-Maximization steps. Given many samples, we can further simplify this to $\mathcal{O}(NK)$. With an inner product decoder, it has a time complexity of $\mathcal{O}(N^2)$. Overall, the final time complexity for one epoch is $\mathcal{O}(NXD) + \mathcal{O}(NK) + \mathcal{O}(N^2)$. In comparison, VGAECD has a time complexity of $\mathcal{O}(NXD^2) + \mathcal{O}(N^2)$ due to its two-layer GCN architecture. Thus, we can conclude that VGAECD-OPT is relatively competitive with VGAECD in terms of time complexity.

We list a summary of the competing method's time complexity in Table 8. Additionally, we note the following notations: $N$—number of nodes, $\gamma$—the number of random walks, $T$—walk length, $W$—window size, $D$—the representation size and $K$—number of communities.

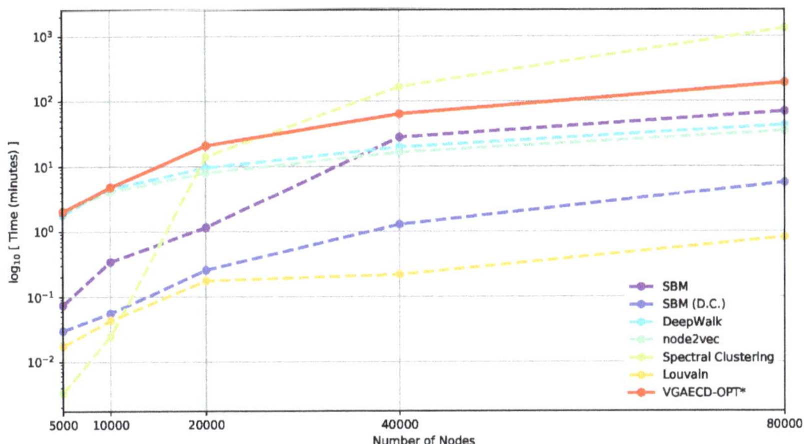


**Figure 6.** Runtime of VGAECD-OPT & baseline methods on LFR benchmark graphs.

*6.5. Limitations of VGAE Framework*

Overall, VGAE and its variants have proven to be an effective algorithm for learning networks with features. In this section, we highlight one shortcoming of VGAE that remains a challenge. In particular, VGAE has scaling difficulties. In VGAE, the inner product decoder uses a cross-entropy loss function. Unfortunately, this requires a dense by dense matrix multiplication, which requires a significant amount of memory for backpropagation purposes. In other literature, methods such as LINE [50] and DeepWalk [12] employs a negative sampling loss function for link prediction. In VGAE's case, such implementation is not trivial as the task differs (reconstruction vs. link prediction). To explain this, let us consider an undirected unweighted graph $\mathcal{G}$. One can observe that negative sampling considers the connectivity of each node by considering edges that are present and absent, $e_{ij} \in \{0,1\}$. Formally, negative sampling can be defined as

$$\log \tau \left( \mathbf{z}_j^\top \cdot \mathbf{z}_i \right) + \sum_{i=1}^{K} E_{v_n \sim p_n(v)} \left[ \log \tau \left( -\mathbf{z}_n^\top \cdot \mathbf{z}_i \right) \right]. \tag{17}$$

Here, we consider positive samples (edges) in the first term of Equation (17) with $\mathbf{z}_i$ being the representation of node $v_i$ and negative samples in the second term. $\tau(\cdot)$ denotes the non-linear function and $K$ defines the number of negative samples which are drawn from some probability distribution $p_n(v)$. The first term of Equation (17) defines the likelihood of positive samples inducing an edge while the second term negates such probability. In other words, the loss function in Equation (17) induces separation of the positive samples from the negative samples, in such a way that the representations of each positive edge would stay further apart from negative edges. In DeepWalk and LINE, the effectiveness of negative sampling is highly dependent on the context (random walker's chain). Under the current VGAE framework, such a context does not exist. Implementing this in VGAE is non-trivial. Hence, to scale our model, we need more efficient decoders, which remains a challenging task.

## 7. Conclusions

In this paper, we demonstrate that VAE and its variants (including VGAE & VGAECD) have a high tendency to favor minimization of the reconstruction loss over a clustering loss. As a result, it performs poorer as training prolongs overtime. To rectify this problem, we propose a dual optimization approach for optimizing VGAECD. We experimentally show the effectiveness of our dual optimization approach on VGAECD, allowing us to outperform its previous achievements. Moreover, to increase the speed of learning, we follow new practices of linearizing the encoder. Although the performance gain is marginal in terms of community detection, it has reduced the number of learnable parameters, which results in faster convergence and training speed.

**Author Contributions:** Conceptualization, J.J.C.; methodology, J.J.C.; software, J.J.C.; validation, J.J.C. and X.L.; formal analysis, J.J.C. and X.L.; investigation, J.J.C. and X.L.; writing—original draft preparation, J.J.C. and X.L.; writing—review and editing, X.L. and T.M.; visualization, J.J.C.; supervision, T.M.; funding acquisition, T.M. All authors have read and agreed to the published version of the manuscript.

**Funding:** We would like to thank JST CREST (Grant Number JPMJCR1687), JSPS Grant-in-Aid for Scientific Research(B) (Grant Number 17H01785) and JSPS Grant-in-Aid for Early-Career Scientists (Grant Number 19K20352).

**Conflicts of Interest:** The authors declare no conflict of interest.

## References


1. Zachary, W.W. An Information Flow Model for Conflict and Fission in Small Groups. *J. Anthropol. Res.* **1977**, *33*, 452–473. [CrossRef]
2. Newman, M.E. The Structure of Scientific Collaboration Networks. *Proc. Natl. Acad. Sci. USA* **2001**, *98*, 404–409. [CrossRef]
3. Harper, F.M.; Konstan, J.A. The Movielens Datasets: History and Context. *ACM Trans. Interact. Intell. Syst. (TiiS)* **2016**, *5*, 19. [CrossRef]
4. Su, X.; Sperlì, G.; Moscato, V.; Picariello, A.; Esposito, C.; Choi, C. An Edge Intelligence Empowered Recommender System Enabling Cultural Heritage Applications. *IEEE Trans. Ind. Inform.* **2019**, *15*, 4266–4275. [CrossRef]
5. Council, N.R. *Network Science*; National Academies Press: Washington, DC, USA, 2006.
6. Barabási, A.L. *Network Science*; Cambridge University Press: Cambridge, UK, 2016.
7. Fortunato, S. Community Detection in Graphs. *Phys. Rep.* **2010**, *486*, 75–174. [CrossRef]
8. Newman, M.E.J. Modularity and Community Structure in Networks. *Proc. Natl. Acad. Sci. USA* **2006**, *103*, 8577–8582. [CrossRef]
9. Bengio, Y.; Courville, A.; Vincent, P. Representation Learning: A Review and New Perspectives. *IEEE Trans. Pattern Anal. Mach. Intell.* **2013**, *35*, 1798–1828. [CrossRef]
10. Yang, Z.; Cohen, W.W.; Salakhutdinov, R. Revisiting Semi-Supervised Learning with Graph Embeddings. In Proceedings of the 33rd International Conference on International Conference on Machine Learning, New York, NY, USA, 19–24 June 2016; Volume 48, pp. 40–48.
11. Grover, A.; Leskovec, J. Node2vec: Scalable Feature Learning for Networks. In Proceedings of the 22nd ACM SIGKDD International Conference on Knowledge Discovery and Data Mining, San Francisco, CA, USA, 13–17 August 2016; Volume 2016, pp. 855–864.
12. Perozzi, B.; Al-Rfou, R.; Skiena, S. DeepWalk: Online Learning of Social Representations. In Proceedings of the 20th ACM SIGKDD International Conference on Knowledge Discovery and Data Mining, New York, NY, USA, 24–27 August 2014; pp. 701–710.
13. Tang, J.; Qu, M.; Wang, M.; Zhang, M.; Yan, J.; Mei, Q. Line: Large-Scale Information Network Embedding. In Proceedings of the 24th International Conference on World Wide Web. International World Wide Web Conferences Steering Committee, Florence, Italy, 18–22 May 2015; pp. 1067–1077.

14. Kipf, T.N.; Welling, M. Semi-Supervised Classification with Graph Convolutional Networks. In Proceedings of the International Conference on Learning Representations, Toulon, France, 24–26 April 2017.

15. Pearl, J. Causal Inference in Statistics: An Overview. *Stat. Surv.* **2009**, *3*, 96–146. [CrossRef]

16. Gal, Y. *Uncertainty in Deep Learning*; University of Cambridge: Cambridge, UK, 2016; Volume 1, p. 3.

17. Kendall, A.; Gal, Y. What Uncertainties Do We Need in Bayesian Deep Learning for Computer Vision? In *Advances in Neural Information Processing Systems 30*; Guyon, I., Luxburg, U.V., Bengio, S., Wallach, H., Fergus, R., Vishwanathan, S., Garnett., R., Eds.; Curran Associates, Inc.: Red Hook, NY, USA, 2017; pp. 5580–5590.

18. Snijders, T.A.B.; Nowicki, K. Estimation and Prediction for Stochastic Blockmodels for Graphs with Latent Block Structure. *J. Classif.* **1997**, *14*, 75–100. [CrossRef]

19. Karrer, B.; Newman, M.E.J. Stochastic Blockmodels and Community Structure in Networks. *Phys. Rev. E* **2011**, *83*, 016107. [CrossRef]

20. Kingma, D.P.; Welling, M. Auto-Encoding Variational Bayes. In Proceedings of the 2nd International Conference on Learning Representations, Banff, AB, Canada, 14–16 April 2014.

21. Kipf, T.N.; Welling, M. Variational Graph Auto-Encoders. In Proceedings of the Bayesian Deep Learning Workshop, 30th Conference on Neural Information Processing Systems (NeurIPS), Centre Convencions Internacional Barcelona, Barcelona, Spain, 5–10 December 2016.

22. Choong, J.J.; Liu, X.; Murata, T. Learning Community Structure with Variational Autoencoder. In Proceedings of the IEEE International Conference on Data Mining, Singapore, 17–20 November 2018; pp. 69–78.

23. Pan, S.; Hu, R.; Long, G.; Jiang, J.; Yao, L.; Zhang, C. Adversarially Regularized Graph Autoencoder for Graph Embedding. In Proceedings of the 27th International Joint Conference on Artificial Intelligence, Stockholm, Sweden, 13–19 July 2018; pp. 2609–2615.

24. Razavi, A.; van den Oord, A.; Poole, B.; Vinyals, O. Preventing Posterior Collapse with Delta-Vaes. In Proceedings of the International Conference on Learning Representations, New Orleans, LA, USA, 6–9 May 2019.

25. Finn, C.; Abbeel, P.; Levine, S. Model-Agnostic Meta-Learning for Fast Adaptation of Deep Networks. In Proceedings of the 34th International Conference on Machine Learning, International Convention Centre, Sydney, Australia, 6–11 August 2017; pp. 1126–1135.

26. Metz, L.; Maheswaranathan, N.; Cheung, B.; Sohl-Dickstein, J. Meta-Learning Update Rules for Unsupervised Representation Learning. In Proceedings of the International Conference on Learning Representations, New Orleans, LA, USA, 6–9 May 2019.

27. Dai, B.; Dai, H.; He, N.; Liu, W.; Liu, Z.; Chen, J.; Xiao, L.; Song, L. Coupled Variational Bayes via Optimization Embedding. In *Advances in Neural Information Processing Systems 31*; Bengio, S., Wallach, H., Larochelle, H., Grauman, K., Cesa-Bianchi, N., Garnett, R., Eds.; Curran Associates, Inc.: Red Hook, NY, USA, 2018; pp. 9690–9700.

28. Greff, K.; van Steenkiste, S.; Schmidhuber, J. Neural Expectation Maximization. In *Advances in Neural Information Processing Systems 30*; Guyon, I., Luxburg, U.V., Bengio, S., Wallach, H., Fergus, R., Vishwanathan, S., Garnett., R., Eds.; Curran Associates, Inc.: Red Hook, NY, USA, 2017; pp. 6691–6701.

29. Jordan, M.I.; Ghahramani, Z.; Jaakkola, T.S.; Saul, L.K. An Introduction to Variational Methods for Graphical Models. *Mach. Learn.* **1999**, *37*, 183–233. [CrossRef]

30. Wu, F.; Zhang, T.; de Souza, A.H., Jr.; Fifty, C.; Yu, T.; Weinberger, K.Q. Simplifying Graph Convolutional Networks. In Proceedings of the International Conference on Learning Representations, New Orleans, LA, USA, 6–9 May 2019.

31. Salha, G.; Hennequin, R.; Vazirgiannis, M. Keep It Simple: Graph Autoencoders Without Graph Convolutional Networks. In Proceedings of the Workshop on Graph Representation Learning, 33rd Conference on Neural Information Processing Systems (NeurIPS), Vancouver Convention Center, Vancouver, BC, Canada, 8–14 December 2019.

32. Murphy, R.C.; Wheeler, K.B.; Barrett, B.W.; Ang, J.A. Introducing the Graph 500. *Cray Users Group (CUG)* **2010**, *19*, 45–74.

33. Ueno, K.; Suzumura, T. Highly Scalable Graph Search for the Graph500 Benchmark. In Proceedings of the 21st International Symposium on High-Performance Parallel and Distributed Computing, Delft, The Netherlands, 18–22 June 2012; pp. 149–160.

34. Hay, M.; Miklau, G.; Jensen, D.; Weis, P.; Srivastava, S. Anonymizing Social Networks. In *Computer Science Department Faculty Publication Series*; University of Massachusetts Amherst: Amherst, MA, USA, 2007; p. 180.

35. Girvan, M.; Newman, M.E.J. Community Structure in Social and Biological Networks. *Proc. Natl. Acad. Sci. USA* **2002**, *99*, 7821–7826. [CrossRef]

36. Pizzuti, C. Ga-Net: A Genetic Algorithm for Community Detection in Social Networks. In Proceedings of the International Conference on Parallel Problem Solving from Nature, Dortmund, Germany, 13–17 September 2008; pp. 1081–1090.

37. Kannan, R.; Vempala, S.; Vetta, A. On Clusterings: Good, Bad and Spectral. *J. ACM* **2004**, *51*, 497–515. [CrossRef]

38. Blondel, V.D.; Guillaume, J.L.; Lambiotte, R.; Lefebvre, E. Fast Unfolding of Communities in Large Networks. *J. Stat. Mech. Theory Exp.* **2008**, *2008*, P10008. [CrossRef]

39. Fortunato, S.; Barthélemy, M. Resolution Limit in Community Detection. *Proc. Natl. Acad. Sci. USA* **2007**, *104*, 36–41. [CrossRef]

40. Good, B.H.; de Montjoye, Y.A.; Clauset, A. The Performance of Modularity Maximization in Practical Contexts. *Phys. Rev. E* **2010**, *81*, 046106. [CrossRef]

41. Raghavan, U.N.; Albert, R.; Kumara, S. Near Linear Time Algorithm to Detect Community Structures in Large-Scale Networks. *Phys. Rev. E* **2007**, *76*, 036106. [CrossRef]

42. Pons, P.; Latapy, M. Computing Communities in Large Networks Using Random Walks. In Proceedings of the International Symposium on Computer and Information Sciences, Istanbul, Turkey, 26–28 October 2005; pp. 284–293.

43. Rosvall, M.; Bergstrom, C.T. Multilevel Compression of Random Walks on Networks Reveals Hierarchical Organization in Large Integrated Systems. *PLoS ONE* **2011**, *6*, e18209. [CrossRef]

44. Agreste, S.; Meo, P.D.; Fiumara, G.; Piccione, G.; Piccolo, S.; Rosaci, D.; Sarné, G.M.L.; Vasilakos, A.V. An Empirical Comparison of Algorithms to Find Communities in Directed Graphs and Their Application in Web Data Analytics. *IEEE Trans. Big Data* **2017**, *3*, 289–306. [CrossRef]

45. Cao, S.; Lu, W.; Xu, Q. GraRep: Learning Graph Representations with Global Structural Information. In Proceedings of the 24th ACM International on Conference on Information and Knowledge Management, Melbourne, Australia, 19–23 October 2015; pp. 891–900.

46. Guo, T.; Pan, S.; Zhu, X.; Zhang, C. CFOND: Consensus Factorization for Co-Clustering Networked Data. *IEEE Trans. Knowl. Data Eng.* **2018**, *31*, 706–719. [CrossRef]

47. Moscato, V.; Picariello, A.; Sperlí, G. Community Detection Based on Game Theory. *Eng. Appl. Artif. Intell.* **2019**, *85*, 773–782. [CrossRef]

48. Krizhevsky, A.; Sutskever, I.; Hinton, G.E. ImageNet Classification with Deep Convolutional Neural Networks. In *Advances in Neural Information Processing Systems 25*; Pereira, F., Burges, C.J.C., Bottou, L., Weinberger, K.Q., Eds.; Curran Associates, Inc.: Red Hook, NY, USA, 2012; pp. 1097–1105.

49. Mikolov, T.; Sutskever, I.; Chen, K.; Corrado, G.S.; Dean, J. Distributed Representations of Words and Phrases and Their Compositionality. In *Advances in Neural Information Processing Systems 26*; Burges, C.J.C., Bottou, L., Ghahramani, Z., Weinberger, K.Q, Eds.; Curran Associates, Inc.: Red Hook, NY, USA, 2013; pp. 3111–3119.

50. Tian, F.; Gao, B.; Cui, Q.; Chen, E.; Liu, T.Y. Learning Deep Representations for Graph Clustering. In Proceedings of the 28th AAAI Conference on Artificial Intelligence, Québec City, QC, Canada, 27–31 July 2014; pp. 1293–1299.

51. Defferrard, M.; Bresson, X.; Vandergheynst, P. Convolutional Neural Networks on Graphs with Fast Localized Spectral Filtering. In *Advances in Neural Information Processing Systems 29*; Lee, D.D., Sugiyama, M., Luxburg, U.V., Guyon, I., Garnett, R., Eds.; Curran Associates, Inc.: Red Hook, NY, USA, 2016; pp. 3844–3852.

52. Qiu, J.; Dong, Y.; Ma, H.; Li, J.; Wang, K.; Tang, J. Network Embedding As Matrix Factorization: Unifying DeepWalk, LINE, PTE, and Node2Vec. In Proceedings of the Eleventh ACM International Conference on Web Search and Data Mining, Los Angeles, CA, USA, 5–9 February 2018; pp. 459–467.

53. Liu, X.; Murata, T.; Kim, K.S.; Kotarasu, C.; Zhuang, C. A General View for Network Embedding as Matrix Factorization. In Proceedings of the Twelfth ACM International Conference on Web Search and Data Mining, Melbourne, Australia, 11–15 February 2019; pp. 375–383.

54. Leskovec, J.; Chakrabarti, D.; Kleinberg, J.; Faloutsos, C.; Ghahramani, Z. Kronecker Graphs: An Approach to Modeling Networks. *J. Mach. Learn. Res.* **2010**, *11*, 985–1042.

*Entropy* **2020**, *22*, 197

55. Seshadhri, C.; Kolda, T.G.; Pinar, A. Community Structure and Scale-Free Collections of Erdős-Rényi Graphs. *Phys. Rev. E* **2012**, *85*, 056109. [CrossRef] [PubMed]
56. Lancichinetti, A.; Fortunato, S.; Radicchi, F. Benchmark Graphs for Testing Community Detection Algorithms. *Phys. Rev. E* **2008**, *78*, 046110.. [CrossRef] [PubMed]
57. Bródka, P. A Method for Group Extraction and Analysis in Multilayer Social Networks. *arXiv* **2016**, arXiv:1612.02377.
58. Airoldi, E.M.; Blei, D.M.; Fienberg, S.E.; Xing, E.P. Mixed Membership Stochastic Blockmodels. *J. Mach. Learn. Res.* **2008**, *9*, 1981–2014.
59. Larremore, D.B.; Clauset, A.; Jacobs, A.Z. Efficiently Inferring Community Structure in Bipartite Networks. *Phys. Rev. E* **2014**, *90*, 012805. [CrossRef]
60. Abbe, E.; Sandon, C. Community Detection in General Stochastic Block Models: Fundamental Limits and Efficient Algorithms for Recovery. In Proceedings of the 2015 IEEE 56th Annual Symposium on Foundations of Computer Science (FOCS), Berkeley, CA, USA, 18–20 October 2015; pp. 670–688.
61. Abbe, E.; Bandeira, A.S.; Hall, G. Exact Recovery in the Stochastic Block Model. *IEEE Trans. Inf. Theory* **2016**, *62*, 471–487. [CrossRef]
62. Fortunato, S.; Hric, D. Community Detection in Networks: A User Guide. *Phys. Rep.* **2016**, *659*, 1–44. [CrossRef]
63. NT, H.; Maehara, T. Revisiting Graph Neural Networks: All We Have Is Low-Pass Filters. *arXiv* **2019**, arXiv:1905.09550.
64. Higgins, I.; Matthey, L.; Pal, A.; Burgess, C.; Glorot, X.; Botvinick, M.; Mohamed, S.; Lerchner, A. Beta-VAE: Learning Basic Visual Concepts with a Constrained Variational Framework. *ICLR* **2017**, *2*, 6
65. Dempster, A.P.; Laird, N.M.; Rubin, D.B. Maximum Likelihood from Incomplete Data via the EM Algorithm. *J. R. Stat. Soc. Ser. B (Methodol.)* **1977**, *39*, 1–22.
66. Kingma, D.P.; Ba, J. Adam: A Method for Stochastic Optimization. In Proceedings of the 3rd International Conference for Learning Representations, San Diego, CA, USA, 7–9 May 2015.
67. Hinton, G.E.; van Camp, D. Keeping the Neural Networks Simple by Minimizing the Description Length of the Weights. In Proceedings of the Sixth Annual Conference on Computational Learning Theory, Santa Cruz, CA, USA, 26-28 July 1993; pp. 5–13.
68. Adamic, L.A.; Glance, N. The Political Blogosphere and the 2004 US Election: Divided They Blog. In Proceedings of the 3rd International Workshop on Link Discovery, Chicago, IL, USA, 21–24 August 2005; pp. 36–43.
69. Sen, P.; Namata, G.; Bilgic, M.; Getoor, L.; Galligher, B.; Eliassi-Rad, T. Collective Classification in Network Data. *AI Mag.* **2008**, *29*, 93. [CrossRef]
70. McCallum, A.K.; Nigam, K.; Rennie, J.; Seymore, K. Automating the Construction of Internet Portals with Machine Learning. *Inf. Retr.* **2000**, *3*, 127–163. [CrossRef]
71. Yang, J.; Leskovec, J. Defining and Evaluating Network Communities Based on Ground-Truth. *Knowl. Inf. Syst.* **2015**, *42*, 181–213. [CrossRef]
72. Danon, L.; Díaz-Guilera, A.; Duch, J.; Arenas, A. Comparing Community Structure Identification. *J. Stat. Mech. Theory Exp.* **2005**, *2005*, P09008. [CrossRef]
73. Newman, M.E.J.; Girvan, M. Finding and Evaluating Community Structure in Networks. *Phys. Rev. E* **2004**, *69*, 026113. [CrossRef]

*Article*

# Properties of the Vascular Networks in Malignant Tumors

Juan Carlos Chimal-Eguía [1,*], Erandi Castillo-Montiel [2] and Ricardo T. Paez-Hernández [3]

[1]   Centro de Investigación en Computación del Instituto Politécnico Nacional, Av. Miguel Othon de Mendizabal s/n. Col. La Escalera, Ciudad de México CP 07738, Mexico
[2]   Department of Técnologias WEB, Instituto Politécnico Nacional (IPN) - Centro Nacional de Cálculo (CENAC), Av. Luis Enrique Erro S/N, Unidad Profesional Adolfo López Mateos, Zacatenco, Gustavo A. Madero, Ciudad de México CP 07738, Mexico; erandicm@gmail.com
[3]   Área de Física de Procesos Irreversibles, Departamento de Ciencias Básicas, Universidad Autónoma Metropolitana, U-Azcapotzalco, Av. San Pablo 180, Col.Reynosa, Ciudad de México CP 02200, Mexico; phrt@correo.azc.uam.mx
*   Correspondence: jchimale@ipn.mx

Received: 4 December 2019; Accepted: 28 January 2020; Published: 31 January 2020

**Abstract:** This work presents an analysis for real and synthetic angiogenic networks using a tomography image that obtains a portrait of a vascular network. After the image conversion into a binary format it is possible to measure various network properties, which includes the average path length, the clustering coefficient, the degree distribution and the fractal dimension. When comparing the observed properties with that produced by the Invasion Percolation algorithm (IPA), we observe that there exist differences between the properties obtained by the real and the synthetic networks produced by the IPA algorithm. Taking into account the former, a new algorithm which models the expansion of an angiogenic network through randomly heuristic rules is proposed. When comparing this new algorithm with the real networks it is observed that now both share some properties. Once creating synthetic networks, we prove the robustness of the network by subjecting the original angiogenic and the synthetic networks to the removal of the most connected nodes, and see to what extent the properties changed. Using this concept of robustness, in a very naive fashion it is possible to launch a hypothetical proposal for a therapeutic treatment based on the robustness of the network.

**Keywords:** complex networks; angiogenesis; network properties

---

## 1. Introduction

The development of cancer has thereby concentrated on an approach which is center around the genetic events which allow cells to escape from growth control and become cancerous. However, even if cancer cells have been generated and can evolve to accumulate more mutations, these cancer cells might not be able to grow beyond a very small size.

One of the most important factors in this respect is the blood supply which provides cancer cells with oxygen, nutrients and necessities required for survival. When the growth of tumoral cells is high enough to consume all supplies in a specific organ or tissue, the tumor stops its growth in order to induce the generation of new blood supply to sustain its growth; this process is called angiogenesis.

Whether a new blood supply can be formed or not appears to be determined by the balance between angiogenesis inhibitors and promoters. When angiogenic cell lines emerge they can shift the balance away from inhibition and in favor of promotion. This induces blood vessels to grow towards the tumor and this process leads to the complete vascularization of the tumor, i.e., a vascular network [1].

Our understanding about the role of angiogenesis in the development of cancers has advanced significantly since the studies of Judah Folkman [2]. Many of these studies have been focused mainly on the understanding of the cells at molecular level. To fully understand the behaviour of an organism, organ or even a single cell, we need to fully comprehend the collective behaviour of the whole system. Recently, the analysis of the behaviours of the biological systems or their emergent properties, that are not apparent from the examination of only a few isolated interactions alone, have emerged as new insights in the study of the systems in biology.

For instance, the use of fractal geometry [3] can describe the pathological structure of tumors, and give us some insights into the mechanisms of tumor growth and angiogenesis, that complement those obtained by modern molecular methods. Another good example is the comparative analysis of the transcription gene regulatory networks of the E. Coli and S. Cervisae made by Santillan et al. [4], or the interesting work made by Abdollahi et al. [5]. In both articles, the authors noticed how the network properties of the gene networks revealed some interesting data from an evolutionary point of view.

These among many others [6–9], are good examples of how emergent properties could help us understand the behaviour of biological systems. In particular, the use of network theory is important because it allows the description of a network structure using graph concepts. Furthermore, the observed network topology gives clues about the evolution, structure, which helps us elucidate the dynamics of hundreds of interacting components [10–13] .

In this work, we present an analysis of two angiogenic networks in patients with Hepato-Cellular Carcinoma (HCC). We used a tomography image (obtained from the National Institute of Nutrition of México INNSZ) in order to obtain the vascular network. After the conversion of the image into a binary skeletonized form, we measured some of the network properties; the performed measurements includes the average path length, the clustering coefficient, the degree distribution, and the fractal dimension.

The observed properties of the tumor vasculature as a whole closely correspond to those produced by a new algorithm of random growth process known as Angiogenesis Random Growth Algorithm (ARGA). ARGA models the expansion of an angiogenic network through randomly heuristic rules. We test the network robustness by subjecting the original angiogenic and the synthetic networks produced by ARGA to the removal of the most connected nodes and seeing to what extent the properties changed (in particular the clustering coefficient). Taking into account this robustness, we proposed a hypothetic therapeutic treatment based on the network robustness.

The paper is organized as follows: In Section 2 we present the analysis of the angiogenic network, after converting into a binary skeletonized form from a tomography image and then analyzed its network properties. Section 3 presents a robustness study of the real network compared with those processed by a new algorithm called ARGA. Finally, Section 4 gives some concluding remarks.

## 2. Analysis of the Angiogenic Network

Angiogenesis is an important natural process that takes into account the growth of new blood vessels that occur in the body, both in health and disease. Angiogenesis is now recognized as one of the critical events required for tumor progression [14], where cancerous growth is dependent on vascular induction and the development of a new vascular supply.

The idea of targeting angiogenesis to inhibit tumor growth was proposed more than three decades ago, and since then, several approaches to block or disrupt tumor angiogenesis have been explored. However, all of these have been focused on the understanding of the molecular behaviour, and only a few in other properties that are not apparent from the molecular point of view.

For instance, some authors have declared: "Angiogenesis in tumors leads to tumor vessels with multiple functional and structural abnormalities. Tumors consist of a chaotic, poorly organized vasculature, with tortuous, irregular shape, and leaky vessels that are often unable to support efficient blood flow and leading to an aberrant vascular system" [15,16]. Furthermore, is it possible to consider that the tumor angiogenesis leads to a poorly organized vasculature without another measurement that the observation in situ. In our opinion, the answer to this question should be supported by some structural analysis. In this context we proposed the following analysis in order to have more elements that show if the vasculature network created by the tumor has this chaotic image, or has some structural elements that are not apparent only from observation that make them in some sense efficient to their purposes.

### 2.1. Creating the Network from a Image Tomography

The first step in our analysis is to obtain the vascular network from a tomography image. In mathematical terms a network is represented by a graph, which is a pair of sets $G = (P, E)$, where $P$ is a set of nodes (or vertices or points) $P_1, P_2, \ldots, P_n$ and $E$ is a set of edges (or links or lines) that connect two elements of $P$. Graphs are usually represented as a set of dots, each corresponding to a node, two of these dots being joined by a line if the corresponding nodes are connected.

From the department of radiology of the National Institute of Medical Sciences "Salvador Zuviran" (INNSZ) of Mexico City, we obtained images used by the INNSZ in order to diagnose the development of the malignant tumors of four patients with Hepato-Cellular Carcinoma (HCC) (see Table 1).

**Table 1.** Characteristics of the four patients studied in the INNSZ, all with Hepato-Cellular Carcinoma (HCC).

| Patient | Sex | Age | Date |
|---------|--------|-----|------------------|
| A | Female | 44 | 23 November 2007 |
| B | Male | 57 | 5 February 2008 |
| C | Female | 63 | 30 October 2007 |
| D | Female | 55 | 8 March 2007 |

The progress in the development of the disease of these patients in the INNSZ is made by a computerized tomogram. This tomograph is not only dedicated to diagnosis of the angiogenesis process, but also other cancerous diseases such as the detection of gliomas, etc. The images taken by the tomogram have a resolution of 960 by 1260 pixels. These images are stored in DICOM format, and using the MedimalView software we can manipulated it to obtain BMP images with a resolution of 960 by 1240 pixels (8 bits for pixel). From the BMP image we obtain different images with different sizes. In our study we obtained four sizes 32 by 32 pixels, 64 by 64 pixels 128 by 128 pixels and 256 by 256 pixels.

Once the image is obtained in a BMP format we proceed to make a digital processing of the image to obtain a binary skeletonized image. The procedure is as follows [17]:

1.  Pre-Processing

    The tomographic images were subjected to a pre-processing stage to obtain the tumor vascular network. Using a representation of the image in 2-D, the first step was to display the image into a gray scale, where each pixel uses an individual value that represents its luminescence, and thus, have greater ease in handling the image. All the tomography images given by the INNSZ were very noisy, making it difficult to identify the blood vessels, so it was decided to make an improvement in the image by adjusting the contrast automatically.

    The representation of an image in an 2-D array is given by the intensity values $f(x, y)$ at each image pixel. The arrangement has $M$ rows and $N$ columns, where $(x, y)$ are discrete coordinates. We used for convenience integer values for discrete coordinates. Then we have for each coordinate $x = 0, 1, 2, \ldots, M - 1$ and $y = 0, 1, 2, \ldots, N - 1$. In a matrix representation obtaining

$$f(x, y) = \begin{bmatrix} f(0,0) & f(0,1) & \cdots & f(0, N-1) \\ \vdots & \vdots & \ddots & \vdots \\ f(M-1, 0) & f(M-1, 1) & \cdots & f(M-1, N-1) \end{bmatrix} \tag{1}$$

The gray scale adjustment consists of multiplying each RGB component by three constants defined by: $\alpha$, $\beta$ and $\gamma$. Subsequently, the intensity obtained in each channel is averaged.

This process subtracts all the color information contained in each pixel and gives a separation of 255 levels between black and white. These three constants are obtained as the separation between the RGB and the black channels as:

- $\alpha$: Division between the red and black. (0.2989)
- $\beta$: Division between the green and black. (0.5870)
- $\gamma$: Division between the blue and black. (0.1140)

Now to obtain the equivalent gray scale value for each pixel we use the following equation:

$$I = \alpha * R + \beta * G + \gamma * B$$

We shall now proceed to the brightness adjustment as the last part of the stage of pre-processing algorithm. Brightness is the percentage of luminescence or darkness of a color. It is possible to go from 0 % which means black, up to 100% which means white. Mathematically, the operation corresponding to the brightness adjustment is: $M + B = C$, where M corresponds to the image matrix, C corresponds to the adjusted image M, and p is the parameter adjusting brightness whose standard ranges from $-100$ to 100.

2. Segmentation

Now proceed to the image segmentation stage in which we obtain the angiogenic network by extracting most of the blood vessels that are connected within the image and store them in a new image. To achieve this we use a threshold which cleaves the image into two classes of objects: blood vessels and background image. Otsu's method [18] calculates this threshold automatically in the following way: in order to find the value of a threshold $T$, for which the variance $\sigma_B^2(T)$ between two regions $C_0$ and $C_1$ (considering only two regions) is maximum (i.e., the point where the two classes are separated), we use following the equation:

$$\sigma_B^2(T) = \frac{[m_G P_1(T) - m(T)]^2}{P_1(T)[1 - P_1(T)]} \tag{2}$$

where, $m_G$ is the average gray level of the entire image and $P_1(T)$ is the occurrence probability into the region.

To separate the blood vessels from the background, the general idea was to label each region of contiguous pixels with a different value, and with this value one can obtain the number of objects in the image which depends on the adjacency used.

3. Obtaining the skeletonized binary form

Skeletonization of an image makes possible the classification, recognition and simplification of the objects within it, and one of its most important applications is that skeletonization reduces the structural form of an image to a graph. The skeleton tries to represent the shape of an object with a relatively small number of pixels and the position, orientation and length of the skeleton lines correspond to those equivalent to the original image.

Once the vascular network is segmented we proceed to represent the image network with a relatively smaller number of pixels using the skeleton of the original image. This process generates a binary image which is stored in an array of $0's$ and $1's$, where the value of 1 corresponds to the image skeleton, while the value of 0 will be considered the image background.

The region skeleton can be defined by the transform of the Median Axis Transformation (MAT) proposed by Blum et al. [19]. To define the MAT for each point p in R (the region), we seek if the point p is a close neighbor to B (edges of the region R).

If $p$ has more than one closed neighbor, it is said to belong to the median axis (i.e., it belongs to the skeleton) of R. It is important to notice that the concept of proximity depends on the definition of distance used. All the procedure is shown in Figure 1.

It is worthwhile to mention that the skeletonized binary form is a 2D representation of the vascular network, this means that we only have 8 possible neighbors with respect to one single node. This apparently limitation can be overcome considering a 3D model and developing the same steps as those mentioned above. In recent years there have appear other models trying to resemble this process [20–22].

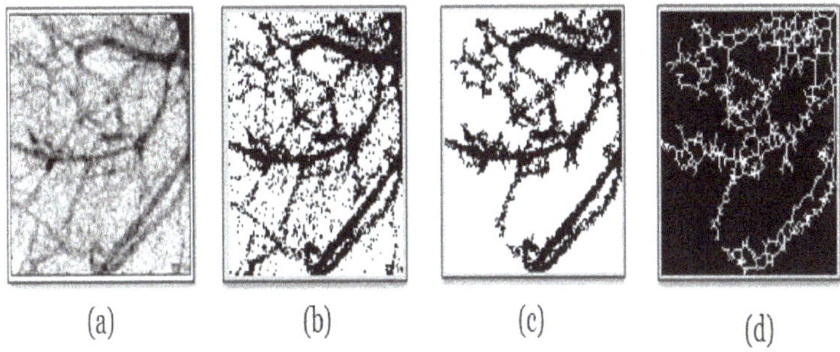

**Figure 1.** Step by Step of the digital processing make to the BMP image to obtain a binary skeletonized form: (**a**) image in gray scale, (**b**) Image in binary form, (**c**) Segmentation procedure, (**d**) Skeletonized binary form.

Once the skeletonized binary form has been obtained, we proceed to get the graph of the vascular network as follows: we postulate that every image pixel represents a single cell or point into a binary matrix. In order to form the network, it is proposed that every cell occupied (value of one) represents a node into the network. If any of these cells have other adjacent cells with value of one, i.e., occupied, it is possible that nodes are connected with their neighbors, so we assigned an edge between these two nodes, creating in this way the edges of the network, as shown in Figure 2.

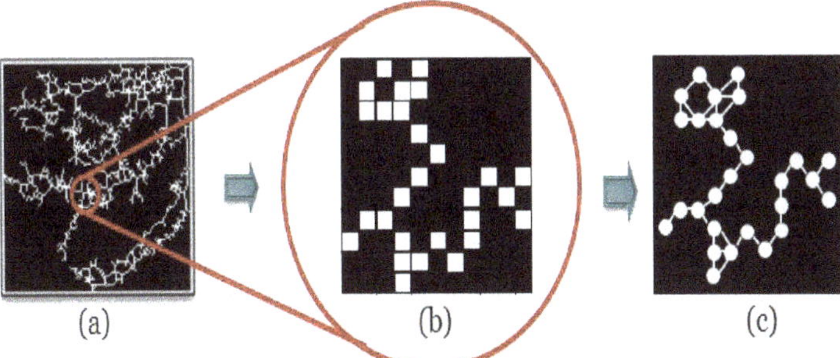

**Figure 2.** Modeling the complex network (**a**) binary skeletonized form. (**b**) Zoom of one part of the skeletonized binary form pixel by pixel (**c**) Network obtained after the assignation of nodes and edges to each pixel.

### 2.2. Structure of the Network

We have built from a tomography image a complex network defined as a graph. With this in mind, we can measure some properties related to the complex network just created and try to understand the system behavior as a whole. Motivated by these ideas and considering that many biological networks share properties of the small world networks, we proceed to perform four measurements [17,23], namely,

- Clustering Coefficient: A common property of complex networks is the cliques that it forms. This inherent tendency to cluster is quantified by the clustering coefficient. Let us analyze briefly the concept; if we focus on a selected node $i$ in the network, having $k_i$ edges which connect it to other $k_i$ nodes. If the nearest neighbours of the original node were part of a clique, there would be $k_i(k_i - 1)/2$ edges between them. The ratio between the number $E_i$ of edges that actually exist between these $k_i$ nodes and the total number $k_i(k_i - 1)/2$ gives the value of the node clustering coefficient $i$, as;

$$C_i = \frac{2E_i}{k_i(k_i - 1)}$$

  The clustering coefficient of the whole network is the average of all individual $C_i$.

- Degree Distribution: The way in which the degree of the nodes is distributed is characterized by the distribution function $P(k)$, which is the probability that a randomly selected node has exactly $k$ edges. For complex networks there are three types of important distributions, which determine different structures or topology of them, namely; Poisson Distribution, Exponential Distribution and Scale-Free Distribution.

  Networks that have a power-type distribution are called scale-free distributions or Power Law distributions. These networks arise in the context of network growth, in which each new node connects preferably to the nodes that are connected to the largest number of nodes in the network. Scale-free networks are also networks of the small world, because they have a coefficient of Clustering larger than a random network and the average of the shortest distance increases logarithmically with the number of nodes N, for this Distribution the probability density function is given by: $P(k) = Ck^{-\alpha}$ [1].

- Average path length: If we consider a unweighted graph $G$ with the set of edges $E$ and let $d(e_1, e_2)$, where $e_1$ and $e_2$, $e_1, e_2 \in E$ denote the shortest distance between $e_1$ and $e_2$. Then, the average path length $l_G$ is defined as;

$$l_G = \frac{1}{n(n-1)} \sum_{i,j} d(e_i, e_j)$$

  where $n$ is the number of vertices of $G$.

- Fractal dimension: The fractal dimension is a statistical quantity that gives an indication of how completely a fractal appears to fill space, as one zooms down to finer scales. In order to obtain the fractal dimension we use the box counting method, this method of counting is used to determine the fractal dimension of an irregular object. It consists of covering the object with a grid and counting how many boxes of the grid contain parts of the object. This process is repeated, several times using boxes with sides equal to $1/2$ of the size of the previous box [24]. The fractal dimension $d$ is then the slope that is obtained from graphing $LogN(r)$ vs $Log(1/r)$ in an equivalent way, the negative of the gradient of graphing $LogN(r)$ vs $Log(r)$;

$$d = \frac{\Delta logN(r)}{\Delta log(r)}$$

Taking into account the aforementioned properties, it is possible to perform these measurements over the binary skeletonized forms obtained from the images of four different patients. We report our results in Table 2.

**Table 2.** General structural properties of the four networks. For each network we have indicated the number of nodes (size), the average degree $k$, the clustering coefficient $C$, the average path length $l$, the fractal dimension $D$ and the distribution exponent $\alpha$ (this exponent was calculated taking into account a power law distribution).

| Image | Size | k | C | l | D | $\alpha$ |
|---|---|---|---|---|---|---|
| A (32 × 32 ) | 98 | 2 | 0.169 | 0.061 | 1.304 | 3.256 |
| A (64 × 64 ) | 340 | 3 | 0.279 | 0.035 | 1.395 | 3.034 |
| A (128 × 128) | 630 | 2 | 0.231 | 0.020 | 1.357 | 4.0624 |
| A (256 × 256) | 2301 | 2 | 0.226 | 0.010 | 1.409 | 4.334 |
| B (32 × 32) | 79 | 2 | 0.122 | 0.066 | 1.278 | 3.302 |
| B (64 × 64) | 234 | 2 | 0.201 | 0.036 | 1.332 | 4.168 |
| B (128 × 128) | 1248 | 2 | 0.180 | 0.009 | 1.470 | 4.481 |
| B (256 × 256) | 2247 | 2 | 0.187 | 0.009 | 1.396 | 4.222 |
| C (32 × 32) | 111 | 2 | 0.156 | 0.063 | 1.342 | 2.552 |
| C (64 × 64) | 211 | 2 | 0.152 | 0.029 | 1.301 | 3.507 |
| C (128 × 128) | 987 | 2 | 0.214 | 0.015 | 1.425 | 2.703 |
| C (256 × 256) | 3570 | 2 | 0.230 | 0.009 | 1.494 | 2.907 |
| D (32 × 32) | 103 | 2 | 0.251 | 0.071 | 1.322 | 3.101 |
| D (64 × 64) | 428 | 2 | 0.207 | 0.026 | 1.463 | 3.320 |
| D (128 × 128) | 894 | 2 | 0.204 | 0.014 | 1.390 | 3.921 |
| D (256 × 256) | 1260 | 2 | 0.169 | 0.010 | 1.291 | 4.250 |

Figure 3 depicts the skeletonized binary form and the degree distribution for the patient A, as an example of the network that we obtain for this special case.

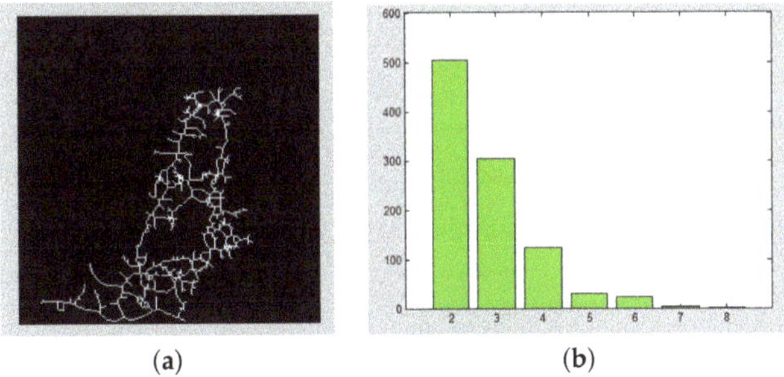

(a)  (b)

**Figure 3.** (a) Example of the Binary skeletonized form for patient B. (b) Degree distribution obtained from patient B.

## 2.3. Robustness Analysis

Angiogenesis is an important natural process that takes into account the growth of new blood vessels that occur in the body, both in healthy and ill hosts. Angiogenesis is now recognized as one of the critical events required for tumor progression [14]. In other words, cancerous growth is dependent on vascular induction and the development of new vascular supplies.

The idea of targeting angiogenesis to inhibit tumor growth was proposed more than three decades ago, and since then, several approaches to block or disrupt tumor angiogenesis have been explored.

However, all of these have been focused on the understanding of the molecular behaviour, and only a few in other properties that are not apparent from the molecular point of view.

Recent studies suggest that a network's connectivity pattern determines its robustness to external perturbations, such as removal of nodes or links [24]. To test this, we measured the effects of directed attacks and random failures on network organization. These measures were carried out as follows:

1. A given fraction of the vascular network nodes was eliminated from the original network. The nodes to be removed were either chosen as the most connected (directed attacks), or at random (random failures).
2. The network's emerging was evaluated by calculating their structural properties, namely, the average path length, the clustering coefficient and the degree distribution.
3. The whole process was repeated for several fractions of removed nodes.

In Tables 3 and 4 we showed the results of the robustness analysis for the real networks for the four patients taken from both directed (Table 3) and at random (Table 4) attacks.

**Table 3.** Robustness analysis for two networks using a random attack. For each network we have indicated the number of nodes ($N$), and the number of nodes eliminated randomly ($EN$), beginning with 1% of the nodes (corresponding to the first row of each patient). It is worthwhile to note that 1% corresponds, in the first case, to 3 disconnected nodes; however, when we disconnect these 3 nodes other adjacent nodes are also disconnected, giving 6 in total disconnected nodes. We did the same for 5%, 10% and finally 15%, the average degree $k$, the average length $l$, the clustering coefficient $C$, the fractal dimension $D$ and the exponent of the distribution $\alpha$. In all the attacks we used images of 64 by 64 pixels.

| PATIENT A | $N$ | $EN$ | $k$ | $l$ | $C$ | $D$ | $\alpha$ |
|---|---|---|---|---|---|---|---|
| | 340 | 6 | 3 | 0.0339 | 0.279 | 1.390 | 0.740 |
| | 340 | 18 | 2 | 0.039 | 0.262 | 1.387 | 0.687 |
| | 340 | 35 | 2 | 0.051 | 0.267 | 1.326 | 0.704 |
| | 340 | 52 | 3 | 0.122 | 0.303 | 1.144 | 0.550 |
| PATIENT B | $N$ | $EN$ | $k$ | $l$ | $C$ | $D$ | $\alpha$ |
| | 234 | 3 | 2 | 0.045 | 0.188 | 1.396 | 0.944 |
| | 234 | 13 | 2 | 0.036 | 0.187 | 1.272 | 1.074 |
| | 234 | 24 | 2 | 0.077 | 0.227 | 1.083 | 0.626 |
| | 234 | 36 | 2 | 0.218 | 0.254 | 1 | 0.548 |

**Table 4.** Robustness analysis for two networks using a direct attack. For each network we have indicated the number of nodes ($N$), and the number of nodes eliminated ($EN$), beginning with the nodes with 7 connections (this corresponds to the first row of each patient and in parenthesis are the remaining nodes), then the nodes with 6 (corresponding to the second row) and so on, the average degree $k$, the average length $l$, the clustering coefficient $C$, the fractal dimension $D$ and the exponent of the distribution $\alpha$. In all the attacks we used images of 64 by 64 pixels.

| PATIENT A | $N$ | $EN$ | $k$ | $l$ | $C$ | $D$ | $\alpha$ |
|---|---|---|---|---|---|---|---|
| | 340 | 6 (334) | 2 | 0.03 | 0.262 | 1.392 | 0.7984 |
| | 340 | 19 (315) | 2 | 0.036 | 0.217 | 1.391 | 1.0072 |
| | 340 | 21 (294) | 2 | 0.048 | 0.226 | 1.237 | 0.78551 |
| | 340 | 64 (230) | 2 | 0.060 | 0.118 | 1.089 | 1.292 |
| PATIENT B | $N$ | $EN$ | $k$ | $l$ | $C$ | $D$ | $\alpha$ |
| | 234 | 1 | 2 | 0.036 | 0.197 | 1.332 | 1.008 |
| | 234 | 1 | 2 | 0.036 | 0.198 | 1.332 | 1.223 |
| | 234 | 7 | 2 | 0.0364 | 0.164 | 1.326 | 0.870 |
| | 234 | 28 | 3 | 0.110 | 0.144 | 1 | 0.778 |

Tables 4 shows that for both patients (patient A and B), when we carried a direct attack, the statistical properties were lost after we eliminated nodes with five connections, i.e., the average path length become higher and the clustering coefficient lower, compared with the original. Our calculations also reveal that random removal nodes (see Table 3) have almost the same effect as in the direct attack when removing almost 15% from all the nodes.

## 3. Computational Modeling of Angiogenic Networks

### 3.1. Invasion Percolation Algorithm

Some years ago, Baish et al. [3] introduced an algorithm called Invasion Percolation in order to show that the fractal dimensions observed in tumor vasculature closely correspond to those produced by a statistical growth process known as Invasion Percolation [25]. In a more technical sense, Invasion Percolation is an algorithm that models the expansion of a network through a medium with randomly distributed heterogeneities. The resulting network always expands into the weakest available sites, yielding structures with voids on a wide range of length scales and pathways that are tortuous over many scales.

The Invasion-Percolation model is motivated by the problem of a fluid to be dispersed in a porous medium. This principle may be applied to any type of invasion process in which the path shows fluid passage resistance [25]. The porous medium may be represented as a network of pores which are connected between the pores. In an ideal medium the network can be viewed as a matrix in which the cells and their neighborhoods represent the pores and the connections between them. It is assigned random numbers to each cells in order to represent the pore size. The simulation of the fluid path through the pores consists of a series of discrete jumps, where each discrete step will be that offering less resistance (low random number). The Invasion-Percolation model involves a single time, in which the jump is generated in the matrix and provides a unique way to traverse the porous medium.

To show how the Invasion-Percolation algorithm works, we performed several experiments in which some networks were generated with this algorithm. Figure 4a depicts a single example of the synthetic network created by the Invasion-Percolation algorithm, for this case we have used a size of the matrix of 128 × 128 cells. Likewise, Figure 4b shows the distribution of nodes generated by the algorithm, in which it is observed how the distributions does not resembles to that obtained using the patient data (see for example Figure 3b for patient B).

The pseudo-code for Percolation is presented as follows:

```
(1) A matrix M of size n x n is initialized with aleatory values between 0 to 1.
(2) The position of the first blood vessel $(x,y)$ will be located on the left
border of the matrix M, where the vertical position (y) of the blood vessel is
chosen by the little value located on the line. The horizontal position (x) will be
equal to 1.
(3) The matrix M is crossed from left to right until the right edge is reached.
(3.1) The number of divisions per blood vessel is chosen between their 8 neighbors.
with  the least value from the above we can have the following
results:
(a) The blood vessel stays the same (this means that the 8 neighbors have the
same value).
(b) The blood vessel have two o more branches.
(4) The matrix M is updated with the new blood vessels.
(5) Then, we return to step 3.
```

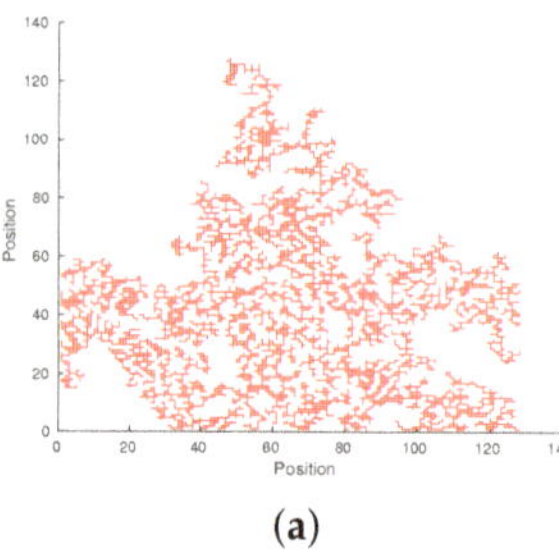

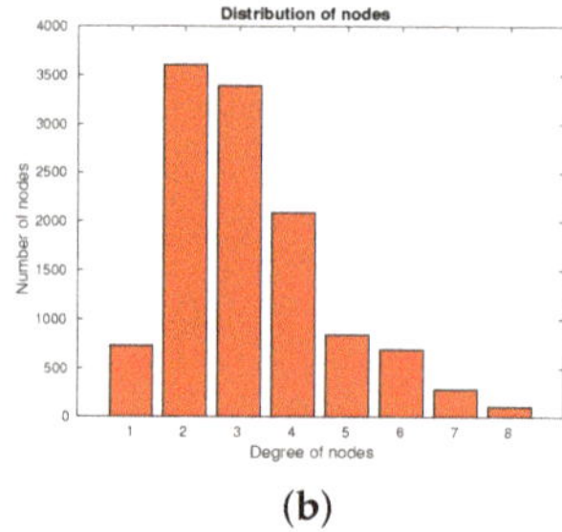


(a)                                              (b)

**Figure 4.** Example of the network generated by Invasion-Percolation algorithm. (**a**) Network generated by the algorithm. (**b**) Distribution of nodes generated by the same algorithm.

As was mentioned previously, the design of the Invasion Percolation algorithm intents to mimic or model the vasculature shown in real tumors. So, in order to have a trustworthy algorithm that could simulate angiogenic networks, we suggest that it is necessary that the Invasion Percolation algorithm could reproduce some of the properties shown in our angiogenic networks. For this purpose, all the synthetic networks produced by the Invasion Percolation were subjected to a structural analysis (i.e., measuring the average path length, the clustering coefficient and the degree of distribution) and a geometry analysis (i.e., measuring its fractal dimension), both previously proposed for studying real networks. The results obtained are presented in Table 5, which shows the study was made using synthetic networks. It is worthwhile to mention that in order to obtain Table 5 we have run our algorithm around five hundred times for different sizes of the matrix ($32 \times 32$, $64 \times 64$, $128 \times 128$, $256 \times 256$ cells) to obtain as many as possible synthetic networks to work with.

**Table 5.** General structural properties for networks created by the Invasion-Percolation algorithm for different matrix sizes. After several simulations (we only have reported the average values for each measure) for each size, we have indicated the average number of nodes (average size) $N$, the average degree $k$, the average clustering coefficient $C$, the average path length $l$ and the average of the fractal dimension $D$, also we have added the standard deviation (in parenthesis) for each size and for each measure.

| Matrix | N | Z | C | l | D |
|---|---|---|---|---|---|
| $32 \times 32$ | 310 (79.14) | 4 | 0.49 (0.02) | 0.007 (0.008) | 1.64 (0.092) |
| $64 \times 64$ | 837 (235.3) | 4 | 0.44 (0.14) | 0.08 (0.014) | 1.62 (0.07) |
| $128 \times 128$ | 4373 (1515.95) | 4 | 0.477 (0.007) | 0.018 (0.002) | 1.72 (0.07) |
| $256 \times 256$ | 16441 (5876) | 4 | 0.47 (0.006) | 0.009 (0.0019) | 1.75 (0.075) |

A detailed analysis of Figure 4 and Table 5, it is shown that the Invasion-Percolation algorithm did not share the structural and geometrical properties shown by those obtained from real networks obtained in patients (see for instance, Table 2 and Figure 3). So, in our opinion the Invasion Percolation algorithm should be transformed in order to share the properties aforementioned. In order to address this point it was necessary to design a new algorithm in which we incorporated some strategies to better simulate not only the fractal properties shown by Baish [3], but also the structural properties revealed by the structural analysis made in real networks.

*3.2. A New Algorithm Called Arga (Angiogenesis Random Growth Algorithm)*

A new algorithm for the generation of angiogenic networks complying with the characteristics of structure and geometry of the modeled network obtained from tomographic images is proposed.

The new algorithm is motivated by the changes that could occur in the medium and responsible for the division of blood vessels. Its aim is to find a pathway through the medium, giving priority to the front feed (from left to right of the matrix) [25].

Similar to the Invasion-Percolation algorithm for network formation an ideal medium is assumed, i.e., it is homogeneous and symmetric. This medium will be represented by a matrix of size $n$ x $n$ and ideally when a change happened in the medium, this will be reflected in a change in the formation behaviour of the synthetic network. However, it is impossible to know these changes in situ. We will simulate these changes using a random variable which varies in time.

Basically, the pseudo-code for ARGA is presented as follows:

```
(1) A matrix A of size 1 x 1 is initialized
(2) The first position of the first blood vessel $(x,y)$ will be located on the left
border of the matrix, where the vertical position (y) of the blood vessel is chosen
in a random way between 0 and 1. The horizontal position (x) will be equal to 1.
(3) As long as there is no position on the edge without dividing and not reaching
position 1, the matrix is traversed untill the right edge is reached.
(3.1) The number of divisions per blood vessel is chosen which can have the following
results:
(a) The blood vessel stays the same
(b) A random number of new blood vessels is produced (5 maximum)
(3.2) If new blood vessels are produced, by each of them a random advance in the
network will occur. Following the following rules:
(a) The highest priority is assigned to the front directions
(b) The direction to advance is selected randomly giving priority to the highest.
(c) The advance for each direction is calculated randomly (maximum size 10).
(4) Changes are made
(5) They are stored in matrix A.
(6) Then, we return to step 3.
```

To show how the ARGA algorithm works, we performed several experiments in which some networks were generated with this algorithm. Figure 5a depicts a single example of the synthetic network created by the ARGA algorithm, for this case we have used a size of the matrix of $128 \times 128$ cells. Likewise, Figure 5b shows the distribution of nodes generated by the algorithm, in which it is observed how the distributions resembles in a better way to those obtained using the patient data (see for example Figure 3b for patient B).

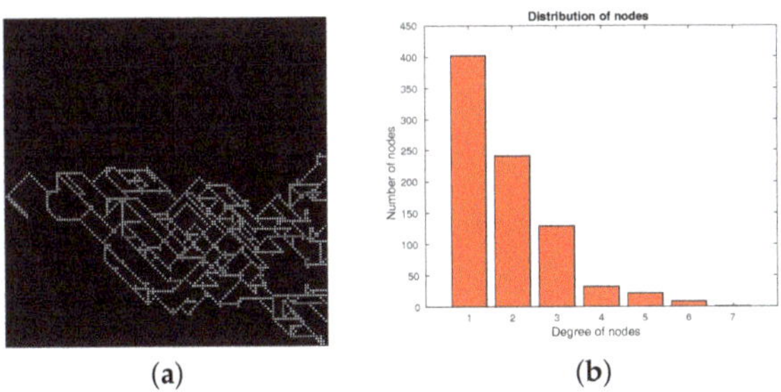

(a)        (b)

**Figure 5.** Example of a single network (with a size of $128 \times 128$ cells ) generated by ARGA algorithm. (**a**) Network generated by the algorithm. (**b**) Distribution of nodes generated by the same algorithm

Once we have our algorithm running, we developed hundreds of simulations (it is worthwhile to mention that in order to obtain Table 6 we have run our algorithm around five hundred times for different sizes of the matrix ($32 \times 32$, $64 \times 64$, $128 \times 128$, $256 \times 256$ cells)) to obtain as many as possible synthetic networks to work with. All these synthetic networks produced by ARGA were subjected to a structural analysis (i.e., measuring the average path length, the clustering coefficient and the degree of distribution) and a geometry analysis (i.e., measuring its fractal dimension), both previously proposed for studying real networks. The results obtained are presented in Table 6, using these synthetic networks, the table only shows the average values and the standard deviation obtained for all the synthetic networks.

**Table 6.** General structural properties for networks created by the ARGA algorithm for different matrix sizes. For each size we have indicated the average number of nodes (average size) $N$, the average degree $k$, the average clustering coefficient $C$, the average path length $l$, the average of the fractal dimension $D$ and the average exponent of the distribution $\alpha$ (this exponent was calculated taking into account a Poisson distribution), also we have added the standard deviation (in parenthesis) for each size and for each measure.

| Matrix | N | Z | C | l | D | $\alpha$ |
|---|---|---|---|---|---|---|
| $32 \times 32$ | 110 (16.2) | 3 | 0.26 (0.08) | 0.032 (0.001) | 1.32 (0.125) | 2.92 (0.67) |
| $64 \times 64$ | 458 (174.8) | 3 | 0.27 (0.04) | 0.03 (0.004) | 1.48 (0.09) | 3.094 (0.51) |
| $128 \times 128$ | 1772 (490.31) | 3 | 0.27 (0.025) | 0.08 (0.002) | 1.56 (0.06) | 3.8 (0.48) |
| $256 \times 256$ | 8522 (3961) | 3 | 0.295 (0.02) | 0.09 (0.001) | 1.6 (0.08) | 3.78 (0.61) |

*3.3. Robustness Analysis for the Arga Algorithm*

Below are the results obtained from the robustness analysis made to the synthetic networks produced by the ARGA algorithm, using the previous cuts of size($32 \times 32$, $64 \times 64$, $128 \times 128$ and $256 \times 256$ cells). As explained in Section 2.3 (for the real networks case), the robustness analysis was carried out by attacking the networks, randomly and in directed way, and then characterized them using the proposed structure and geometry analyzes. The full results of the robustness analysis for the case of the networks created by the ARGA algorithm can be seen in Tables 7 and 8. In these we have shown the results of the robustness analysis for the synthetic networks for four different sizes taken from both directed (Table 7) and random (Table 8) attacks.

**Table 7.** Robustness analysis for synthetic networks using a random attack. For each network we have indicated the number of nodes ($N$), the number of nodes eliminated randomly ($EN$) beginning with 1% of the nodes (corresponding to the first row to the size of the network), for 5%, 10% and finally 15% respectively, the average degree $k$, the average length $l$, the clustering coefficient $C$ and the fractal dimension $D$.

| Size ($32 \times 32$ cells) | N | EN | k | l | C | D |
|---|---|---|---|---|---|---|
| 1% | 108 | 1 | 3 | 0.0839 | 0.3754 | 1.4235 |
| 5% | 108 | 7 | 2 | 0.0819 | 0.3870 | 1.415 |
| 10% | 108 | 12 | 2 | 0.0737 | 0.3606 | 1.3853 |
| 15% | 108 | 18 | 3 | 0.1030 | 0.3545 | 1.2682 |
| **Size ($64 \times 64$ cells)** | **N** | **EN** | **k** | **l** | **C** | **D** |
| 1% | 483 | 6 | 3 | 0.0412 | 0.3345 | 1.4860 |
| 5% | 483 | 25 | 3 | 0.0409 | 0.3352 | 1.402 |
| 10% | 483 | 49 | 3 | 0.0456 | 0.3185 | 1.450 |
| 15% | 483 | 73 | 3 | 0.346 | 0.3580 | 1.340 |
| **Size ($128 \times 128$ cells)** | **N** | **EN** | **k** | **l** | **C** | **D** |
| 1% | 1463 | 32 | 2 | 0.017 | 0.2708 | 1.5544 |
| 5% | 1463 | 837 | 2 | 0.026 | 0.2767 | 1.3553 |
| 10% | 1463 | 911 | 2 | 0.0298 | 0.2668 | 1.378 |
| 15% | 1463 | 1216 | 2 | 0.0319 | 0.2780 | 1.19 |
| **Size ($256 \times 256$ cells)** | **N** | **EN** | **k** | **l** | **C** | **D** |
| 1% | 4231 | 1904 | 2 | 0.009 | 0.2443 | 1.4980 |
| 5% | 4231 | 617 | 2 | 0.0094 | 0.2592 | 1.5412 |
| 10% | 4231 | 2060 | 2 | 0.0064 | 0.2757 | 1.4808 |
| 15% | 4231 | 3767 | 2 | 0.011 | 0.2732 | 1.200 |

**Table 8.** Robustness analysis for the synthetic networks using a direct attack. For each network we have indicated the number of nodes ($N$), the number of nodes eliminated ($EN$) beginning with the nodes with 7 connections (this corresponds to the first row), then the nodes with 6 (corresponding to the second row) and so on, the average degree $k$, the average length $l$, the clustering coefficient $C$ and the fractal dimension $D$.

| Size (32 × 32 cells) | $N$ | $EN$ | $k$ | $l$ | $C$ | $D$ |
|---|---|---|---|---|---|---|
| 7 | 108 | 0 | 3 | 0.0839 | 0.3754 | 1.4253 |
| 6 | 108 | 7 | 2 | 0.07020 | 0.2799 | 1.409 |
| 5 | 108 | 6 | 2 | 0.0807 | 0.313 | 1.411 |
| 4 | 108 | 21 | 2 | 0.0758 | 0.1760 | 1.557 |
| Size (64 × 64 cells) | $N$ | $EN$ | $k$ | $l$ | $C$ | $D$ |
| 7 | 483 | 5 | 3 | 0.0411 | 0.3248 | 1.4867 |
| 6 | 483 | 26 | 2 | 0.03834 | 0.3016 | 1.4763 |
| 5 | 483 | 48 | 2 | 0.0334 | 0.2709 | 1.4685 |
| 4 | 483 | 126 | 2 | 0.1023 | 0.1031 | 1.05 |
| Size (128 × 128 cells) | $N$ | $EN$ | $k$ | $l$ | $C$ | $D$ |
| 7 | 1463 | 4 | 2 | 0.0180 | 0.2614 | 1.5592 |
| 6 | 1463 | 21 | 2 | 0.0177 | 0.2472 | 1.5577 |
| 5 | 1463 | 59 | 2 | 0.0168 | 0.2310 | 1.5540 |
| 4 | 1463 | 1354 | 2 | 0.0356 | 0.068 | 1.1899 |
| Size (256 × 256 cells) | $N$ | $EN$ | $k$ | $l$ | $C$ | $D$ |
| 7 | 4231 | 14 | 2 | 0.0105 | 0.2591 | 1.5716 |
| 6 | 4231 | 96 | 2 | 0.0102 | 0.2439 | 1.5697 |
| 5 | 4231 | 329 | 2 | 0.0093 | 0.2260 | 1.5610 |
| 4 | 4231 | 3971 | 2 | 0.022 | 0.2239 | 1.03 |

## 4. Concluding Remarks

An analysis has been carried out on the vascular angiogenic network on four patients with Hepato-Cellular Carcinoma (HCC). This analysis consisted of measuring a number of statistical properties of a vascular network obtained from four digital tomographies and digitalized until a binary skeletonized scheme is obtained. From this, we generated a network of nodes and edges which represent the original angiogenetic vascular network. Some interesting observations arising from these measurements are:

- The clustering coefficient in all 16 generated networks is less than 0.4. This indicates that they were well connected networks.
- The degree distribution in all the networks have an exponential tail with the distribution exponent, between 0.6 and 1.1.
- The average path length is small in all the networks being between 0.009 and 0.071.
- The fractal dimension is found to be around 1.4.

Many authors have considered that: *"Tumor consists of a chaotic, poorly organized vasculature, with tortuous, irregularly shape, and leaky vessels that are often unable to support efficient blood flow and leading to an aberrant vascular system"*. From our observations the above consideration is incorrect because we have shown that there is a well connected network (high clustering coefficient); besides, the network has an efficient communication. This is reflected in a small average path length. So, when observing in situ a poorly organized shape, it does not take into account that there are good structural properties that offer support for various dynamical processes, i.e., it is thought that the network topology plays a crucial role, which supports an efficient blood flow among other dynamical properties.

The high interest in scale-free networks in literature might give the impression that all complex networks in nature have power-law degree distributions. It is true for several complex networks of highest interest in the scientific community, such as the World Wide Web, social networks among others, that in all of them the degree distribution has a power-law tail. However, some other networks such as neural and power grid showed exponential degree distributions in literature, and these are called evolving networks [3]. In our case, we have exponential distributions for all the vascular angiogenic networks generated. This means having and evolving a network with aging effects and growth constraints that leads to this exponential decay.

A study of the robustness of the generated angiogenic vascular network has been carried out. We studied the robustness of the network analyzing the connectivity pattern subjected to external perturbations, such as the removal of nodes or links. To test this, the effects of directed attacks and random failures on the network organization were measured.

This study shows that both patients (patient A and B), when we carried a direct attack, their statistical properties were lost after we eliminated nodes with five connections, i.e., the average path length become higher and the clustering coefficient lower, compared with the original. Our calculations also reveal that random removal nodes have almost the same effect as in the direct attack when removing almost 15% from all the nodes.

Taking into account a new kind of algorithm (ARGA) it was shown that this algorithm simulates in a better way the growth of the vascular network than the Invasion-Percolation algorithm. This is clear because with this algorithm it was possible to reproduce, in a better way, the structural and geometric measurements for the real network than with the Invasion-Percolation algorithm. Some interesting observations arising from these measurements (see Table 6) are:

- The clustering coefficient in all the generated networks is less than 0.3. This indicate that they were well connected networks, as in the case of real data.
- The degree distribution in all the networks have an exponential tail with the distribution exponent, between 2.9 and 3.78.
- The average path length is small in all the synthetic networks being between 0.03 and 0.09, as in the case of real data.
- The fractal dimension is found to be around 1.4, as in the case of real data.

Furthermore, both algorithms were subjected to a variety of direct attacks and random failures and, for the ARGA algorithm the same effect as in the patients A and B was observed, i.e., all the statistical properties were lost in the same manner as the real networks (see Tables 7 and 8). However, the Invasion Percolation did not share the behaviour (the experiments are not included in this article) of real networks as the ARGA algorithm did for the case of direct and random attacks. This indicates that in order to simulate in a better way real networks the use of the ARGA algorithm will produce synthetic networks more similar to the real ones.

Taking into account the considerations of the last two paragraphs, and based on the statistical properties of the network, some possible therapies could be suggested. We think that more clinical research related to the structure and robustness of the vascular network for the angiogenic process could be done in order to prove the latter hypothesis.

**Author Contributions:** Conceptualization, J.C.C.-E.; Investigation, J.C.C.-E., E.C.-M. and R.T.P.-H.; Software, E.C.-M.; Validation, R.T.P.-H.; Writing—original draft, J.C.C.-E. All authors have read and agreed to the published version of the manuscript.

**Funding:** This research was funded by "Consejo Nacional de Ciencia y Tecnología" (CONACyT), "Comisión de Operación y Fomento de Actividades Académicas del Instituto Politécnico Nacional" (COFAA-IPN, project number 20196225) and "Estimulos al Desempeño de los Investigadores del Instituto Politécnico Nacional"(EDI-IPN).

**Acknowledgments:** The authors wish to thanks "Consejo Nacional de Ciencia y Tecnología" (CONACyT), "Comisión de Operación y Fomento de Actividades Académicas del Instituto Politécnico Nacional" (COFAA-IPN, project number 20196225) and "Estimulos al Desempeño de los Investigadores del Instituto Politécnico Nacional" (EDI-IPN) for the support given for this work.

**Conflicts of Interest:** The authors declare no conflict of interest.

## References

1. Wodarz, D.; Komarova, N. *Computational Biology of Cancer: Lecture Notes and Mathematical Modeling*; World Scientific: Singapore, 2005.
2. Folkman, J. Tumor angiogenesis: Therapeutic implications. *N. Engl. J. Med.* **1971**, *285*, 1182–1186. [PubMed]
3. Baish, J.W.; Jain, R.K. Fractals and cancer. *Cancer Res.* **2000**, *60*, 3683–3688. [PubMed]

4. Guzmán-Vargas, L.; Santillán, M. Comparative analysis of the transcription-factor gene regulatory networks of E. coli and S. cerevisiae. *BMC Syst. Biol.* **2008**, *2*, 13. [CrossRef] [PubMed]

5. Abdollahi, A.; Schwager, C.; Kleeff, J.; Esposito, I.; Domhan, S.; Peschke, P.; Hauser, K.; Hahnfeldt, P.; Hlatky, L.; Debus, J.; et al. Transcriptional network governing the angiogenic switch in human pancreatic cancer. *Proc. Natl. Acad. Sci. USA* **2007**, *104*, 12890–12895. [CrossRef]

6. Vogelstein, B.; Kinzler, K.W. *The Genetic Basis of Human Cancer*; McGraw-Hill: New York, NY, USA, 2002.

7. Moolgavkar, S.H.; Knudson, A.G. Mutation and Cancer: A Model for Human Carcinogenesis 2. *JNCI J. Natl. Cancer Inst.* **1981**, *66*, 1037–1052. [CrossRef]

8. Preziosi, L. *Cancer Modelling and Simulation*; CRC Press: Boca Raton, FL, USA, 2003.

9. Gatenby, R.A.; Gawlinski, E.T. The glycolytic phenotype in carcinogenesis and tumor invasion. *Cancer Res.* **2003**, *63*, 3847–3854.

10. Balazsi, G.; Barabási, A.L.; Oltvai, Z. Topological units of environmental signal processing in the transcriptional regulatory network of Escherichia coli. *Proc. Natl. Acad. Sci. USA* **2005**, *102*, 7841–7846. [CrossRef]

11. Dobrin, R.; Beg, Q.K.; Barabási, A.L.; Oltvai, Z.N. Aggregation of topological motifs in the Escherichia coli transcriptional regulatory network. *BMC Bioinform.* **2004**, *5*, 10. [CrossRef]

12. Guelzim, N.; Bottani, S.; Bourgine, P.; Képès, F. Topological and causal structure of the yeast transcriptional regulatory network. *Nat. Genet.* **2002**, *31*, 60. [CrossRef]

13. Albert, R.; Jeong, H.; Barabási, A.L. Error and attack tolerance of complex networks. *Nature* **2000**, *406*, 378–382. [CrossRef]

14. Hanahan, D.; Weinberg, R.A. The hallmarks of cancer. *Cell* **2000**, *100*, 57–70. [CrossRef]

15. Jain, R.K. Molecular regulation of vessel maturation. *Nat. Med.* **2003**, *9*, 685–693. [CrossRef] [PubMed]

16. Morikawa, S.; Baluk, P.; Kaidoh, T.; Haskell, A.; Jain, R.K.; McDonald, D.M. Abnormalities in pericytes on blood vessels and endothelial sprouts in tumors. *Am. J. Pathol.* **2002**, *160*, 985–1000. [CrossRef]

17. Castillo Montiel, E. Estudio de Neoplasias Malignas Utilizando Dinámica No Lineal. Ph.D. Thesis, Instituto Politécnico Nacional, Centro de Investigación en Computación, Ciudad de México, Mexico, 2009.

18. Otsu, N. A threshold selection method from gray-level histograms. *IEEE Trans. Syst. Man Cybern.* **1979**, *9*, 62–66. [CrossRef]

19. Blum, H. A transformation for extracting new descriptors of shape. *Models Percept. Speech Vis. Form* **1967**, *19*, 362–380.

20. Stephanou, A.; McDougall, S.R.; Anderson, A.R.; Chaplain, M.A. Mathematical modelling of flow in 2D and 3D vascular networks: Applications to anti-angiogenic and chemotherapeutic drug strategies. *Math. Comput. Model.* **2005**, *41*, 1137–1156. [CrossRef]

21. Macklin, P.; McDougall, S.; Anderson, A.R.; Chaplain, M.A.; Cristini, V.; Lowengrub, J. Multiscale modelling and nonlinear simulation of vascular tumour growth. *J. Math. Biol.* **2009**, *58*, 765–798. [CrossRef]

22. Dawson, T.H. Modeling of vascular networks. *J. Exp. Biol.* **2005**, *208*, 1687–1694. [CrossRef]

23. Newman, M.E. The structure and function of complex networks. *SIAM Rev.* **2003**, *45*, 167–256. [CrossRef]

24. Albert, R.; Barabási, A.L. Statistical mechanics of complex networks. *Rev. Mod. Phys.* **2002**, *74*, 47. [CrossRef]

25. Wilkinson, D.; Willemsen, J.F. Invasion percolation: A new form of percolation theory. *J. Phys. A Math. Gen.* **1983**, *16*, 3365. [CrossRef]

*Communication*

# Complex Network Construction of Univariate Chaotic Time Series Based on Maximum Mean Discrepancy

**Jiancheng Sun**

School of Software and Internet of Things Engineering, Jiangxi University of Finance and Economics, Nanchang 330013, China; sunjc@jeufe.edu.cn; Tel.: +86-0791-8384-5702

Received: 11 December 2019; Accepted: 23 January 2020; Published: 24 January 2020

**Abstract:** The analysis of chaotic time series is usually a challenging task due to its complexity. In this communication, a method of complex network construction is proposed for univariate chaotic time series, which provides a novel way to analyze time series. In the process of complex network construction, how to measure the similarity between the time series is a key problem to be solved. Due to the complexity of chaotic systems, the common metrics is hard to measure the similarity. Consequently, the proposed method first transforms univariate time series into high-dimensional phase space to increase its information, then uses Gaussian mixture model (GMM) to represent time series, and finally introduces maximum mean discrepancy (MMD) to measure the similarity between GMMs. The Lorenz system is used to validate the correctness and effectiveness of the proposed method for measuring the similarity.

**Keywords:** complex network; chaotic time series; Gaussian mixture model; maximum mean discrepancy

## 1. Introduction

Chaotic time series exist widely in many fields, such as economics, physics, hydrology and so on [1]. In chaotic systems, the "butterfly effect" is a typical phenomenon in which small causes can have large effects [2]. Therefore, a chaotic system usually has highly complex behaviours, and the relevant analysis is a challenging task.

In recent years, the application of complex network theory to time series analysis is increasing rapidly. Firstly, the time series was transformed into a network, and then, various complex network tools were used for analysis [3–7]. There are three kinds of network reconstruction methods: recurrence network based on phase space and visibility graphs and transition network based on Markov chain [6]. Regardless of the network construction method, a key problem to be solved is how to measure the similarity between nodes. For example, in order to measure the similarity between nodes, Euclidean distance, visual distance, and transition probability were applied to recurrence network [8], visibility graphs [9], and transition network [10], respectively.

In this communication, we focus on the construction of a complex network of univariate chaotic time series, which is an effective way to analyse the time series [11]. Similarly, in this task, a core problem to be solved is how to measure the similarity between time series. In the community of time series analysis, some commonly used metrics, such as Euclidean distance [12], correlation coefficient [13], and dynamic time warping distance (DTW) [14], were used to measure similarities between time series. Especially for DTW, its outstanding advantage lies in the ability to find the optimal nonlinear alignment between two given sequences. However, most of the metrics cannot effectively measure the similarity in the case of chaotic time series. For example, time series from the same chaotic system are completely different in the form of local sub-sequences, which makes the pair matching metric, such as Euclidean distance, unable to measure the similarity between them. Even with statistical metrics such as the correlation coefficient, due to the rich structure of chaotic system, its effect is not as expected. In addition, the probability distribution of chaotic time series usually presents mixed distribution [15], which also brings challenges to measuring the similarity between sequences using statistical distances.

Considering the characteristics of chaotic time series mentioned above, in this communication, we improved the performance of similarity measurement from two aspects: (1) transform univariate time series (UTS) into a high-dimensional space to describe time series more accurately; (2) in the high-dimensional space, Gaussian mixture model (GMM) is used as the representation of time series, and distance metric is introduced to catch the similarity between GMMs.

## 2. Approach of Constructing Complex Networks

In the complex network constructed, nodes represent the time series themselves, and the edges between nodes are determined by the strength of similarity between the time series. The process of constructing a network can be divided into the following sections.

### 2.1. Representation of Univariate Chaotic Time Series

We can realize the representation of UTS by using the idea of phase space reconstruction. Although a complex system is usually described by multiple variables, in most cases we can only observe a univariate (scalar) time series $T = \{x_1, x_2, \ldots, x_n\}$ from the system, where $n$ is the length of the time series. Fortunately, using the embedded theorem [16], we can reconstruct the original space of the system by unfolding the scalar time series into higher dimensional phase space. With the help of phase space reconstruction, we can investigate the geometric and dynamic properties of the original phase space as well as unobserved variables. In other words, it provides a new approach, which can transform UTS into state vectors in higher dimensional space, so as to describe and understand the characteristics of the system more accurately. This is exactly the motivation for representation of univariate time series.

By choosing an appropriate embedding dimension $m$ and a time delay $\tau$, we can transform a UTS $T = \{x_1, x_2, \ldots, x_n\}$ into a state vector in the phase space as

$$x_i = \left(x_i, x_{i+\tau}, x_{i+2\tau}, \ldots, x_{i+(m-1)\tau}\right)^T \tag{1}$$

where $m$ can also be regarded as the number of variables in the original phase space. Therefore, the phase space can be described by a $m \times (n - \tau)$ matrix $X$, where each column represents the state point $x_i$ at time $i$, and each row represents a subsequence of the UTS. On the other hand, each row in $X$ can also be viewed as observation of a variable. Consequently, $X$ is multivariable time series (MTS) with $m$ variables, which is converted from the UTS.

To illustrate the phase space reconstruction clearly, the well-known Lorenz system is used as an example to illustrate the reconstruction process [2]. The Lorenz system is described by three ordinary differential equations:

$$\begin{aligned} \frac{dx}{dt} &= \sigma(y - x), \\ \frac{dy}{dt} &= x(\rho - z) - y, \\ \frac{dz}{dt} &= xy - \beta z \end{aligned} \tag{2}$$

where $x$, $y$, and $z$ are system variables; $t$ is the time, and $\sigma, \rho, \beta$ are the system parameters. Two time series $T_1$ and $T_2$ ($x$ component of the system) shown at the top of Figure 1a are generated using (2) with $\sigma = 8$, $\rho = 28$ and $\beta = 8/3$. All system parameters are kept the same here, except that the initial conditions for generating $T_1$ and $T_2$ are slightly different by $10^{-2}$. The difference between the two-time series is shown at the bottom of Figure 1a. It can be seen from Figure 1a, in the beginning, the two-time series kept the same shape, but as time went on, the differences became larger and larger, and this phenomenon is known as the "butterfly effect". In other words, although the two time series are generated from the same system with slightly different initial values, they are very different in local characteristics. Therefore, it is usually not feasible to measure the similarity between chaotic time series by general metric. In Figure 1b, pairs of time series values $x_i = (x_i, x_{i+\tau})^T$ are plotted with black dots; this is the state vector in the reconstructed phase space of the $x$ component of $T_1$ using (1). In other words, a system with two variables is reconstructed from the observation of one variable (a scalar time series). As can be seen from Figure 1b, more abundant geometric structure of the time series can be seen by using the phase space reconstruction, thus it can provide more information about the time series.

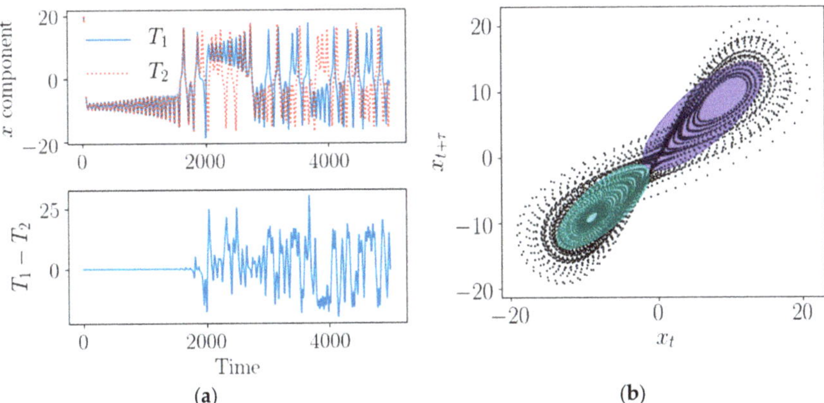

**Figure 1.** Chaotic time series of Lorenz system and its reconstructed phase space. (a) $x$ component of Lorenz system. (b) Reconstructed phase space of $T_1$.

When the phase space is constructed, the next problem to be solved is how to represent the time series in the space. As mentioned above, the general point-to-point metrics are hard to measure the similarity between chaotic time series due to the complexity. Therefore, it is a more reasonable choice to calculate the statistical characteristics of time series and then measure the similarity between them. An intuitive way is to estimate the covariance matrix of the multivariable time series (MTS) in phase space and then calculate geodesic distance between them [17]. However, considering the complex structure of MTS in phase space, a single covariance matrix usually cannot accurately describe the data distribution of a chaotic system. For example, in Figure 1b, at least two independent Gaussian distribution models are required to accurately describe the Lorenz systems. Thus, a natural way here is to adopt multivariate GMM to represent MTS generated by chaotic systems. The GMM is given by

$$G = \sum_{i=1}^{n} \alpha_i g_i = \sum_{i=1}^{n} \alpha_i \mathcal{N}(\boldsymbol{\mu}_i, \boldsymbol{\Sigma}_i) \tag{3}$$

where the $i$-th component is characterized by normal distributions $g_i = \mathcal{N}(\boldsymbol{\mu}_i, \boldsymbol{\Sigma}_i)$ with weights $\alpha_i$, means $\boldsymbol{\mu}_i$, and covariance matrices $\boldsymbol{\Sigma}_i$. In other words, GMM is a linear combination of several Gaussian distribution functions. In theory, GMM can fit any type of distribution, which is usually used to solve the case that one data set contains many different distributions. As shown in Figure 1b, GMM with two components is used to model MTS, and it (denoting as the ellipses) can well describe the double scroll structure of Lorenz system.

### 2.2. Complex Network Construction with Similarity Metric

Once the time series is represented by GMM, the similarity between time series is converted into the similarity between GMMs. Kullback–Leibler divergence is a commonly used solution to measure the distance between two probability distributions. However, it has no closed form solution in the case of GMM, and the implementation of Monte Carlo simulation becomes computationally expensive [18]. Therefore, we introduce maximum mean discrepancy (MMD) in reproducing kernel Hilbert spaces to quantify the similarity between GMMs [19]. Suppose we have two GMMs in $\mathbb{R}^d$:

$$
\begin{aligned}
P &= \sum_{i=1}^{m} \alpha_i p_i = \sum_{i=1}^{m} \alpha_i \mathcal{N}(\mu_i, \Sigma_i) \\
Q &= \sum_{j=1}^{n} \beta_j q_j = \sum_{j=1}^{n} \beta_j \mathcal{N}\left(\mu_j', \Sigma_j'\right)
\end{aligned}
\tag{4}
$$

where $p_i = \mathcal{N}(\mu_i, \Sigma_i)$; $q_j = \mathcal{N}\left(\mu_j', \Sigma_j'\right)$; and $m$, $n$ is the components number of $P$ and $Q$, respectively. Given a kernel function $k(x, y) = \langle \varphi(x), \varphi(y) \rangle_{\mathcal{H}}$, the reproducing kernel Hilbert space (RKHS) $\mathcal{H}$ corresponding to $k(x, y)$ can be defined, where $\varphi(x)$ is a feature mapping [20]. Given that we are in an RKHS, the mean map kernel can be defined as

$$
K(P, Q) = \mathbb{E}_{x \sim P, y \sim Q} k(x, y) = \left\langle \mathbb{E}_{x \sim P}[\varphi(x)], \mathbb{E}_{y \sim Q}[\varphi(y)] \right\rangle
\tag{5}
$$

Then MMD can be easily defined as

$$
\begin{aligned}
\mathrm{MMD}(P, Q) &= \|\mathbb{E}_{x \sim P}[\varphi(x)] - \mathbb{E}_{y \sim Q}[\varphi(y)]\| \\
&= \sqrt{K(P, P) + K(Q, Q) - 2K(P, Q)}
\end{aligned}
\tag{6}
$$

In the case of insufficient data, we can approximate the kernel function $K(P, Q)$ by empirical estimation [21]:

$$
K_{\mathrm{emp}}(P, Q) = \frac{1}{n_P \cdot n_Q} \sum_{i=1}^{n_P} \sum_{j=1}^{n_Q} k\left(x_i, y_j\right)
\tag{7}
$$

where $\{x_i\}_{i=1}^{n_P}$ and $\{y_j\}_{j=1}^{n_Q}$ are random samples. However, the approximation obtained with (7) introduces errors with high probability. Fortunately, when enough data is available, we can estimate the true distribution of the data; when GMM is used to approximate the distribution of the data, $K(P, Q)$ has a closed solution:

$$
K(P, Q) = \sum_{i,j} \alpha_i \beta_j K\left(\mathcal{N}(\mu_i, \Sigma_i), \mathcal{N}\left(\mu_j', \Sigma_j'\right)\right)
\tag{8}
$$

With (8), the form of $K(P, P)$ and $K(Q, Q)$ can be derived similarly. It turns out, introducing the Gaussian RBF kernel $k(x, y) = exp\left(-\gamma\|x - y\|^2/2\right)$, the product kernel of the Gaussian distribution is derived as:

$$
\left(\mathcal{N}(\mu_i, \Sigma_i), \mathcal{N}\left(\mu_j', \Sigma_j'\right)\right) = 1/\left|\gamma\Sigma_i + \gamma\Sigma_i + \mathbf{I}\right|^{\frac{1}{2}} \exp\left(-\frac{1}{2}\left(\mu_i - \mu_j'\right)^T \left(\Sigma_i + \Sigma_j' + \gamma^{-1}\mathbf{I}\right)^{-1}\left(\mu_i - \mu_j'\right)\right)
\tag{9}
$$

With (6), (8), and (9), we can obtain the analytic form of $\mathrm{MMD}(P, Q)$ by introducing the Gaussian RBF kernel.

Once similarity measures are in place, the construction of complex networks is straightforward. First, each UTS is represented by a GMM, and then, MMD in (6) is used to calculate the distance between each pair of GMM to form a distance matrix $D = \left(\mathrm{MMD}\left(P_i, Q_j\right)\right)$, where $i$ and $j$ denote different UTS. With a critical threshold $r_c$, $D$ can be converted into adjacent matrix whose elements indicate whether pairs of nodes are connected or not in the network. An adjacent matrix $A = \left(a\left(P_i, Q_j\right)\right)$, here $a\left(P_i, Q_j\right) = 1$ if $\mathrm{MMD}\left(P_i, Q_j\right) \le r_c$ and $a\left(P_i, Q_j\right) = 0$ if $\mathrm{MMD}\left(P_i, Q_j\right) > r_c$.

## 3. Experiments and Results

The Lorenz system in (2) has highly complex behaviors with the variation of the system parameters. With the change of system parameters, Lorenz system presents highly complex behavior. We randomly generate 150 time series of $x$ components by keeping $\sigma = 10.0$ and $\beta = 8/3$ while varying $\rho \in [28, 45]$. The reason is that $(\sigma, \rho, \beta)$ form a vast three-dimensional parameter space. Considering the complexity of the Lorenz system, its characteristics have not been fully studied when $\sigma$ and $\beta$ take other values [22]. To simplify the problem, many researchers fix $\sigma$ and $\beta$. while changing $\rho$. That is, each set of $(\sigma, \rho, \beta)$ corresponds to a UTS and different parameter $\rho$ corresponds to different class of time series. The length of each time series is 6000 data points, and the first 1500 points are removed to reduce the initialization effect of the system. The differential equation (2) is solved by scipy.integrate.odeint() in Python package SciPy [23], and the time point step is 0.01.

Firstly, with $m = 3, \tau = 8$, each UTS is transformed into MTS in phase space by (1). Then, the GMM corresponding to each MTS is estimated, where the components number is 3. Finally, the MMD between the GMMs is calculated and eventually converted to the adjacency matrix. In addition, to evaluate the proposed method, three other metrics (geodesic distance, DTW and correlation coefficient) are also used to construct the network for comparison. By estimating the covariance matrix of MTS in phase space, the geodesic distance can be obtained and then a network formed [5]. For DTW and correlation coefficient, the metrics can be calculated directly between UTS.

The spring layout method in NetworkX package [24] was used to plot the network, and the results were shown in Figure 2. In the network, each node corresponds to a UTS, and the connection between nodes is determined by the adjacency matrix. The selected threshold $r_c$. enables 20% of the edges to be preserved to highlight the geometric structure of the network. The validity of network construction can be evaluated from two aspects: one is to see whether the similarity between nodes can be effectively captured; the other is to see whether the geometry of the network is conducive to the analysis of time series. In the first aspect, geodesic distance (Figure 2a) and MMD (Figure 2b) are better metrics of similarity because nodes with similar $\rho$ are clustered together. In contrast, in Figure 2c,d, nodes with different $\rho$ are mixed together, especially in Figure 2d, the nodes are completely confused and indistinguishable, like a random network, indicating that the metric used cannot effectively measure similarity. From the second aspect, the MMD is superior to the geodesic distance in the geometry of the network because the nodes in Figure 2a are squeezed together to make it difficult to distinguish. This phenomenon also indicates that MMD is more sensitive to measure similarity, which results in a looser network. In the following description, we will explain why a loose network structure is better than a tight one.

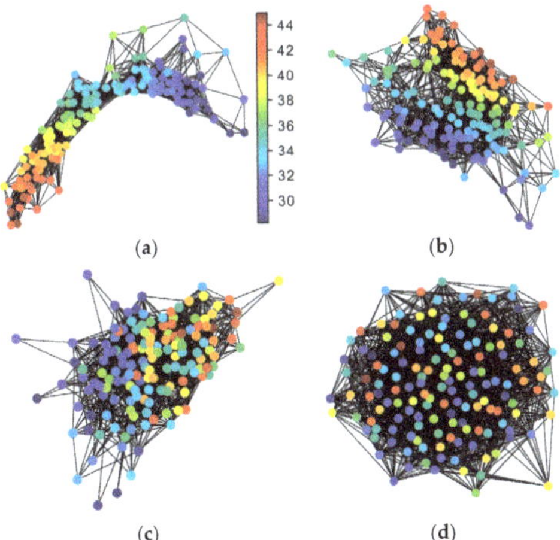

**Figure 2.** Construction of complex network based on different metric (colour bar denotes the value of $\rho$). (**a**) Network construction based on geodesic distance; (**b**) network construction based on maximum mean discrepancy (MMD); (**c**) network construction based on dynamic time warping distance (DTW); (**d**) network construction based on correlation coefficient.

To analyse the characteristics of MMD and geodesic distance in detail, we show the heat map of the related distance matrix in Figure 3, which corresponds to the network structure. The size of the heat map is $150 \times 150$, corresponding to 150 nodes, and each pixel denotes the distance between a pair of nodes (time series). The nodes are arranged in ascending order according to the value of $\rho$. From Figure 3a,b, it can be seen that the distance near the diagonal is small (high similarity), otherwise the distance is large, which means that the node pairs with similar $\rho$ have high similarity. To investigate this point more clearly, we set a certain threshold $r_c$ and retained the 20% of the edge (mentioned in Section 2.2), as shown in Figure 3c,d. As you can see, the reserved edges are centered diagonally and gradually spread to both sides. Therefore, if more edges are retained by increasing $r_c$, the topology of the network (the relationship between adjacent nodes) can still remain stable to some extent.

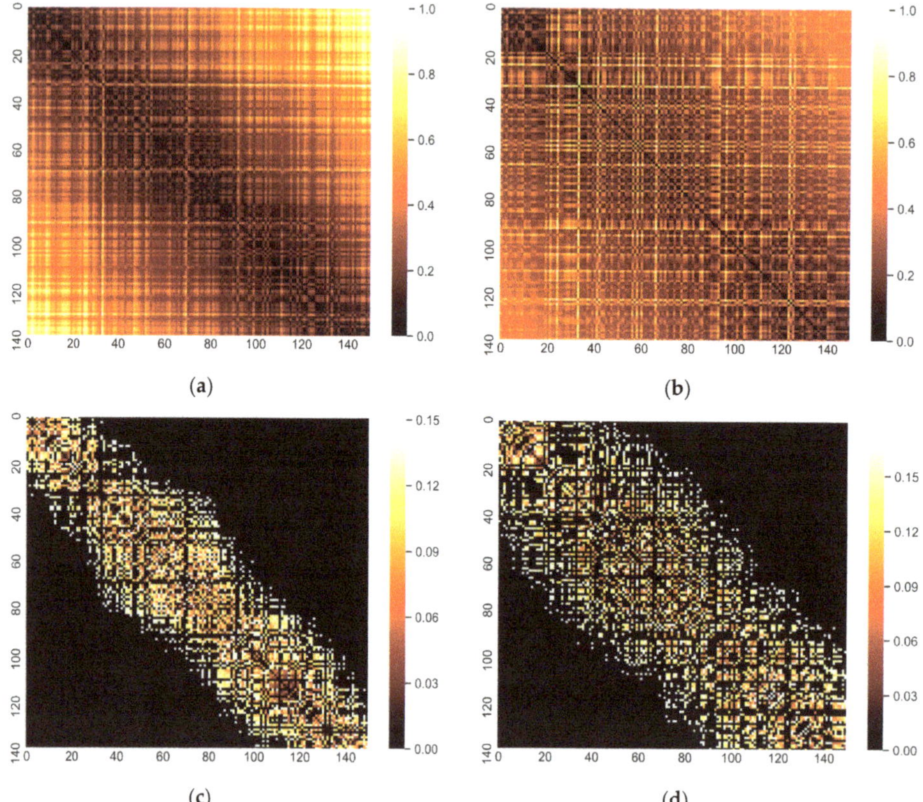

**Figure 3.** Heat map of distance matrix based on MMD and geodesic distance (coordinate label indicates the number of nodes and colour bar denote the value of distance between two nodes). (**a**) Heat map based on geodesic distance; (**b**) heat map based on MMD; (**c**) heat map in (**a**) with 20% of the edges to be preserved; (**d**) heat map in (**b**) with 20% of the edges to be preserved.

By comparing Figure 3c,d (similar to Figure 2a,b), we can find that the network structure based on MMD is looser. From the characteristics of Lorenz system, the loose network structure is more reasonable. This is because small changes in $\rho$ do not exactly correspond to smooth changes in the properties of time series. Although time series with similar $\rho$ usually have similar properties, time series with different $\rho$ sometimes have similar behaviors [25]. That is, a node should be similar to a node with a similar $\rho$, but it may also be similar to a node with a different $\rho$, which results in a loose network structure. Compared with geodesic method, MMD can capture the two similarities more effectively, and this results in a looser network. This is because the geodesic method is a special case of MMD in some ways. The deeper reason is that geodesic method uses only ONE covariance matrix (Gaussian distribution with a zero mean vector) to represent the data, while the MMD method uses GMM (linear combination of multiple Gaussian distributions) to fit the data. In contrast, MMD can capture more detailed information to find more neighbor nodes.

## 4. Conclusions

In this communication, a method was proposed for constructing a complex network of univariate chaotic time series. Compared with the commonly used metric, the introduced MMD can capture the similarity between GMMs more effectively, which is the key problem of constructing complex networks of the chaotic time series. Although the proposed method is specific to chaotic time series, it can also be applied to time series in other fields. In addition, it can be directly generalized to the case of multivariate time series by omitting phase space reconstruction.

**Funding:** This research was funded by the National Natural Science Foundation of China, grant number 61362024, 61761020.

**Conflicts of Interest:** The author declares no conflict of interest.

## References


1. Kantz, H.; Thomas, S. *Non-Linear Time Series Analysis*, 2nd ed.; Cambridge University Press: London, UK, 2004; ISBN 978-0521529020.
2. Lorenz, E.N. Deterministic Nonperiodic Flow. *J. Atmos. Sci.* **1963**, *20*, 130–141. [CrossRef]
3. Bezsudnov, I.V.; Snarskii, A.A. From the time series to the complex networks: The parametric natural visibility graph. *Phys. A Stat. Mech. Its Appl.* **2014**, *414*, 53–60. [CrossRef]
4. Zhao, Y.; Weng, T.; Ye, S. Geometrical invariability of transformation between a time series and a complex network. *Phys. Rev. E* **2014**, *90*, 012804. [CrossRef] [PubMed]
5. Sun, J.; Yang, Y.; Xiong, N.N.; Dai, L.; Peng, X.; Luo, J. Complex Network Construction of Multivariate Time Series Using Information Geometry. *IEEE Trans. Syst. Man, Cybern. Syst.* **2019**, *49*, 107–122. [CrossRef]
6. Zou, Y.; Donner, R.V.; Marwan, N.; Donges, J.F.; Kurths, J. Complex network approaches to nonlinear time series analysis. *Phys. Rep.* **2019**, *787*, 1–97. [CrossRef]
7. Gao, Z.-K.; Small, M.; Kurths, J. Complex network analysis of time series. *Europhysics Lett.* **2016**, *116*, 50001. [CrossRef]
8. Donner, R.V.; Zou, Y.; Donges, J.F.; Marwan, N.; Kurths, J. Recurrence networks—A novel paradigm for nonlinear time series analysis. *New J. Phys.* **2010**, *12*, 033025. [CrossRef]
9. Lacasa, L.; Luque, B.; Ballesteros, F.; Luque, J.; Nuño, J.C. From time series to complex networks: The visibility graph. *Proc. Natl. Acad. Sci.* **2008**, *105*, 4972–4975. [CrossRef] [PubMed]
10. Ruan, Y.; Donner, R.V.; Guan, S.; Zou, Y. Ordinal partition transition network based complexity measures for inferring coupling direction and delay from time series. *Chaos An Interdiscip. J. Nonlinear Sci.* **2019**, *29*, 043111. [CrossRef] [PubMed]
11. Zhang, J.; Small, M. Complex Network from Pseudoperiodic Time Series: Topology versus Dynamics. *Phys. Rev. Lett.* **2006**, *96*, 238701. [CrossRef] [PubMed]
12. Berthold, M.R.; Höppner, F. On Clustering Time Series Using Euclidean Distance and Pearson Correlation. *arXiv* **2016**, arXiv:1601.02213.
13. Zhang, J.; Luo, X.; Small, M. Detecting chaos in pseudoperiodic time series without embedding. *Phys. Rev. E* **2006**, *73*, 016216. [CrossRef] [PubMed]
14. Jeong, Y.-S.; Jeong, M.K.; Omitaomu, O.A. Weighted dynamic time warping for time series classification. *Pattern Recognit.* **2011**, *44*, 2231–2240. [CrossRef]
15. Shekofteh, Y.; Jafari, S.; Sprott, J.C.; Hashemi Golpayegani, S.M.R.; Almasganj, F. A Gaussian mixture model based cost function for parameter estimation of chaotic biological systems. *Commun. Nonlinear Sci. Numer. Simul.* **2015**, *20*, 469–481. [CrossRef]
16. Takens, F. Detecting strange attractors in turbulence. In *Dynamical Systems and Turbulence, Warwick 1980*; Springer: Berlin/Heidelberg, Germany, 1981; pp. 366–381.

17. Pennec, X.; Fillard, P.; Ayache, N. A Riemannian Framework for Tensor Computing. *Int. J. Comput. Vis.* **2006**, *66*, 41–66. [CrossRef]

18. Hershey, J.R.; Olsen, P.A. Approximating the Kullback Leibler Divergence Between Gaussian Mixture Models. In Proceedings of the 2007 IEEE International Conference on Acoustics, Speech and Signal Processing—ICASSP '07, Honolulu, HI, USA, 15–20 April 2007; pp. IV-317–IV-320.

19. Gretton, A.; Borgwardt, K.M.; Rasch, M.J.; Schölkopf, B.; Smola, A. A kernel two-sample test. *J. Mach. Learn. Res.* **2012**, *13*, 723–773.

20. Reed, M.; Simon, B. *Methods of Modern Mathematical Physics*; Academic Press: San Diego, CA, USA, 1980; ISBN 9780125850506.

21. Muandet, K.; Fukumizu, K.; Dinuzzo, F.; Schölkopf, B. Learning from distributions via support measure machines. In Proceedings of the Advances in Neural Information Processing Systems, Lake Tahoe, NV, USA, 3–6 December 2012; pp. 10–18.

22. Strogatz, S.H. *Nonlinear Dynamics and Chaos: With Applications to Physics, Biology, Chemistry, and Engineering*; Westview Press: Cambridge, MA, USA, 1994.

23. Scipy.integrate.odeint. Available online: https://docs.scipy.org/doc/scipy/reference/generated/scipy.integrate.odeint.html (accessed on 24 January 2020).

24. Software for Complex Networks. Available online: http://networkx.github.io/ (accessed on 24 January 2020).

25. Sparrow, C. *The Lorenz Equations: Bifurcations, Chaos, and Strange Attractors*; Springer: New York, NY, USA, 1982.

*Article*

# Analyzing Uncertainty in Complex Socio-Ecological Networks

**Ana D. Maldonado [1,*], María Morales [2], Pedro A. Aguilera [3] and Antonio Salmerón [2,4]**

[1]  Data Analysis Research Group, University of Almería, 04120 Almería, Spain
[2]  Department of Mathematics, University of Almería, 04120 Almería, Spain; maria.morales@ual.es (M.M.); antonio.salmeron@ual.es (A.S.)
[3]  Department of Biology and Geology, University of Almería, 04120 Almería, Spain; aguilera@ual.es
[4]  Department of Mathematics and Center for the Development and Transfer of Mathematical Research to Industry (CDTIME), University of Almería, 04120 Almería, Spain
*  Correspondence: ana.d.maldonado@ual.es

Received: 20 December 2019; Accepted: 17 January 2020; Published: 19 January 2020

**Abstract:** Socio-ecological systems are recognized as complex adaptive systems whose multiple interactions might change as a response to external or internal changes. Due to its complexity, the behavior of the system is often uncertain. Bayesian networks provide a sound approach for handling complex domains endowed with uncertainty. The aim of this paper is to analyze the impact of the Bayesian network structure on the uncertainty of the model, expressed as the Shannon entropy. In particular, three strategies for model structure have been followed: naive Bayes (NB), tree augmented network (TAN) and network with unrestricted structure (GSS). Using these network structures, two experiments are carried out: (1) the impact of the Bayesian network structure on the entropy of the model is assessed and (2) the entropy of the posterior distribution of the class variable obtained from the different structures is compared. The results show that GSS constantly outperforms both NB and TAN when it comes to evaluating the uncertainty of the entire model. On the other hand, NB and TAN yielded lower entropy values of the posterior distribution of the class variable, which makes them preferable when the goal is to carry out predictions.

**Keywords:** Bayesian networks; entropy; socio-ecological system

---

## 1. Introduction

Socio-ecological systems (SESs) constitute an outstanding example of complex systems, where multiple social and ecological components interact with each other in space and time [1,2]. SESs are complex adaptive systems whose interactions might change as a response to external events or endogenous changes [3,4]. As a consequence, the state of the SES evolves to a new one to adapt to these changes [5]. This brings about challenges not only from the modeling perspective but also when it comes to making predictions and diagnosing problems. An example of such complex socio-ecological systems is *cultural landscapes*, which are the outcome of the interaction of humans and nature over time [6]. Cultural landscapes [7] are typically heterogeneous systems providing diverse *ecosystem services* as the result of a complex relationship between human cultural management and the ecosystem.

Furthermore, there is a strong relationship between cultural landscapes and the socio-economy [8–10] and this relationship must be appropriately modeled in order to make well founded decisions on, for instance, implementing suitable landscape conservation policies [9]. Traditional analysis methods have been applied to this problem [11–13] but they sometimes fail to capture the complexity of the cultural landscape elements, connections and cause-effect relations, specially when ecosystem services are taken into account [14].

Another key issue is handling the uncertainty in data and in the predictions made by the models. In this sense, Bayesian networks (BNs) [15], provide a sound approach for handling complex domains endowed with uncertainty. The underlying formalism for uncertainty treatment is probability theory, which entails to quantify the uncertainty associated with the decisions made from BNs using measures as, for instance, Shannon entropy [16].

BNs have been widely used in the last decade as a modeling tool in environmental problems in general [17] and in cultural landscapes applications in particular [18]. A recent example employs the so-called object-oriented Bayesian networks (OOBNs) which are basically a structured way of representing Bayesian networks taking advantage of repeated and hierarchical components [19] so that the modeling task is simplified [20].

In this paper, we analyze the resulting model uncertainty when complex socio-ecological systems are modeled using Bayesian networks. More precisely, we investigate the impact of different network structures on the value of Shannon entropy from an experimental point of view. This analysis is relevant for practitioners when making decisions, since less uncertain models are potentially more reliable when making predictions using the model.

## 2. Materials and Methods

From now on, we will use uppercase letters to denote random variables and lowercase letters to denote a value of a random variable. Boldfaced characters will be used to denote random vectors (i.e., multidimensional random variables). The set of all possible values of a random vector $\mathbf{X}$ (also called its *support*) is denoted as $\Omega_{\mathbf{X}}$. A *Bayesian network* [15] with variables $\mathbf{X} = \{X_1, \ldots, X_n\}$ is a directed acyclic graph with $n$ nodes where each one corresponds to a variable in $\mathbf{X}$. Attached to each node $X_i \in \mathbf{X}$, there is a conditional distribution of $X_i$ given its parents in the network, $Pa(X_i)$, so that the joint distribution of random vector $\mathbf{X}$ factorizes as

$$p(x_1, \ldots, x_n) = \prod_{i=1}^{n} p(x_i | pa(x_i)), \tag{1}$$

where $pa(x_i)$ denotes a configuration of the values of the parents of $X_i$.

A simple example of a Bayesian network representing the joint distribution of variables $X_1, \ldots, X_5$ is shown in Figure 1. It encodes the factorization

$$p(x_1, x_2, x_3, x_4, x_5) = p(x_1)p(x_2|x_1)p(x_3|x_1)p(x_5|x_3)p(x_4|x_2, x_3). \tag{2}$$

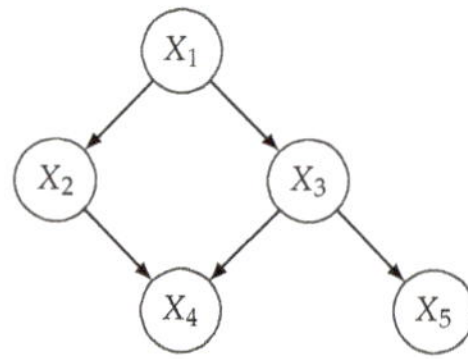

**Figure 1.** An example of a Bayesian network structure with 5 variables.

From a modeling perspective, one advantage of Bayesian networks is that the induced factorization avoids the specification of large multivariate distributions that are replaced by a set of smaller ones, which are more easily specified, since the number of parameter is lower. For example, the factorization in Equation (2) replaces the specification of a joint distribution over 5 variables by the specification of 5 smaller distributions, each one of them with at most 3 variables. Another advantage is that the network structure describes the interaction between the variables in the model, in a way that can be easily interpretable.

One of the most successful areas of application of Bayesian networks is *classification* [21], which is a prediction task in which there is a discrete target variable $C$, called the *class*, whose value is to be forecasted from the values of a set of *feature* variables $\mathbf{X} = \{X_1, \ldots, X_n\}$. The predicted value $c^*$ of $C$ is computed as the one that maximizes the posterior distribution of $C$ given the observed values of the features, that is,

$$c^* = \arg\max_{c \in \Omega_C} p(c|x_1, \ldots, x_n).$$ (3)

Note that

$$p(c|x_1, \ldots, x_n) = \frac{p(c) \times p(x_1, \ldots, x_n|c)}{p(x_1, \ldots, x_n)} \propto p(c) \times p(x_1, \ldots, x_n|c),$$ (4)

which means that solving the classification problem requires the specification of an $n$-dimensional distribution for $X_1, \ldots, X_n$ given $C$. The problem can be simplified by representing the joint distribution using a Bayesian network and taking advantage of the factorization encoded by its structure. The strongest simplification is achieved when the network is forced to adopt a naive Bayes (NB) structure, where the feature variables are assumed to be conditionally independent given the class. The BN structure is depicted in Figure 2a.

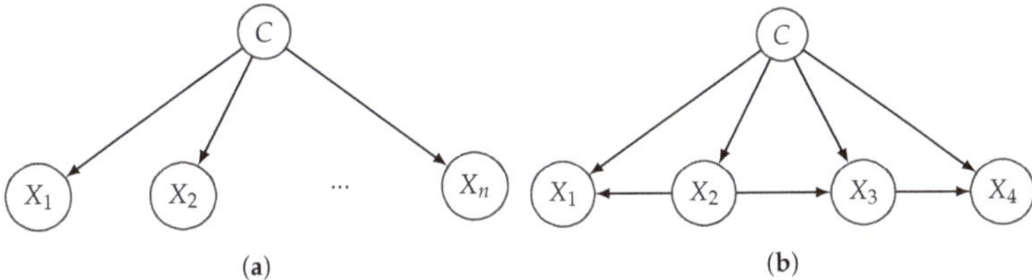

**(a)**          **(b)**

**Figure 2.** Structure of a naive Bayes model with $n$ features (**a**) and a tree augmented network (TAN) model with 4 features (**b**).

Adopting an NB structure actually means a strong independence assumption, but in practice it is compensated by the low number of parameters that need to be specified. Notice that, in this case the factorization results in

$$p(c|x_1, \ldots, x_n) \propto p(c) \prod_{i=1}^{n} p(x_i|c),$$ (5)

meaning that $n$ one-dimensional conditional distributions must be specified, instead of one $n$-dimensional conditional distribution.

The independence assumption underlying NB models can be relaxed, resulting in more expressive models that still keep a reduced number of parameters. This is the motivation of the *tree augmented network* (TAN) structure [21], where each feature variable is allowed to have another feature as a parent, besides the class, as long as the resulting subgraph containing the features is a tree (i.e., it contains no directed cycles). An example of a TAN model is given in Figure 2b, corresponding to the factorization

$$p(c|x_1, \ldots, x_n) \propto p(c)p(x_1|x_2, c)p(x_1|c)p(x_3|x_2, c)p(x_4|x_3, c).$$ (6)

Given that there are multiple structures that one can choose when facing classification problems, ranging from NB to unrestricted Bayesian networks, a natural question is to know whether this choice has an impact on the performance of the classification model. This problem has been analyzed from the point of view of the accuracy of the classification model [21]. In this paper we are more interested in analyzing the impact of the model structure on the uncertainty over the predictions, which in this context can be evaluated as the uncertainty of the used Bayesian network.

After all, a Bayesian network represents a probability distribution and a well known approach to quantifying the uncertainty of a probability distribution is to use Shannon entropy [16]. The Shannon entropy of a discrete random variable $X$ is

$$H(X) = - \sum_{x \in \Omega_X} p(x) \log p(x). \tag{7}$$

Analogously, it can be defined over a random vector $\mathbf{X} = \{X_1, \ldots, X_n\}$ as

$$H(\mathbf{X}) = - \sum_{\mathbf{x} \in \Omega_{\mathbf{X}}} p(\mathbf{x}) \log p(\mathbf{x}), \tag{8}$$

which in the case of a Bayesian network can be written as

$$\begin{aligned} H_{\mathrm{BN}}(\mathbf{X}) &= - \sum_{\mathbf{x} \in \Omega_{\mathbf{X}}} \prod_{i=1}^{n} p(x_i | pa(x_i)) \log \prod_{j=1}^{n} p(x_j | pa(x_j)) \\ &= - \sum_{\mathbf{x} \in \Omega_{\mathbf{X}}} \prod_{i=1}^{n} p(x_i | pa(x_i)) \left( \sum_{j=1}^{n} \log p(x_j | pa(x_j)) \right). \end{aligned} \tag{9}$$

Particularly, for a Bayesian network with NB structure and variables $\mathbf{X} = \{C, X_1, \ldots, X_n\}$, the entropy can be computed as

$$H_{\mathrm{NB}}(\mathbf{X}) = - \sum_{\mathbf{x} \in \Omega_{\mathbf{X}}} p(c) \prod_{i=1}^{n} p(x_i | c) \left( \log p(c) + \sum_{j=1}^{n} \log p(x_j | c) \right). \tag{10}$$

Shannon entropy is usually preferred to other entropies as a measure of uncertainty within the context of Bayesian networks due to its decomposability properties, which allow to efficiently compute it by taking advantage of the factorization of the distribution induced by the Bayesian network.

### 2.1. Experimental Analysis

In order to study the impact of the Bayesian network structure on the model uncertainty, we have conducted an experiment taking as a basis a Bayesian network that models a complex socio-ecological system. More precisely, we use the network described in [20]. It models the entire region of Andalusia (southern Spain) which contains a wide variety of scenarios from an ecological point of view.

The variables in the network describe social and economic indicators taken from the Multiterritorial Information System of Andalusia (SIMA) (http://www.juntadeandalucia.es/institutodeestadisticaycartografia/sima/) as well as environmental information collected from the Andalusian Environmental Information Network (http://www.juntadeandalucia.es/medioambiente/site/rediam). The network contains a total of 75 variables, described in the on-line material (https://w3.ual.es/personal/amg457/Downloads_protected/Experimentos.zip).

We conducted two experiments:

### 2.1.1. Experiment 1

The goal of this experiment is to assess the impact of the Bayesian network structure on the entropy of the model. The starting point was the Bayesian network in [20], that will be referred to as Original BN. Its structure is displayed in Figure 3 and it gives an idea of the complexity of the described system. Out of Original BN, we generated samples of sizes ranging from 500 to 100,000. From each sample, we constructed 9 networks with NB structure, each one of them with a different class variable, 9 networks with TAN structure, with the same class variables as NB and 1 network where we imposed no restriction on its structure. NB and TAN networks were built using package `bnlearn` in R [22] while the other network was constructed using the greedy search (GSS) method implemented in Hugin (http://www.hugin.com).

Instead of computing the entropy of each of the obtained networks using Equations (9) and (10), we decided to estimate them. The reason is that a straight application of those formulas requires summing over a number of terms that grows exponentially with the number of variables. For instance, in the case of Original BN, that contains 75 variables, assuming that all of them had only 2 possible values, evaluating the entropy would require summing over $2^{75}$ terms (approximately $3.8 \times 10^{22}$).

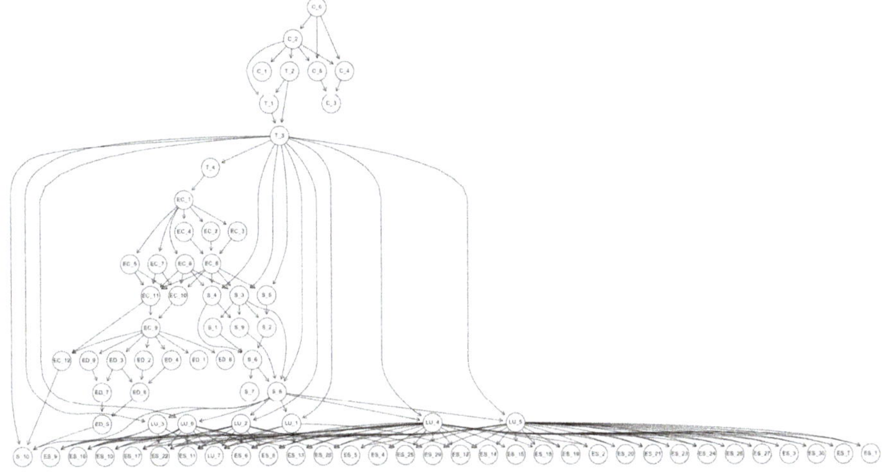

**Figure 3.** Structure of the Bayesian network used as reference in the experiments.

The estimation of the entropy was carried out using the same samples utilized for constructing the Bayesian networks. For a sample of size $m$, $\{\mathbf{x}^{(1)}, \ldots, \mathbf{x}^{(m)}\}$, we estimated $H_{\mathrm{BN}}(\mathbf{X})$ as

$$\hat{H}_{\mathrm{BN}}(\mathbf{X}) = -\frac{1}{m}\left(\sum_{j=1}^{n} \log p(x_j^{(r)}|pa(x_j^{(r)}))\right), \tag{11}$$

where $x_j^{(r)}$ denotes the value of variable $X_j$ in the $r$-th element of the sample and $pa(x_j^{(r)})$ is the value of the parent variables of $X_j$ in the $r$-th element of the sample.

Similarly, we estimated $H_{\mathrm{NB}}(\mathbf{X})$ as

$$\hat{H}_{\mathrm{NB}}(\mathbf{X}) = -\frac{1}{m}\left(\log p(c^{(r)}) + \sum_{j=1}^{n} \log p(x_j^{(r)}|c^{(r)})\right). \tag{12}$$

Note that $\hat{H}_{\mathrm{BN}}(\mathbf{X})$ and $\hat{H}_{\mathrm{NB}}(\mathbf{X})$ are, respectively, unbiased estimators of $H_{\mathrm{BN}}(\mathbf{X})$ and $H_{\mathrm{NB}}(\mathbf{X})$. It can be easily proved taking into account that

$$\begin{aligned}
H_{\mathrm{BN}}(\mathbf{X}) &= -\sum_{\mathbf{x}\in\Omega_{\mathbf{X}}} \prod_{i=1}^{n} p(x_i|pa(x_i)) \left(\sum_{j=1}^{n} \log p(x_j|pa(x_j))\right) \\
&= \mathbb{E}_p\left[-\sum_{j=1}^{n} \log p(X_j|pa(X_j))\right],
\end{aligned}$$

where $\mathbb{E}_p$ denotes the expectation computed with respect to distribution $\prod_{i=1}^{n} p(x_i|pa(x_i))$. Therefore, $\hat{H}_{\mathrm{BN}}(\mathbf{X})$ is just the sample mean estimator of $H_{\mathrm{BN}}(\mathbf{X})$, which is known to be unbiased. Likewise,

$\hat{H}_{\mathrm{NB}}(\mathbf{X})$ is the sample mean estimator of $H_{\mathrm{NB}}(\mathbf{X})$. Since both estimators are unbiased, their accuracy can be measured using their variance or equivalently, their standard deviation, as variance coincides with mean squared error for unbiased estimators.

### 2.1.2. Experiment 2

In this experiment we used the same networks as in Experiment 1. Then we generated three scenarios in the socio-ecosystem described by the Bayesian network. Each scenario corresponds to a particular configuration of values of some variables in the network. For each scenario, we computed the *posterior* distribution of the class variable—see Equation (4)—from each one of the nine networks in Experiment 1 and estimated the entropy of the posterior distribution as we describe next. The *prior* distribution of the class variable corresponds to the marginal distribution of variable $C$ in the corresponding network in Experiment 1, without taking into account the data corresponding to the three scenarios analyzed here. This is equivalent to adopting a parametric empirical Bayes approach, where the parameters of the prior distribution are estimated by maximum likelihood. This is the usual way of approaching prediction problems with Bayesian networks when we have an initial sample with a high number of elements and without missing values. If we denote by $q(c)$ the posterior distribution of the class variable for one particular scenario, then the entropy in this experiment is calculated as

$$H(C) = - \sum_{c \in \Omega_C} q(c) \log q(c). \tag{13}$$

Note that in this case there is no need to estimate the entropy from the sample, as we only need to sum over the values of the class variable.

## 3. Results and Discussion

The results of Experiment 1 are reported in Figure 4. The dashed line corresponds to the Original BN, that constitutes the ground truth. The dots represent the estimated entropy values, while the bars centered on each point represent the standard deviation, and thus the accuracy of the estimated value. It can be seen how in this case the network with unrestricted structure (GSS), consistently outperforms both NB and TAN. In fact, the entropy of the GSS network converges to the exact one (Original BN) when the sample size increases. Focusing on the classification-oriented networks, the uncertainty is clearly lower (lower entropy) for TAN compared to NB. This comes to no surprise, as the structure of the NB is the most simple one and therefore it is more unlikely that it is able to capture the exact model accurately and this is reflected in the model uncertainty. In the case of NB and TAN, the increase in sample size does not lead to a reduction in the entropy. This is also consistent with the lack of ability to fit the right model of both structures, due to the independence assumptions.

With respect to Experiment 2, the results for the three scenarios considered is similar, as can be inferred from Figures 5–7. The comparison carried out in this experiment is more fair with respect to NB and TAN because it refers to prediction scenarios, in which case we are only interested in the distribution over the target variable and not the entire model. In the three scenarios, the entropy corresponding to NB and TAN, likewise GSS, also converges to the entropy of the class posterior distribution computed with the original network.

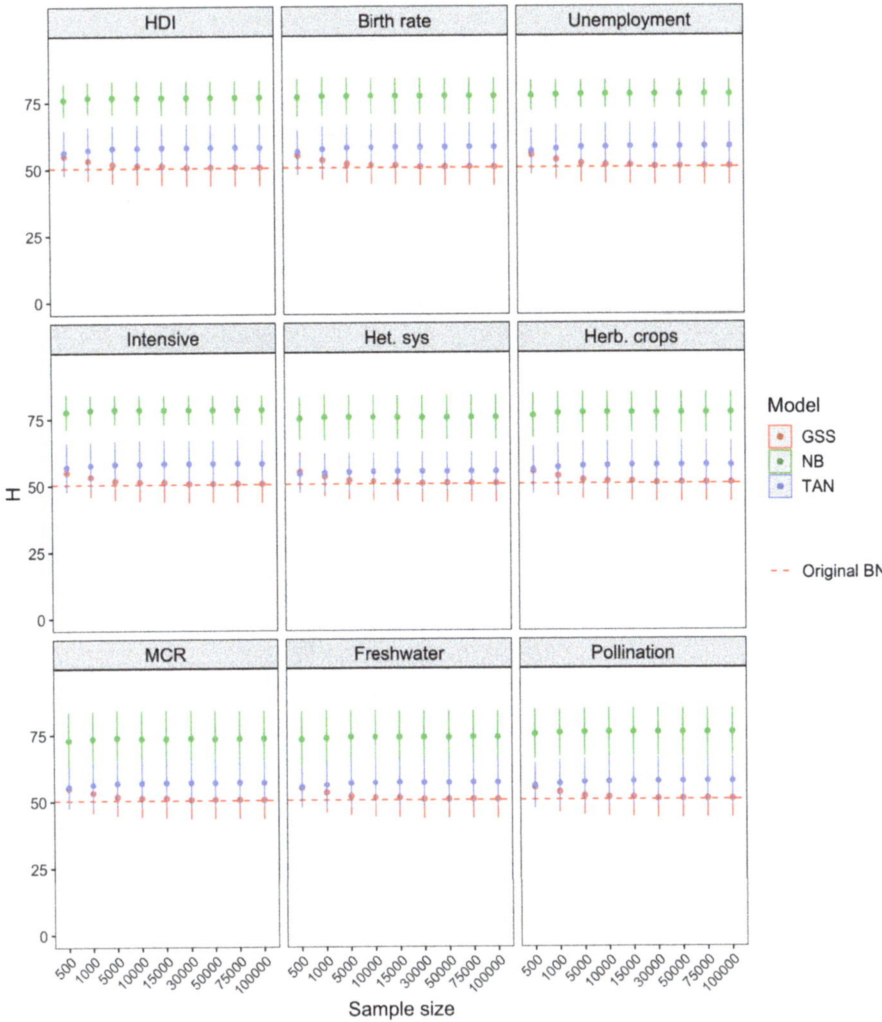

**Figure 4.** Shannon entropy vs. sample size for the Bayesian networks used in Experiment 1.

For smaller sample sizes, the uncertainty of GSS is typically higher than the exact one, which is in-line with the result obtained in Experiment 1 for this network. However, the uncertainty of the class posterior obtained from NB and TAN structures is often below the entropy of the Original BN and, in general, clearly below the uncertainty obtained from GSS. The extreme case is the posterior of variable MCR in scenario 1 computed from NB (bottom left panel of Figure 5).

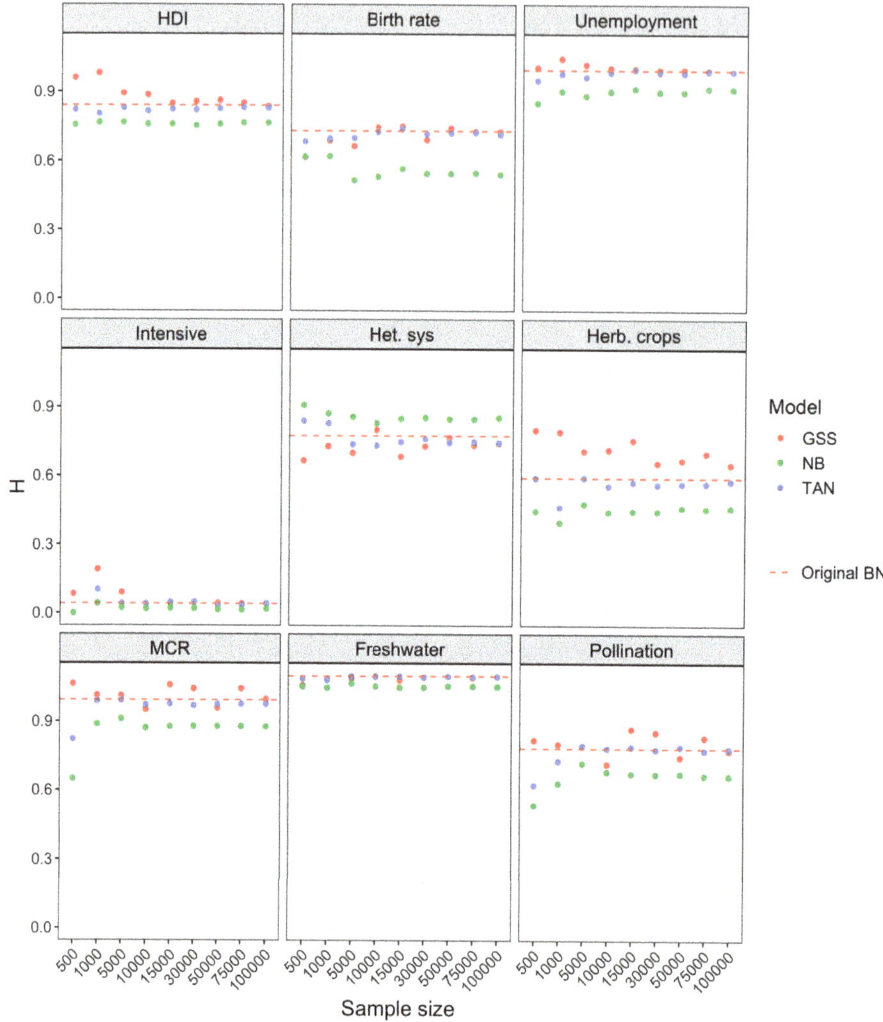

**Figure 5.** Shannon entropy of the class posterior distribution vs. sample size for scenario 1 in Experiment 2.

The observed behavior of the analyzed models support the idea of using NB and TAN for classification instead of unrestricted Bayesian network structures. The fact that the uncertainty is lower means that the class posterior distribution is less smooth. In other words, it better discriminates the most probable value of the class, which is in fact the value that corresponds to the outcome of the prediction model, as seen in Equation (3). This is precisely the effect that is sought by NB and TAN, which are focused on being accurate in the predictions rather than in goodness of fit.

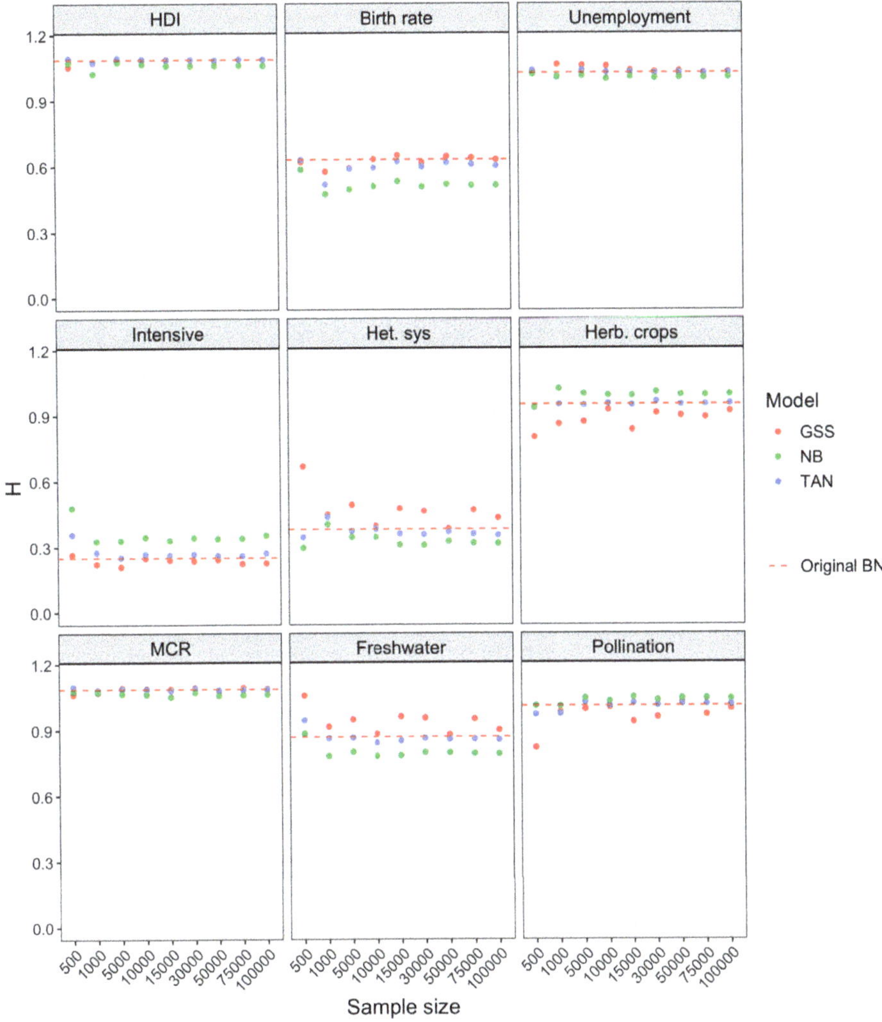

**Figure 6.** Shannon entropy of the class posterior distribution vs. sample size for scenario 2 in Experiment 2.

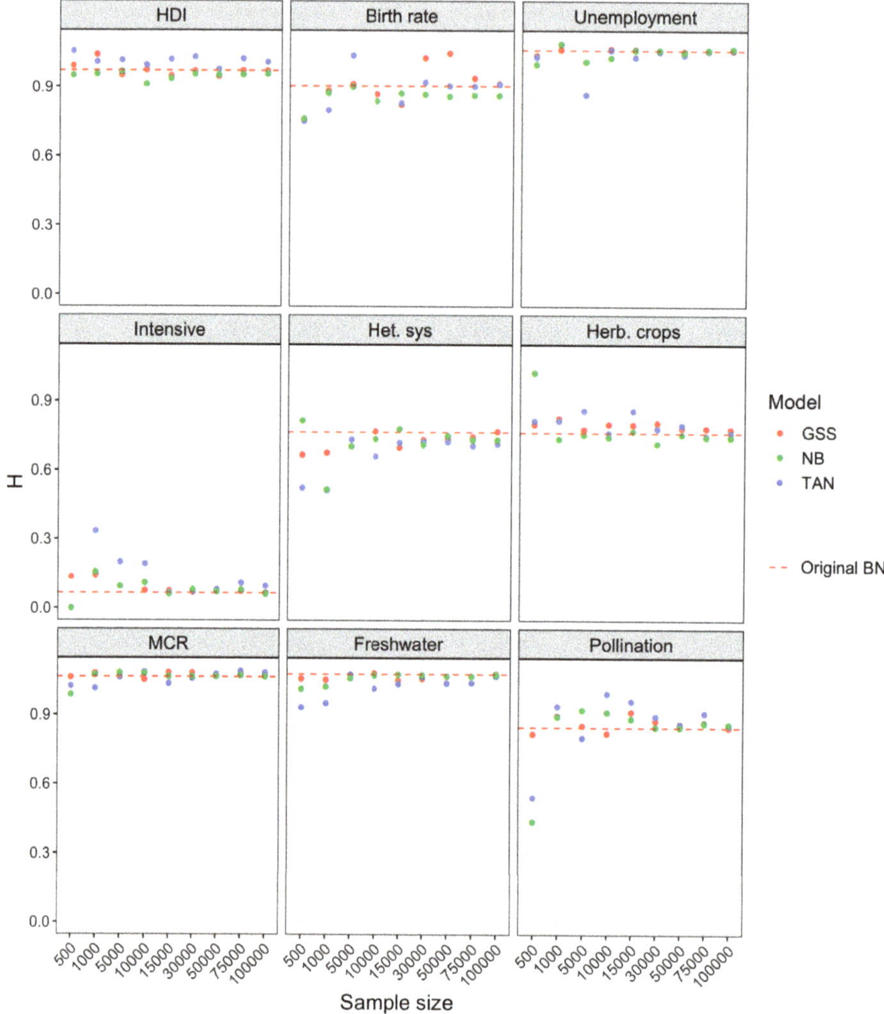

**Figure 7.** Shannon entropy of the class posterior distribution vs. sample size for scenario 3 in Experiment 2.

## 4. Conclusions

In this paper we have carried out two experiments analyzing the uncertainty in various Bayesian network structures representing complex environmental networks. More precisely, we have tested unrestricted structure, NB and TAN models representing a complex socio-economic system with 75 variables.

According to the results of the experiments, the conclusion is that, from the point of view of uncertainty, unrestricted structures are preferable when the goal is the representation of the entire complex system, that is, the full model. However, if the goal is to carry out predictions, then NB and TAN yield less uncertain results.

**Author Contributions:** Conceptualization, A.S. and P.A.A.; methodology, A.S.; software, A.D.M. and M.M.; validation, A.D.M., M.M., P.A.A. and A.S.; formal analysis, A.D.M., M.M. and A.S.; investigation, A.D.M. and M.M.; resources, A.D.M. and P.A.A.; data curation, A.D.M. and M.M.; writing—original draft preparation, A.S.; writing—review and editing, A.D.M., M.M., P.A.A. and A.S.; visualization, A.D.M.; supervision, A.S.; funding acquisition, A.S. All authors have read and agreed to the published version of the manuscript.

**Funding:** This research was funded by the Spanish Ministerio de Ciencia, Innovación y Universidades, through grant TIN2016- 77902-C3-3-P and by ERDF-FEDER funds.

**Conflicts of Interest:** The authors declare no conflict of interest.

## References


1. Liu, J.; Dietz, T.; Carpenter, S.R.; Folke, C.; Alberti, M.; Redman, C.L.; Schneider, S.H.; Ostrom, E.; Pell, A.N.; Lubchenco, J.; et al. Coupled human and natural systems. *AMBIO* **2007**, *36*, 639–650. [CrossRef]

2. Rescia, A.; Pérez-Corona, M.E.; Arribas-Ureña, P.; Dover, J.W. Cultural landscapes as complex adaptive systems: The cases of northern Spain and northern Argentina. In *Resilience and the Cultural Landscape: Understanding and Managing Change in Human-Shaped Environments*; Cambridge University Press: Cambridge, UK, 2012; pp. 126–145.

3. Parrott, L.; Meyer, W.S. Future landscapes: Managing within complexity. *Front. Ecol. Environ.* **2012**, *10*, 382–389. [CrossRef]

4. Ropero, R.F.; Rumí, R.; Aguilera, P.A. Modelling uncertainty in social-natural interactions. *Environ. Model. Softw.* **2016**, *75*, 362–372. [CrossRef]

5. Schlüter, M.; Haider, L.J.; Lade, S.J.; Lindkvist, E.; Martin, R.; Orach, K.; Wijermans, N.; Folke, C. Capturing emergent phenomena in social-ecological systems. *Ecol. Soc.* **2019**, *24*, 11. [CrossRef]

6. Blondel, J. The design of Mediterranean landscapes: A millennial story of humans and ecological systems during the historical period. *Hum. Ecol.* **2006**, *34*, 713–729. [CrossRef]

7. García-Llorente, M.; Martín-López, B.; Iniesta-Arandia, I.; López-Santiago, C.A.; Aguilera, P.A.; Montes, C. The role of multi-functionality in social preferences toward semi-arid rural landscapes: An ecosystem service approach. *Environ. Sci. Policy* **2012**, *19*, 136–146. [CrossRef]

8. Moreira, F.; Rego, F.C.; Ferreira, P.G. Temporal (1958–1995) pattern of change in a cultural landscape of northwestern Portugal: Implications for fire occurrence. *Landsc. Ecol.* **2001**, *16*, 557–567. [CrossRef]

9. Schmitz, M.F.; De Aranzabal, I.; Aguilera, P.A.; Rescia, A.; Pineda, F.D. Relationship between landscape typology and socioeconomic structure: Scenarios of change in Spanish cultural landscapes. *Ecol. Model.* **2003**, *168*, 343–356. [CrossRef]

10. Peña, J.; Bonet, A.; Bellot, J.; Sánchez, J.; Eisenhuth, D.; Hallett, S.; Aledo, A. Driving Forces of Land-Use Change in a Cultural Landscape of Spain. In *Modelling Land-Use Change: Progress and Applications*; Koomen, E., Stillwell, J., Bakema, A., Scholten, H.J., Eds.; Springer: Dordrecht, The Netherlands, 2007; pp. 97–116. [CrossRef]

11. De Aranzabal, I.; Schmitz, M.F.; Aguilera, P.; Pineda, F.D. Modelling of landscape changes derived from the dynamics of socio-ecological systems: A case of study in a semiarid Mediterranean landscape. *Ecol. Indic.* **2008**, *8*, 672–685. [CrossRef]

12. Álvarez Martínez, J.; Suárez-Seoane, S.; De Luis Calbuig, E. Modelling the risk of land cover change from environmental and socio-economic drivers in heterogeneous and changing landscapes: The role of uncertainty. *Landsc. Urban Plan.* **2011**, *101*, 108–119. [CrossRef]

13. Schmitz, M.; Matos, D.; De Aranzabal, I.; Ruíz-Labourdette, D.; Pineda, F. Effects of a protected area on land-use dynamics and socioeconomic development of local populations. *Biol. Conserv.* **2012**, *149*, 122–135. [CrossRef]

14. Müller, F.; Burkhard, B. The indicator side of ecosystem services. *Ecosyst. Serv.* **2012**, *1*, 26–30. [CrossRef]

15. Pearl, J. *Probabilistic Reasoning in Intelligent Systems*; Morgan-Kaufmann: San Mateo, CA, USA, 1988.

16. Shannon, C.E. A mathematical theory of communication. *Bell Syst. Tech. J.* **1948**, *27*, 623–666. [CrossRef]

17. Aguilera, P.A.; Fernández, A.; Fernández, R.; Rumí, R.; Salmerón, A. Bayesian networks in environmental modelling. *Environ. Model. Softw.* **2011**, *26*, 1376–1388. [CrossRef]

18. Landuyt, D.; Broekx, S.; D'hondt, R.; Engelen, G.; Aertsens, J.; Goethals, P. A review of Bayesian belief networks in ecosystem service modelling. *Environ. Model. Softw.* **2013**, *46*, 1–11. [CrossRef]

19. Korb, K.B.; Nicholson, A.E. *Bayesian Artificial Intelligence*; Chapman & Hall: London, UK, 2003.

20.  Maldonado, A.D.; Aguilera, P.A.; Salmerón, A.; Nicholson, A.E. Probabilistic modeling of the relationship between socioeconomy and ecosystem services in cultural landscapes. *Ecosyst. Serv.* **2018**, *33*, 146–164. [CrossRef]

21.  Friedman, N.; Geiger, D.; Goldszmidt, M. Bayesian Network Classifiers. *Mach. Learn.* **1997**, *29*, 131–163. [CrossRef]

22.  Scutari, M. Learning Bayesian Networks with the bnlearn R Package. *J. Stat. Softw.* **2010**, *35*, 1–22. [CrossRef]

*Article*

# Multi-Type Node Detection in Network Communities

**Chinenye Ezeh** [1,2]**, Ren Tao** [1,*]**, Li Zhe** [1]**, Wang Yiqun** [1] **and Qu Ying** [1]

[1]  Software College, Northeastern University, Shenyang 110000, China; noblenenye@gmail.com (C.E.);
    gislzneu@163.com (L.Z.); neuwangyiqun@163.com (W.Y.); qybefore1727@163.com (Q.Y.)

[2]  Department of Computer Engineering, Michael Okpara University of Agriculture, Umudike 440109, Nigeria

*  Correspondence: chinarentao@163.com

Received: 31 October 2019; Accepted: 27 November 2019; Published: 17 December 2019

**Abstract:** Patterns of connectivity among nodes on networks can be revealed by community detection algorithms. The great significance of communities in the study of clustering patterns of nodes in different systems has led to the development of various methods for identifying different node types on diverse complex systems. However, most of the existing methods identify only either disjoint nodes or overlapping nodes. Many of these methods rarely identify disjunct nodes, even though they could play significant roles on networks. In this paper, a new method, which distinctly identifies disjoint nodes (node clusters), disjunct nodes (single node partitions) and overlapping nodes (nodes binding overlapping communities), is proposed. The approach, which differs from existing methods, involves iterative computation of bridging centrality to determine nodes with the highest bridging centrality value. Additionally, node similarity is computed between the bridge-node and its neighbours, and the neighbours with the least node similarity values are disconnected. This process is sustained until a stoppage criterion condition is met. Bridging centrality metric and Jaccard similarity coefficient are employed to identify bridge-nodes (nodes at cut points) and the level of similarity between the bridge-nodes and their direct neighbours respectively. Properties that characterise disjunct nodes are equally highlighted. Extensive experiments are conducted with artificial networks and real-world datasets and the results obtained demonstrate efficiency of the proposed method in distinctly detecting and classifying multi-type nodes in network communities. This method can be applied to vast areas such as examination of cell interactions and drug designs, disease control in epidemics, dislodging organised crime gangs and drug courier networks, etc.

**Keywords:** bridging centrality; community detection; disjoint nodes; disjunct nodes; node similarity; overlapping nodes

---

## 1. Introduction

Over the years, numerous research works have been devoted to identification and description of community with respect to networks or graphs without a consensus on its definition [1]. Some characteristic features can easily be extracted from the nodes in a graph to describe a community [2,3]. Intuitively, communities are usually acquired from the removal of bridges (edges), bridge-nodes or articulation points (cut vertexes) from a graph. Identification and removal of these sets of nodes and edges can effectively disintegrate a network naturally into densely connected subgroups [4–11]. A community can effectively be described as clusters of densely connected nodes that are revealed along disconnected lines of weak connections of bridge-nodes.

Communities are very useful in detecting hierarchical clusters in various fields such as cells interaction, epidemic/disease control in natural and biological sciences, design of power grid and road networks in engineering, collaboration networks, social networks in social sciences and so on [6,7,11–13]. Most networks reveal hierarchical structures, i.e., they reveal smaller clusters contained within larger clusters. One of the most popular clustering methods is the hierarchical clustering method, which is further divided into two categories namely agglomerative algorithms and divisive algorithms. In agglomerative algorithms, clusters of nodes with high similarity are merged together in successive iterations to achieve better clusters, whereas in divisive algorithms, nodes with low similarity values are disconnected in successive iterations to reveal better clusters of nodes with higher similarity [1,14].

In recent years, existing community detection algorithms reported in the literature were specifically designed to either detect only disjoint nodes or overlapping nodes. Disjoint nodes, also known as node clusters, are nonoverlapping groups of densely connected subgraphs of a network [1,12,14–18]. Overlapping nodes are nodes shared by two or more communities at the same time, thereby creating overlapping communities [1,14–16,19–22]. Previous methods rarely take into consideration disjunct nodes (isolated or neutral nodes) [23]. However, when critically examined, real complex networks reveal the existence of multi-type nodes [1]. For example, Peel et al. [24] reported that the majority of community detection algorithms cannot recover the metadata of a certain node or often mislabelled this node (person number 9) in the popular Zachary's karate club network, which, most likely, had a neutral political support during the feud that eventually divided the karate club. Nodes of this type can only be discovered by suitably designed algorithms that are capable of distinguishing the different node types on a network.

There has been proliferation of different community detection algorithms over the past few years, with each algorithm being designed to achieve what has already been attained in the past with little or no difference. The idea of implementing these algorithms differently on datasets for set purposes not only consume much resources but take quite precious amount of time. We set out to achieve a unified process of community detection which focuses on and reveals the various node types, and therefore we propose a method that detects multi-type nodes in network communities that disintegrate a network into communities. This method ensures that various node types are recovered and duly classified. In other words, when an overlapping node is identified, it is easier to distinguished the communities been overlapped by it. Also, the disjoint nodes are clearly separated whereas the disjunct nodes do not adhere to any clusters. Some of the foremost community detection algorithms were proposed by Girvan and Newman [4,5]. In these algorithms, the edge with the highest betweeness centrality value is iteratively disconnected until the network disintegrates into modules. It is reported that these algorithms cannot discover overlapping nodes, as each node is assigned to a cluster [1]. However, we know that most real networks often share nodes between communities, resulting in community overlap and sometimes disjunct nodes are discovered [1,22]. In their work [5], Newman and Girvan introduced a quality measure known as modularity measure, which is used to determine the strength of community structures found by the algorithm. This measure further inspired other community detection algorithms based on modularity optimisation methods. Newman [25] proposed a fast optimisation of the quality function modularity. In this method, at the initial stage, there are $|N|$ communities formed by each node. At every successive iteration, communities are merged only if it improves the value of the quality function modularity [1,25]). Even though Newman's method is quite fast and detected quality communities on networks, Clauset et al. [26] pointed out that it consumed much storage space and time in the computation of adjacency matrix. As a result, they proposed a more efficient method known as greedy modularity optimisation algorithm, which uses data structures to compute and retain only significant improvements in the value of the quality function modularity [1,26]). Similar to the greedy modularity optimisation techniques of Newman [25] and Clauset et al. [26] is the very popular Louvain algorithm [27]. This method is suitable for both weighted and unweighted networks. In the first phase, each node is assigned to its own community. Nodes are joined to form supernodes only if there is gain in the value of modularity. The second phase involves fusion of connected supernodes on the condition that the value of modularity increased. The entire process is repeated recursively until gain in the value of modularity is no longer possible. The Louvain algorithm is reported to be one of the fastest community detection algorithms and is capable of handling networks with millions of nodes and edges [1]. The modularity optimisation methods fall under the category of hierarchical agglomeration community detection algorithms, and they detect only disjoint or none overlapping clusters. Unlike the modularity optimisation based methods, Label propagation algorithm (LPA) proposed by Raghavan et al. [28] uses only structural information of networks to detect communities. At the initial stage, each node obtains a unique identifier or label and subsequently adopts the majority label of its neighbours after every successive propagation iteration. The propagation process terminates when a convergence point is reached, i.e., when every node adopts the majority label of its neighbours or the preassigned number of iterations is attained. At this stage, densely connected clusters of nodes assume same label thereby forming communities [1,28]. The Spectral algorithm is a matrix-based clustering method that uses eigenvectors for clustering. Here, the nodes on a network form data points and the edges between nodes form distances. The eigenvector of these points is calculated from the generated affinity matrix, and a clustering method such as the $k$-means clustering technique is used to partition these points [1,29,30]. As noted earlier, complex networks have the tendency to allow multi-membership of nodes to two or more communities per time and, consequently, this brings about node overlaps and overlapping communities in networks [22,31,32]. To capture such distinctive characteristics of networks, researchers proposed and designed community detection algorithms that are capable of capturing the overlapping structures of complex networks. Yuan et al. [19] proposed a constraint model that necessitates recursive edge-cuts that meet the constraint condition. This algorithm detects overlapped communities at the end of the process.

Note that the majority of the previously proposed algorithms can only detect disjoint nodes (node clusters) or overlapping nodes (nodes binding overlapping communities) and rarely disjunct/neutral nodes (single node partitions). We propose a new method which distinctly identifies disjoint nodes, disjunct nodes and overlapping nodes following a natural pattern of network division. Our approach rather focuses on identifying the various node types, as when these node types are identified, network communities are naturally recovered. The procedure involves iteratively finding nodes with the highest bridging centrality value and subsequently its neighbours that yield the least node similarity value are determined and the links joining them disconnected [33]. The process is sustained until a stoppage criterion condition is met. Our approach focuses on revealing the node types and this ensures that nodes are distinctly identified as well as classified into communities with high value of modularity. Singleton nodes with a degree value of one are ignored to avoid the possibility of cutting them off during network division, so as not to mix them up with what we classify as disjunct nodes in this work. Additionally, the properties that characterise disjunct/neutral nodes are highlighted and clearly demonstrated. The proposed algorithm was tested and compared with other community detection algorithms on artificial and real-world datasets, and the results indicated impressive performance against the compared algorithms.

The outline for the rest of this paper is as follows. In Section 2, we define some relevant terms and design and implement an algorithm to detect disjoint nodes, disjunct nodes and overlapping nodes. We further highlight some of the properties that characterise disjunct nodes. We analyse the experimental results, discuss our findings and offer recommendations in Section 3. Finally, we conclude in Section 4.

## 2. Methodology

Bearing in mind the usefulness of communities in studying and understanding patterns of node connectivity on networks, we propose a new method to discover disjoint nodes, disjunct nodes and overlapping nodes. Our method iteratively identifies bridge-nodes using the Bridging centrality metric [6] to compute the nodes with the highest bridging centrality value. Furthermore, the node similarity value between the identified bridge-node and all of its neighbours is calculated. We rank the node similarity values in decreasing order and detach the edges/links with the least node similarity value. Intuitively, the bridge-node forms a community by aligning with its neighbours that return high node similarity values unless there is anything to the contrary [3,34]. The edge/link which has the least node similarity value is the edge between the bridge-node and another community. If the node similarity values between the bridge-node and its neighbours return a value equal to zero, then the bridge-node would most certainly be isolated upon network division and we classify this node to be a disjunct node without any community. This signifies that the isolated nodes do not share any nodes in common with any of their neighbours. Some of the bridge-nodes which seem to be isolated are actually overlapping nodes. The proposed algorithm identifies them by cutting them out just like the isolated nodes, but they differ from isolated nodes in the sense that they have paths linking back to them from their neighbours, they share some common nodes and can form communities with their neighbours.

The proposed algorithm is designed to be implemented on a typical undirected and unweighted graph $G = (V, E)$, in which $V = \{v_1, v_2 \cdots v_n\}$ is of $n$ nodes and $E = \{e_1, e_2 \cdots e_m\}$ is a set of edges denoted by $m$. The $n$ nodes and their connections are represented by an adjacency matrix $= [A_{ij}](n \times n)$ where $A_{ij} = 1$ if $v_i$ is connected to $v_j$, and $A_{ij} = 0$ otherwise.

### 2.1. Definition of Important Measures and Terms

#### 2.1.1. Similarity Measure

The node similarity measure is used to compute the level of relationship between nodes. This measure is equally used to ascertain if nodes can be grouped together into the same community [1,3,16]. We determine the similarity between nodes via the structural similarity, which computes the intersections between the neighbourhood sets of any two nodes. There are a couple of node similarity measures but we adopt the Jaccard similarity coefficient because of its intuitive appeal. The model is shown in Equation (1).

$$\frac{|n_i \cap n_j|}{|n_i \cup n_j|} \tag{1}$$

$n_i$ is the neighbourhood set of node $i$ and $n_j$ is the neighbourhood set of the neighbours of node $i$.

#### 2.1.2. Modularity

Modularity is an optimisation function that is used to evaluate the quality of a graph partition, which was designed by Newman and Girvan [5]. The larger the value of the modularity function, the better the quality of the detected communities [17,18]. The model is given in Equation (2).

$$Q = \sum e_{ii} - a_i^2 \tag{2}$$

$e_{ii}$ is the fraction of edges included in the community $i$ and $a_i$ is the fraction of nodes' degree included in the community $i$.

$$e_{ii} = E_i/m \tag{3}$$

where $E_i$ is the number of edges contained inside the community $i$ and $m$ is the total number of edges in $G$.

$$a^2 = \frac{\sum_{v \in C_i} d_v}{\sum_{v \in G} d_v} \tag{4}$$

where $C_i$ is the community $i$ and $d_v$ is the degree of node $v$.

#### 2.1.3. Betweeness Centrality

The Betweeness centrality of a node $v$, first designed by Freeman [35], is given in Equation (5):

$$C_B(v) = \sum_{\substack{s \neq v \neq t \\ s,v,t \in V}} \frac{\rho_{st}(v)}{\rho_{st}} \tag{5}$$

where $\rho_{st}(v)$ is the number of shortest paths from node $s$ to node $t$ that pass through node $v$, and $\rho_{st}$ is the number of shortest paths from node $s$ to node $t$.

### 2.1.4. Bridging Coefficient and Bridging Centrality

The Bridging coefficient is defined as

$$BC(v) = \frac{d(v)^{-1}}{\sum_{i \in N(v)} \frac{1}{d(i)}} \tag{6}$$

where $d(v)$ is the degree of node $v$ and $N(v)$ is the set of neighbours of node $v$. Bridging centrality, on the other hand, is used to quantitatively measure the extent of bridging capability of all nodes in a network. Comparatively to other components on the same network, the bridge-nodes are identified on the basis of their high value of bridging centrality [6,7]. The bridging centrality $C_R(v)$ of a node $v$ is defined by

$$C_R(v) = BC(v) \times C_B(v) \tag{7}$$

where $BC(v)$ is the Bridging coefficient and $C_B(v)$ is the Betweeness centrality.

### 2.1.5. Clustering Coefficient

Clustering coefficient measures the degree of clustering that exists between node $v$ and its direct neighbours [6]. The model is given in Equation (8).

$$Cl(v) = \frac{2L}{d_v(d_v - 1)} \tag{8}$$

where $d_v$ is the degree of node $v$ and $L$ is the number of links between $d_v$ neighbours of node $v$.

### 2.2. The Algorithm

The steps involved in the implementation of the proposed method for detecting disjoint nodes, disjunct nodes and overlapping nodes are stated in Algorithm 1. First, assign the desired number of partitions $P$ to be detected. Initialise modularity $Q = 0$ and create a copy of the network $G' \leftarrow G$. Then, compute the bridging centrality value $C_{BR_i}$ of all nodes in the network $G$. Select the node $B_{r_i}$ with the highest bridging centrality value. Compute the node similarity values between $B_{r_i}$ and all of its neighbours. Select the nodes that return the least node similarity value and delete the links/edges connecting them to $B_{r_i}$. Repeat the cycle until the number of connected components, modules or partitions of $G' == P$. In other words, the algorithm loops and keeps count of the number of modules/partitions until the network is divided up into total number of desired partitions $P$ which was assigned at the beginning of the experiment. Assign all partitions with components greater than 1 to cluster nodes $C_{cluster}$. Find all single node partitions $SP$ and compute their clustering coefficient $Cl_{coeff}$ from the original network $G$. Classify $SP$ as neutral node $C_{neutral}$ if $Cl_{coeff} = 0$, or overlapping node $C_{overlap}$ otherwise. Compute the quality of the resultant communities' modularity, $Q$, and display the cluster nodes $C_{cluster}$, neutral node $C_{neutral}$ and overlapping node $C_{overlap}$.

### 2.3. Properties of an Isolated Bridge-Node

From the synthetic graph displayed in Figure 1a, we note that node $v_4$ has the highest bridging centrality value contained in Table 1. Further computations of the node similarity values between node $v_4$ and its neighbours nodes $v_3$ and $v_5$ returned the value 0, i.e., $sim(v_4, v_3) = sim(v_4, v_5) = 0$. When the links connecting these nodes are disconnected, the network $G$ disintegrates. This makes node $v_4$ become an isolated node as it has no similarity with any of its neighbours, yet it is very vital in bridging communities. From Table 2, we note that edges $G(4,5); G(5,4)$ and $G(4,3); G(3,4)$ returned the highest edge-betweeness values, respectively. These are the edges which link node $v_4$ with its

---

**Algorithm 1** Multi-type Node Detection Algorithm

---

    **Input:** Network $G$; desired number of partitions $P$
    **Output:** $C_{cluster}$, $C_{neutral}$, $C_{overlap}$, $Q$

1:  **initialize** $Q = 0$, **copy** $G' \leftarrow G$;
2:  **compute** $C_{BR_i} = bridgingcentrality(G')$                                   ▷ use Equation (7);
3:  **select** $B_{r_i} \leftarrow \max(C_{BR_i})$                        ▷ nodes with max. bridging centrality value;
4:  $Neb_{B_{r_i}} \leftarrow$ **find** $(neighbours(B_{r_i}))$;
5:  **if** $Neb_{B_{r_i}} \leq 1$ **then**
6:     continue;
7:  **end if**
8:  **compute** $sim(B_{r_i}, Neb_{B_{r_i}})$,                            ▷ node similarity, use Equation (1);
9:  **find** $\min(sim(B_{r_i}, Neb_{B_{r_i}}))$                                    ▷ remove links;
10: **repeat**
11:     2–11
12: **until** number_connected_components$(G') == P$
13: $C_{cluster} ==$ **find**$(connected\_components(G') > 1))$; $SP ==$ **find**$(connected\_components(G') ==$
    1$))$;                                      ▷ SP refers to Single Node Partitions
14: **compute** $Cl_{coeff} = clusteringcoeff(G, SP)$; **calculate** $Q$;
15: **if** $Cl_{coeff} = 0$ **then**
16:     $C_{neutral} \leftarrow SP$
17: **else**
18:     $C_{overlap} \leftarrow SP$
19: **end if**
20: **print** $C_{cluster}$, $C_{neutral}$, $C_{overlap}$, $Q$

---

neighbour's nodes $v_5$ and $v_3$, respectively. Even though these edges have the highest edge-betweeness values, they are linked to an isolated bridge-node, which cannot form a community with any of its neighbours because it has zero node similarity values with them. The network $G$ is disconnected into two distinct communities, with node $v_4$ not belonging to any particular community. Therefore, we designate node $v_4$ as a disjunct node without any community. This also demonstrates that, with respect to bridge-nodes, the link that yields the least node similarity value is same link with the highest edge-betweeness centrality value. In other words, node similarity has an inverse correlation with edge-betweeness centrality.

We can summarise the properties of an isolated-bridge node as follows.

- They are bridge-nodes.
- They have degree $k_i > 1$.
- They have no path linking back to them. In other words, they do not share common nodes with any other node on the network. i.e., $|n_i \cap n_j| = \varnothing$. Therefore, they have zero node similarity values with all of their neighbours.

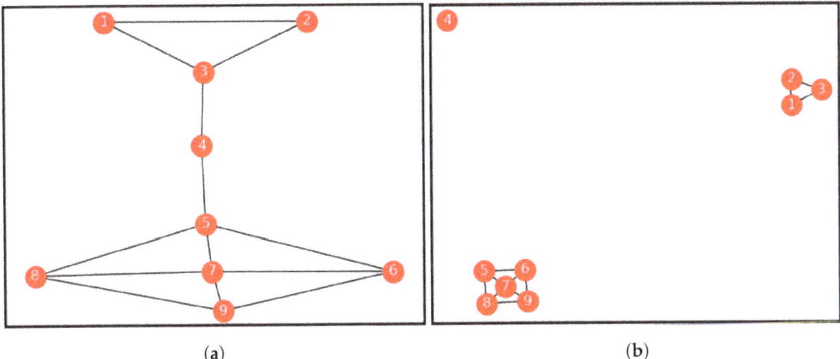

(a)  (b)

**Figure 1.** Example synthetic network. (**a**) Full network. (**b**) Fragmented network.

**Table 1.** Bridging centrality and node similarity values of nodes in network $G$.

| Iteration Count | Node ID. | Bridge Centrality Value | Neighbours. | Node Similarity Value |
|---|---|---|---|---|
| 1st | 4 | 0.4592 | 3 | 0 |
|  |  |  | 5 | 0 |
| 2nd | 7 | 0.0045 | 5 | 0.2857 |
|  |  |  | 6 | 0.2857 |
|  |  |  | 8 | 0.2857 |
|  |  |  | 9 | 0.2857 |

**Table 2.** Edge-Betweeness values of links/edges with the highest values in network $G$.

| Edge | Edge-Betweeness Value |
|---|---|
| $G(4,5); G(5,4)$ | 0.2778 |
| $G(4,3); G(3,4)$ | 0.2500 |
| $G(1,3); G(3,1)$ | 0.0972 |
| $G(2,3); G(3,2)$ | 0.0972 |

## 3. Results, Evaluation and Discussion

The algorithm is implemented with PYTHON3.7 and related packages (Networkx [36], Numpy [37,38], Matplotlib [39] and Scipy [40]) and run on a computer with Windows 7 OS (64-bits), Intel (R) Core$^{(TM)}$ i7-4790 CPU (3.60 GHz) and 4 GB RAM.

### 3.1. Tests on Artificial Networks

The proposed algorithm was tested on Lancichinetti–Fortunato–Radicchi (LFR) benchmark [1,41] against the greedy algorithm of Clauset, Newman and Moore (CNM) [26]; Linear Propagation algorithm (LPA) [28]; Louvain algorithm (Louvain) [27]; Spectral Clustering algorithm (SPA) [29,30]; and Girvan-Newman algorithm (GN) [4]. The algorithm implemented in the work of Yuan et al. [19] was not included in any of the experiments in this work as we could not re-implement it. In the LFR benchmark, $N$ is the number of nodes rendered in the network by the benchmark. $\tau_1$ and $\tau_2$ represent the power law exponent of the degree distribution and the power law exponent of the community size distribution produced in the network, respectively. $<k>$ is the average degree of nodes in the network, and the mixing parameter $\mu$ is the fraction of intra-community links or edges connecting each node. $min_C$ and $max_C$ are the minimum size of communities and the maximum size of communities, respectively. The results obtained from the LFR benchmark, as shown in Figure 2a,b, indicate that the quality of communities detected by all the algorithms, except for the proposed algorithm deteriorates sharply at mixing parameter $\mu = 0.2$. The proposed algorithm decline steadily in contrast to LPA, GN and SPA algorithms until $\mu = 0.3$. The implication is that from $\mu \leq 0.3$ qualities of communities detected are very good, but from $\mu > 0.3$, the qualities of the communities detected deteriorate. In any case, the proposed algorithm performs better than the other compared algorithms. For the LFR benchmark experiment in Figure 2a, we set $N = 1000$ nodes, $\tau_1 = 5, \tau_2 = 1.5, < k > = 10, \min_C = 20, \max_C = 50$. The number of communities to be detected was set at 100 for the proposed algorithm, GN and SPA. Likewise, In Figure 2b, we set $N = 2000$ nodes, $\tau_1 = 5, \tau_2 = 1.5, < k > = 10, \min_C = 20, \max_C = 60$. The number of communities to be detected was set at 200 for the proposed algorithm, GN and SPA. Due to the high CPU time in computing GN and the proposed algorithms, we did one iteration only.

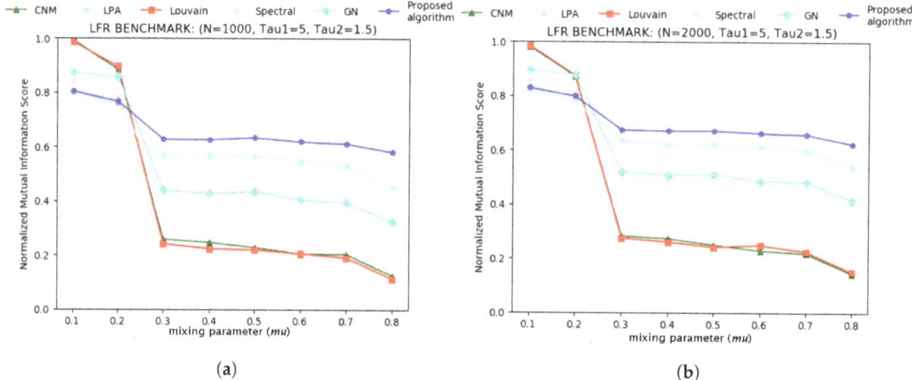

**Figure 2.** (**a**) Normalised mutual Information performance comparison of the proposed algorithm using Lancichinetti–Fortunato–Radicchi (LFR) benchmark. Number of nodes $N = 1000, \tau_1 = 5, \tau_2 = 1.5, < k > = 10, \min_C = 20, \max_C = 50$. (**b**) Normalised mutual information performance comparison of the proposed algorithm using LFR benchmark. Number of nodes $N = 2000, \tau_1 = 5, \tau_2 = 1.5, < k > = 10, \min_C = 20, \max_C = 60$. The mixing parameter $mu$ ranges from 0 to 0.8 with a step increment of 0.1.

## 3.2. Tests on Real-World Network Datasets

We further demonstrate the efficiency of the proposed algorithm with real-world datasets such as Zachary's karate club network (Karate), Dolphins network (Dolphins), American football club network (Football), Kreb's network of political books (Polbooks) and email data from European research institution (Email). Nodes and edges are indicated as $n$ and $m$, respectively, whereas ground-truth represents the number of communities in the original network as shown in Table 3. The performance of the proposed algorithm is tested on real datasets against CNM, LPA, Louvain, SPA and GN algorithms using modularity measure and F1-score, which is an average of precision and recall computed from ground-truth community dataset and detected community dataset [32]. For modularity measure comparison among the stated algorithms, the number of communities to be detected for karate club network was set at 3 for SPA, GN and the proposed algorithm. For the dolphins network, the number of communities to be detected was set at 4 for SPA, GN and the proposed algorithm. For football network, the number of communities was set at 12 for SPA and GN. The proposed algorithm detected at most nine communities in the football network. Therefore, the number of communities was set at 9. For the polbooks network, the number of communities were set at 4 for SPA, GN and the proposed algorithm. For the email network, the number of communities was set at 42 for SPA and GN. Just like in the case of football network, the proposed algorithm detected at most 30 communities in the email network. Therefore, the number of communities was set at 30. As shown in Figure 3a, the proposed algorithm outperformed the compared algorithms in karate club network, dolphins network, football network and polbooks network. In the email network, the proposed algorithm performed marginally above the other algorithms. In Figure 3b, the proposed algorithm performed better than the other algorithms in Karate network and Dolphins network. Expectedly, LPA and Spectral algorithms performed better ahead of the proposed algorithm, CNM, Louvain and GN algorithms in the football network. This could be as a result of the proposed algorithm detecting at most nine communities in this network. In the polbooks network, the performance of the proposed algorithm is good but less than the performance of CNM and GN algorithms. The email network was not considered for the F1-score computation due to unavailability of its ground-truth dataset.

**Table 3.** Properties and description of network datasets used.

| Network | $n/m$ | Ground-Truth | Description | Ref |
|---|---|---|---|---|
| Karate Club | 34/78 | 2 | Friendship network of karate club members | [12] |
| Dolphin | 62/159 | 2 | Association network of bottlenose dolphins | [43] |
| Polbooks | 105/441 | 3 | A co-purchasing network of political books | [44] |
| Football | 115/613 | 12 | A game-scheduling network of teams | [45] |
| Email EU | 1005/16706 | 42 | European research institution's email data | [46,47] |

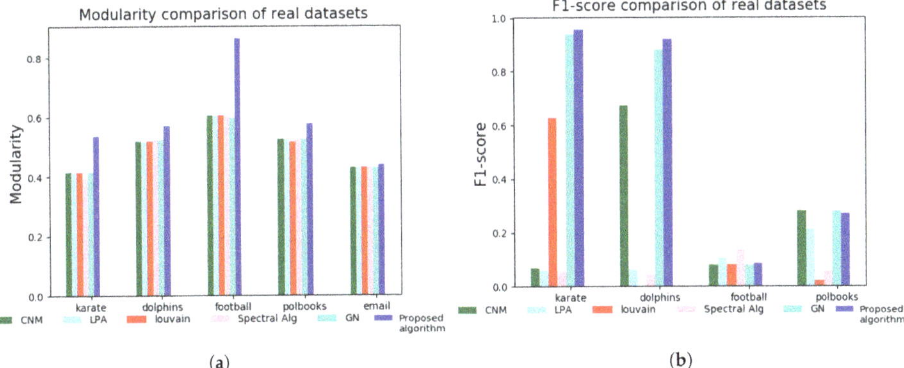

(a)                    (b)

**Figure 3.** (a) Modularity measure comparison among CNM, LPA, Louvain, SPA, GN and the proposed algorithm. (b) F1-score comparison among CNM, LPA, Louvain, SPA, GN and the proposed algorithm. The email network is ommitted in the F1-score computation due to unavailability of its ground-truth data.

### 3.2.1. Zachary's Karate Club Network

The results obtained show that the proposed algorithm is quite efficient in identifying disjoint nodes, disjunct nodes and overlapping nodes. In Zachary's karate club network, shown in Figure 4a, the proposed algorithm detected three partitions (two cluster node partitions and one single node partition). The two cluster node partitions (disjoint nodes) are the two main communities whereas the single node partition (node 9) is a disjunct node. The ground-truth community of this network comprises two main partitions, as indicated in Table 3, but some useful clusters can be found at sub-modular levels as indicated in Figure 4b. The proposed algorithm was able to recover the metadata of node 9 as a disjunct node. This corresponds to what is reported in the work of Peel et al. [24], where person number 9 is indicated to likely have possessed neutral political inclination neither towards the karate club president nor the club instructor during the feud between these two persons that eventually resulted in the split of the karate club into two. Often, most algorithms fail to recover this particular node or they mislabel it [24]. In Figure 4b, the proposed algorithm detected four main communities with one disjunct node (node 9) and one overlapping node (node 28). The partitions overlapped by node 28 are overlapping communities. Information revealed at sub-modular levels of partitions can be very useful in situations where one needs to examine the connections and relationships among nodes at sub-modular structures. Node 9 (displayed in green) in Figure 4a,b and node 28 (displayed in cyan) in Figure 4b are shown as being isolated, but a careful examination shows that only node 9 meets the requirements to be classified as a disjunct node. Node 28 is an overlapping node as it has at least an edge linking back to it and it shares clusters with two of its neighbours (nodes 31 and 33), which are in different communities that form the overlapping communities. Yuan et al. [19] correctly classified this node as an overlapping node which corresponds to node 29 in their work. Also, the proposed algorithm achieved modularity value of 0.5789 at three communities as indicated in Table 4, which is greater than SPA and GN's modularity values of 0.4188 and 0.4188 respectively at three communities each. CNM and LPA returned three communities each with modularity values of 0.4198 and 0.4198, respectively. At 4 communities, the proposed algorithm achieved modularity value of 0.5940 which is greater than the modularity value of 0.4156 achieved by Louvain algorithm at four communities. It is very apparent that the modularity values achieved by the proposed algorithm on the Karate club network are higher than those of the other algorithms considered for comparison as can be seen in Table 4. This is a clear indication that the proposed algorithm attains better clustering quality than the compared algorithms.

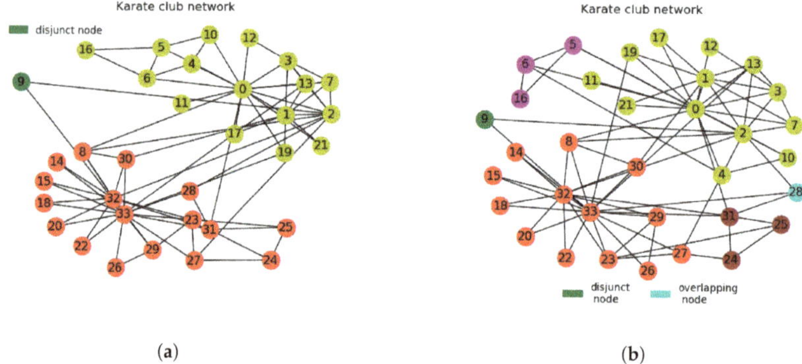


**Figure 4.** (a) Zachary's karate club network partitioned into 2 communities with 1 disjunct node. (b) Zachary's karate club network partitioned into 4 communities with 1 disjunct node and 1 overlapping node. The partitions overlapped by node 28 are overlapping communities. The rest of the nodes not indicated on the legends in Figure 4a,b represent different communities according to their respective colours.

**Table 4.** Modularity values and number of communities gotten from real complex networks. Number of communities indicated against CNM, LPA and Louvain are auto-generated since they do not need prior parameters before execution. The proposed algorithm could detect at most 9 communities for the football network and 30 communities for the Email network. The modularity values shown against SPA, GN and the proposed algorithms for Karate, Dolphin and Polbooks networks are based on the smallest number of communities returned among CNM, LPA and Louvain algorithms.

| | Modularity $Q$ and Number of Communities (C) | | | | | |
|---|---|---|---|---|---|---|
| Network | CNM | LPA | Louvain | SPA | GN | Proposed Algorithm |
| Karate | 0.4198 $C = 3$ | 0.4198 $C = 3$ | 0.4156 $C = 4$ | 0.4188 $C = 3$ | 0.4188 $C = 3$ | 0.5789 $C = 3$ |
| Dolphin | 0.5188 $C = 4$ | 0.5196 $C = 6$ | 0.5268 $C = 6$ | 0.5188 $C = 4$ | 0.4156 $C = 4$ | 0.6989 $C = 4$ |
| Polbooks | 0.5266 $C = 4$ | 0.5268 $C = 8$ | 0.5270 $C = 4$ | 0.5270 $C = 4$ | 0.5266 $C = 4$ | 0.5905 $C = 4$ |
| Football | 0.6046 $C = 6$ | 0.6043 $C = 11$ | 0.6044 $C = 10$ | 0.6046 $C = 12$ | 0.6043 $C = 12$ | 0.8641 $C = 9$ |
| Email | 0.4324 $C = 44$ | 0.4306 $C = 38$ | 0.4322 $C = 28$ | 0.4314 $C = 42$ | 0.4328 $C = 42$ | 0.4415 $C = 30$ |

### 3.2.2. Dolphins Network

The proposed algorithm can choose the number of partitions to be returned. This way, modular structures at lower hierarchies are revealed. In the dolphins network, shown in Figure 5a, the two larger communities (disjoint nodes) are clearly indicated with one disjunct node (node 39). Yuan et al. [19] reported node 40, which corresponds to node 39 in our work, as an overlapping node rather than as a disjunct node, but we understand that this is as a result of differences in methods implemented in the respective algorithms. The proposed algorithm achieved modularity value of 0.6989 at four communities, which is higher than the modularity values of 0.5188 for CNM and SPA each and 0.4156 for GN at four communities. LPA and Louvain achieved modularity values of 0.5196 and 0.5268, respectively, at six communities each. These values are less than the modularity value of 0.6989 achieved by the proposed algorithm as indicated in Table 4.

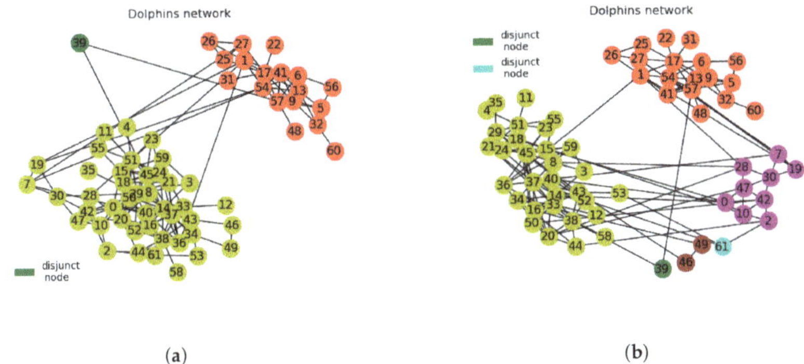

(a)       (b)

**Figure 5.** (a) Dolphins network partitioned into 2 communities with 1 disjunct node. (b) Dolphins network partitioned into 4 communities with 2 disjunct nodes. The rest of the nodes not indicated on the legends in Figure 5a,b represent different communities according to their respective colours.

### 3.2.3. The Other Networks

In Kreb's network of political books, the proposed algorithm achieved modularity value of 0.5905 at four communities (all disjoint nodes) in comparison to CNM, Louvain, SPA and GN's modularity values of 0.5266, 0.5270, 0.5270 and 0.5266, respectively, at four communities each. At 8 communities, LPA algorithm achieved modularity value of 0.5268 as against the proposed algorithm's modularity value of 0.6964 at eight communities. Yuan et al. [19] classified nodes 30 and 86 as overlapping nodes at four communities. Our results show that these nodes which correspond to nodes 29 and 85 in our work as shown in Figure 6 are members of clusters.

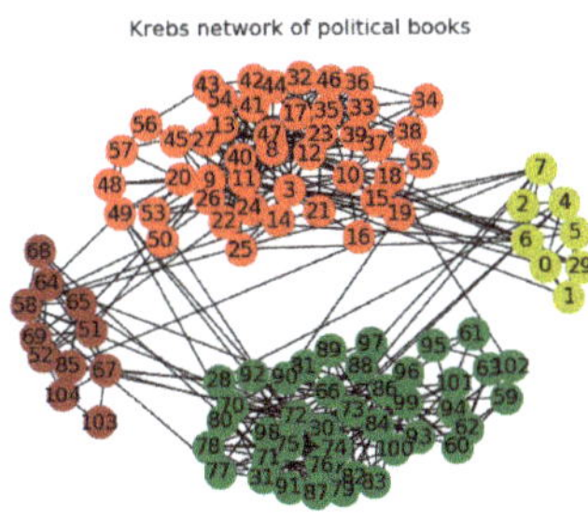

**Figure 6.** Kreb's network of political books at 4 communities.

In American college football network, the proposed algorithm could detect at most nine communities, contrary to the ground-truth of 12 communities indicated in Table 3 and what others reported in the literature. The quality of the communities detected by the proposed algorithm is still quite high in comparison to other methods with modularity value of 0.8641. We noticed that six of the conferences combined to form three bigger conferences. Clauset et al. [26] reportedly detected six communities with modularity value of 0.6046. Yuan et al. [19] reportedly detected 10 communities with node 37 as overlapping node, whereas the proposed algorithm does not have any overlapping node.

In the email data network of European research institution, the proposed algorithm detected at most 30 communities with modularity value of 0.4415. CNM algorithm returned 44 communities with modularity value of 0.4324. LPA algorithm returned 38 communities with modularity value of 0.4306. Louvain algorithm returned 28 communities with modularity value of 0.4322. SPA and GN algorithms' number of communities were fixed at 42 each and they achieved modularity values of 0.4314 and 0.4328, respectively. These values are presented in Table 4.

The method developed in this paper leads the way in multi-type node detection on networks contrary to previous methods that detect either only cluster nodes or overlapping nodes. Most of the methods often rarely identify disjunct nodes, which are integral parts of complex networks that play various significant roles. We further highlighted the unique properties of disjunct nodes which prior to this time had not been properly characterised by any other work. From our observation, the disjunct nodes can have several connections to their direct neighbours but when the network is disintegrated, they are shown to be isolated. In other words, they do not belong to any community. Discovery of these types of nodes could be very useful in certain instances to determine the actual impact they may have on the network and their neighbours. For example, a protein molecule in a network of protein–protein interactions (PPI) can connect other modular protein clusters and could be revealed to be a disjunct protein molecule at a sub-modular level when the network is divided up. One can investigate the significant roles such protein molecules play and the possible effects their malfunction can have on the surrounding protein molecule clusters. With an understanding of something of this nature, careful study of biological cells can help in designing drugs for disease treatment and epidemic controls. In computer networks, this can be very helpful in the design of network configuration of computers. Also, in the fight against drugs and related crimes, a drug mule or courier who works for drug cartels, but is not necessarily a member of any of the drug cartels, can be intercepted and the cartels infiltrated. Another possible area of interest might be in the design of power grid networks.

To actualise our set objectives, we used the bridging centrality metric [6] as a tool to help us determine bridge-nodes. We also used the Jaccard similarity coefficient to help determine the level of similarity or relationship between the bridge-nodes and their neighbours. This helped us to distinctly identify and classify the node types. A clear distinction was made between the disjunct nodes and the overlapping nodes. It is imperative we point out that our method and objectives are quite different from the method and objectives in [7]. Hwang et al. [7] proposed bridge-cut algorithm which is based on bridging centrality of edges. We have not compared the performance of these two methods as it's not part of the scope of this present work.

Additionally, we set the number of desired output partitions ahead of time before executing this algorithm. This allows one to adjust the number of partitions to be returned so as to ensure careful study of the multi-level hierarchical structures in networks. Such information as this can be very useful in disease control by deletion of certain edges connected to isolated or overlapping nodes. Some existing studies also support this point of view [7,48]. Differentiating multi-type nodes in a natural way on networks can equally be helpful in critical examination of cell interactions and drug designs, protein–protein networks, etc. [6,7]. It can also give insight to future studies and understanding of terrorist cells operations, illegal transfer of funds among terrorists, drug courier networks, organised crime gangs, power grids, internet infrastructure designs, road network designs and so on.

### 3.2.4. Computational Complexity Analysis

The bridging centrality metric is bounded by the time complexity of betweeness centrality based on Brande's betweeness algorithm, which is what is implemented in the Networkx python package used in this work. It is calculated in $\mathcal{O}(nm)$ time, where $n$ and $m$ are the total number of nodes and edges on a network, respectively [7,49]. Its space complexity takes $\Theta(n^2)$ to be computed. The bridging coefficient consumes approximately $\mathcal{O}(n(\log n)^2)$ time [7]. The Jaccard similarity coefficient takes $\mathcal{O}(m^2)$ time to be computed [50]. Due to the recomputation of bridging centrality and Jaccard similarity coefficient after every iteration; therefore, our algorithm can be computed in a total time and space complexity of $\mathcal{O}((nm) + (m^2))^2$ and $\Theta(n^2)$, respectively. The processing time expended on executing each algorithm on different networks is give in Table 5. The proposed algorithm only performs better than GN with respect to small networks and performs poorly in large networks.

**Table 5.** CPU execution time of the algorithms in seconds.

| Network | CNM | LPA | Louvain | SPA | GN | Proposed Algorithm |
|---|---|---|---|---|---|---|
| Karate | 0.0037 | 0.0012 | 0.0110 | 0.0110 | 0.0467 | 0.0311 |
| Dolphin | 0.0147 | 0.0101 | 0.0301 | 0.0604 | 0.1264 | 0.0960 |
| Polbooks | 0.0232 | 0.0061 | 0.0400 | 0.0712 | 1.3444 | 0.8653 |
| Football | 0.0513 | 0.0290 | 0.0655 | 0.1937 | 5.5780 | 3.5012 |
| Email EU | 2.3331 | 0.1486 | 1.3961 | 1.4739 | 324.79 | 6804.38 |

### 3.2.5. Limitations and Future Works

In future works, we hope to design an autonomous divisive algorithm that needs no parameters to stop the iteration. We also hope to make the algorithm scalable for very large networks because the betweeness centrality metric, as a global metric, has a high computational efficiency as indicated from the processing time in Table 5. This algorithm will be deployed in various application domains to explore further studies in these areas.

## 4. Conclusions

We designed a new algorithm that distinctly identifies and classifies multi-type nodes in network communities. Bridging centrality metric was used to calculate and select nodes with the highest bridging centrality value. Jaccard similarity coefficient was used to determine the level of similarity or relationship between the bridge-nodes and all of their neighbours. The nodes with the least similarity value were disconnected iteratively after which the bridging centrality of all nodes are recomputed until the stopping condition was met. We also validated the existence of disjunct/neutral nodes and highlighted the properties that characterise them. The results from extensive experiments done with real-world datasets show that this algorithm is efficient in distinctly discovering and classifying disjoint nodes, overlapping nodes and disjunct nodes, which are shown to be neutral nodes in terms of community membership. These results demonstrate the effectiveness of the proposed method and we believe that it will be of significant use in various application domains of community detection as well as arouse interests in future designs of an all inclusive community detection algorithms. This way, node connectivity relations can be revealed and studied better at sub-modular levels of different complex systems.

**Author Contributions:** Conceptualisation, C.E.; Formal analysis, C.E. and L.Z.; Funding acquisition, R.T.; Investigation, C.E.; Methodology, C.E.; Software, C.E., W.Y. and Q.Y.; Supervision, R.T.; Validation, C.E., R.T. and L.Z.; Writing—original draft, C.E.

**Funding:** This work was partially supported by the National Natural Science Foundation of China (61473073, 61433014), the Fundamental Research Funds for the Central Universities (N161702001, N171706003, 182608003, 181706001) and Program for Liaoning Excellent Talents in Universities (LJQ2014028).

**Acknowledgments:** The authors would like to thank members of Complex Network group under the watch of Ren Tao for their wonderful and insightful discussions. Chinenye Ezeh offers gratitude to Patrice Monkam who took the time to read the first draft of the manuscript and offered valuable suggestions. We are equally grateful to the editors and anonymous reviewers for their insight and helpful remarks.

**Conflicts of Interest:** The authors declare no conflicts of interest.

## References and Notes


1. Fortunato, S. Community detection in graphs. *Phys. Rep.* **2010**, *486*, 75–174, doi:10.1016/j.physrep.2009.11.002. [CrossRef]
2. Sonia, C.; Gilles, C.; Pierre, H.; Sylvain, P.; Alberto, C. Finding communities in networks in the strong and almost-strong sense. *Phys. Rev. E* **2012**, *85*, doi:10.1103/PhysRevE.85.046113. [CrossRef]
3. Zarandi, F.D.; Rafsanjani, M.K. Community detection in complex networks using structural similarity. *Phys. A* **2018**, *503*, 882–891, doi:10.1016/j.physa.2018.02.212. [CrossRef]
4. Girvan, M.; Newman, M.E. Community structure in social and biological networks. *Proc. Natl. Acad. Sci. USA* **2002**, *99*, 7821–7826, doi:10.1073/pnas.122653799. [CrossRef] [PubMed]
5. Newman, M.E.J.; Girvan, M. Finding and evaluating community structure in networks. *Phys. Rev. E* **2004**, *69*, 026113, doi:10.1103/PhysRevE.69.026113. [CrossRef] [PubMed]
6. Hwang, W.; Cho, Y.; Zhang, A.; Ramanathan, M. Bridging centrality: Identifying bridging nodes in scale-free networks. In Proceedings of the KDD-06, Philadelphia, PA, USA, 20–23 August 2006.
7. Hwang, W.; Ramanathan, M.; Kim, T.; Zhang, A. Bridging centrality: Graph mining from element level to group level. In Proceedings of the 14th ACM SIGKDD International Conference on KDD, Las Vegas, NV, USA, 24–27 August 2008.
8. Nanda, S.; Kotz, D. Localized bridging centrality. In *Handbook of Optimization in Complex Networks*; Thai, M., Pardalos, P., Eds.; SOIA: New York, NY, USA, 2012; pp. 197–218.
9. Yanqing, H.; Hongbin, C.; Zhang, P.; Menghui, L.; Zengru, D.; Ying, F. Comparative definition of community and corresponding identifying algorithm. *Phys. Rev. E* **2008**, *78*, 026121, doi:10.1103/PhysRevE.78.026121. [CrossRef]
10. Enugala, R.; Rajamani, L.; Ali, K.; Kurapati, S. Community detection in dynamic social networks: A survey. *IJRA* **2015**, *2*, 278–285. [CrossRef]
11. Baruah, A.K.; Bora, T. Bridging centrality: Identifying bridging nodes in transportation networks. *IJANA* **2018**, *9*, 3669–3673.
12. Aloise, D.; Caporossi, G.; Hansen, P.; Liberti, L.; Perron, S.; Ruiz, M. Modularity maximization in networks by variable neighborhood search. In Proceedings of the 10th DIMACS Implementation Challenge Workshop, Atlanta, GA, USA, 13–14 February 2012; p. 113, doi:10.1090/conm/588/11705. [CrossRef]
13. Chen, M.; Kuzmin, K.; Szymanski, B.K. Community detection via maximization of modularity and its variants. *IEEE Trans. Comp. Soc. Syst.* **2014**, *1*, 46–65, doi:10.1109/TCSS.2014.2307458. [CrossRef]
14. Greeshma, V.; Vani, K.S. Community detection in networks using page rank vectors. *IJBB* **2015**, *5*, doi:10.5121/ijbb.2015.5401. [CrossRef]
15. Scripps, J.; Tan, P. Clustering in the presence of bridge-nodes. In Proceedings of the 2006 SIAM International Conference on Data Mining, Bethesda, MD, USA, 20–22 April 2006, doi:10.1137/1.9781611972764.24. [CrossRef]
16. Saoud, B.; Moussaoui, A. Node similarity and modularity for finding communities in networks. *Phys. A* **2018**, *492*, 1958–1966, doi:10.1016/j.physa.2017.11.110. [CrossRef]
17. De Montgolfier, F.; Soto, M.; Viennot, L. Asymptotic modularity of some graph classes. In *Algorithms and Computation*; Asano, S.N., Okamoto, Y., Watanabe, O., Eds.; Springer: Berlin/Heidelberg, Germany, 2011; pp. 435–444.
18. Chen, M.; Kuzmin, K.; Szymanski, B.K. Extension of modularity density for overlapping community structure. In Proceedings of the IEEE/ACM ASONAM, Beijing, China, 17–20 August 2014; pp. 856–863, doi:10.1109/ASONAM.2014.6921686. [CrossRef]
19. Yuan, C.; Chai, Y.; Wei, S.B. Feature analysis and modeling of the network community structure. *CTP* **2012**, *58*, 604–612, doi:10.1088/0253-6102/58/4/27. [CrossRef]
20. Jiang, Y.; Jia, C.; Yu, J. An efficient community detection method based on rank centrality. *Phys. A* **2013**, *392*, 2182–2194, doi:10.1016/j.physa.2012.12.013. [CrossRef]

21. Zalik, K.R.; Zalik, A.B. Framework for detecting communities of unbalanced sizes in networks. *Phys. A* **2018**, *490*, 24–37, doi:10.1016/j.physa.2017.07.028. [CrossRef]

22. Ahn, Y.Y.; Bagrow, J.P.; Lehmann, S. Link communities reveal multiscale complexity in networks. *Nature* **2010**, *466*, 761–764, doi:10.1038/nature09182. [CrossRef]

23. We use interchangeably disjoint nodes for cluster nodes and disjunct nodes for isolated or neutral nodes. In this context, disjunct nodes refer to nodes that do not belong to any communities after network divisions. They appear to be neutral in adhering to clusters or communities. What we refer to as disjunct nodes in this paper is quite different from singleton nodes with degree value of 1.

24. Peel, L.; Larremore, D.B.; Clauset, A. The ground truth about metadata and community detection in networks. *Sci. Adv.* **2017**, *3*, doi:10.1126/sciadv.1602548. [CrossRef]

25. Newman, M. Fast algorithm for detecting community structure in networks. *Phys. Rev. E Stat. Nonlinear Soft. Matter Phys.* **2004**, *69*, 066133, doi:10.1103/PhysRevE.69.066133. [CrossRef]

26. Clauset, A.; Newman, M.E.J.; Moore, C. Finding community structure in very large networks. *Phys. Rev. E* **2004**, *70*, 066111, doi:10.1103/PhysRevE.70.066111. [CrossRef]

27. Blondel, V.D.; Guillaume, J.L.; Lambiotte, R.; Lefebvre, E. Fast unfolding of communities in large networks. *J. Stat. Mech. Theory Exp.* **2008**, *2008*, P10008, doi:10.1088/1742-5468/2008/10/p10008. [CrossRef]

28. Raghavan, U.N.; Albert, R.; Kumara, S. Near linear time algorithm to detect community structures in large-scale networks. *Phys. Rev. E* **2007**, *76*, doi:10.1103/physreve.76.036106. [CrossRef]

29. Shi, J.; Malik, J. Normalized cuts and image segmentation. *IEEE Trans. Pattern Anal. Mach. Intell.* **2000**, *22*, 888–905, doi:10.1109/34.868688. [CrossRef]

30. Ng, A.Y.; Jordan, M.I.; Weiss, Y. On Spectral Clustering: Analysis and an Algorithm. In Proceedings of the 14th International Conference on Neural Information Processing Systems: Natural and Synthetic, Vancouver, BC, Canada, 3–8 December 2001; MIT Press: Cambridge, MA, USA, 2001; pp. 849–856.

31. Javed, M.A.; Younis, M.S.; Latif, S.; Qadir, J.; Baig, A. Community detection in networks: A multidisciplinary review. *J. Netw. Comput. Appl.* **2018**, *108*, 87–111, doi:10.1016/j.jnca.2018.02.011. [CrossRef]

32. Malliaros, F.; Vazirgiannis, M. Clustering and Community Detection in Directed Networks: A Survey. *Phys. Rep.* **2013**, *533*, doi:10.1016/j.physrep.2013.08.002. [CrossRef]

33. The bridging centrality of a node is the product of the betweeness centrality of the node and its bridging coefficient [6,7].

34. Radicchi, F.; Castellano, C.; Cecconi, F.; Loreto, V.; Parisi, D. Defining and identifying communities in networks. *Proc. Natl. Acad. Sci. USA* **2004**, *101*, 2658–2663, doi:10.1073/pnas.0400054101. [CrossRef]

35. Freeman, L. A set of measures of centrality based on betweenness. *Sociometry* **1977**, *40*, 35–41, doi:10.2307/3033543. [CrossRef]

36. Hagberg, A.A.; Schult, D.A.; Swart, P.J. Exploring network structure, dynamics, and function using networkx. In Proceedings of the 7th Python in Science Conference, Pasadena, CA, USA, 19–24 August 2008; Varoquaux, T.V., Millman, J., Eds.; Pasadena: California, CA, USA, 2008; pp. 11–15.

37. Oliphant, T.E. A Guide to NumPy. 2006. Available online: https://www.scipy.org/citing.html (accessed on 3 December 2019).

38. Walt, S.V.; Colbert, S.C.; Varoquaux, G. The numpy array: A structure for efficient numerical computation. *MCSE* **2011**, *13*, 22, doi:10.1109/MCSE.2011.37. [CrossRef]

39. Hunter, J.D. Matplotlib: A 2D graphics environment. *MCSE* **2007**, *9*, 90–95, doi:10.1109/MCSE.2007.55. [CrossRef]

40. Jones, E.; Oliphant, E.; Peterson, P. Scipy: Open Source Scientific Tools for Python. Available online: https://www.bibsonomy.org/bibtex/24b71448b262807648d60582c036b8e02/neurokernel (accessed on 29 November 2019).

41. Lancichinetti, A.; Fortunato, S.; Radicchi, F. Benchmark graphs for testing community detection algorithms. *Phys. Rev. E* **2008**, *78*, 046110, doi:10.1103/PhysRevE.78.046110. [CrossRef]

42. Zachary, W. An information flow model for conflict and fission in small groups. *JAR* **1976**, *33*, 473, doi:10.1086/jar.33.4.3629752. [CrossRef]

43. Lusseau, D.; Schneider, K.; Boisseau, O.J.; Haase, P.; Slooten, E.; Dawson, S.M. The bottlenose dolphin community of Doubtful Sound features a large proportion of long-lasting associations. *Behav. Ecol. Sociobiol.* **2003**, *54*, 396–405, doi:10.1007/s00265-003-0651-y. [CrossRef]

44. Krebs, V. *Krebs Amazon Political Books Dataset*; Unpublished work, 2019.

45. Available online: http://www-personal.umich.edu/~mejn/netdata/football.zip (accessed on 3 December 2019).
46. Yin, H.; Benson, A.; Leskovec, J.; Gleich, D. Local Higher-Order Graph Clustering. In Proceedings of the 23rd ACM SIGKDD International Conference on Knowledge Discovery and Data Mining, Halifax, NS, Canada, 13–17 August 2017; pp. 555–564.
47. Leskovec, J.; Kleinberg, J.; Faloutsos, C. Graph Evolution: Densification and Shrinking Diameters. *arXiv* **2007**, arXiv:physics/0603229.
48. Kovacs, I.; Barabasi, A.L. Destruction Perfected. *Nature* **2015**, *524*, 38–39. [CrossRef] [PubMed]
49. Akabane,A.T.; Immich, R.; Pazzi, R.W.; Madeira, E.R.M.; Villas, L.A. Distributed Egocentric Betweenness Measure as a Vehicle Selection Mechanism in VANETs: A Performance Evaluation Study. *Sensors* **2018**, *18*, 2731, doi:10.3390/s18082731. [CrossRef] [PubMed]
50. Butcher, N. Jaccard Coefficients. Available online: https://www3.nd.edu/~kogge/courses/cse60742-Fall2018/Public/StudentWork/KernelPaperFinal/jaccard-butcher3.pdf (accessed on 29 November 2019).

*Article*

# Predicting the Evolution of Physics Research from a Complex Network Perspective

**Wenyuan Liu** [1,2,*,†], **Stanisław Saganowski** [3,*,†], **Przemysław Kazienko** [3] **and Siew Ann Cheong** [1,2]

[1] School of Physical and Mathematical Sciences, Nanyang Technological University, 21 Nanyang Link, Singapore 637371, Singapore; cheongsa@ntu.edu.sg
[2] Complexity Institute, Nanyang Technological University, 61 Nanyang Drive, Singapore 637335, Singapore
[3] Department of Computational Intelligence, Faculty of Computer Science and Management, Wrocław University of Science and Technology, Ignacego Łukasiewicza 5, 50-371 Wrocław, Poland; kazienko@pwr.edu.pl
* Correspondence: wenyuan.liu@ntu.edu.sg (W.L.); stanislaw.saganowski@pwr.edu.pl (S.S.)
† These authors contributed equally to this work.

Received: 13 October 2019; Accepted: 19 November 2019; Published: 26 November 2019

**Abstract:** The advancement of science, as outlined by Popper and Kuhn, is largely qualitative, but with bibliometric data, it is possible and desirable to develop a quantitative picture of scientific progress. Furthermore, it is also important to allocate finite resources to research topics that have the growth potential to accelerate the process from scientific breakthroughs to technological innovations. In this paper, we address this problem of quantitative knowledge evolution by analyzing the APS data sets from 1981 to 2010. We build the bibliographic coupling and co-citation networks, use the Louvain method to detect topical clusters (TCs) in each year, measure the similarity of TCs in consecutive years, and visualize the results as alluvial diagrams. Having the predictive features describing a given TC and its known evolution in the next year, we can train a machine learning model to predict future changes of TCs, i.e., their continuing, dissolving, merging, and splitting. We found the number of papers from certain journals, the degree, closeness, and betweenness to be the most predictive features. Additionally, betweenness increased significantly for merging events and decreased significantly for splitting events. Our results represent the first step from a descriptive understanding of the science of science (SciSci), towards one that is ultimately prescriptive.

**Keywords:** SciSci; knowledge evolution; machine learning

## 1. Introduction

We all become scientists because we want to create an impact and make a difference to the lives of those around us and also to the many generations that are to come. We all strive to make choices in the problems we study, but not all choices lead to breakthroughs. There is actually much more about scientific breakthroughs that we can try to understand. For one, science is an ecosystem of scholars, ideas, and papers published. In this ecosystem, scientists can form strongly interacting groups over a particular period to solve specific problems, but later drift apart as their interests diverge, or due to the availability or paucity of funds, or other factors. The evolution of these problem driven groups is more or less completely documented by the papers published as outcomes of their research. By analyzing groups of closely related papers, researchers could extract rich information about knowledge processes [1–4]. The potential to map scientific progress using publication data has attracted enormous interest recently [5–7]. However, compared to the study of science at the level of individual papers [8–10] and at the level of the whole citation network [11–15], where much work has already been done, the research on science at the community level is still limited [1,3,16,17].

In a recent paper, Liu et al. demonstrated the utility of visualizing and analyzing scientific knowledge evolution for physics at the aggregated mesoscale through the use of alluvial diagrams [3]. In this picture, papers are clustered into groups (or communities), and these groups can grow or shrink, merge or split, new groups may arise, while the others may dissolve. This shares a very strong parallel with what some researchers discovered in social group dynamics [18]. More importantly, many breakthroughs were made by scientists absorbing knowledge from other fields, often in a very short time. On the alluvial diagrams, these knowledge transformations manifest themselves as merging and splitting events. Clearly, funding agencies, universities, and research institutes would want to promote growing research fields, and particularly those where breakthroughs are imminent. This is why it is important to be able to predict future events. Liu et al. [3] attempted this in their paper by analyzing the correlation between event types and several network metrics. Unfortunately, such predictions are very noisy. While merging events are highly correlated with interconnections between communities, the correlation between splitting events and the internal structure of communities are much more complex; besides, the predictions of forming, dissolving, growing, and shrinking were not considered at all.

Given the recent successes in the area of machine learning and artificial intelligence in a variety of prediction problems [19,20], as well as having developed and validated a general framework to predict social group evolution in Saganowski et al. [21], we decided to utilize machine learning techniques to fill the gap in predicting scientific knowledge events [22–24]. The overall idea behind the group evolution prediction (GEP) method is to build a classification model trained with historical observations in order to predict the future group changes based on their current characteristics, such as size, density, the average degree of nodes, etc. A single historical observation consists of a set of features describing the group at a given point in time and an event type that this group just experienced. The profile of the group may reflect its structure (e.g., density), dynamics (e.g., average age of its member articles), or context (e.g., the journals from which the articles (group members) come). In total, we used over 100 features, some of which were already known to the literature, whereas the others focusing on the dynamics and context were the new, unique features proposed in this paper. Indeed, when we ranked the most valuable features contributing to the successful prediction of knowledge evolution events, the new features were among the best ones. In order to be able to perform the prediction of future group changes, we have to track and learn the model on the historical cases. For that purpose, the group changes from the past (historical evolution) need to be defined and discovered using the methods successfully applied in the social network analysis field, e.g., the GED method [25], Tajeuna et al.'s method [26], or others [27]. Most of the methods consider the similarity between the groups in the consecutive time windows as a major factor to match similar groups and further to identify the evolution event type between them. In our work, we apply the GED method, which facilitates both the group quantity (the number of common members) and the group quality (the importance of common members), in order to match related groups. This allows us to enrich the co-citation evolution network with information about member relations, which is depicted in the social position measure [28].

In this study, we extract groups (topical clusters (TCs)) from the bibliographic coupling networks (BCNs) and independently from the co-citation networks (CNs) for the period 1981–2010. Next, the GED method is utilized to label four types of evolution events (changes of TCs): continuing, dissolving, merging, and splitting. Then, we use an auto-adaptive mechanism to find the most predictive machine learning model, together with its parameters for each network. Additionally, two scenarios were considered for each network: when the number of events of each kind is imbalanced (the original case) and balanced by equal sampling. In general, the prediction quality was satisfactory and good for all event types, with F-measures substantially exceeding 0.5. Such values are significantly greater than the baseline F-measures of 0.14–0.21 for both networks. The feature ranking tells us that the most informative features are context based like the number of PRE, PRB, and RMP papers belonging to the group and the structural features like the degree, closeness, and betweenness. While looking more carefully at the betweenness of papers from two merging TCs, we found significantly higher betweenness for papers that are linked across these two TCs than those connected inside the TCs. No such enhancement in betweenness was found for continuing TCs, while a significant decrease in average betweenness was found for splitting TCs. In summary, our findings suggest that evolutionary events in the landscape of physics research can be predicted accurately using various machine learning models, and understanding this predictive power in terms of important features is a worthwhile future research direction.

## 2. Materials and Methods

The entire analytical process consists of several steps that are primarily defined by the group evolution prediction (GEP) framework. First, the bibliographic coupling network (BCN) and co-citation network (CN) are extracted from the references placed in the papers from a given time window (see Figure 2), and this is carried out separately for each period.

As a result, we get a time series of BCNs/CNs. Next, paper groups called topical clusters (TCs) are extracted using the Louvain clustering methods, independently for each BCN/CN in the time series. Each group is described by the set of predictive features. Having TCs for consecutive periods, we were able to identify changes in TC evolution using the group evolution discovery (GED) method that appropriately labels the TC changes; see below.

Independently, the features' ranking and its validation were performed to find the most valuable TC measures. Based on this ranking, a structural measure node betweenness was selected for the more in-depth studies as the early signal for splitting or merging. The above-mentioned steps are summarized and visualized in Figure 1.

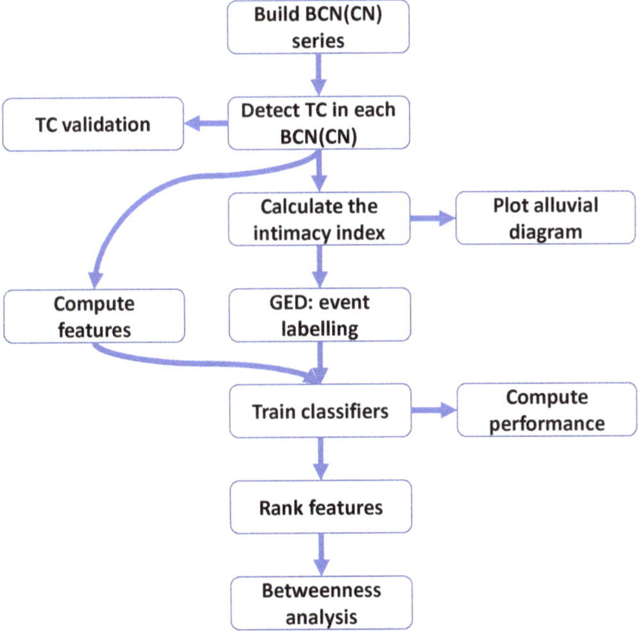


**Figure 1.** The workflow of this paper. TC, topical cluster.

### 2.1. GEP Method

The group evolution prediction (GEP) method is the first generic approach for the prediction of the evolution of groups [21]; in our case, groups correspond to TCs. The GEP process consists of six main steps: (1) time window definition, (2) temporal network creation, (3) group detection, (4) group evolution tracking, (5) evolution chain identification and feature calculation, and (6) classification using machine learning techniques. Thanks to its adaptable character, we were able to apply it to the BCN and CN differently. For the group (TC) detection in both networks, we applied the Louvain method [29]. The group evolution tracking was performed with the GED method (see below), but we used different similarity measures for each network BCN and CN (see below). The set of features describing the group at a given time window was adjusted to our networks, as some of the features defined in the GEP method were not applicable in our case. We also introduced some new, dedicated measures appropriate for bibliographical data; see SI (Supplementary Information) for the complete list. Finally, we applied the Auto-WEKA tool to find the best predictive model and its parameters from a wide range of all possible solutions. The commonly known average F-measure was used as a prediction performance measure. The stratified sampling and 10-fold cross-validation techniques were used to validate the model. The feature selection technique was applied to prevent model overfitting.

### 2.2. Bibliographic Coupling Network and Co-Citation Network

In the BCN and CN, nodes represent papers, and undirected but weighted edges denote the bibliographic coupling strengths and co-citation strengths, respectively. That is, if two papers share $w$ common references, the BCN edge between them would have a weight of $w$. For example, Papers 1 and 2 in Figure 2 share three citations: A, B, and C, whereas Papers 3 and 4 commonly cite only one paper: E. On the other hand, if two papers are cited together by $w'$ papers, the edge between them in the CN receives weight $w'$. Papers A and B are cited together by two other papers: 1 and 2, but Papers B and C by three, i.e., additionally by Paper 3. Both BCN and CN are temporal networks, in which the nodes are all papers published (BCN) or papers cited (CN) within a specific time window. We assumed that the reasonable time window for bibliographical data was one year to facilitate the analysis of changes in scientific knowledge, i.e., changes in topical clusters year-by-year. For the BCN, only the giant component, which in most cases occupied 99% of the whole BCN, would be considered for the TC detection and evolution analysis. For the CN, we did not use all papers cited in the given time window because most of them were cited only a small number of times, and thus, they had little influence on the broader knowledge evolution. Therefore, we ranked all available $N$ papers $p_1, p_2, \ldots, p_N$ in descending order by the number of times they were cited in this time window (year): $f_1, f_2, \ldots, f_N, f_1 \geq f_2 \geq \ldots \geq f_N$. Next, we chose the top $n$ papers $p_1, p_2, \ldots, p_n$ that totally gathered $\frac{1}{4}$ of all citations, i.e., such that $n < N$ was the smallest integer to satisfy $\sum_{i=1}^{n} f_i \geq \frac{1}{4} \sum_{j=1}^{N} f_j$. The data we used in this paper were the APSdataset, consisting of about half a million publications between 1893 and 2013 and six million citation relations among them [30].

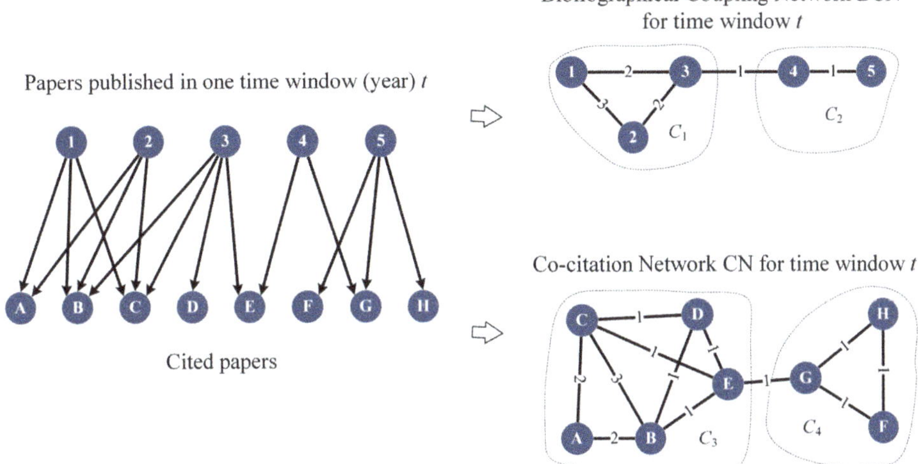



**Figure 2.** The process of building a bibliographical coupling network (BCN) and co-citation network (CN) from the citation bipartite network for a given period: year $t$. Both BCN and CN are undirected and weighted; the weights denote the number of shared citations (BCN) or co-citing papers (CN). Separate topical clusters are extracted for BCN ($C_1$, $C_2$) and CN ($C_3$, $C_4$). Nodes with numbers are papers from a given period being considered, and nodes with letters are their references.

## 2.3. Community Detection and Validation

There are many approaches to community detection, including modularity based algorithms, hierarchical clustering, non-negative matrix factorization, principal component analysis, link partitioning, and others [31]. In this work, we used the Louvain method [29] to extract community structure from BCNs and CN. The community partitions we obtained in BCNs and CNs had considerably high modularities (about 0.75), which suggest clear and robust community structures. Furthermore, a different community detection algorithm was also used, i.e., Infomap, which gave very similar results as the Louvain method. For instance, the normalized mutual information between community partitions of BCN in 1991 from the Louvain method and Infomap algorithm was 0.66, which confirmed the existence and robustness of community structure in BCNs and CNs. In this study, we only used the Louvain method; however, the results were similar if we switched to Infomap or other community detection algorithms.

To verify that the communities extracted were really focused on closely related questions, we checked the Physics and Astronomy Classification Scheme (PACS) numbers of members of the communities. This cross-validation was independent of network structure; therefore, it provided more evidence for the robustness of TCs. In our study, we only used the first two digits of the PACS numbers, as a balance between accuracy and coverage. To test whether the PACS numbers appearing in the communities could have occurred due to randomness, we chose one year $t$, built its BCN, extracting the community structure with sizes $\{s_1, s_2, \ldots, s_n\}$, and then randomly assigned papers in year $t$ into $n$ pseudo-communities of the same sizes, to remove any potential size effects. The results showed that the papers in the same community significantly focused on a small number of PACS numbers compared with a null model; see Figure 3. Interested readers can get more details on the systematic validation of TC in Liu et al. [3].

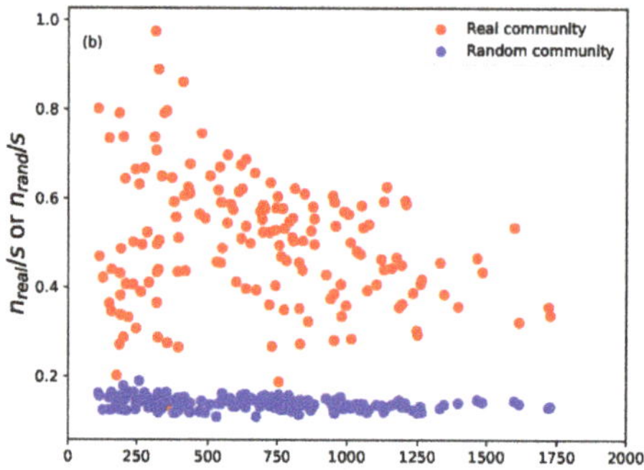



**Figure 3.** Comparison of PACS homogeneity between real BCN TCs, which are between 1991 and 2000 and have more than 100 papers, and their corresponding random collections. The fraction of the largest subset of papers sharing at least one PACS number as a function of *s* for real communities in the BCN and random collections. For clarity, the error bars are not shown in the figures because they are smaller than the marker size.

*2.4. Intimacy Indices*

To analyze the evolution of TCs, we needed to match them from consecutive years. The set of cited papers to a large extent overlapped year-by-year, so for the CN, we could use the regular approach proposed together with the GED method; see below and Brodka et al. [25]. For BCN, however, there was no overlap at all between papers published in the successive years because every paper could be published only once and in only one year. Even if we did not have the corresponding papers in TCs from two BCNs, i.e., two years, the papers' references overlapped each another. Therefore, we could measure the similarity of their reference pools to reflect their inheritance. For that purpose, we introduced the forward intimacy index and backward intimacy index in Liu et al. [3]. The idea behind intimacy indices is that the references related to a particular topic change gradually. The forward intimacy index $I_{mn}^f$ and the backward intimacy index $I_{mn}^b$ between TCs $C_m^t$ in year $t$ and $C_n^{t+1}$ in year $t+1$ are defined as follows:

$$I_{mn}^f = \sum_i \frac{N\left(R_i, \mathcal{R}_n^{t+1}\right)}{N\left(R_i, \mathcal{R}^{t+1}\right)} \frac{N\left(R_i, \mathcal{R}_m^t\right)}{L\left(\mathcal{R}_m^t\right)},$$

$$I_{mn}^b = \sum_i \frac{N\left(R_i, \mathcal{R}_m^t\right)}{N\left(R_i, \mathcal{R}^t\right)} \frac{N\left(R_i, \mathcal{R}_n^{t+1}\right)}{L\left(\mathcal{R}_n^{t+1}\right)}. \tag{1}$$

Here, the TCs at $t$ and $t+1$ are $\mathcal{C}^t = \left\{C_1^t, ..., C_m^t, ..., C_u^t\right\}$ and $\mathcal{C}^{t+1} = \left\{C_1^{t+1}, ..., C_n^{t+1}, ..., C_v^{t+1}\right\}$, and we denote the references cited by papers in $C_m^t$ and $C_n^{t+1}$ as $\mathcal{R}_m^t = \mathcal{R}(C_m^t) = \left[R_{m1}, ..., R_{mp}\right]$ and $\mathcal{R}_n^{t+1} = \mathcal{R}(C_n^{t+1}) = \left[R_{n1}, ..., R_{nq}\right]$; $\mathcal{R}^t = \left\{\mathcal{R}_1^t, ..., \mathcal{R}_m^t, ...\right\}$. $N(element, list)$ is the number of times *element* occurs in *list*, and $L(list)$ is the length of *list*. For more details and examples of intimacy indices, please refer to Liu et al. [3].

*2.5. GED Method*

The group evolution discovery (GED) method [25] was used for tracking group evolution for historical cases to learn the classifier and for testing cases to validate classification results. The GED method makes use of the similarity between groups in the following years, as well as their sizes to label one of six event types: continuing, dissolving, merging, splitting, growing, and shrinking. However, we have adapted the GED method to label only four types of events: continuing, dissolving, merging, and splitting, as these were the most important to us. The other two (growing and shrinking) were covered by continuing. In general, the GED method allowed us to use various metrics as a similarity measure between groups. Therefore, the intimacy indices defined in Equation (1) were used for the BCN to match similar groups in the consecutive time windows. However, the original GED inclusion measures were used for the CN. This means that the similarity between two groups from two successive time windows was reflected by the inclusion measure, which was calculated for two scenarios: inclusion $I(C_n^t, C_m^{t+1})$ of a group $C_n^t$ from time window $t$ in another group $C_m^{t+1}$ from time window $t+1$ (forward; Equation (2)) and inclusion $I(C_m^{t+1}, C_n^t)$ of this second group $C_m^{t+1}$ from $t+1$ in the first group $C_n^t$ from $t$ (backward; Equation (3)). The inclusion measure makes use of the social position $SP(p)$, which is a kind of weighted PageRank. It denotes the importance of paper $p$ being cited among all other papers [28]. The inclusions for CN are defined as follows:

$$I(C_n^t, C_m^{t+1}) = \overbrace{\frac{\|C_n^t \cap C_m^{t+1}\|}{\|C_n^t\|}}^{\text{group quantity}} \cdot \underbrace{\frac{\sum\limits_{p \in (C_n^t \cap C_m^{t+1})} SP(p)}{\sum\limits_{p \in (C_n^t)} SP(p)}}_{\text{group quality}} \cdot 100\%, \tag{2}$$

$$I(C_m^{t+1}, C_n^t) = \overbrace{\frac{\|C_m^{t+1} \cap C_n^t\|}{\|C_m^{t+1}\|}}^{\text{group quantity}} \cdot \underbrace{\frac{\sum\limits_{p \in (C_m^{t+1} \cap C_n^t)} SP(p)}{\sum\limits_{p \in (C_m^{t+1})} SP(p)}}_{\text{group quality}} \cdot 100\%. \tag{3}$$

If both inclusions (CN) or both intimacy indices (BCN) are greater than the percentage thresholds alpha and beta (the only parameters in this method), the method labels the event continuing. If at least one inclusion or one intimacy index exceeds one of the thresholds, the splitting and merging events considered, the proper event is assigned depending on the number of similar groups in $t$ and $t+1$. If both inclusions or both intimacy indexes are below the thresholds, i.e., the group has no corresponding group in the next time window, the dissolving event is assigned.

### 2.6. Feature Ranking

Rankings of the most prominent features were obtained by repeating the feature selection 1000 times using a basic evolutionary algorithm [32], as proposed in Saganowski et al. [21]. The alternative approach would be to use the forward or backward feature elimination technique, but our own implementation gave us more flexibility and control over the experiment. The rankings were received for the 30 year span (1981–2010). Next, only the top 10 features were selected to described TCs in two additional years (2010–2012) and predict TC evolution. The results revealed the superiority of feature selection compared to the raw approach with all features' engagement.

## 3. Results

### 3.1. Physics Research Evolution for 1981–2010

We begin with studying how scientific knowledge evolved in terms of communities of research papers and how these communities changed over time. There were several studies on the evolution of knowledge within the set of whole journals [2], which was considered as the analysis on the macroscopic level. Furthermore, some research was carried out for the collection of papers, usually involving some subjective criterion provided by the authors, e.g., only papers cited at least 100 times [1]. As a result, they focused only on a small subset: the most prominent, frequently cited papers, which do not represent the whole diverse domain knowledge. This kind of analysis was considered as microscopic. In our approach, we assumed that the most informative way was to analyze neither the entire journal, nor the most cited papers, but whole communities of closely related papers. These communities emerged naturally since they shared the same citation patterns. The analysis at such a level provided a better balance between high and low granularity. We called this kind of analysis mesoscopic because it was in between the macroscopic scale of journals and the microscopic scale of individual papers. However, if we performed community detection directly on the citation network, we might end up with communities consisting of both old and recent papers simultaneously. In such a case, it is difficult to interpret how scientific knowledge has evolved from the past to the present. We should be able to explain that such and such communities represent scientific knowledge from an earlier year, whereas the other communities correspond to scientific knowledge from another consecutive year. This enabled us to compare them and to distil a picture of how scientific knowledge has evolved from past to present. It required, however, constructing the networks from research papers that were published in a given year (bibliographic coupling) or papers that were cited in a given year (co-citation). The bibliographic coupling network (BCN) reflected the relation between present publications, while the co-citation network (CN) represented the relation between papers that had a strong influence on recent publications. In this way, we could detect communities over the years and study how they evolved year-by-year; see the Methods Section for details on BCN and CN.

After building BCN and CN, the Louvain method was used to extract the community structures. By checking the Physics and Astronomy Classification Scheme (PACS) numbers of the papers in these communities, we showed that the BCN communities were meaningful and reflected the real structure of the scientific communities. These results suggested that the papers in the same community were very similar to each other in terms of research topic. These results suggest papers in the same community had high similarity to each other in terms of research topic. The method and results of the validation are briefly reviewed in the Methods Section; the interested reader is referred to Liu et al. for details [3]. For the CN communities, this validation is tricky because of two problems: (i) the old physics review papers had no PACS numbers, and (ii) PACS was revised several times, so the same numbers in different versions can potentially refer to different topics, or the same topics are referred to by different numbers in different versions. Nevertheless, systematic validation seemed to be impossible, although a quick check on some CN communities after 2010 suggested that the CN community structure also reliably reflected the actual scientific community. We refer to these validated units of knowledge evolution as topical clusters (TCs) in this paper.

In Figure 4, we provide the alluvial diagram that depicts the evolution of TCs within the BCNs for the period between 1981 and 2010. The equivalent alluvial diagram for the CNs is shown in Figure S2 in the SI. In both alluvial diagrams, we visualize the sequences of TCs, their inheritance relations, which can be intimacy indices (for the BCN communities), a fraction of common members or inclusion measures (for the CN communities), and the evolution processes they undergo; see the Methods Section for more details. The events (changes) that we can discern from the alluvial diagram (shown in Figure 4) are analogous to those recognized in social group evolution [18]. They represent forming, dissolving, growing, shrinking, merging, and splitting. We found in Liu et al. that the prediction of such events was hard since the correlation between them was nonlinear and complex. This challenge is addressed in the following section by tapping into the power of machine learning.

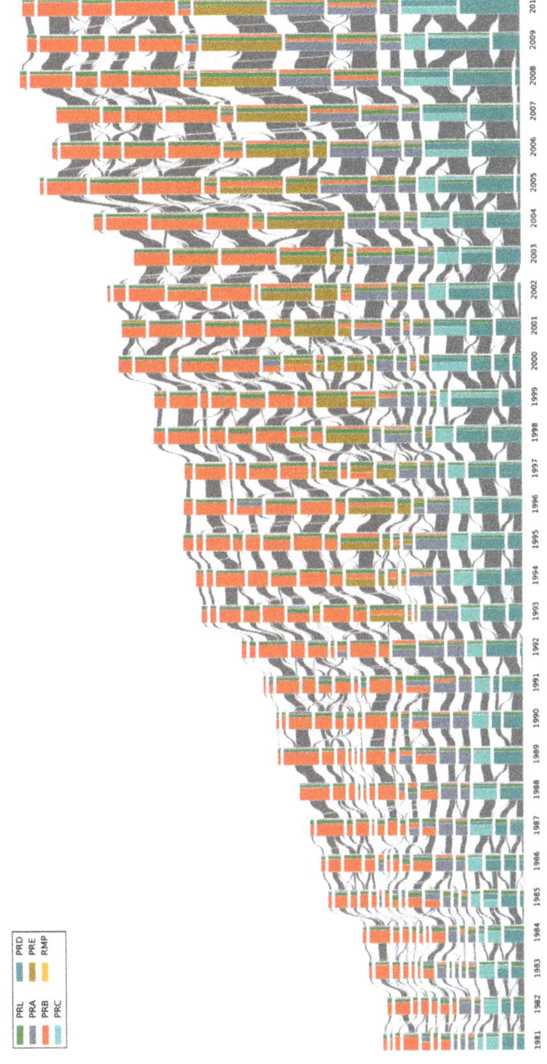



**Figure 4.** The alluvial diagram of APSpapers from 1981 to 2010 for the BCNs. Each block in a column represents a TC, and the height of the block is proportional to the number of papers in the TC. For clarity reason, only TCs comprising more than 100 papers are shown. TCs in successive years are connected by streams whose widths at the left and right ends are proportional to the forward and backward intimacy indices. The colors inside a TC represent the relative contributions from different journals.

### 3.2. Event Labeling

The GED method takes into account the size and the similarity between groups (TCs) in the consecutive time frames in order to label groups' changes (assign event type). There are four events considered in this work:

- Continuing: A research field is said to be continuing when the problems identified and solutions obtained from one year to another are of an incremental nature. It is likely to correspond to the repeated hypothesis testing picture of the progress of science proposed by Karl Popper [33].

Therefore, in the CN, this would appear as a group of papers that are repeatedly cited together year-by-year. In the BCN, this shows up as groups of articles from successive years sharing more or less the same reference list.

- Dissolving: A research field is thought to disappear in the following year if the problems are solved or abandoned, and no new significant work is done after this. For the CN, we will find a group of papers that are cited up to a given year, but receiving very few new citations afterwards. In the BCN, no new relevant papers are published in the field; hence, the reference chain terminates.

- Splitting: A research field splits in the following year, when the community of scientists who used to work on the same problems starts to form two or more sub-communities, which are more and more distant from one another. In terms of the CN, we will find a group of papers that are almost always cited together up till a given year, breaking up into smaller and disjoint groups of papers that are cited together in the next year. In the BCN, we will find the transition between new papers citing a group of older papers to new papers citing only a part of this reference group.

- Merging: Multiple research fields are considered to have merged in the following year when the previously disjoint communities of scientists found a mutual interest in each other's field so that they solve the problems in their own domain using methods from another domain. In the CN, we find previously distinct groups of papers that are cited together by papers published after a given year. In the BCN, newly published papers will form a group commonly citing several previously disjoint groups of older papers.

The GED method has two main parameters (alpha and beta), which are the levels of inclusion that groups in the consecutive years have to cross in order to be considered as matching groups. We applied the GED method with a wide range of these parameters from 5 to 100%. The characteristics of the considered networks required us to set the alpha and beta thresholds to very low values, i.e., 30% for the BCN and 10% for the CN; see SI for more details. In total, we obtained 479 various events for the BCN and 492 events for the CN, which were our observations and the labels in the prediction part of our study. In both networks, the distribution of the events was imbalanced with the continuing event dominating over all other types; see Figure 5(A1,B1).

*3.3. Future Events' Prediction*

The machine learning approach to prediction requires dividing the data into two parts: the training dataset and test dataset. The training data are used to train the classifier, which can then label events in the test data. The labeled values are compared with the event labels, and the prediction performance is calculated. More than 450 observations were used to train the classifiers. Each observation contained 77 normalized features (preselected from the initial 100) divided into three categories: microscopic features (related to nodes in the group, e.g., node degree), mesoscopic features (related to the entire group, e.g., the group size), and macroscopic features (related to the whole network, e.g., network density). Mesoscopic features calculated for individual nodes are commonly aggregated for all nodes from the group, e.g., average node degree or betweenness in the group. See SI for the complete list of features used.To select the best classification algorithm (model) automatically, as well as its hyper-parameter settings to maximize the prediction performance, the Auto-WEKA software package [34] was utilized. For each network, we ran the Auto-WEKA for 48 h, which allowed us to validate nearly 20,000 configurations per network. The metric being maximized was the F-measure, commonly used for multi-class classification. The overall classification quality was calculated as the average F-measure for all event types, treating them as equally important.

The predicted output variable (event labels) had an imbalanced distribution. Commonly, classifiers tend to focus on the dominant event type (class), which is very well predicted, but at the expense of the minority event types. For the imbalanced BCN dataset, the best performance was achieved with the attribute selected classifier (with the SMOas a base classifier), which performed feature selection [35]. The percentage of the correctly classified instances was 80.6%, while the average F-measure was only 0.50 due to classifier focusing on continuing, which was the most frequently occurring event type; see Figure 5A. For this event, the F-measure value was equal to 0.89, and only seven events out of 352 were incorrectly classified. The worst classified was the splitting event, whose F-measure was only 0.11. Most of the splitting events were incorrectly classified as continuing (31 out of 33 events). The second worst was merging, with the F-measure of 0.35. Again, the majority of the merging events were wrongly classified as continuing events: 38 out of 56. Interestingly, the splitting and merging events were never cross-classified mistakenly. For the imbalanced CN dataset, the best performance was achieved with a lazy classifier, which used locally weighted learning [36]. The percentage of the correctly classified instances was 73.3%, while the average F-measure was only 0.53, again due to the classifier concentrating on the dominating continuing event type; see Figure 5B. The F-measure value for the continuing event was only 0.83; however, as many as 50 continuing events (out of 337) were wrongly classified as dissolving. Similar to BCN, many splitting and merging events were incorrectly classified as continuing: 17 out of 22 events and 24 out of 46 events, with the F-measure equal to 0.30 and 0.42, respectively.

By balancing the imbalanced training datasets (i.e., by under-sampling them), we forced the classifiers to pay more attention to the features rather than to the number of occurrences of the particular majority event type. Please note that the test set was untouched, i.e., left imbalanced. As a result of balancing datasets, the previously minor event types (dissolving, merging, and splitting) were predicted much better, but with a significant drop in performance of the continuing event classification. More importantly, by balancing the datasets, we increased the overall prediction quality by over 20%. For the balanced BCN dataset, the best performance was achieved by means of the boosting based classifier AdaBoost with Bayes net as the base model. The percentage of the correctly classified instances was 62.0%, and the average F-measure was 0.61. The biggest sources of errors were merging events, which were wrongly classified as continuing and dissolving, as well as continuing wrongly classified as splitting. The best classified event was dissolving (only four mistakes in 27 classifications; the overall score was 0.79), followed by the splitting event (six mistakes in 27 classifications; overall F-measure of 0.70). For the balanced CN dataset, the attribute selected classifier (with the PART [37] as a base classifier) provided the best results: the percentage of the correctly classified instances was 69.32%, while the average F-measure was 0.69. The dissolving, merging, and splitting events were classified very well with the F-measure values equal to 0.79, 0.82, and 0.75, respectively. Most of the continuing events were wrongly classified as splitting (13 out of 22), which resulted in a lower F-measure value of 0.40.

What is interesting for us to note is that the prediction results for the CN were slightly better than for the BCN. A possible explanation is that for the CN, we used a richer similarity measure containing users' importance information. Thus, the event tracking and, therefore, the ground truth could be more accurate. Overall, the prediction quality expressed by the average F-measure was very good for the imbalanced, as well as for the balanced datasets, as the baseline results obtained with the ZeroR classifier were much worse: F-measure of 0.21 for both BCN and CN imbalanced datasets, 0.18 for the balanced BCN, and 0.14 for the balanced CN. For each dataset, different classifiers turned out to be the best; however, most models were wrapped with the boosting or meta classifiers.

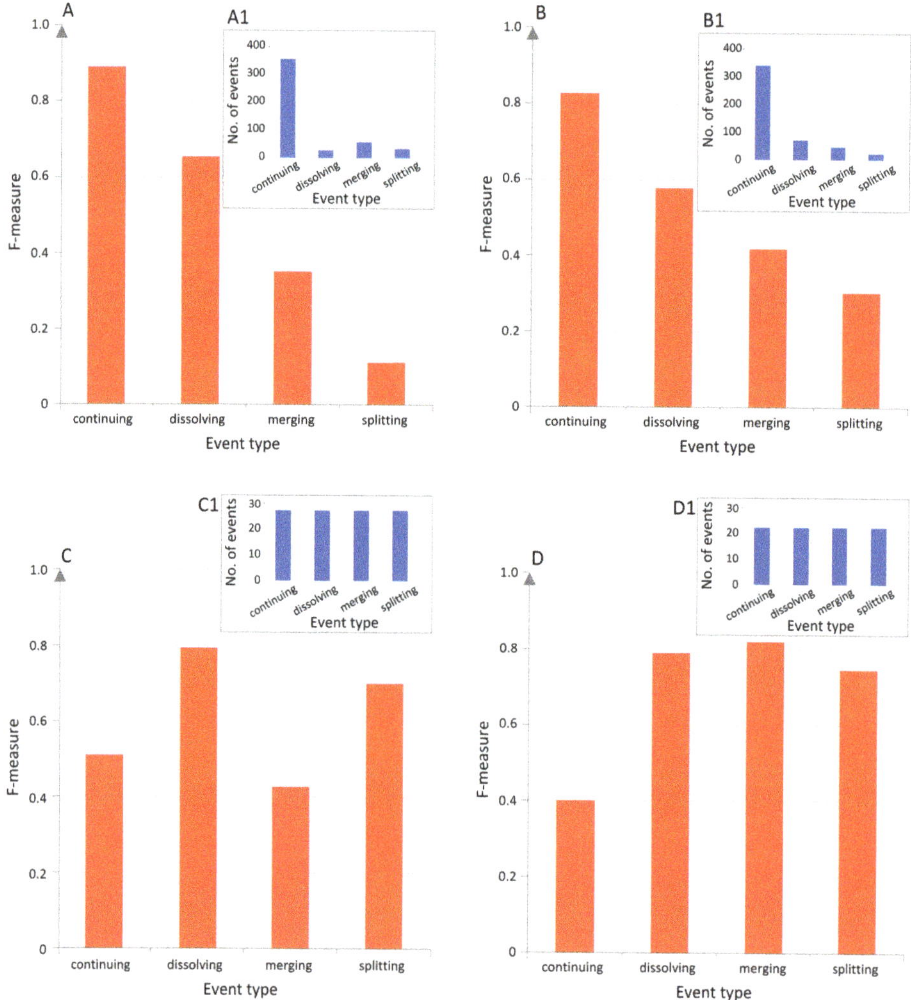

**Figure 5.** The prediction quality of classification results. The F-measure values for the imbalanced BCN (**A**) and CN (**B**) datasets, as well as the balanced BCN (**C**) and CN (**D**) datasets. The distribution of classes in the training sets is provided for each dataset: **A1**, **B1**, **C1**, **D1**, respectively. For the imbalanced datasets, the classifier focused on the dominating continuing event. Balancing the datasets increased the overall prediction quality by over 20%.

### 3.4. Predictive Feature Ranking

The feature selection technique is used in machine learning to find the most informative features, to avoid classifier overfitting, to eliminate (or at least to reduce) the noise in the data, as well as to provide some explanations about phenomena [32]. By repeating the feature selection 1000 times, we obtained 1000 sets of selected features. Next, we calculated how many times each feature was selected, thus receiving the ranking of the most often selected features. For the BCN, the context-based features dominated the ranking. It referred especially to the number of papers from the Physical Review E, Physics Review B, and Physical Review A; see Figure 6A. Besides the context, the network features based on degree, betweenness, size, and closeness measures were most informative, which tells us that the structural properties were as important as context awareness. The context based feature, i.e., the number of papers published in the Review of Modern Physics, was the most often selected for the CN dataset. It was followed by closeness and degree based features in the ranking; see Figure 6B. For both networks, macroscopic features were ranked rather low, which suggests that the general network profile was not very important, perhaps because of the smooth changes in the entire network. Surprisingly, the dynamic features, e.g., related to the average age of references (for BCN) and age of articles (for CN), did not show an informative value and were ranked very low for both networks. The rankings were validated in the additional two years of data available (2010–2012). The prediction was performed twice: (i) using all features and (ii) using the top 10 ranked features only. Selecting only the top 10 features boosted the quality of the prediction by 11% for the CN and by 2% for the BCN, which underlined the necessity of the feature selection process.

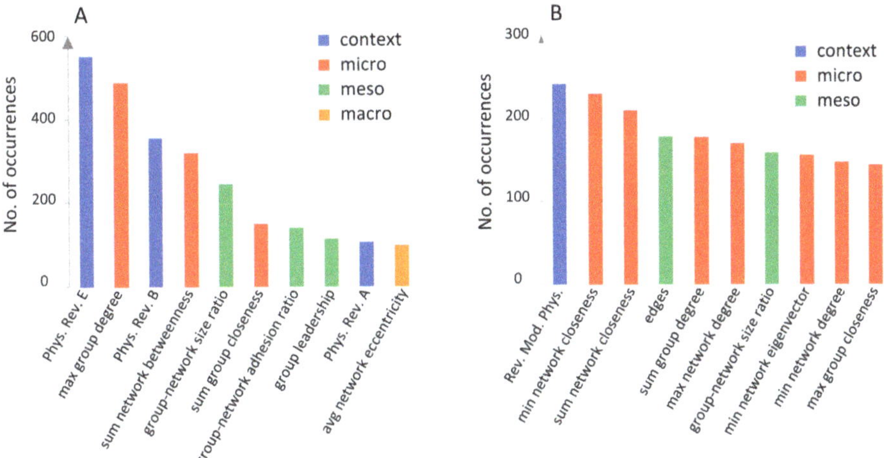

**Figure 6.** Feature ranking. The most frequently selected features in 1000 iterations for the BCN (**A**) and CN (**B**) datasets. The context based features (number of papers published in a given journal) turned out to be the most informative, followed by the microscopic structural measures, especially closeness, degree, and betweenness.

### 3.5. Changes to the Betweenness Distributions Associated with Merging and Splitting Events in BCN

Having the list of best predictive features (Figure 6), we can analyze some of them more in-depth to look for early warning signals. Basically, we believe that scientific knowledge evolves slowly, and this slow evolution drives the evolution of citation patterns. Therefore, there must be specific changes in citation patterns that precede merging and splitting events. Besides the number of PRE papers in a TC, sum_network_betweennessis also a strongly predictive feature; see Figure 6A. This suggests that we should look at the betweenness of papers in the BCN more carefully. The betweenness of the node denotes what percentage of the shortest paths between all pairs of nodes in the network passes a given node. Values of nodes' betweenness can be aggregated (sum, average, max, min) for all nodes in the TC, as we list in Table S1 in SI. However, in this section, we only focus on the distribution of the original node betweenness. Naively, when we considered the part of the BCN adjacency matrix corresponding to two TCs that ultimately merged, we expected to find few links between TCs at first. However, as the number of links between TCs increased over time, the modularity-maximizing Louvain method would eventually merge the two TCs into a single TC. This is shown schematically in Figure 7, where in general, betweenness would increase on average with time as the two TCs merge.

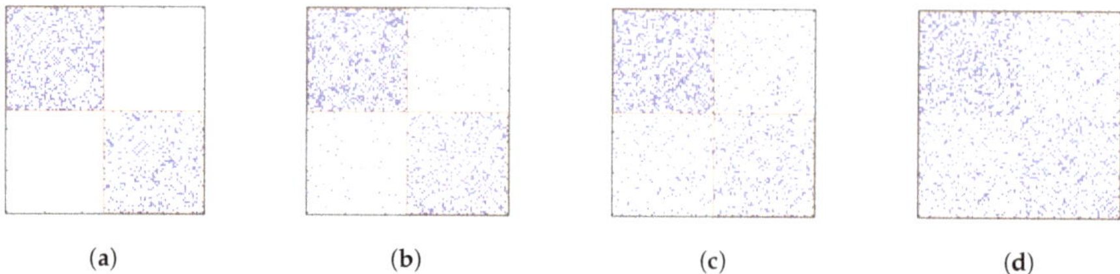

(a)   (b)   (c)   (d)

**Figure 7.** Part of the BCN adjacency matrix for two TCs (red boxes) that ultimately merged. (**a**) No links between the two TCs at first. (**b**) Few links between the two TCs. (**c**) More links between the two TCs. (**d**) Many links between the two TCs, leading to their identification as a single merged TC (big red box) by the Louvain method.

In reality, there are always links between TCs, and the numbers and strengths of these links fluctuate over time. To develop a more quantitative description of the merging events outlined in Figure 4, as well as splitting and continuing events, we focused on five events going from 1999 to 2000, shown in Table 1.

**Table 1.** The five evolution events from 1999 to 2000 in the BCN alluvial diagram Figure 4 that we will study quantitatively. The naming convention for TC is that four digits before '.' is the year of TC, two digits after '.' is the position of the TC in the diagram, starting with 00 for the bottom TC; the one just above bottom is 01, and so on. In the left panel, we highlight the related TCs.

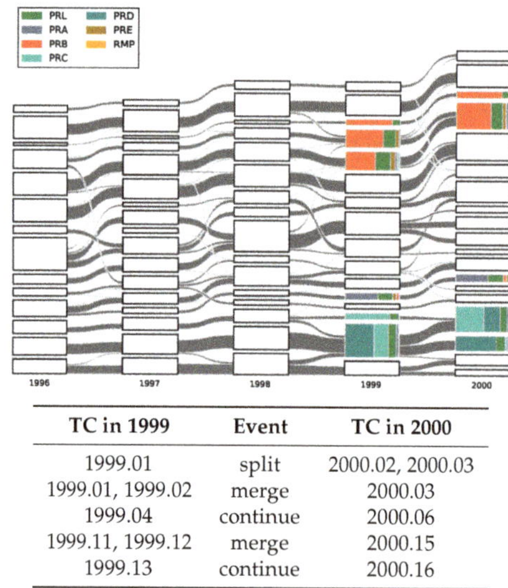

| TC in 1999 | Event | TC in 2000 |
|---|---|---|
| 1999.01 | split | 2000.02, 2000.03 |
| 1999.01, 1999.02 | merge | 2000.03 |
| 1999.04 | continue | 2000.06 |
| 1999.11, 1999.12 | merge | 2000.15 |
| 1999.13 | continue | 2000.16 |

### 3.5.1. **1999.01 + 1999.02 → 2000.03**

Let us consider the part of the BCN associated with the TCs. For example, for 1999.01 and 1999.02, we can see from Figure 8a that connections within 1999.01 and 1999.02 were very dense, but there were also some links between the two TCs. In fact, we found 164 out of 1849 papers in 1999.01 with non-zero bibliographic coupling to 144 papers in 1999.02 (344 papers).

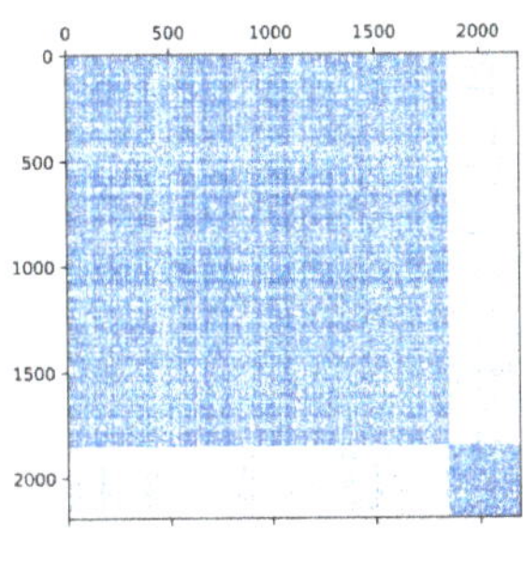 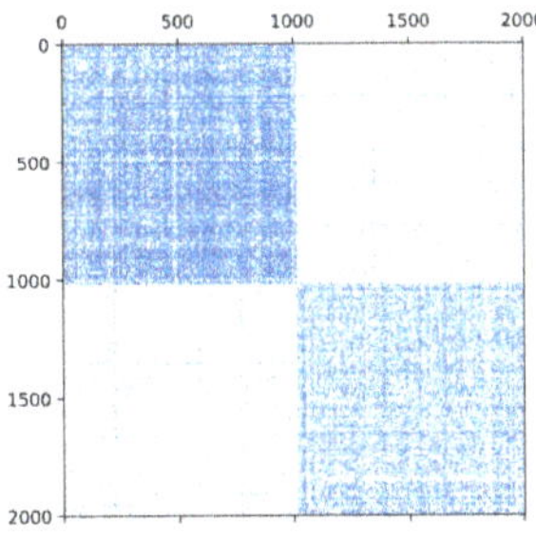

(a)                       (b)

**Figure 8.** (**a**) The adjacency matrix of the BCN associated with the TCs 1999.01 (top dense block) and 1999.02 (bottom dense block). (**b**) The adjacency matrix of the BCN associated with the TCs 1999.11 (top dense block) and 1999.12 (bottom dense block).

The natural question we then ask is: are the betweennesses of the 164 papers in 1999.01 that are coupled to 1999.02 larger, equal to, or smaller than the betweenness of the rest of the 1685 papers in 1999.01 not coupled to 1999.02? Alternatively, if we think of the 164 papers as randomly sampled from the 1849 papers in 1999.01, are we sampling the 164 betweenness in an unbiased fashion? To distinguish the different parts of the TC, we call all papers in 1999.01 that have coupling with papers in 1999.02 as 1999.01*a* and the rest of the papers as 1999.01*b*. For more detail analysis, we will divide 1999.01*a* and 1999.01*b* into 1999.01*aα*, 1999.01*aβ*, 1999.01*bα*, and 1999.01*bβ*. 1999.01*aα* consists of 17 papers in 1999.01*a* that do not have references in common with papers in 1999.01*b*; 1999.01*aβ* consists of 147 papers in 1999.01*a* that have references in common with papers in 1999.01*b*; 1999.01*bα* are 907 papers in 1999.01*b* that have references in common with papers in 1999.01*a*; and 1999.01*bβ* represents 778 papers in 1999.01*b* that do not have references in common with papers in 1999.01*a*.

In Table 2, we show the 25th, 50th, and 75th percentiles of the papers in these smaller groups, compared to those of the 1849 papers in 1999.01 and the 344 papers in 1999.02. As we can see, the 25th, 50th, and 75th percentile betweenness in the connecting parts (1999.01*a* and 1999.02*a*) were all higher than the 25th, 50th, and 75th percentile betweenness in the non-connecting parts (1999.01*b* and 1999.02*b*). More importantly, these percentile betweenness were higher than the 25th, 50th, and 75th percentile betweenness of the TCs 1999.01 and 1999.02 themselves. To test how significant these quartiles were in 1999.01*a*, we randomly sampled 164 betweenness values from 1999.01 $10^6$ times and measured the quartiles of these samples. When we draw random samples from a TC, the 25th, 50th, and 75th percentiles depend on the size of the TC. There as more variability in these quartiles in smaller samples than in larger samples. Therefore, in the test for statistical significance, the observed quartile had to be tested against different null model quartiles for samples of different sizes. To do this, we drew samples with a range of sizes from the same set of betweenness and, for a given quartile (25%, 50%, or 75%), fit the minimum quartile value against the sample size to a cubic spline and the maximum quartile value against sample size to a different cubic spline. With these two cubic splines, we could then check whether the observed quartile value for a sample of size $n$ was more than or less than the null model minimum or maximum using cubic spline interpolation. From the histograms shown in Figure 9a, we see that the betweenness quartiles of 1999.01*a* were statistically larger than random samples of the same size from 1999.01, at the level of $p < 10^{-6}$, which means the papers in 1999.01*a* had significantly larger betweenness than other papers in 1999.01.

**Table 2.** The 25th, 50th, and 75th percentiles of the betweenness of 1849 papers in 1999.01, the 164 papers in 1999.01*a*, the 17 papers in 1999.01*a*$\alpha$, the 147 papers in 1999.01*a*$\beta$, the 1685 papers in 1999.01*b*, the 907 papers in 1999.01*b*$\alpha$, the 778 papers in 1999.01*b*$\beta$, the 344 papers in 1999.02, the 144 papers in 1999.02*a*, the 200 papers in 1999.02*b*, the 1014 papers in 1999.11, the 299 papers in 1999.11*a*, the 715 papers in 1999.11*b*, the 988 papers in 1999.12, the 347 papers in 1999.12*a*, and the 641 papers in 1999.12*b*.

|  | Percentile | | |
| --- | --- | --- | --- |
|  | 25 | 50 | 75 |
| 1999.01 | $8.06 \times 10^{-6}$ | $5.73 \times 10^{-5}$ | $2.05 \times 10^{-4}$ |
| 1999.01*a* | $5.90 \times 10^{-5}$ | $1.58 \times 10^{-4}$ | $4.67 \times 10^{-4}$ |
| 1999.01*a*$\alpha$ | $7.77 \times 10^{-6}$ | $1.95 \times 10^{-5}$ | $2.44 \times 10^{-4}$ |
| 1999.01*a*$\beta$ | $5.29 \times 10^{-6}$ | $4.96 \times 10^{-5}$ | $2.48 \times 10^{-4}$ |
| 1999.01*b* | $6.22 \times 10^{-6}$ | $5.04 \times 10^{-5}$ | $1.88 \times 10^{-4}$ |
| 1999.01*b*$\alpha$ | $8.59 \times 10^{-6}$ | $6.00 \times 10^{-5}$ | $2.14 \times 10^{-4}$ |
| 1999.01*b*$\beta$ | $7.97 \times 10^{-6}$ | $5.32 \times 10^{-5}$ | $1.83 \times 10^{-4}$ |
| 1999.02 | $2.47 \times 10^{-6}$ | $5.54 \times 10^{-5}$ | $2.13 \times 10^{-4}$ |
| 1999.02*a* | $3.08 \times 10^{-5}$ | $1.13 \times 10^{-4}$ | $3.17 \times 10^{-4}$ |
| 1999.02*b* | $2.14 \times 10^{-7}$ | $1.44 \times 10^{-5}$ | $1.60 \times 10^{-4}$ |
| 1999.11 | $1.73 \times 10^{-5}$ | $9.04 \times 10^{-5}$ | $2.81 \times 10^{-4}$ |
| 1999.11*a* | $6.38 \times 10^{-5}$ | $1.98 \times 10^{-4}$ | $4.61 \times 10^{-4}$ |
| 1999.11*b* | $9.91 \times 10^{-6}$ | $6.17 \times 10^{-5}$ | $2.17 \times 10^{-4}$ |
| 1999.12 | $6.56 \times 10^{-6}$ | $4.54 \times 10^{-5}$ | $1.62 \times 10^{-4}$ |
| 1999.12*a* | $2.74 \times 10^{-5}$ | $9.08 \times 10^{-5}$ | $2.33 \times 10^{-4}$ |
| 1999.12*b* | $2.52 \times 10^{-6}$ | $2.69 \times 10^{-5}$ | $1.20 \times 10^{-4}$ |

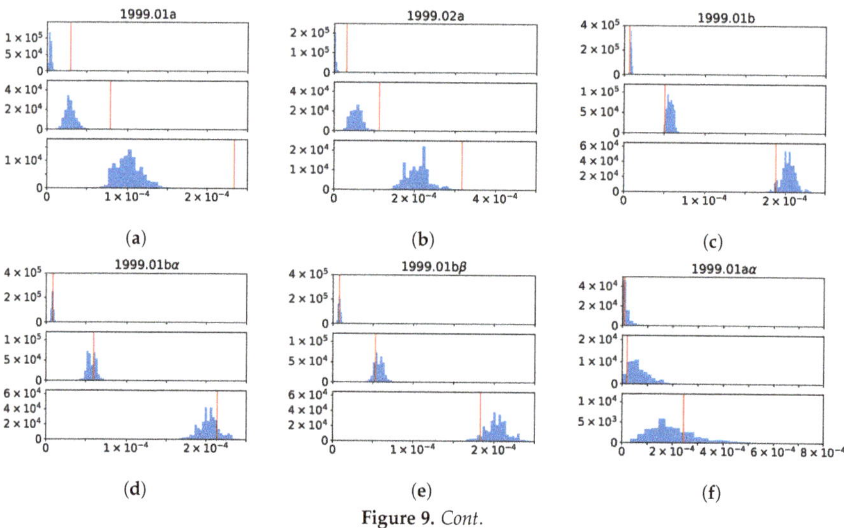

**Figure 9.** *Cont.*

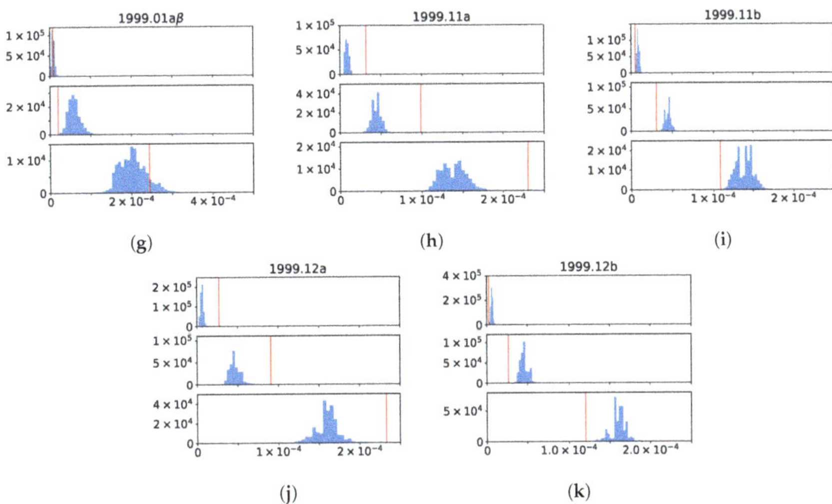

**Figure 9.** The lower (top), median (middle), and top quartile (bottom) of the betweennesses in
(**a**) 1999.01*a*, (**b**) 1999.02*a*, (**c**) 1999.01*b*, (**d**) 1999.01*bα*, (**e**) 1999.01*bβ*, (**f**) 1999.01*aα*, (**g**) 1999.01*aβ*,
(**h**) 1999.11*a*, (**i**) 1999.11*b*, (**j**) 1999.12*a*, and (**k**) 1999.12*b* shown as red vertical lines and $10^6$ random
samples of the same number of betweennesses from 1999.01 (**a**,**c**–**g**), or 1999.02 (**b**), or 1999.11 (**h**,**i**),
or 1999.12 (**j**,**k**) shown as blue histograms. The x-axes are "quartile value", and y-axes are "null
model density".

### 3.5.2. 1999.01 → 2000.02 + 2000.03

When a TC splits into two in the next year, we expect the links between two parts $a$ and $b$ in the
TC to have thinned out to the point that the modularity $Q$ of the whole is lower than the modularities
$Q_a$ and $Q_b$ of the two parts. However, in general, we would not know how to separate the TC into
the two parts $a$ and $b$. Fortunately, for the 1999.01 → 2000.02 + 2000.03 splitting event, we also
knew the part 1999.01*a*, which merged with 1999.02*a*, became 2000.03. Therefore, we might naively
expect 1999.01*b* to be the part that split from 1999.01 to become 2000.02. If we test the quartiles of
1999.01*b*, against random samples of the same size from 1999.01, we find the histograms shown in
Figure 9c. As we can see, the betweenness quartiles of 1999.01*b* were quite a bit lower than the typical
values in 1999.01, but this difference was statistically not as significant as the quartiles of 1999.01*a*.
Thinking about this problem more deeply, we realized that while papers in 1999.01*b* had no references
in common with 1999.02, some of them did share common references with 1999.01*a*. Let us call these
sets of papers 1999.01*aα* (papers do not have references in common with papers in 1999.01*b*), 1999.01*aβ*
(papers have references in common with papers in 1999.01*b*), 1999.01*bα*(papers have references in
common with papers in 1999.01*a*), and 1999.01*bβ* (papers that do not have references in common with
papers in 1999.01*a*). In Figure 9d, we learn from the histograms that the betweenness quartiles of
1999.01*bα* are indistinguishable with random samples of the same size from 1999.01. On the other
hand, from the histograms in Figure 9e, we find out that while the lower betweenness quartile of
1999.01*bβ* is indistinguishable with the random samples of the same size from 1999.01, its median and
the upper quartile are both on the low sides of the random sample distributions. This suggests a split
of 1999.01 to (1999.01a + 1999.01*bα*) + 1999.01*bβ*.

Just to be safe, we also checked the betweenness quartiles of 1999.01*aα* and 1999.01*aβ*,
against random samples of the same sizes from 1999.01. As we can see from Figure 9f,g, the lower
quartiles and medians are lower than those obtained from random samples, but the upper quartiles
are decidedly higher. However, the difference between 1999.01*aα* and 1999.01*aβ* was not as obvious as
the difference between 1999.01*bα* and 1999.01*bβ*, and one possible reason was the smaller sample size

(17, 147 vs. 907, 778). Again, these results were consistent with the picture that the rise in betweenness in 1999.01*a* was driving the merging with 1999.02*a*, while the fall in betweenness in 1999.01*b*$\beta$ was driving a splitting inside 1999.01.

### 3.5.3. **1999.11 + 1999.12 → 2000.15**

Although a small part split off from each of 1999.11 and 1999.12, the main event associated with the two TCs was a symmetric merging. Looking again into the relevant parts of the BCN, we found 299 out of 1014 papers in 1999.11 coupled to 347 out of 988 papers in 1999.12, and we called them 1999.11*a* and 1999.12*a*, respectively. As we can see from the histograms in Figure 9h,j, the betweenness quartiles in 1999.11*a* and 1999.12*a* were significantly higher than one would expect from random samples of 1999.11 and 1999.12. Simultaneously, the betweenness quartiles in 1999.11*b* and 1999.12*b* were significantly lower than in random samples of 1999.11 and 1999.12 (see Figure 9i,k). Therefore, what we see here might be the early warning signals of merging, as well as that of asymmetric splitting.

### 3.5.4. **1999.04 → 2000.06** and **1999.13 → 2000.16**

So far, we have learned that a decrease in betweenness within a TC signals a possible split, whereas an increase in betweenness of the part of the TC coupled to another TC signals a merger between the two TCs. For this story to be consistent, we must not see these signals in the continuing events 1999.04 → 2000.06 and 1999.13 → 2000.16. However, if we go through the full BCN, we find that 370 out of 389 papers in 1999.04 and 308 out of 319 papers in 1999.13 are coupled to papers outside of these TCs, which suggests the possibility of merging or splitting.

However, as we can conclude from Table 3, while the lower betweenness quartiles of the coupling parts of 1999.04 and 1999.13 with other TCs may be significantly larger than those of random samples of the two TCs, the highest betweenness quartiles were never significantly larger. Therefore, at the same level of confidence that we have set for the precursors of merging between 1999.01 and 1999.02, as well as between 1999.11 and 1999.12, we have to say that there were no significant precursors for 1999.04 and 1999.13 to merge with other TCs.

What about splitting then? A TC is likely to split into two if at least one of two parts has reduced betweenness. We see in Table 3 that betweenness in the coupling parts of 1999.04 and 1999.13 was not significantly lower than that of random samples. Therefore, we looked at the non-coupling part, i.e., papers in 1999.04 and 1999.13, which had no references in common with papers in other TCs, but they may have common references with papers in the same TCs. We called these non-coupling parts 1999.04*b* and 1999.13*b*, respectively (the bottom row in Table 3). Only the top betweenness quartile of 1999.04*b* fell below that of random samples from 1999.04 in Table 3. Therefore, the early warning for a splitting event in the next year is not strong enough. For 1999.13*b*, on the other hand, all three betweenness quartiles fell below that of random samples from 1999.13, even after we accounted for the small size of 1999.13*b*. This suggests that the probability of a splitting event next year is high, but 1999.13 continued on to 2000.16, which thereafter continued to 2001 without merging or splitting. This might be because additional conditions, like the size of TC being large, must be satisfied before a splitting can occur.

**Table 3.** The distributions of the betweennesses of papers in 1999.04 and 1999.13 that share common references with the other TCs in 1999 (1999.00 to 1999.15). The four columns below 1999.04 and 1999.13 denote the following: the first column shows how many papers have common references with the other TCs, while the second, third, and fourth columns show the lower, median, and upper quartile values of betweennesses of these papers, respectively. For example, there are 25 papers in 1999.04 that share common references with papers in 1999.03, and the betweennesses of these papers have a lower quartile value of $1.6 \times 10^{-5}$, a median value of $4.3 \times 10^{-4}$, and an upper quartile value of $8.1 \times 10^{-4}$. Similarly, there are 254 papers in 1999.13 that share common references with papers in 1999.10, and the betweennesses of these papers have a lower quartile value of $3.6 \times 10^{-5}$, a median value of $8.8 \times 10^{-5}$, and an upper quartile value of $2.7 \times 10^{-4}$. The bottom row "b" represents 1999.04b and 1999.13b, respectively, which are papers in 1999.04 and 1999.13 that have no references in common with papers in other TCs. A betweenness value in red means that it is larger than the maximum of the corresponding quartile distribution of $10^6$ random samples, and a betweenness value in blue denotes it is smaller than the minimum of the corresponding $10^6$ random samples.

| | | 1999.04 | | | | 1999.13 | | |
| --- | --- | --- | --- | --- | --- | --- | --- | --- |
| | Size | Percentile | | | Size | Percentile | | |
| | | 25 | 50 | 75 | | 25 | 50 | 75 |
| 1999.00 | 12 | $9.0 \times 10^{-5}$ | $1.1 \times 10^{-3}$ | $2.3 \times 10^{-3}$ | 1 | - | - | $1.8 \times 10^{-3}$ |
| 1999.01 | 56 | $1.6 \times 10^{-4}$ | $4.2 \times 10^{-4}$ | $1.0 \times 10^{-3}$ | 6 | $2.0 \times 10^{-4}$ | $4.9 \times 10^{-4}$ | $6.5 \times 10^{-4}$ |
| 1999.02 | 6 | $3.0 \times 10^{-4}$ | $5.1 \times 10^{-4}$ | $7.4 \times 10^{-4}$ | 2 | $6.0 \times 10^{-4}$ | - | $2.6 \times 10^{-4}$ |
| 1999.03 | 25 | $1.6 \times 10^{-5}$ | $4.3 \times 10^{-4}$ | $8.1 \times 10^{-4}$ | 0 | - | - | - |
| 1999.04 | - | - | - | - | 8 | $1.5 \times 10^{-4}$ | $4.8 \times 10^{-4}$ | $8.0 \times 10^{-4}$ |
| 1999.05 | 179 | $4.9 \times 10^{-5}$ | $1.7 \times 10^{-4}$ | $4.5 \times 10^{-4}$ | 4 | $2.2 \times 10^{-4}$ | $4.3 \times 10^{-4}$ | $6.5 \times 10^{-4}$ |
| 1999.06 | 110 | $8.7 \times 10^{-5}$ | $2.0 \times 10^{-4}$ | $6.2 \times 10^{-4}$ | 40 | $5.9 \times 10^{-5}$ | $1.6 \times 10^{-4}$ | $4.5 \times 10^{-4}$ |
| 1999.07 | 29 | $1.7 \times 10^{-4}$ | $5.6 \times 10^{-4}$ | $1.2 \times 10^{-3}$ | 44 | $1.4 \times 10^{-4}$ | $3.1 \times 10^{-4}$ | $5.5 \times 10^{-4}$ |
| 1999.08 | 63 | $1.1 \times 10^{-4}$ | $3.2 \times 10^{-4}$ | $8.6 \times 10^{-4}$ | 17 | $2.2 \times 10^{-4}$ | $5.2 \times 10^{-4}$ | $8.5 \times 10^{-4}$ |
| 1999.09 | 49 | $7.8 \times 10^{-5}$ | $2.6 \times 10^{-4}$ | $8.0 \times 10^{-4}$ | 99 | $8.0 \times 10^{-5}$ | $2.5 \times 10^{-4}$ | $4.8 \times 10^{-4}$ |
| 1999.10 | 53 | $1.2 \times 10^{-4}$ | $3.8 \times 10^{-4}$ | $8.2 \times 10^{-4}$ | 254 | $3.6 \times 10^{-5}$ | $8.8 \times 10^{-5}$ | $2.7 \times 10^{-4}$ |
| 1999.11 | 89 | $1.0 \times 10^{-4}$ | $3.2 \times 10^{-4}$ | $9.2 \times 10^{-4}$ | 71 | $1.4 \times 10^{-4}$ | $3.4 \times 10^{-4}$ | $5.7 \times 10^{-4}$ |
| 1999.12 | 53 | $8.7 \times 10^{-5}$ | $2.9 \times 10^{-4}$ | $9.3 \times 10^{-4}$ | 39 | $1.3 \times 10^{-4}$ | $2.7 \times 10^{-4}$ | $4.6 \times 10^{-4}$ |
| 1999.13 | 9 | $1.3 \times 10^{-4}$ | $4.2 \times 10^{-4}$ | $1.1 \times 10^{-3}$ | - | - | - | - |
| 1999.14 | 62 | $1.4 \times 10^{-4}$ | $4.8 \times 10^{-4}$ | $1.0 \times 10^{-3}$ | 210 | $4.2 \times 10^{-5}$ | $1.0 \times 10^{-4}$ | $2.7 \times 10^{-4}$ |
| 1999.15 | 17 | $1.8 \times 10^{-4}$ | $3.6 \times 10^{-4}$ | $9.7 \times 10^{-4}$ | 176 | $5.1 \times 10^{-5}$ | $1.3 \times 10^{-4}$ | $3.1 \times 10^{-4}$ |
| b | 88 | $2.1 \times 10^{-6}$ | $2.2 \times 10^{-5}$ | $5.8 \times 10^{-5}$ | 27 | $9.1 \times 10^{-11}$ | $4.3 \times 10^{-6}$ | $1.8 \times 10^{-5}$ |

## 4. Discussion and Conclusions

During the past two decades, researchers have made many efforts to understand the system of science. Many problems have been solved; however, the understanding of interactions between different fields is still limited. Investigating the temporal network (BCN, CN) and its community structures, we were able to measure and quantify the complex interaction between different fields, particularly in physics, over time. Naturally, we would like to have a predictive power based on this picture. However, the correlation between network structure and evolution events is nonlinear and complex. Therefore, we turned to machine learning techniques, which have shown a great power to solve predictive problems that are hard to solve using traditional statistical methods. To our knowledge, this is the first study that utilized both machine learning and network science approaches to predict the future of science at the community level.

To be able to identify changes in TCs, we needed to define time windows used for network creation and community detection. The natural choice for bibliographical data was the usage of single years, since the publishing process may last many months. Obviously, another detail may be considered like multiple years, e.g., two or five years. In our approach, i.e., both for BCN and CN, every citation had the same importance. However, there were some other concepts like fractional counting of citations [38]. It assumes that the impact of each citation is proportionate to the number of references in the citing document. Additionally, it can be differentiated depending on, e.g., the quality of the journal. For the CN, we calculated the similarity between groups in the consecutive time windows in two ways: (i) using the plain relative overlap measure and (ii) using the inclusion measure based on social position. The idea was to enrich evolution data with the structural information occurring between the nodes. It turned out that both measures provided similar labeling, but the evolution tracking with the social position information produced a slightly better initial prediction. Therefore, the study was continued only for the inclusion measure; see SI for more information.

We decided to analyze more in-depth only one feature describing the structural profile of TCs, namely node betweenness. It was primarily caused by the limited amount of resources and the complexity of the analyses. The entire process required much human assistance and could not have been easily automated. In our experiments, we utilized the raw, imbalanced or artificially flattened, balanced datasets. However, depending on the study purpose, we could bias some classes we were more interested in, e.g., split. This could be achieved either by means of appropriate balancing (sampling for the learning set or reformulating the problem into the binary question: Is split expected (true) or not (false)?) As of now, the betweenness analysis was still limited to several case studies; in the future, a more rigorous framework would be desired. The idea of analyzing science by the discovery of knowledge changes is general and can be applied to all bibliographical data containing citations. We focused solely on APS journals; however, also papers indexed by PubMed, Web of Science, or Google Scholar may be studied.

**Supplementary Materials:** The following are available online at http://www.mdpi.com/1099-4300/21/12/1152/s1, Figure S1: The scatter plots for simple overlap measure and inclusion measure for CNs between 1981 to 2010, Figure S2: The alluvial diagram of APS papers' references (CNs) from 1981 to 2010, Figure S3: Bibliographic coupling network in 1991, Table S1: List of all features used in the study.

**Author Contributions:** Conceptualization, W.L., S.S., S.A.C., and P.K.; methodology, W.L., S.S., S.A.C., and P.K.; software, W.L. and S.S.; validation, W.L. and S.S.; formal analysis, W.L. and S.S.; investigation, W.L. and S.S.; resources, W.L. and S.S.; data curation, W.L. and S.S.; writing, original draft preparation, W.L. and S.S.; writing, review and editing, W.L., S.S., S.A.C., and P.K.; visualization, W.L. and S.S.; supervision, S.A.C. and P.K.; project administration, S.A.C. and P.K.; funding acquisition, S.A.C. and P.K.

**Funding:** This research was funded by the Singapore Ministry of Education Academic Research Fund Tier 2 under Grant Number MOE2017-T2-2-075, the National Science Centre, Poland, Project No. 2016/21/B/ST6/01463, the European Union's Marie Skłodowska-Curie Program under Grant Agreement No. 691152 (RENOIR), and the Polish Ministry of Science and Higher Education under Grant Agreement No. 3628/H2020/2016/2, and the statutory funds of the Department of Computational Intelligence, Wrocław University of Science and Technology.

**Conflicts of Interest:** The authors declare no conflict of interest. The funders had no role in the design of the study; in the collection, analyses, or interpretation of data; in the writing of the manuscript; nor in the decision to publish the results.

## Abbreviations

The following abbreviations are used in this manuscript:

| | |
|---|---|
| PR | Physical Review |
| PRA | Physical Review A |
| PRB | Physical Review B |
| PRC | Physical Review C |
| PRD | Physical Review D |
| PRE | Physical Review E |
| PRL | Physical Review Letters |
| RMP | Reviews of Modern Physics |
| BCN | Bibliographic coupling network |
| CN | Co-citation network |
| GEP | Group evolution prediction |
| GED | Group evolution discover |
| SI | Supplementary Information |

## References


1. Chen, P.; Redner, S. Community structure of the physical review citation network. *J. Inf.* **2010**, *4*, 278–290. [CrossRef]
2. Rosvall, M.; Bergstrom, C.T. Mapping Change in Large Networks. *PLoS ONE* **2010**, *5*, e8694. [CrossRef] [PubMed]
3. Liu, W.; Nanetti, A.; Cheong, S.A. Knowledge evolution in physics research: An analysis of bibliographic coupling networks. *PLoS ONE* **2017**, *12*, e0184821. [CrossRef] [PubMed]
4. Helbing, D.; Brockmann, D.; Chadefaux, T.; Donnay, K.; Blanke, U.; Woolley-Meza, O.; Moussaid, M.; Johansson, A.; Krause, J.; Schutte, S.; et al. Saving Human Lives: What Complexity Science and Information Systems can Contribute. *J. Stat. Phys.* **2015**, *158*, 735–781. [CrossRef] [PubMed]
5. Zeng, A.; Shen, Z.; Zhou, J.; Wu, J.; Fan, Y.; Wang, Y.; Stanley, H.E. The science of science: From the perspective of complex systems. *Phys. Rep.* **2017**, *714–715*, 1–73. [CrossRef]
6. Fortunato, S.; Bergstrom, C.T.; Börner, K.; Evans, J.A.; Helbing, D.; Milojević, S.; Petersen, A.M.; Radicchi, F.; Sinatra, R.; Uzzi, B.; et al. Science of science. *Science* **2018**, *359*, 185. [CrossRef]
7. Hicks, D.; Wouters, P.; Waltman, L.; de Rijcke, S.; Rafols, I. Bibliometrics: The Leiden Manifesto for research metrics. *Nature* **2015**, *520*, 429–431. [CrossRef]
8. Radicchi, F.; Fortunato, S.; Castellano, C. Universality of citation distributions: Toward an objective measure of scientific impact. *Proc. Natl. Acad. Sci. USA* **2008**, *105*, 17268–17272. [CrossRef]
9. Wang, D.; Song, C.; Barabási, A.L. Quantifying Long-Term Scientific Impact. *Science* **2013**, *342*, 127–132. [CrossRef]
10. Ke, Q.; Ferrara, E.; Radicchi, F.; Flammini, A. Defining and identifying Sleeping Beauties in science. *Proc. Natl. Acad. Sci. USA* **2015**, *112*, 7426–7431. [CrossRef]
11. Small, H. Visualizing science by citation mapping. *J. Am. Soc. Inf. Sci.* **1999**, *50*, 799–813. [CrossRef]
12. Boyack, K.W.; Klavans, R.; Börner, K. Mapping the backbone of science. *Scientometrics* **2005**, *64*, 351–374. [CrossRef]
13. Bollen, J.; Van de Sompel, H.; Hagberg, A.; Bettencourt, L.; Chute, R.; Rodriguez, M.A.; Balakireva, L. Clickstream Data Yields High-Resolution Maps of Science. *PLoS ONE* **2009**, *4*, e4803. [CrossRef]
14. Perc, M. Self-organization of progress across the century of physics. *Sci. Rep.* **2013**, *3*, 1720. [CrossRef]
15. Kuhn, T.; Perc, M.; Helbing, D. Inheritance Patterns in Citation Networks Reveal Scientific Memes. *Phys. Rev. X* **2014**, *4*, 041036. [CrossRef]
16. Zhang, Y.; Chen, H.; Lu, J.; Zhang, G. Detecting and predicting the topic change of Knowledge-based Systems: A topic-based bibliometric analysis from 1991 to 2016. *Knowl.-Based Syst.* **2017**, *133*, 255–268. [CrossRef]
17. Van Eck, N.J.; Waltman, L. Software survey: VOSviewer, a computer program for bibliometric mapping. *Scientometrics* **2010**, *84*, 523–538. [CrossRef]
18. Palla, G.; Barabási, A.L.; Vicsek, T. Quantifying social group evolution. *Nature* **2007**, *446*, 664–667. [CrossRef]

19. Carrasquilla, J.; Melko, R.G. Machine learning phases of matter. *Nat. Phys.* **2017**, *13*, 431–434. [CrossRef]
20. Ahneman, D.T.; Estrada, J.G.; Lin, S.; Dreher, S.D.; Doyle, A.G. Predicting reaction performance in C–N cross-coupling using machine learning. *Science* **2018**, *360*, 186–190. [CrossRef]
21. Saganowski, S.; Bródka, P.; Koziarski, M.; Kazienko, P. Analysis of group evolution prediction in complex networks. *PLoS ONE* **2019**, *14*, 1–18. [CrossRef] [PubMed]
22. Saganowski, S.; Gliwa, B.; Bródka, P.; Zygmunt, A.; Kazienko, P.; Koźlak, J. Predicting Community Evolution in Social Networks. *Entropy* **2015**, *17*, 3053–3096. [CrossRef]
23. İlhan, N.; Öğüdücü, G. Feature identification for predicting community evolution in dynamic social networks. *Eng. Appl. Artif. Intell.* **2016**, *55*, 202–218. [CrossRef]
24. Pavlopoulou, M.E.G.; Tzortzis, G.; Vogiatzis, D.; Paliouras, G. Predicting the evolution of communities in social networks using structural and temporal features. In Proceedings of the 2017 12th International Workshop on Semantic and Social Media Adaptation and Personalization (SMAP), Bratislava, Slovakia, 9–10 July 2017; pp. 40–45. [CrossRef]
25. Bródka, P.; Saganowski, S.; Kazienko, P. GED: The method for group evolution discovery in social networks. *Soc. Netw. Anal. Min.* **2013**, *3*, 1–14. [CrossRef]
26. Tajeuna, E.G.; Bouguessa, M.; Wang, S. Tracking the evolution of community structures in time-evolving social networks. In Proceedings of the 2015 IEEE International Conference on Data Science and Advanced Analytics (DSAA), Paris, France, 19–21 October 2015; pp. 1–10. [CrossRef]
27. Alhajj, R.; Rokne, J. (Eds.) *Encyclopedia of Social Network Analysis and Mining*; Springer: New York, NY, USA, 2014; doi:10.1007/978-1-4614-6170-8. [CrossRef]
28. Brodka, P.; Musial, K.; Kazienko, P. A Performance of Centrality Calculation in Social Networks. In Proceedings of the 2009 International Conference on Computational Aspects of Social Networks, Fontainebleau, France, 24–27 June 2009; pp. 24–31. [CrossRef]
29. Blondel, V.D.; Guillaume, J.L.; Lambiotte, R.; Lefebvre, E. Fast unfolding of communities in large networks. *J. Stat. Mech. Theory Exp.* **2008**, *2008*, P10008. [CrossRef]
30. APS Data Sets for Research. Available online: https://journals.aps.org/datasets (accessed on 26 December 2019).
31. Javed, M.A.; Younis, M.S.; Latif, S.; Qadir, J.; Baig, A. Community detection in networks: A multidisciplinary review. *J. Netw. Comput. Appl.* **2018**, *108*, 87–111. [CrossRef]
32. Yang, J.; Honavar, V. Feature Subset Selection Using a Genetic Algorithm. In *Feature Extraction, Construction and Selection: A Data Mining Perspective*; Liu, H., Motoda, H., Eds.; The Springer International Series in Engineering and Computer Science; Springer: Boston, MA, USA, 1998; pp. 117–136. [CrossRef]
33. Popper, K.R. *All Life Is Problem Solving*; Routledge: London, UK, 2010.
34. Kotthoff, L.; Thornton, C.; Hoos, H.H.; Hutter, F.; Leyton-Brown, K. Auto-WEKA 2.0: Automatic model selection and hyperparameter optimization in WEKA. *J. Mach. Learn. Res.* **2017**, *18*, 1–5.
35. Platt, J. Fast Training of Support Vector Machines Using Sequential Minimal Optimization. In *Advances in Kernel Methods—Support Vector Learning*; MIT Press: Cambridge, MA, USA, 1998.
36. Atkeson, C.G.; Moore, A.W.; Schaal, S. Locally Weighted Learning. *Artif. Intell. Rev.* **1997**, *11*, 11–73. [CrossRef]
37. Frank, E.; Witten, I.H. Generating Accurate Rule Sets Without Global Optimization. In Proceedings of the Fifteenth International Conference on Machine Learning, Madison, WI, USA, 24–27 July 1998; Morgan Kaufmann Publishers Inc.: San Francisco, CA, USA, 1998; pp. 144–151.
38. Leydesdorff, L.; Opthof, T. Scopus's source normalized impact per paper (SNIP) versus a journal impact factor based on fractional counting of citations. *J. Am. Soc. Inf. Sci. Technol.* **2010**, *61*, 2365–2369. [CrossRef]

*Article*

# Uncovering the Dependence of Cascading Failures on Network Topology by Constructing Null Models

Lin Ding [1], Si-Yuan Liu [2], Quan Yang [1] and Xiao-Ke Xu [2,*]

[1] School of Computer, University of South China, Hengyang 421001, China; linding@usc.edu.cn (L.D.); yangquan19941024@163.com (Q.Y.)
[2] College of Information and Communication Engineering, Dalian Minzu University, Dalian 116600, China; liusiyuan0311@foxmail.com
* Correspondence: xuxiaoke@foxmail.com

Received: 10 October 2019; Accepted: 14 November 2019; Published: 15 November 2019

**Abstract:** Cascading failures are the significant cause of network breakdowns in a variety of complex infrastructure systems. Given such a system, uncovering the dependence of cascading failures on its underlying topology is essential but still not well explored in the field of complex networks. This study offers an original approach to systematically investigate the association between cascading failures and topological variation occurring in realistic complex networks by constructing different types of null models. As an example of its application, we study several standard Internet networks in detail. The null models first transform the original network into a series of randomized networks representing alternate realistic topologies, while taking its basic topological characteristics into account. Then considering the routing rule of shortest-path flow, it is sought to determine the implications of different topological circumstances, and the findings reveal the effects of micro-scale (such as degree distribution, assortativity, and transitivity) and meso-scale (such as rich-club and community structure) features on the cascade damage caused by deliberate node attacks. Our results demonstrate that the proposed method is suitable and promising to comprehensively analyze realistic influence of various topological properties, providing insight into designing the networks to make them more robust against cascading failures.

**Keywords:** complex networks; cascading failures; network topology; null models

---

## 1. Introduction

Complex networks, involving interactive specific nodes abstracted from the real-world systems, have attracted much attention in recent decades [1–3]. Many man-made infrastructure systems such as the Internet, transportation networks, and power grids, are examples of complex networks playing essential roles in our modern society. Understanding their robustness concerning random failures and deliberate attacks is of utmost importance and has an increasing interest. Early studies have concentrated on the static failures of a network and the impact of random and deliberate removal of a node (or edge) or group of nodes altogether [4–6], while in some cases, the networks undergoing failures may experience a more catastrophic condition as soon as cascading failures take place [7]. For instance, Hub nodes may fail due to targeted attacks. Taking into consideration the inherent dynamics of network flow, the initial removal of only a few nodes may trigger a cascade of overload failures and eventually propagating the failure to a large fraction of the network, leading to a much more devastating result than the case of static failure [8]. Indeed, such cascading failures were found to be particularly relevant for large-scale breakdowns in various infrastructure networks, such as the Internet collapses [9] and huge blackouts in some countries [10]. As these catastrophic incidents can induce excessive losses in a short period, they alarm the whole world and bring serious concern on the dynamics of cascading failure [11–14].

From the perspective of complex networks, different models were built to imitate cascading phenomena [15,16]. Our point of interest is the type of load models, in which deriving the flow distribution in a network is one of the key issues. Notably Motter et al. built a cascading load model, where node betweenness based on shortest paths is used to represent the flow of physical quantities [8]. Owing to the fact that the shortest-path flow is common in realistic networks such as the Internet and power grids, their model has enjoyed extensive adoption as the basis of various studies [17–23]. Also, this basic model was extended to consider network information conditions [24], network weights [25], and system laws [26].

Because it is shown that the dynamic behavior of a network largely depends on its topological structure, based on these cascading models, many efforts were made to analyze the topological impact on the cascade robustness of complex networks. Previous methods adopted for the analysis can be divided into two main types: the empirical approach and modeling approach. The two approaches focus on examining the cascade consequences on different real-life networks and traditional model networks respectively, where the preferential attachment model [16–18], the small-world model [19,27], and many others [28,29] are examples of such model networks. Although studies such as these have clarified that network topological properties show some relations with the cascade robustness, including degree distribution [18,29], interdependence characteristics [19–21], community structure [22–24], assortativity [25], and transitivity effects [26–28], hardly any attention is paid to enumerate a real-life network robustness to cascading failures in terms of its multi-scale topological features.

As we all know, for a large-scale real-life network, its topological data is generally difficult to be acquired, and once acquired, the network topology is fixed so that it is hard to study the impact of topological variations. Due to the inflexibility of the purely empirical approach, the modeling approach is usually preferable. However, the traditional network modeling can handle only simple microscopic dynamics driving the formation of the network, and thus the resulting networks are universal, which are difficult to approximate full topological properties of a real network. Moreover, when a certain topological parameter of the model network is adjusted to study its influence, accordingly other parameters are often changed simultaneously. Since statistical parameters which define topological properties are not dimensioned, and network size and structure vary widely, the research results from empirical and traditional modeling approaches cannot be carefully compared. Therefore, with the current methods, it is hard to have a sound understanding of the dependence of cascading failures on underlying network topology. So far, for a specific network such as the Internet, the relationship between topological metrics, such as degree distribution, assortativity, transitivity, rich-club, and community structure with respect to the cascading evolution is still unclear. It is thus desirable to develop a novel approach to give the quantitatively accurate evaluation, which helps to take appropriate measures to establish a stronger system.

Recently, null models for real-life networks were increasingly used to analyze structural complexity [30,31], link prediction [32,33], and community detection [34]. In general, a random network, with certain characteristics of a real network, is called a null network of the original one. The null models (networks) may accurately reflect the non-trivial properties of the original network, and can arrange for a precise reference of the original network together with statistical measurements. Therefore, different from the current empirical and modeling approaches, applying null models allow us to comprehensively and explicitly exploit topological features of a real network and systematically study how the modification of these features can affect the cascade robustness. However, to the best knowledge of ours, there is a lack of studies on cascading failures with null models.

In this study, we aim to close this gap by suggesting a novel approach founded on null models to investigate cascading failures on the Internet, and the effect of the cascade with varying topological structures is explored for a given network. To this end, we first construct various null models to generate realistic alternate networks derived from the standard Internet, where different topological properties are considered including degree distribution, assortativity, transitivity, rich-club, and community structure. Considering the distribution of the actual shortest-path flow, the cascading failure propagation triggered by deliberate node attacks is modeled under varying topological conditions of the Internet. For each of network topologies, the size of the largest connected component of the attacked network is monitored. Then the results are used to establish the relationship between the cascading failures and the variations in different topological features. Based on three Internet AS-level networks, our study validates the proposed method and clearly show how micro-scale properties (i.e., degree distribution, assortativity, and transitivity) and meso-scale properties (i.e., rich-club and community structure) exert impacts on the network robustness against cascading failure, where there are substantially different results from those in traditional model networks.

Furthermore, it should be remarked on that both the perception of the cascades with shortest-path flow and null models were deeply studied, but the approach of integration of the two fields is novel and adopted to investigate the relationship between cascading failures and the topological variations occurring in a given realistic network, which is the key contribution of our paper. Although the study is performed in the framework of propagating failures on the Internet, we believe that the proposed approach can be applicable to studying the robustness of other kinds of real-life networks with reasonable modification because the basic models involved are easily extended.

The remainder of the article is arranged as follows. Section 2 introduces topological parameters and null models engaged in generating distinct topological structures of the Internet. Section 3 states the cascading model with the routing rule of shortest-path flow. Section 4 discusses the procedure involved to explore the topological effect on the cascade robustness and tests it on various Internet AS-level networks with results. Finally, Section 5 concludes this work.

## 2. Constructing Null Models of the Internet

The Internet can be represented as a complex graph with $N$ nodes and $E$ edges, where the nodes can be routers or ASs, and the edges are the physical connections between nodes. The network topology defines how nodes within the network are arranged and connected to each other. A minor shift in the topology, such as edge swapping, can initiate varying the properties of the network that accordingly affect its dynamical behaviors and functions.

### 2.1. Network Parameters

There are many metrics or parameters to describe statistical properties of network topology, but in this study, we restrict ourselves to consider five basic ones: degree distribution, assortativity coefficient, clustering coefficient, rich-club coefficient, and modularity coefficient, which were widely studied in traditional model networks and exported important impacts on a variety of network-based dynamical processes [1–3]. The degree distribution $p(k)$ represents the probability with which a node in the network chosen randomly has degree $k$. Scale-free networks widely observed in reality have a power-law degree distribution, namely $p(k) \sim k^{-\lambda}$, where $\lambda$ is the scaling exponent. The specific definitions of other four parameters are as follows:

1.  Assortativity Coefficient

$$r = \frac{E^{-1}\sum\limits_{e_{ij}} k_i k_j - \left[ E^{-1}\sum\limits_{e_{ij}} \frac{1}{2}(k_i + k_j) \right]^2}{E^{-1}\sum\limits_{e_{ij}} \frac{1}{2}(k_i^2 + k_j^2) - \left[ E^{-1}\sum\limits_{e_{ij}} \frac{1}{2}(k_i + k_j) \right]^2}, \tag{1}$$

where $e_{ij}$ is an edge connecting nodes $i$ and $j$; $k_i$ and $k_j$ denote the degrees of nodes $i$ and $j$, respectively.

2. Clustering Coefficient

$$c = \frac{1}{N} \sum_{i=1}^{N} c_i,$$ (2)

where the node clustering coefficient $c_i = \frac{2\lambda_i}{k_i(k_i-1)}$, where $\lambda_i$ is the number of the edges existing among $k_i$ neighbors of node $i$. The clustering coefficient can be used to measure the transitivity property of a real-life network.

3. Rich-club Coefficient

$$\phi = \frac{M}{n(n-1)/2} = \frac{2M}{n(n-1)},$$ (3)

where $n$ and $M$ are the numbers of rich nodes and the edges existing among these rich nodes, respectively.

4. Modularity Coefficient

$$Q = \sum_i \left( h_{ii} - a_i^2 \right),$$ (4)

where $a_i = \sum_w h_{iw}$ signifies the row (or column) sums, symbolizing the fraction of edges that link to nodes in community $i$ and $h_{iw}$ is the fraction of edges in the original network that connect nodes in the subset $i$ with nodes in the subset $w$.

*2.2. Synthetic Networks Generated by Null Models*

Given a real network, its topological structure is settled. To investigate the influence of the above topological properties (e.g., degree distribution, assortativity, transitivity, rich-club, and community structure) on cascading dynamics in detail, several null models based on randomized algorithms are considered to generate alternate realistic topologies of the original network. Here, the randomized algorithms cannot just rewire edges of the original network but also randomize some factors on the condition of precisely keeping some original connection properties. It is clear that the topologies thus created can rigorously grasp real topological characteristics as they are derived from the original network.

2.2.1. dK-Series of Null Networks

We consider two approaches with randomized algorithms for constructing null models. One is the dK-series of prospect distributions, where all degree correlations are indicated in d-sized subgraphs of a specified graph [35,36]. This approach can produce null models of different orders, including 0K, 1K, 2K, and 3K that are applied to approach the original network progressively and then spot its micro-scale features at multi-levels. Null networks of all these orders are interconnected, i.e., 0K $\supseteq$ 1K $\supseteq$ 2K $\supseteq$ 3K. Any higher-order null network embraces the features of lower-order null network.

Figure 1a illustrates the process of constructing the properties $Pd$, which we call the dK-series of null networks. The $d = 0, \cdots, 4$ corresponds to different order of dK-series [34]. We use the total number of corresponding subgraphs to represent all the values of $P$. That is to say, $P(2,2) = 1$ means that the network has one edge between two 2-degree nodes. 0K null network in Figure 1b is the simplest and the most randomized version of the original network, which only retains the number of nodes and the average degree of it. 1K null network maintains the degree distribution of the original network, but it has randomly rewired the link relationship as shown in Figure 1c. 2K null network holds the identical joint degree distribution of the original network in Figure 1d, which means they have the same degree values for the end nodes of each edge. That is to say, they have the same values of assortativity coefficient as the original one. The rewiring procedure of 3K null network is demonstrated in Figure 1e. 3K null and the original networks hold identical clustering coefficient for each node.

Thus, with the increase of null model orders (i.e., the increase of the constraints for generating null models), the null networks can be gradually approaching the original network theoretically.

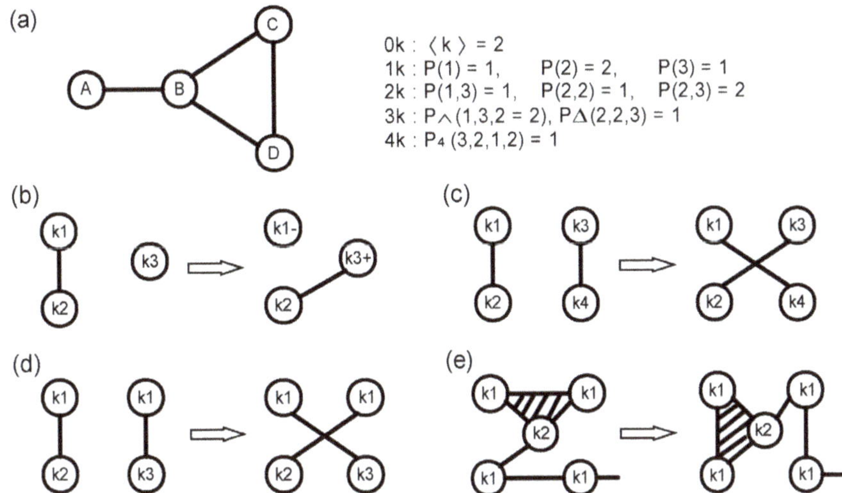


**Figure 1.** The summary of dK-series null models. (**a**) The properties $Pd$, $d = 0, \cdots, 4$, calculated for a given toy network of size 4. The rewiring procedure of (**b**) 0K null network, (**c**) 1K null network, (**d**) 2K null network, and (**e**) 3K null network.

### 2.2.2. Null Networks of Tunable Properties

Although the above four null models of different orders are useful in understanding the behaviors of the original network, they cannot capture how to control it more efficiently. Therefore, we also consider the approach of the targeted edge-swapping [37], which can create null models with tunable micro-scale properties, such as assortativity, transitivity and meso-scale properties, such as rich-club, community structure. We refer to them as null models with tunable properties.

To obtain the increased and decreased assortativity $r$, we consider null models of increasing and decreasing assortativity respectively, where $r$ is tunable [37,38]. Such null networks are constructed as follows. The process of edge swapping is conducted on the original network with preferred $r$. First, two edges $e_{st}$ and $e_{uv}$ are randomly picked up from the network, where the rank of the degrees of nodes $s$, $t$, $u$, and $v$ is denoted by $k_s > k_u > k_t > k_v$, and $e_{su}$, $e_{tv}$, $e_{sv}$, and $e_{ut}$ do not exist. Then, $e_{st}$ and $e_{uv}$ are removed. For generating the null network of increasing assortativity, we add the edges $e_{su}$ and $e_{tv}$; while for generating the null network of decreasing assortativity, we add the edges $e_{sv}$ and $e_{ut}$. The swapping procedure goes on iteratively until the error between the observed and preferred value is within a very tiny value, such as 0.005.

Similarly, we consider null models of increasing and decreasing transitivity, rich-club property [33,39], and community structure [34], combined with well-controlled parameters $c$, $\phi$, and $Q$ respectively. Besides, to gain further insights about community structures, we consider null models of rewiring edges within a community and between two communities [34]. The former only varies the inner topology of each community, at the same time maintaining the structural features between communities and the number of communities. The latter only changes the links between two communities but maintains the structural features inside each community.

Note that for the null models of tunable assortativity, when the assortativity coefficient is adjusted, its lower-order property (i.e., the degree distribution) can remain the same as that of the original network. In a similar way, the null models of tunable clustering coefficients can keep its lower-order properties (i.e., the degree distribution and assortativity) unchanged. Moreover, when one meso-scale property, such as the rich-club coefficient or modularity coefficient is adjusted, because the desired value of the property can be obtained only by exchanging a small number of edges in the original network, micro-scale properties of most connectivity between network nodes, including the degree distribution, assortativitity, and transitivity, all can remain almost unchanged. Therefore, the null models in this study allow us to identify how the modification of one meso-scale property affects cascading dynamics in the case of keeping micro-scale properties unchanged in the network.

## 3. Cascading Failure Model

With the desired null networks at hand, these networks are then enriched with data flow. We follow the routing rule of shortest-path flow described in [8]. In the rule, at each time step, one packet is exchanged between every couple of network nodes and transferred along the shortest paths linking them. Under this situation, the load at a node can be denoted by its betweenness [18,19]. This definition method on the load was widely applied to different kinds of realistic networks, including communication networks such as the Internet, transmission systems such as power grids, and transportation networks [17].

The capacity of a node is the highest load that it can handle. Generally, the capacity is restricted by the cost in a real-life network. Therefore, it is sensible to suppose that the capacity $C_i$ of node $i$ is proportional to its initial load $L_i(0)$ [21]:

$$C_i = (1 + \alpha)L_i(0), \tag{5}$$

where $\alpha(\alpha \geq 0)$ is a redundancy parameter. $\alpha \geq 0$ guarantees that initially, all nodes work properly (i.e., without overload).

Here, the potential cascading failure is considered to be triggered by removing a single node with the highest load, because many prior studies concerning both model networks and real networks have shown that such a node failure can affect loads at other nodes considerably and thus cause severe loss to the network. Assume such deliberate attack arises at $t = 1$. The removal of the attacked node in general changes the distribution of shortest paths. The traffic load used to go through this node has to be rerouted. Therefore, the loads of some nodes may increase beyond their capacities. Consequently, the corresponding nodes are overloaded and thus fail. Once more nodes fail, the shortest paths among all node pairs and the loads are then recalculated based on the topological modification. The process of node cascading failure and the load redistribution is iterated until no node fails, at which point the cascading propagation course is considered as being accomplished.

The dynamical function of a real-life network depends on the node capability to communicate efficiently with each other. Suppose the number of nodes in the largest component before and after the cascade to be $N$ and $N'$ respectively, without loss of generality, the damage triggered by a cascade is calculated by the relative size $G$ of the largest connected component, i.e., $G = N'/N$. As $G$ of the attacked networks is checked, the profile of this parameter variation can display the network invulnerability and robustness against cascading failure. Obviously, the greater the value of the index $G$, the better the network robustness against cascading failure.

## 4. Main Results

This section adopts real network data to identify the topological impact on the cascade damage in detail. We consider three Internet AS-level networks [40]. They contains 3015 nodes, 530 nodes and 493 nodes, respectively. The data were collected from online data and reports of the University of Oregon Route Views Project. For each of the original networks, their multiple alternate networks are generated by employing four null models of different orders as well as ten null models with tunable properties, and the cascading model is applied to them. Extensive simulations are implemented to reveal the potential relationship between the cascade robustness with topological variations occurring in the Internet network. In the simulations, each curve for null networks is averaged over 30 realizations. It should be noted that although the following analysis is performed by using the real Internet networks, our framework and proposed approach involved are applicable to other kinds of realistic networks.

### 4.1. Cascading Failures in Null Networks of Different Orders

Let us first investigate cascading failures in null networks of different orders (0K–3K). Figure 2 compares the results of cascading failures in these null networks to that in the original Internet with 3015 nodes. The curves demonstrate the association between the relative size $G$ and the redundancy parameter $\alpha$ under topological variation. As expecting in Figure 2, $G$ monotonically increases with the increase of $\alpha$ for each curve. Based on the definition of the cascading model, increasing $\alpha$ means each node has more capacity redundancy to receive the redistributed load from failed nodes, which will reduce the likelihood of subsequent overload failures. Then the robustness of the whole network becomes stronger (i.e., the robustness measurement index $G$ increases). Moreover, we observe that the network robustness against cascading failure is gradually weakened with increasing the order of null networks from 0K to 3K, leading to getting closer to the original one. This can be explained by the different role of each topological feature in the network robustness.

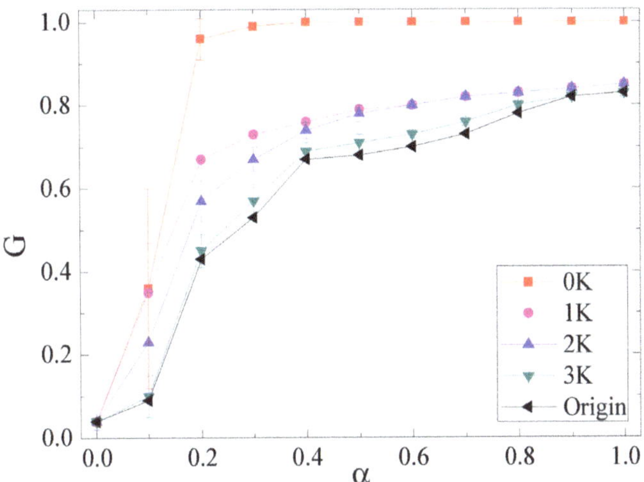

**Figure 2.** The relationship between $G$ and $\alpha$ in null networks of different orders for the Internet with 3015 nodes and 5156 edges.

Table 1 shows the variation of topological parameters for the Internet with 3015 nodes and its corresponding 0K–3K null networks. As it is seen, 0k null networks, which are obtained after sufficient randomization on the original network, only have the same number $N$ of nodes and average degree $<k>$ as the origin. Moreover, they display a Poisson degree distribution rather than a power-law (scale-free) distribution which the original network shows. Such homogeneous structure makes them much more robust than the original heterogeneous network. This is in line with prior studies on cascading failures in traditional model networks [8,23,29], where cascading failure was shown to occur less likely in a homogeneous network than in a heterogeneous one. Because it was shown that cascading dynamics are strongly related to the degree distribution of a network, it also raises a new question whether it is certain that both of them have the similar ability in resisting cascading failures if a network rigorously grasps the degree distribution of the original network.

In this work, 1K null networks are such examples but clearly perform very different ability from the original network. Compared with 1K null networks, the robustness of 2K null networks is further worse, and then their curve is closer to that of the original network, because 2K null networks maintain more micro-scale structures (i.e., assortativity). Though, there is yet a great difference between 2K null networks and the original one. Furthermore, the evolution of 3K null networks more resembles that of the original network, because they can preserve the transitivity characteristics of the original network. The above findings, hence, confirm that different topological features of networks all have great impact on the outcomes of cascading dynamics taking place on them. Actually, most previous studies based on network models pay attention to the degree distribution to evaluate the robustness of a network against cascading failures [23,29]. This study clearly shows that the degree distribution is not enough to guarantee the robustness of a real-life network, and 2K and 3K micro-scale structures (i.e., assortativity and transitivity) are also significant properties. Therefore, further studies are needed to reveal how the modification of assortativity, transitivity and higher-order properties affects cascading failures in case of keeping its lower-order properties of the original network.

**Table 1.** The variation of topological parameters for the Internet with 3015 nodes and its corresponding 0K–3K null networks. Each data point for 0K–3K null networks is averaged over 30 different network realizations.

| Network | $N$ | $<k>$ | $p(k)$ | $\gamma$ | $r$ | $c$ |
|---|---|---|---|---|---|---|
| Origin | 3015 | 3.42 | Power-law | 2.5 | −0.23 | 0.18 |
| 0K | 3015 | 3.42 | Poisson | − | −0.008 | 0.0009 |
| 1K | 3015 | 3.42 | Power-law | 2.5 | −0.22 | 0.10 |
| 2K | 3015 | 3.42 | Power-law | 2.5 | −0.23 | 0.12 |
| 3K | 3015 | 3.42 | Power-law | 2.5 | −0.23 | 0.18 |

*4.2. Cascading Failures in Null Networks of Tunable Assortativity*

Figure 3a shows the relationship between $G$ and $\alpha$ in null networks of increasing assortativity ($r = -0.20$) and decreasing assortativity ($r = -0.26$) for the Internet with 3015 nodes, where the $r$ value of the original network is $-0.23$. In contrast to the case of $r = -0.26$, for the curve of $r = -0.20$, the values of the error bars are so small(about 0.01) that we cannot see them. This can be explained by the fact that under the algorithm of the edge-swapping of increasing assortativity, the difference of connectivity patterns of all generated networks is small due to the constraint of a few nodes with high degree in the original network. Meanwhile, the structure (i.e., a large number of nodes with low degree) of the network can make all generated null networks of decreasing assortativity show a relatively large difference between their connectivity patterns. The difference of connectivity patterns accordingly affect the difference of cascade results, leading to the difference of the error bars of the two curves. In addition, from the curve of $r = -0.20$, we can see that $\alpha$ exerts a relatively small impact on the cascade results, which is different from the case of the other two curves. The reason is also related to the network structure.

More importantly, Figure 3a illustrates that $r$ has a strong effect on the network robustness. The networks of increasing assortativity take more robustness and resistance to the damage of the cascade as compared to those of decreasing assortativity for any choice of parameter $\alpha$. However, compared with the origin, there exist crossover points $\alpha_c$ ($\alpha_c \approx 0.6$) for the networks of increasing assortativity and $\alpha_s$ ($\alpha_s \approx 0.2$) for the networks of decreasing assortativity. When $\alpha_s < \alpha < \alpha_c$, the cascade robustness of the network with higher assortativity indeed increases. In contrast, when $\alpha > \alpha_c$ or $\alpha < \alpha_s$, the network with lower assortativity performs better. This means that the network robustness does not always increase monotonically with its assortativity, which is sensitive to the capacity redundancy. Similar evolution with different cross points can be observed in Figure 3b,c. It can also further confirm that the capacity redundancy is considered to determine how the assortativity affects the network robustness. This is different from the results obtained in traditional model networks [25], which indicate that the robustness of complex networks against cascading failures varied monotonously with the variations of its assortativity regardless of the capacity redundancy.

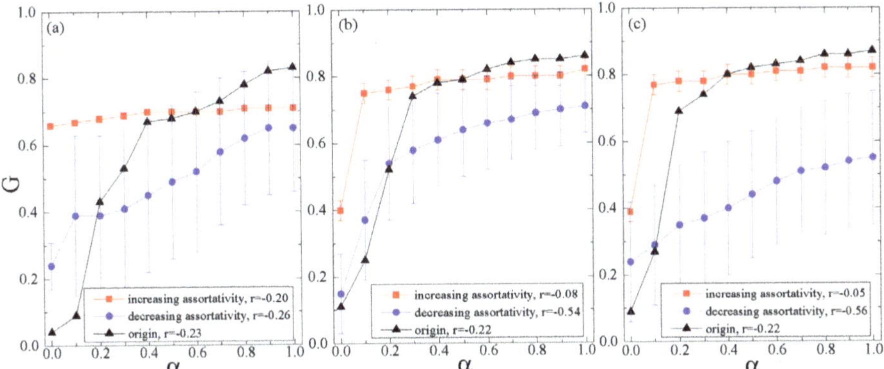

**Figure 3.** The relationship between $G$ and $\alpha$ in null networks of (**a**) increasing assortativity ($r = -0.20$) and decreasing assortativity ($r=-0.26$) for the Internet with 3015 nodes and 5156 edges ($r = -0.23$), (**b**) increasing assortativity ($r = -0.08$) and decreasing assortativity ($r = -0.54$) for the Internet with 530 nodes and 1289 edges ($r = -0.22$), and (**c**) increasing assortativity ($r = -0.05$) and decreasing assortativity ($r = -0.56$) for the Internet with 493 nodes and 1234 edges ($r = -0.22$).

### 4.3. Cascading Failures in Null Networks of Tunable Clustering Coefficient

Figure 4 plots the relationship between $G$ and $\alpha$ in null networks of increasing transitivity ($c = 0.31$) and decreasing transitivity ($c = 0.03$) for the Internet with 3015 nodes, where the $c$ value of the original network is 0.18. It can be seen that the networks of both increasing transitivity and decreasing transitivity are more invulnerable to cascading failure as compared to the original network, and the impact of increasing transitivity on promoting the network robustness is more prominent. For example, in the case of $\alpha = 0.3$, increasing transitivity and decreasing transitivity make the value of $G$ increase from 0.53 to 0.65 and 0.58, respectively.

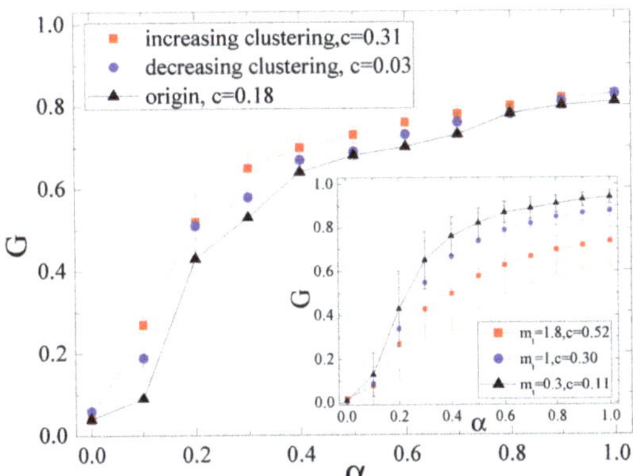

**Figure 4.** The relationship between $G$ and $\alpha$ in null networks of increasing transitivity ($c = 0.31$) and decreasing transitivity ($c = 0.03$) for the Internet. For comparison, the inset shows the relationship between $G$ and $\alpha$ in traditional networks (i.e., HK scale-free networks) with different values of clustering parameter $m_t$. Each data is averaged over 50 individual runs.

Such results also indicate a non-monotonic behavior between the transitivity property and the cascade robustness. This phenomenon is different from the result obtained in those traditional model networks as shown in the subgraph of Figure 4. Here we use a typical model, namely the HK scale-free network model proposed by Holm and Kim [26], in which the clustering coefficient can be adjusted by a particular control factor $m_t$. In the case of $m_t = 0$, the HK model degenerates into the well-known Barabási-Albert(BA) model [41]. The larger the $m_t$ value, the larger the clustering coefficient $c$. When applying our cascading model to the scale-free networks with tunable clustering coefficient $c$, the subgraph of Figure 4 shows that the robustness of such networks is monotonously reduced with the increase of $c$. Obviously, this is not in accordance with the result of our proposed null networks. Coupled with the results in Figure 3, we can confirm that due to the structural complexity of real-life networks, from which deviations of traditional model networks can bring great impacts on understanding and controlling cascading behaviors, and hence constructing null models for empirical analysis of cascading failures is of great significance.

### 4.4. Cascading Failures in Null Networks of Tunable Rich-Club Property

Until now, we have shown the effects of micro-scale features (i.e., assortativity and transitivity) on the cascade. However, we do not know what the effects of meso-scale features are on that. The rich-club and community structure are two typical meso-scale features of the Internet. In the following, the impact of them each on cascading behaviors will be also investigated.

Figure 5a shows the relationship between $G$ and $\alpha$ in null networks of increasing rich-club property ($\phi = 0.08$) and decreasing rich-club property ($\phi = 0.02$) for the Internet with 3015 nodes, where the $\phi$ value of the original network is 0.05. Clearly, one can see that the $\phi$ value is positively related to the network invulnerability against cascading failures, i.e., the higher $\phi$, the more desirable network behavior in resisting cascading failures. The similar evolutions can also be observed in Figure 5b,c. This suggests a plausible way to enhance network robustness by increasing rich-club property.

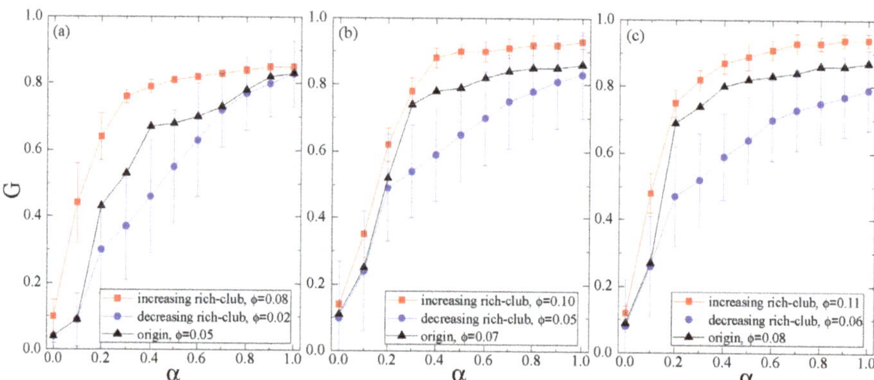

**Figure 5.** The relationship between $G$ and $\alpha$ in null networks of (**a**) increasing rich-club ($\phi = 0.08$) and decreasing rich-club ($\phi = 0.02$) for the Internet with 3015 nodes and 5156 edges ($\phi = 0.05$), (**b**) increasing rich-club ($\phi = 0.10$) and decreasing rich-club ($\phi = 0.05$) for the Internet with 530 nodes and 1289 edges ($\phi = 0.07$), and (**c**) increasing rich-club ($\phi = 0.11$) and decreasing rich-club ($\phi = 0.06$) for the Internet with 493 nodes and 1234 edges ($\phi = 0.08$).

### 4.5. Cascading Failures in Null Networks of Tunable Community Structure

In Figure 6, we plot the relationship between $G$ and $\alpha$ in null networks of increasing community structure ($Q = 0.63$) and decreasing community structure ($Q = 0.60, 0.57$, respectively) for the Internet with 3015 nodes, where the original $Q$ value is 0.62. With decreasing $Q$ of the original network, the cascade robustness of the null networks becomes stronger. However, when we increase $Q$, the robustness is almost the same as that of the original one. This can be explained by the fact that the original version of the Internet has already a clear community characteristic. Using the algorithm of increasing community structure, the $Q$ value can be adjusted to the maximum value ($Q = 0.63$), which is only increased by 0.01 as compared to the original value. Such enhancement of the community structure has no obvious effect on the robustness of the original network.

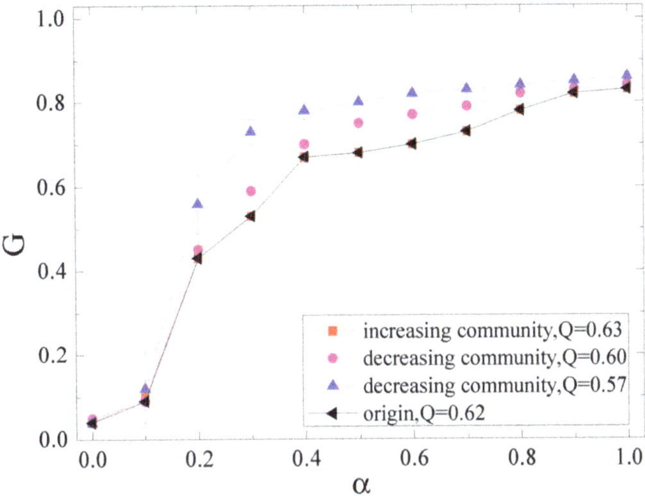


**Figure 6.** The relationship between $G$ and $\alpha$ in null networks of increasing community structure ($Q = 0.63$) and decreasing community structure ($Q = 0.60, 0.57$, respectively) for the Internet.

The characteristic of community structure is that the density of inner edges among communities is relatively greater than the density of external edges. To disclose the community structure effect in detail, Figure 7 displays the relationship between $G$ and $\alpha$ in null networks of rewiring edges within a community and between communities for the Internet with 3015 nodes. For generating these two null networks, we consider the edge-swapping algorithm of the 1K null model to only destroy the inward structure of each community and the external structure of all communities, respectively. It should be noticed that both of them are different from the classical 1K null network, which is based on rewiring edges within the whole original network and thus destroys the meso-scale characteristics completely.

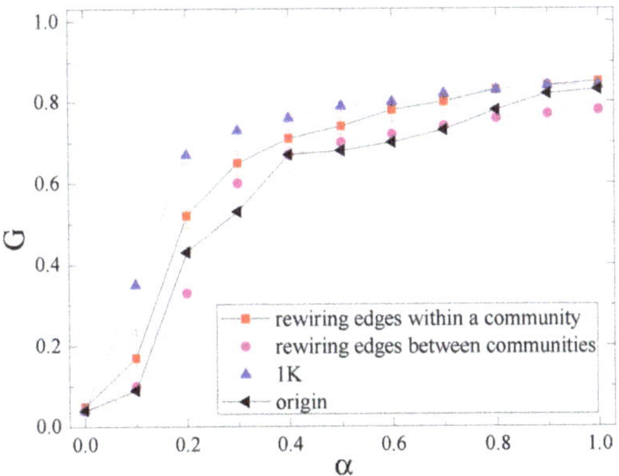


**Figure 7.** The relationship between $G$ and $\alpha$ in null networks of edge swapping within a community and between communities for the Internet.

As seen in Figure 7, different kinds of edges play different roles in the effects of cascading failure. Rewiring edges inside a community (i.e., the random exchange of inner edges) makes the network stronger to resist cascading failures, yet for rewiring edges among communities (i.e., the random exchange of exterior edges), the effect on promoting the network invulnerability is not clear enough because of their fluctuations. Furthermore, the network with the modification of inner edges is still more vulnerable than the 1K null network. Because the 1K null network completely destroys community structures of the original network, its modularity is lower than the other three cases, and the smaller modularity is in favor of the cascade robustness.

## 5. Conclusions

In this work, Internet networks were researched in terms of topology and the association with the cascade robustness was established. Considering realistic topological characteristics including degree distribution, assortativity, transitivity, rich-club coefficient, and modularity, we generated multiple null networks derived from three Internet AS-level networks. The methods used for the generation of null networks offer feasible network configurations inherently. By considering the routing rule of shortest-path flow, the effect of cascade-based attacks was investigated in each of the original networks and its various null networks with the topological variation.

By comparing the largest connected sizes of these attacked networks, our results reveal that the degree distribution is not enough to identify the robustness of a network, and micro-scale structures (i.e., assortativity and transitivity) and meso-scale structures (i.e., rich-club and community structure) are also important for that. In detail, at the micro-scale level, both assortativity and clustering coefficient show a non-monotonic behavior with network robustness. Moreover, the impact of the assortativity on network robustness is related to the capacity redundancy of nodes. In comparison to decreasing transitivity, increasing transitivity contributes more to the promotion of network robustness. At the meso-scale level, the rich-club structure is positively related to network robustness, indicating that increasing it leads to stronger robustness. In contrast, the modularity is inversely related to network robustness, indicating that the network robustness increases with reducing modularity characteristic. In addition, the topological structure within every community plays a more significant role in improving the robustness as compared to that between communities.

The results obtained here are meaningful in guiding the construction or optimization of the Internet to resist the propagation of cascading failure effectively. Inspired by this work, we will aim to understand the relationship between complex characteristics of the Internet and cascading failure more comprehensively in our future studies. For example, an interesting challenge is to analyze interdependence of the systems in such a way to launch a robust network design. Our results also demonstrate the significance of constructing null networks for understanding and analysis of cascading failure in more real-world networks, because it can produce substantially different results from those in traditional model networks. We expect the analysis method proposed here is promising for extended applications in studying robust systems with different network structures in the real-world other than the Internet as well.

**Author Contributions:** Conceptualization, L.D. and X.-K.X.; Methodology, L.D. and X.-K.X.; Data curation, S.-Y.L. and Q.Y.; Formal analysis, L.D. and X.-K.X.; Writing—original draft preparation, L.D.; Validation, all authors; Writing—review and editing, X.-K.X.

**Funding:** This work was supported by the National Natural Science Foundation of China (61403183 and 61773091), the Key Research and Development Plan of Liaoning Province (2018104016), the LiaoNing Revitalization Talents Program (XLYC1807106), and the Program for the Outstanding Innovative Talents of Higher Learning Institutions of Liaoning (LR2016070).

**Conflicts of Interest:** The authors declare no conflicts of interest.

## References


1. Dorogovtsev, S.N.; Goltsev, A.V.; Mendes, J.F.F. Critical phenomena in complex networks. *Rev. Mod. Phys.* **2008**, *80*, 1275–1335. [CrossRef]
2. Newman, M. J. *Networks: An Introduction*; Oxford University Press: Oxford, UK, 2010.
3. Giulio, C.; Tizano, S.; Fabio, S.; Diego, G.; Andrea, G.; Guido, C. The statistical physics of real-world networks. *Nat. Rev. Phys.* **2019**, *1*, 58–71.
4. Albert, R.; Jeong, H.; Barabási, A.L. Error and attack tolerance in complex networks. *Nature* **2000**, *406*, 6794. [CrossRef] [PubMed]
5. Holme, P.; Kim, B.J.; Yoon, C.N.; Han, S.K. Attack vulnerability of complex networks. *Phys. Rev. E* **2002**, *65*, 056109. [CrossRef] [PubMed]
6. Xia, Y.X.; Hill, D.J. Attack vulnerability of complex communication networks. *IEEE Trans. Circuits Syst. II Exp. Briefs* **2008**, *55*, 65–69. [CrossRef]
7. Fu, C.; Wang, Y.; Wang, X.; Gao, Y. Multi-node attack strategy of complex networks due to cascading breakdown. *Chaos Solitons Fract.* **2018**, *106*, 61–66.
8. Motter, A.E.; Lai, Y.C. Cascade-based attacks on complex networks. *Phys. Rev. E* **2002**, *66*, 065102. [CrossRef]
9. Zheng, J.F.; Gao, Z.Y.; Zhao, X.M. Modeling cascading failures in congested complex networks. *Phys. A* **2007**, *385*, 700–706. [CrossRef]
10. Chu, C.; Lu, H.H. Complex networks theory for modern smart grid applications: A survey. *IEEE J. Emerg. Sel. Top. Circuits Syst.* **2017**, *7*, 177–191. [CrossRef]
11. Zhang, Y.; Arenas, A.; Yağan, O. Cascading failures in interdependent systems under a flow redistribution model. *Phys. Rev. E* **2018**, *97*, 022307. [CrossRef]
12. Schäfer, B.; Witthaut, D.; Timme, M.; Latora, V. Dynamically induced cascading failures in power grids. *Nat. Commun.* **2018**, *9*, 1975. [CrossRef] [PubMed]
13. Ozel, O.; Sinopoli, B.; Yağan, O. Uniform redundancy allocation maximizes the robustness of flow networks against cascading failures. *Phys. Rev. E* **2018**, *98*, 042306. [CrossRef]
14. Dey, P.; Mehra, R.; Kazi, F.; Wagh, S.; Singh, N.M. Impact of topology on the propagation of cascading failure in power Grid. *IEEE Trans. Smart Grid* **2016**, *7*, 1970–1978. [CrossRef]
15. Tu, H.; Xia, Y.X.; Iu, H.H.; Chen, X. Optimal robustness in power grid from a network science perspective. *IEEE Trans. Circuits Syst. II Exp. Briefs* **2019**, *66*, 126–130. [CrossRef]
16. Wang, J.W.; Jiang, C.; Cheng, J.F. Robustness of Internet under targeted attack: A cascading failure perspective. *J. Netw. Comput. Appl.* **2014**, *40*, 97–104. [CrossRef]
17. Hong, S.; Wang, B.Q.; Ma, X.M.; Wang, J.H.; Zhao, T.D. Cascading failure analysis and restoration strategy in an interdependent network. *J. Phys. A Math. Theor.* **2015**, *48*, 485101. [CrossRef]
18. Zhang, X.; Liu, D.; Zhan, C.J.; Tse, C.K. Effects of cyber coupling on cascading failures in power systems. *IEEE J. Emerg. Sel. Top. Circuits Syst.* **2017**, *7*, 228–238. [CrossRef]
19. Chen, Z.; Wu, J.J.; Xia, Y.X.; Zhang, X. Robustness of interdependent power grids and communication Networks: A Complex Network Perspective. *IEEE Trans. Circuits Syst. II Exp. Briefs* **2018**, *65*, 115–119. [CrossRef]
20. Min, B.; Zheng, M. Correlated network of networks enhances robustness against catastrophic failures. *PLoS ONE* **2018**, *13*, e0195539. [CrossRef]
21. Chen, Z.; Du, W.B.; Cao, X.B.; Zhou, X.L. Cascading failure of interdependent networks with different coupling preference under targeted attack. *Chaos Solitons Fract.* **2015**, *80*, 7–12. [CrossRef]
22. Babaei, M.; Ghassemieh, H.; Jalili, M. Cascading failure tolerance of modular small-world networks. *IEEE Trans. Circuits Syst. II Exp. Briefs* **2011**, *58*, 527–531. [CrossRef]
23. Wu, J.J.; Gao, Z.Y.; Sun, H.J. Cascade and breakdown in scale-free networks with community structure. *Phys. Rev. E* **2006**, *74*, 066111. [CrossRef]
24. Ren, W.; Wu, J.; Zhang, X.; Lai, R.; Chen, L. A stochastic model of cascading failure dynamics in communication Networks. *IEEE Trans. Circuits Syst. II Exp. Briefs* **2018**, *65*, 632–636. [CrossRef]
25. Yang, Z.; Liu, J. Robustness of scale-free networks with various parameter against cascading failures. *Phys. A* **2018**, *492*, 628–638. [CrossRef]
26. Zheng, J.F.; Gao, Z.Y.; Zhao, X.M. Clustering and congestion effects on cascading failures of scale-free networks. *Eur. Lett.* **2007**, *79*, 58002. [CrossRef]

27. Ghanbari, R.; Jalili, M.; Yu, X. Correlation of cascade failures and centrality measures in complex networks. *Future Gener. Comp. Syst.* **2018**, *83*, 390–400. [CrossRef]
28. La, R. Influence of clustering on cascading failures in interdependent systems. *IEEE Trans. Netw. Sci. Eng.* **2018**, *2*, 2805720. [CrossRef]
29. Sun, S.W.; Wu, Y.F.; Ma, Y.L.; Wang, L.; Gao, Z.K.; Xia, C.Y. Impact of degree heterogeneity on attack vulnerability of interdependent networks. *Sci. Rep.* **2016**, *6*, 32983. [CrossRef]
30. Maslov, S. Specificity and stability in topology of protein networks. *Science* **2002**, *296*, 910–913. [CrossRef]
31. Orsini, C.; Dankulov, M.M.; Colomer-de-Simón, P.; Jamakovic, A.; Mahadevan, P.; Vahdat, A.; Bassler, K.E.; Toroczkai, Z.; Boguná, M.; Caldarelli, G.; et al. Quantifying randomness in real networks. *Nat. Commun.* **2015**, *6*, 8627. [CrossRef]
32. Shang, K.K.; Small, M.; Xu, X.K.; Wen, W.S. The role of direct links for link prediction in evolving networks. *Europhys. Lett.* **2017**, *117*, 28002. [CrossRef]
33. Liu, B.; Xu, S.; Li, T.; Xiao, J.; Xu, X.K. Quantifying the effects of topology and weight for link prediction in weighted complex Networks. *Entropy* **2018**, *20*, 363–367. [CrossRef]
34. Cui, W.K.; Shang, K.K.; Zhang, Y.J.; Xiao, J.; Xu, X.K. Constructing null networks for community detection in complex networks. *Eur. Phys. J. B* **2018**, *91*, 145–153. [CrossRef]
35. Mahadevan, P.; Krioukov, D.; Fall, K.; Vahdat, A. Systematic topology analysis and generation using degree correlations. *ACM SIGCOMM Comput. Commun. Rev.* **2006**, *36*, 135–146. [CrossRef]
36. Mahadevan, P.; Hubble, C.; Krioukov, D. Orbis: Rescaling degree correlations to generate annotated Internet topologies. *ACM SIGCOMM Comput. Commun. Rev.* **2007**, *37*, 325–336. [CrossRef]
37. Zhou, S.; Mondragon, R.J. Structural constraints in complex networks. *New J. Phys.* **2007**, *9*, 173. [CrossRef]
38. Xu, X.K.; Zhang, J.; Sun, J.; Small, M. Revising the simple measures of assortativity in complex networks. *Phys. Rev. E* **2009**, *80*, 56106. [CrossRef]
39. Xu, X.K.; Zhang, J.; Small, M. Rich-club connectivity dominates assortativity and transitivity of complex networks. *Phys. Rev. E* **2010**, *82*, 046117. [CrossRef]
40. Leskovec, J. Stanford Network Analysis Project. 2018. Available online: http://snap.standford.edu/data/as.html (accessed on 15 November 2019).
41. Barabási, A.L.; Albert, R. Emergence of scaling in random networks. *Science* **1999**, *286*, 509–512. [CrossRef]

*Article*

# Service-Oriented Model Encapsulation and Selection Method for Complex System Simulation Based on Cloud Architecture

**Siqi Xiong, Feng Zhu *, Yiping Yao, Wenjie Tang and Yuhao Xiao**

College of Systems Engineering, National University of Defense Technology, Changsha 410073, China; siqi@mail.ustc.edu.cn (S.X.); ypyao@nudt.edu.cn (Y.Y.); tangwenjie@nudt.edu.cn (W.T.); xiaoyuhao19@nudt.edu.cn (Y.X.)
* Correspondence: zhufeng@nudt.edu.cn

Received: 15 August 2019; Accepted: 13 September 2019; Published: 14 September 2019

**Abstract:** With the rise in cloud computing architecture, the development of service-oriented simulation models has gradually become a prominent topic in the field of complex system simulation. In order to support the distributed sharing of the simulation models with large computational requirements and to select the optimal service model to construct complex system simulation applications, this paper proposes a service-oriented model encapsulation and selection method. This method encapsulates models into shared simulation services, supports the distributed scheduling of model services in the network, and designs a semantic search framework which can support users in searching models according to model correlation. An optimization selection algorithm based on quality of service (QoS) is proposed to support users in customizing the weights of QoS indices and obtaining the ordered candidate model set by weighted comparison. The experimental results showed that the parallel operation of service models can effectively improve the execution efficiency of complex system simulation applications, and the performance was increased by 19.76% compared with that of scatter distribution strategy. The QoS weighted model selection method based on semantic search can support the effective search and selection of simulation models in the cloud environment according to the user's preferences.

**Keywords:** complex system simulation; cloud computing architecture; service-oriented modeling; semantic search framework; QoS-based service selection

---

## 1. Introduction

The continuous evolution of complex systems (e.g., social systems, ecosystems, and war systems) has had a tremendous impact on people's daily life and social development. Due to the limitation of existing theoretical analysis methods and the difficulty of experimental analysis methods in some real-world complex systems (e.g., geological changes, nuclear explosions, economic growth [1], and ecosystem evolution), complex system simulation technology has gradually become an attractive approach for the research on complex systems and their complexity [2].

Complex system simulation applications often contain a large number of simulation model entities, and there are complex interactions between these entities, also the entities and the external environment. Such system simulations usually have a large computational load [3]. With the increase in scale and complexity of complex system simulation applications, there are increasingly requirements for the composite mode of simulation models, the computational capabilities of simulation architectures, and the execution efficiency of simulation applications. The popularity of cloud computing technology provides a new approach, platform architecture, and efficient computing power for the research and development of complex system simulations. Simulation users can use the computing resources in the cloud environment on demand at different terminals and invoke the simulation model services stored in the cloud center to assemble complex system simulation applications. Therefore, the development of service-oriented simulation models has gradually become a prominent topic in the field of complex system simulation [4]. Service-oriented technology is mainly directed at models with large computational requirements, such as an electromagnetic environment calculation model, a ballistic path planning model, a radar detection model, and so forth. These models are expected to be provided outward as a shared service. Then, their integration and code porting could be eliminated, and the construction efficiency of complex system simulation applications and the utilization of related models could be improved. The interoperability between heterogeneous simulation models and the distributed collaborative calculation of simulation models on multiple computing nodes could be realized, which could improve the execution efficiency of simulation applications. Therefore, it is necessary to carry out the research on how to construct and select the service-oriented complex simulation models based on cloud computing environment.

To make the complex system simulation model into a shared service in cloud, firstly, models with large computational requirements need to be encapsulated into simulation services in the cloud environment and be parallel processed in the execution of complex system simulation applications under a cloud-based simulation model service framework. Because the simulation model services released and stored in the cloud center have differences in attributes, functions, and quality of service (QoS), it is necessary to find and select the appropriate simulation model that accurately meets the user's requirements in terms of function and can provide high QoS for building complex system simulation applications [5]. Reusable model development (RUM) specification [6] cannot support network communication between a simulation model and a simulation engine under the cloud architecture. Ontology web language [7] (OWL)-based simulation model search methods lack the mechanism to search simulation models through the correlations between models. Also, current simulation model optimization selection algorithms lack the induction for QoS [8] attributes of simulation models in the cloud environment and cannot provide a selection mechanism that satisfies the user's preference for a model's QoS.

In order to solve the abovementioned problems in the existing studies, this paper proposes a service-oriented model encapsulation and selection method for complex system simulation based on cloud architecture. The novelty and contribution of this method includes that it designs a cloud-service-oriented reusable model development (C-RUM) specification to encapsulate the simulation model into a shareable simulation service in the cloud, and then devises a cloud-based simulation model service framework, which solves the problem of network communication in the former RUM specification. This method also uses a knowledge graph [9] to describe the simulation model services and establishes a model semantic search framework in the constructed model description knowledge graph, which supports users in setting correlations between models to obtain the required model. A QoS weighted-based optimization selection algorithm is also proposed, which can select the optimal simulation model that satisfies the user's preference for QoS according to a weighted comparison of QoS indices.

The organization of this paper is as follows: Section 2 discusses related works. Section 3 introduces the C-RUM specification and the cloud-based simulation model service framework. Section 4 introduces the selection method of simulation models based on semantic search. Section 5 describes a case study of the service-oriented model encapsulation and selection method. Section 6 is the summary and the outlook for future work.

## 2. Related Works

### 2.1. High-Level Architecture (HLA)-Based Simulation Model Development Specification

The basic idea of HLA is to use an object-oriented method to design, develop, and implement object models of different levels and granularities and to obtain high-level interoperability and reusability of simulation models and simulation systems. The object model template (OMT) is a standardized description of the properties of simulation models and their interaction formats, but it is not a standard for establishing the object model. With the development of complex system simulation, there are higher requirements for the efficiency, flexibility, and openness of simulation model development. HLA-based simulation model development specification gradually exposes some problems in the application process, such as efficiency, ease of use, fault tolerance, dynamic compatibility, and so forth [10].

### 2.2. RUM Specification

In order to realize the interapplication and interplatform reuse of simulation models and the rapid development of simulation applications, many researchers have proposed reusable and composable development specifications and methods for simulation models. Lee et al. applied the product line engineering concept to the development of simulation model components [11]. Feng et al. proposed a reusable component model development approach for parallel and distributed simulation, requiring that the simulation model have self-contained features; that is, the model can be packaged and released independently, without relying on other models, and is separate from the simulation engine [6]. Jianbo and Yiping proposed a reusable component model framework (RCMF) model development tool called SuKit, which can be used to regenerate models and guide model integration [12].

A patent for RUM specification [13] was proposed by Yiping and Feng and revised in 2017, which has been widely used. RUM specification encapsulates the simulation model into a separate service entity, and the model and the outside world can only interact through the "service interface". RUM specification enables local reuse and composition of simulation models, realizing invocation and communication of simulation models by passing local parameters. However, in the cloud environment, the user terminal and the cloud server are connected by the network, and RUM specification does not support communication between the simulation model and the specific simulation engine framework in the network. Therefore, the simulation model developed by RUM specification cannot be provided as a shared service released and stored in the cloud environment.

### 2.3. OWL-Based Simulation Model Search Method

Ontologies in the Semantic Web can describe simulation models at the semantic level. Web service ontology description language (OWL-S) was designed to make the Web service an entity which computers can understand based on the description of ontology. OWL-S describes Web services in three aspects: (1) service profile, (2) service model, and (3) service grounding [14,15]. Ontology can improve the accuracy of simulation model search by describing simulation models based on semantics [16]. In order to support the composite modeling of complex system simulation applications, some experts have carried out research on simulation service description methods based on semantics and have proposed description ontologies of simulation model resources (e.g., OWL-SS [17] and OWL-SM [18]). At present, OWL-based simulation model description methods generally lack descriptions of the characteristics of simulation models in the cloud environment [19] and lack expression of the correlations between simulation models, which are not effective enough to support users in searching and selecting relevant models conveniently through the correlations between models in the cloud environment.

### 2.4. QoS-Based Simulation Model Selection Method

Similar to Web services, QoS is a key factor in choosing simulation models that are stored in the cloud environment as a service [20]. At present, many researchers have defined suitable QoS indices for simulation services and have proposed model selection mechanisms based on QoS [21]. However, current descriptions of simulation models lack the induction of QoS attributes of simulation models in the cloud environment [22]. Current selection algorithms lack a selection mechanism that can select models in the cloud environment according to users' preferences for QoS indices and thus cannot meet users' specific QoS requirements when constructing complex system simulation applications.

In summary, current HLA-based simulation model development specification and RUM specification cannot support simulation models as a shared service to be invoked and operated on cloud architecture by developers. Existing OWL-based model search methods and QoS-based model selection methods cannot support users in searching for relevant models through the correlations between models and selecting simulation models that meet their QoS requirements to construct complex system simulation applications in the cloud environment.

## 3. C-RUM Specification and Cloud-Based Simulation Model Service Framework

### 3.1. C-RUM Specification

Complex system simulation applications usually contain some simulation models which have a large computational load. The operation of such a model requires an immense amount of computing resources, making other simulation models in the same process fall into a long queue, thus delaying the advancement of the simulation timing and reducing the execution efficiency of the simulation application. If such simulation models are encapsulated in the form of shared services and are distributed and stored in the cloud environment, the construction efficiency of complex system simulation applications and the utilization of related models could be improved. Cloud computing resources can be used to realize interoperability between heterogeneous simulation models and distributed collaborative computing of simulation models on multiple computing nodes, so as to improve the execution efficiency of simulation applications.

The RUM specification can implement local invocation and communication of simulation models. However, in the cloud architecture, the cloud server where the simulation service model is located and the user terminal are interconnected through the network. The RUM specification does not support communication between the simulation model and the specific simulation engine framework on the network. This paper proposes C-RUM specification by transforming the RUM specification. The purpose is to invoke the simulation model encapsulated by C-RUM specification as a form of shared service in the cloud architecture, to make the model service transmit the data through the network protocol to interact with the simulation engine framework, and to implement distributed collaborative computing and heterogeneous execution of simulation applications. The original RUM specification specifies seven standard (service) interfaces for the simulation model to interact with the outside world—model initialization, parameter input, model status recovery, parameter and status adjustment, data output, model status acquisition, and model calculating interfaces—to provide seven standard operations, as shown in Figure 1.

In the cloud architecture, the service interfaces in the original RUM specification cannot identify or parse the network data. Therefore, the C-RUM specification defines the network data input interface and network data output interface. These two interfaces are used to encapsulate the seven service interfaces in order to implement data conversion between the network and the original interfaces. According to the execution flow of the simulation model under the original RUM specification [6], the C-RUM specification divides the network interaction into three types: model invoke command, model calculate command, and calculating data and model status output, as shown in Figure 1. The detailed function of the two new interfaces and the network interaction of the C-RUM-encapsulated simulation model is discussed below.

Network data input interface. This interface is mainly used for parsing network data transmitted based on the socket network transmission protocol. The purpose of the parsing is to obtain the target standard service interface of the network data packet and convert the input data into the specified standard data format of the corresponding service interface. Finally, the parsed instructions or parameters are passed to the target standard service interface. The model invoke and calculate commands need to be analyzed by the network data input interface.

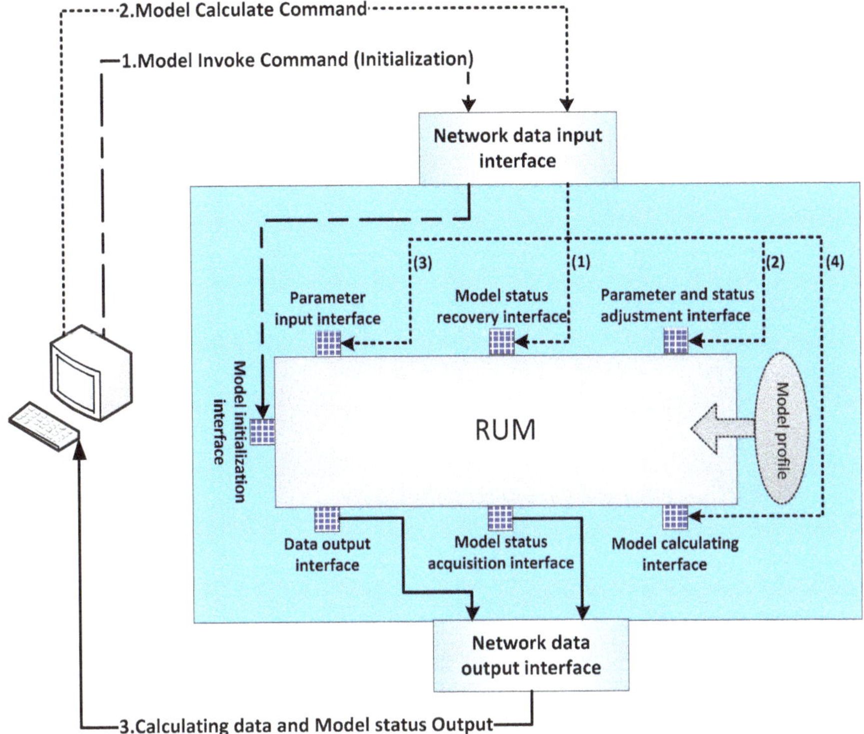



**Figure 1.** Cloud-service-oriented reusable model development (C-RUM) specification and its execution flow.

Model invoke command: When the simulation system of the user terminal needs to be initialized, the terminal will send a model invoke command to the simulation model in the cloud server, and the distributed invocation system in the cloud server will mount the simulation service model into a process. Then, the model initialization command and related parameters are passed to the model initialization interface through the network data input interface to complete the initialization operation of the simulation service model.

Model calculate command: When the simulation system in the user terminal needs the simulation service model to calculate, the model calculate command will be sent to the simulation model. After the command is parsed by the network data input interface, the model will (1) first recover its status through the model status recovery interface, (2) then check whether there is a working parameter adjustment instruction to adjust the working parameters and status, (3) then set the input data through the parameter input interface, and (4) finally start the simulation model calculating operation through the model calculating interface.

Network data output interface. This interface is used to encapsulate the data output from the simulation model after calculation and the status information of the simulation model. That includes indicating the standard interface source of the data and the destination address of the transmission and converting them into the format for the socket network communication protocol. Before output to the network, the calculating result and model status need to be encapsulated by the network data output interface.

Calculating result and model status output: After the simulation service model finishes its calculation, the network data output interface will obtain the data after calculating from the simulation model output interface and acquire the model status from the simulation model status acquisition interface and encapsulate the data into a socket communication protocol package. The package will be forwarded to the simulation system in the corresponding user terminal through the distributed architecture in the cloud environment.

### 3.2. Cloud-Based Simulation Model Service Framework

Most of the current complex simulation system application frameworks need to integrate or migrate simulation models into specific simulation platforms, making it difficult to separate the models from the platform. For different simulation platforms, the operation mechanism of the simulation engine is quite different, and it is not easy to bind the service simulation model of the cloud center to a specific simulation platform. In order to invoke the service-oriented simulation model in the cloud architecture without relying on the simulation platform, running under any simulation engine framework [23,24], this paper proposes a cloud-based simulation model service framework, as shown in Figure 2.

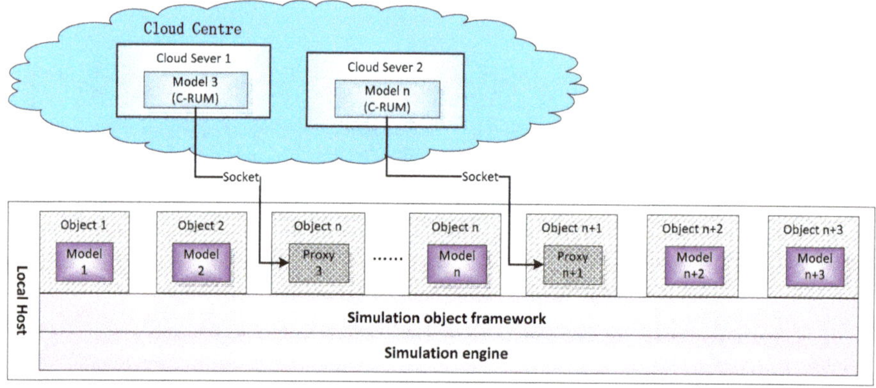


**Figure 2.** Cloud-based simulation model service framework.

The complex system simulation application consists of a large number of simulation object instances, and there are complex interactions and collaborative calculations between the instances. The simulation object framework is built on a specific simulation engine, and the simulation object instances are defined by the simulation object framework (including the declarations of these simulation object instances, their roles in the simulation application, the interaction between them, etc.). Each object instance is implemented by a specific simulation model. The local simulation model can be directly mounted or integrated into the object framework, while the simulation model service stored in the cloud environment relies on the communication with its proxy model in the corresponding object framework. The proxy model does not have the specific function of the simulation service model; it only takes the place of the simulation service model in the entire simulation object framework, defining the interaction relationship with other models. When the simulation application needs to interact with the simulation service model in the cloud server or obtain its status, the simulation object framework will accept and transmit the corresponding parameters and data through the socket communication between the proxy model and the simulation service model. The cloud-based simulation model service framework is applicable to the invocation and operation of the C-RUM-encapsulated simulation model in the cloud environment and does not depend on a specific simulation platform.

## 4. Simulation Model Selection Method Based on Semantic Search in Cloud Environment

### 4.1. Semantic Search Framework

The traditional Web services description language (WSDL)-based [25] simulation model search mechanism uses keyword matching to find simulation model description texts with the same keywords. A knowledge graph uses a more expressive way to describe simulation models by semantic description, and a search method based on a knowledge graph can find simulation models at the semantic level through the link relations between data and things [26]. Compared with ontology description language, a knowledge graph stores resource description framework (RDF) [27] triples in the graph database directly, which means the correlations between simulation models can be described in a simple and intuitive way in the form of graphs.

In this study, a description method of cloud simulation model resources based on a knowledge graph [28] was used to describe simulation models, which describe the characteristics of cloud simulation models and their QoS indices. Then, a simulation model semantic search framework was proposed based on the simulation model description knowledge graph. This search framework provides two patterns for simulation model search: (1) users can associate the required simulation model by attribute information such as the name, domain, type, time scale, and model granularity of the simulation model; or (2) users can search for the required simulation model by the correlations between models. According to the search conditions input by the user, simulation models that meet the search conditions can be found in the knowledge graph stored in the graph database, as shown in Figure 3.

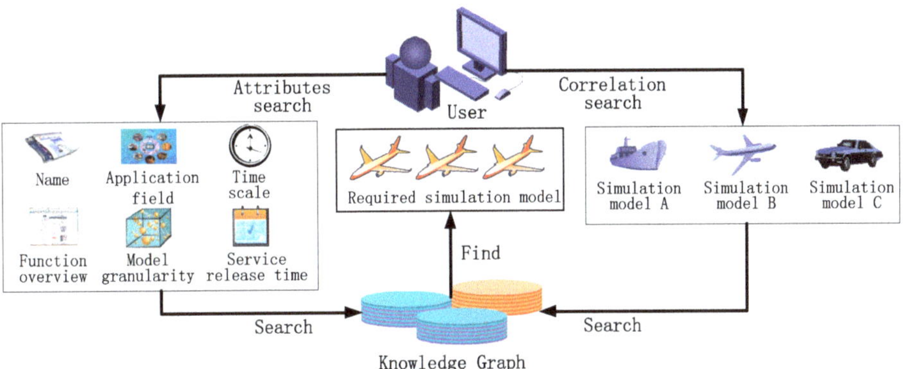


**Figure 3.** Simulation model semantic search framework.

Under the search framework proposed in this paper, users input model attribute requirements as semantic search conditions stored in the array Attributes_conditions [M]. Each item of the array corresponds to 1 to M attribute requirements of the simulation model. The user can input one or more attribute requirements (e.g., model name, domain, category, time scale, model granularity, etc.) as semantic search conditions to search for simulation models that meet the requirements of these attributes. The user can also input the required association model and specific association relationships (e.g., command relationship, equipment-carrying relationship, etc.) as semantic search conditions stored in the two variables Correlated model and Relationship, respectively, to search for simulation models that have a certain relationship with the correlated model. The input of correlated models is necessary in this search pattern. Algorithm 1 shows the semantic search algorithm.

Data_Base represents a knowledge graph database that stores simulation model description information and correlation relationships. model $\nleqslant \alpha$ indicates that the simulation model does not satisfy the attribute requirement $\alpha$ by the judgment method of fuzzy search combined with synonym expansion. relationship (Correlated model, model) indicates the correlation between correlated model and present model. Relationship $\nleq \beta$ indicates that the specified association relationship does not satisfy the correlation between correlated model and present model by means of fuzzy search combined with synonym expansion. push_into_list (model, $\Omega$) indicates adding the simulation model into the model initial set $\Omega$.

---

**Algorithm 1** Semantic_Search

---

**Input:** *Attributes_conditions* [M], the vector for storing model attribute requirements;
*Correlated model*; *Relationship*, the relationship with correlated model;
**Output:** $\Omega$, simulation model initial set;
1: Boolean flag1 ← true, flag2 ← true;
2:  **if** (*Attributes_search_conditions* ≠ *null*)||(*Relationship_search_conditions* ≠ *null*) then
3:     **for each** *model* ∈ *Data_Base* **do**   // Loop traversal of the simulation model
4:         **for** *i* ← 0 **to** *M* **do**   // Loop traversal of model attribute requirement condition
5:             **if** *model* $\nless$ *Attributes_search_conditions* [*i*] **then** flag1 ← false;
6:         **end for**
7:         **if** *relationship* (*Correlation model, model*) ≠ *NULL*
8:             **if** *Relationship* $\nleq$ *relationship* (*Correlation model, model*)
**then** flag2 ← false;
9:             **else** flag2 ← false;
10:         **if** (*flag1* & *flag2*) **then** *push_into_list* (*model*, $\Omega$);
11:     **end for**
12: **end if**
13: **return** $\Omega$

---

*4.2. QoS Weighted-Based Simulation Model Selection Method in Cloud Environment*

The simulation model that the user needs to use has to not only meet the requirements of its function but also have high QoS to reach the quality requirements of building complex system simulation applications. The simulation models obtained under the semantic search framework proposed in Section 4.1. are not unique, and they have similar functions and attributes, but they differ in terms of QoS. In order to build higher-quality complex system simulation applications, after the initial set that meets the search conditions is acquired under the semantic search framework, it is necessary to order that set through a QoS-based selection mechanism and select the optimal simulation models from the ordered candidate set, as shown in Figure 4.

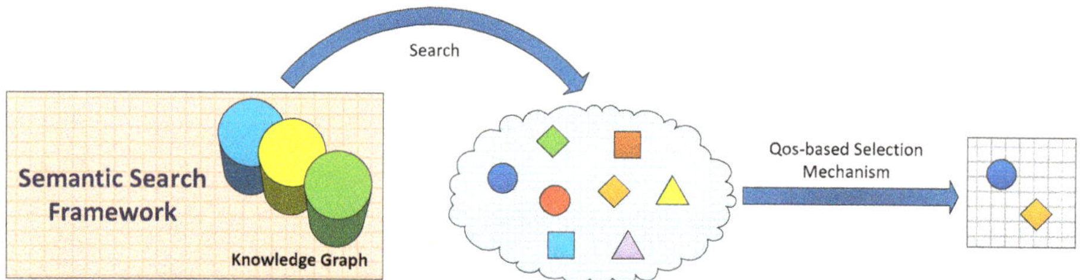


**Figure 4.** Search and selection process of simulation model.

The QoS weighted-based simulation model selection mechanism proposed in this paper can support users in customizing QoS index weights and selecting the simulation model that satisfies their QoS preference from the initial set according to the weighted comparison of QoS indices. The specific method is discussed below.

Definition of QoS indices. Referring to the QoS indices of Web services and considering the uniqueness of the simulation model as a kind of special Web service [21,29], the QoS indices of the simulation model in the cloud environment can be summarized as follows:

1.  Model performance ($Q_M$) is determined by the computation of the model. A simulation model with more computations has lower model performance.
2.  Communication capability ($Q_C$) reflects the communication capability of the link between the user's terminal node and the cloud server.
3.  Availability ($Q_A$) indicates the probability that the simulation model can be called and used. It is defined by the mean time between failures and the mean time to repair.
4.  Reliability ($Q_R$) is defined by the execution success rate of the service, which refers to the probability of obtaining the correct response to the user's requirements within the maximum expected time range.
5.  Security ($Q_S$) is measured by the data management capability of a model service, which mainly depends on the user's historical experience. Terminal users should be given a [0, 10] range to score the service (regarding the confidentiality, integrality, realness, etc., of data) after using it. Then, the value of $Q_S$ is the average score; with the increase and accumulation of evaluations, this value becomes reliable.

QoS weighted-based selection algorithm. The above five attributes ($Q_M$, $Q_C$, $Q_A$, $Q_R$, and $Q_S$) are all positive metrics; that is, the higher the value, the higher the quality. In order to eliminate the gap between the different QoS indices, we used the following formula [22] to limit their values to the range of [0, 1]:

$$V\left(Q_i^k\right) = \frac{maxQ_i^k - Q_i^k}{maxQ_i^k - minQ_i^k}. \tag{1}$$

These five QoS indices are assigned numbers 1–5. $Q_i^k$ indicates the value of the $i$th QoS index of the $k$th model in the candidate set, $maxQ_i^k$ and $minQ_i^k$ indicate the maximum and minimum values, respectively, that the QoS index may reach, and $V\left(Q_i^k\right)$ indicates the value after standardization of this QoS index.

After entering the search condition under the search framework, the simulation user also needs to provide a QoS preference, which is expressed by a weight vector as the following formula:

$$W = (w_i\,, 1 \le i \le 5, \sum w_i = 1). \tag{2}$$

That is, the percentage each QoS index should be accounted for. According to the weight vector given by the user, the total QoS index of the $k$th model in the candidate set is

$$Q^k = \sum_{i=1}^{5} w_i \leftarrow V\left(Q_i^k\right). \tag{3}$$

The model that meets the user's search conditions under the semantic search framework will be added to the initial set. According to the weight vector representing the QoS preference provided by the user, the target QoS index Q of each model in the initial set is obtained by the above formulas. Finally, the candidate set of simulation models ordered by Q will be provided to the user for selection. Algorithm 2 shows the QoS weighted-based model selection process.

---

**Algorithm 2** QoS Weighted-Based_Selection

---

**Input:** W, Simulation model QoS index weight vector;
$\Omega$, Simulation model initial set (from Algorithm 1);
**Output:** $\Phi$, Ordered model candidate set;
1:  **if** $\Omega \neq$ null **then**
2:      **for each** *model* $\in \Omega$ **do**            // Loop traversal of model initial set
3:          **for each** $i \leftarrow 0$ to 5 **do**            // Loop through 5 QoS indices
4:              $V(Q_i) \leftarrow \frac{maxQ_i - Q_i}{maxQ_i - minQ_i}$            // Calculate the standard value of the QoS index
5:              $Q \leftarrow \sum_{i=1}^{5} w_i \cdot V(Q_i)$       // Calculate the target QoS value of the simulation model
6:              *push_into_list*(<model,Q>, $\Phi$)       // Insert the binary <model, Q> into the set $\Phi$
7:          **end for**
8:      **end for**
9:      *rank_list_by* $(\Phi, Q)$        // Sort the elements in $\Phi$ by Q
10: **end if**
11: **return** $\Phi$

---

## 5. Case Study: Airport Operation Control System Simulation

An airport operation control system simulation is mainly used to simulate the control and arrangement of an airport control center in different dispatching strategies. By simulating a period in the real world, simulation results of airliners' punctuality rates and average delay times can be obtained. This complex system simulation provides an effective research method for the scheduling and control of airliners in airports. The airport operation control simulation system mainly includes airliner, airport runway, and air traffic control center (ATC) models. The airliner model records the delay time and has three statuses: taking off, landing, and waiting. The airport runway model records the idling and queuing status of runway. The ATC model needs to do many complex calculations based on relevant strategy, queue waiting of runways, and delay time of airliners, and then schedule the relevant airliners to wait on specified runways. Therefore, it takes much more time to calculate than the other two models.

The abovementioned airport operation control system simulation was used as an experimental case to analyze the efficiency and practicability of the service-oriented model encapsulation and selection method for complex system simulation based on cloud architecture proposed in this paper. The simulation platform used in the case study was SUPE, and all experiments were run on two computing nodes with a Linux (centos7) operating system. Each node was equipped with a 3.40 GHz Intel (R) Core (TM) i7-6700 quad core CPU processor. Docker (version 1.13.0) container technology was used as a virtualization method to build a two-node cloud architecture, in which the distributed operation of the airport operation control simulation system was implemented. In the experiment, a simulation time was set up corresponding to the physical time of 10 min to study the actual system of 1 week (7 days), so each simulation promoted the logical simulation time of 1008. The time that was measured in the test was the execution time when the simulation application finished the 1008 simulation steps (logical simulation time). Each piece of experimental data in the analysis chart was the average value after 10 test runs. The experimental configuration is shown in Table 1.

**Table 1.** Experimental configuration.

| Experimental Parameters | Description/Value |
| --- | --- |
| Number of airliners | (50, 250) |
| Number of airport runways | 5 |
| Number of air traffic control centers | 1 |
| Scheduling policy | Punctuality prioritized |
| Simulation run time | 1008 |
| Model distribution mode | Scatter, model servitization |
| Degree of parallelism | 1, 2, 4 |

### 5.1. Performance Evaluation

#### 5.1.1. Cloud-Architecture-Based Distributed Simulation (CDS)

In order to test the performance of the CDS (which refers to the architecture, the computing nodes of which communicate by a cloud architecture network), this experiment tested the execution time of the airport operation control simulation application under three operation modes: serial simulation on a single process in a single node (S-1P), traditional distributed simulation (TDS, which refers to the architecture, the computing nodes of which communicate by local connection) on two processes (TDS-2P, per process per node), and CDS on two processes (CDS-2P, per process per node) based on the above experimental parameter settings. TDS-2P and CDS-2P used the scatter distribution method (each type of simulation model was distributed to each process in turn), and each process ran in one node.

As shown in Figure 5, when the number of airliner instances was 50, 100, 150, 200, and 250, compared with the running time of the simulation application using the S-1P operation mode, both the CDS and TDS could reduce the running time of the simulation application and improve the execution efficiency. As the number of airliner instances increased, because the model distribution mode was scatter, the amount of computation assigned to each process would get closer to being equal. So, the running time of TDS-2P and CDS-2P would get closer to half that of S-1P. Compared with TDS, the performance of the CDS lost an average of 5.37% in five sets of experiments. This is because in the cloud architecture, Docker container technology uses virtualization to isolate interprocess resources. In CDS, processes at different nodes need to communicate through virtual addresses, which leads to higher communication latency than TDS. However, cloud computing has the advantages of computing resources being used on demand and service models being shareable, which would balance such performance loss. Therefore, it is feasible to use CDS architecture to run complex system simulations.

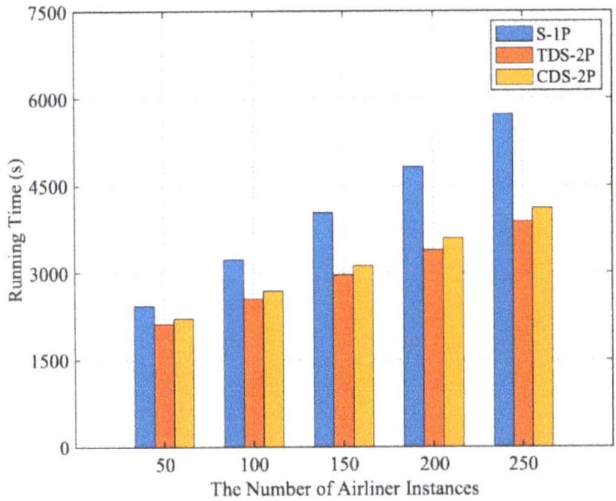


**Figure 5.** Running times of serial simulation on a single process in a single node (S-1P), traditional distributed simulation on two processes (TDS-2P), and cloud-architecture-based distribution on two processes (CDS-2P).

#### 5.1.2. Model Servitization

This experiment packaged the simulation models with large computational requirements into a shareable simulation service through the C-RUM specification. The service model ran in parallel on a single process of a cloud server node, participating in the execution of a complex system simulation application in the cloud-based simulation model service framework (model servitization, MS). In order to study its performance, the case study encapsulated the ATC model, which has a greater number of calculations than the other models, into a service model based on the C-RUM specification. Then, we tested the effect of using scatter and MS distribution methods under CDS (Scatter-CDS and

MS-CDS) on simulation execution time. The specific operation and distribution modes are shown in Figure 6. The experiment used two-process (per process per node) and four-process (two processes per node) parallel simulation to test the performance of Scatter-CDS and MS-CDS (Scatter-CDS-2P, MS-CDS-2P, Scatter-CDS-4P, and MS-CDS-4P). The MS distribution method operated the service model separately in one process, and the remaining models were distributed to the rest of the processes using scatter distribution.

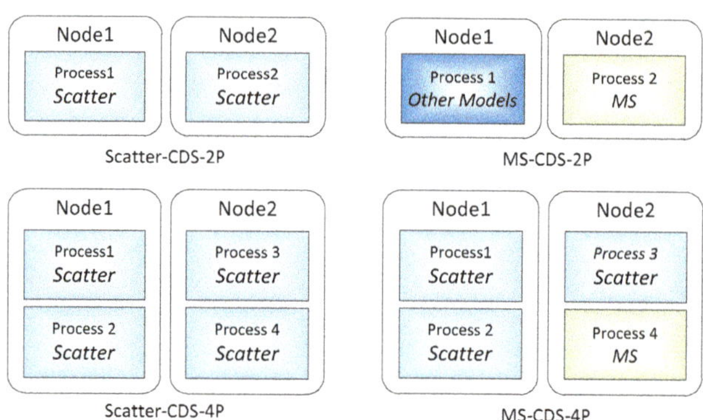


**Figure 6.** Operation and distribution modes.

The results of the test are shown in Figures 7 and 8, compared with the execution time of the simulation application by S-1P operation mode: (1) In the two-process parallel operation mode, when the number of airliner instances was less than 150, MS-CDS-2P was better able to reduce the execution time of simulation applications than Scatter-CDS-2P and had a higher running time speed-up ratio. However, when the number of airliner instances exceeded 150, the computation load was more unbalanced on two computing nodes, and the speed-up ratio of MS-CDS-2P gradually decreased and became even lower than that of Scatter-CDS-2P. (2) In the four-process parallel operation mode, MS-CDS-4P was better able to reduce the execution time of the simulation application and had a higher speed-up ratio (execution performance) than Scatter-CDS-4P when instantiating the number of airliners from 50 to 250. When the number of airliners was 250, the execution performance of MS-CDS-4P improved by 35.28% compared with Scatter-CDS-4P.

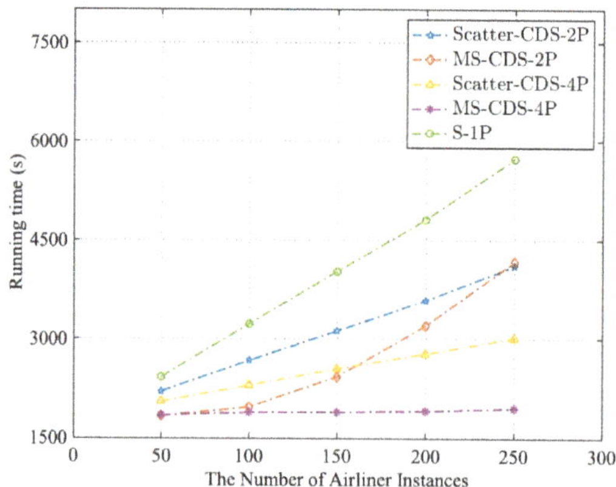


**Figure 7.** Running times of S-1P, Scatter-CDS-2P, model servitization (MS)-CDS-2P, Scatter-CDS-4P, and MS-CDS-4P.

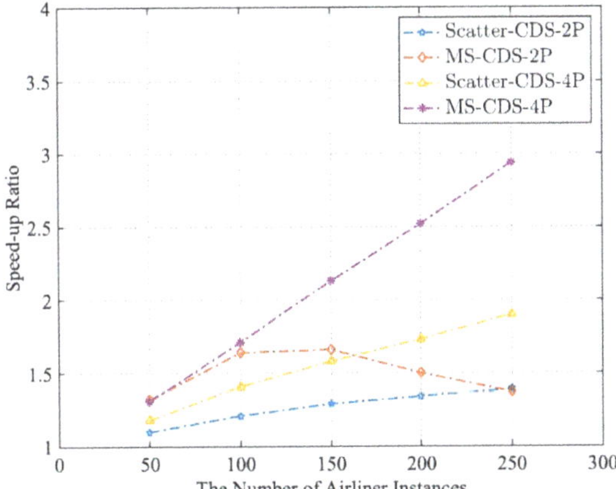

**Figure 8.** Speed-up ratios of S-1P, Scatter-CDS-2P, MS-CDS-2P, Scatter-CDS-4P, and MS-CDS-4P.

Through the experimental results and analysis, we found that packaging models with large computational loads into a shareable service, on the one hand, can provide support for quickly constructing the simulation system in the form of service combination. On the other hand, it can effectively improve the execution efficiency of the simulation system. Also, when the total calculation of the remaining models is gradually increased, the degree of parallelism should be increased to ensure load balancing, so as to maximize the effect of MS on increasing the execution efficiency of simulation applications.

### 5.1.3. Simulation Model Selection Method Based on Semantic Search

In order to prove the practicability of the simulation model selection method based on semantic search proposed in this paper, five kinds of ATC service models with different QoS attribute characteristics were constructed by C-RUM specification. A simulation model description method based on a knowledge graph [28] was used to describe the simulation models of the airport operation control system simulation. The description information was added to the database that stored the model description knowledge graph (a knowledge graph that contained the description information of a large number of different models in various fields). Algorithm 1 was implemented by Cypher query language [30], and the semantic search framework was built in the model description knowledge graph database to find the required models. Then, based on Algorithm 2, according to different QoS index weight vectors, the simulation model candidate set with optimization order could be obtained for users to choose.

Under the simulation model semantic search framework, the five ATC service models (ATC-A, ATC-B, ATC-C, ATC-D, and ATC-E) could be accurately found by correlation with the airliner or runway model or by the attributes of the ATC model. These five models were added to the initial model set, and then based on the QoS index values of the five simulation models and QoS index weight vector, the target QoS value Q of each model could be obtained. The simulation model candidate set that was obtained by sorting Q was available for users to select. The experiment assumed that the user wants to select the service model that can optimize the execution efficiency of the simulation application. Directed at two operation modes, two QoS index weight vectors for different experimental methods were provided to select simulation models. The effectiveness of the semantic search framework and the QoS weighted-based model selection method was verified by running and testing the performance of the simulation application that was assembled by the selected ATC models

(1) When the entire simulation system is running on a single node, the external network communication capability Qc does not affect the execution efficiency of the simulation application.

The model performance $Q_M$ dominates the effect (the other three QoS indices may have little effect on the execution efficiency), so the QoS index weight was set to W = (0.7, 0, 0.1, 0.1, 0.1), and the specific selection process is shown in Table 2.

**Table 2.** Selection process 1.

|  | QoS1 [0,100] | QoS2 [0,100] | QoS3 [0,1] | QoS4 [0,1] | QoS5 [0,10] | Q |
| --- | --- | --- | --- | --- | --- | --- |
| ATC-A | 85 (0.85) | 83 (0.83) | 0.98 | 0.9 | 9 (0.9) | 0.873 |
| ATC-B | 65 (0.65) | 55 (0.55) | 0.96 | 0.92 | 9 (0.9) | 0.733 |
| ATC-C | 92 (0.92) | 52 (0.52) | 0.95 | 0.92 | 10 (1) | 0.931 |
| ATC-D | 71 (0.71) | 70 (0.7) | 0.97 | 0.93 | 10 (1) | 0.787 |
| ATC-E | 74 (0.74) | 65 (0.65) | 0.95 | 0.91 | 9 (0.9) | 0.794 |
|  | W = (0.7, | 0, | 0.1, | 0.1, | 0.1) | |
| Ordered candidate set: {ATC-C, ATC-A, ATC-E, ATC-D, ATC-B} | | | | | | |

The QoS weight vector gave a large weight to $Q_M$, and the optimized candidate model set {ATC-C, ATC-A, ATC-E, ATC-D, ATC-B} could be obtained through calculation. The five ATC service models were assembled, respectively, to the same five airport operation control simulation systems, and the airliner instance was set to 100. The execution times of the five simulation applications operated by S-1P are shown in Figure 9.

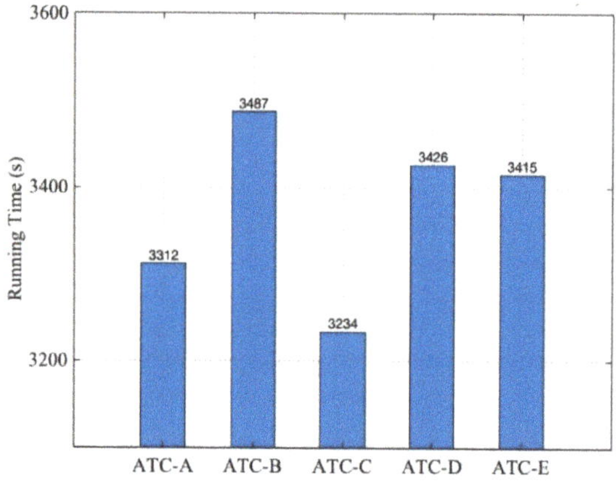


**Figure 9.** Running times of the five simulation systems (1).

These applications could run effectively, and it can be seen that in S-1P operation mode, the order of the ATC models corresponding to the execution efficiency of the five simulation applications was consistent with the optimization order in the model candidate set. The simulation application assembled by the model ATC-C, which ranked first in the candidate set, had the shortest running time (3234 s).

(2) In cloud architecture, shareable simulation services are often stored in the cloud center. The service model and simulation engine framework need to communicate through network interconnection. Both $Q_C$ and $Q_M$ of the simulation service model affect the execution efficiency of the simulation application, so the QoS preference weight vector was set to W = (0.35, 0.35, 0.1, 0.1, 0.1). The specific selection process is shown in Table 3.

**Table 3.** Selection process 2.

|  | QoS1 [0,100] | QoS2 [0,100] | QoS3 [0,1] | QoS4 [0,1] | QoS5 [0,10] | Q |
|---|---|---|---|---|---|---|
| ATC-A | 85 (0.85) | 83 (0.83) | 0.98 | 0.9 | 9 (0.9) | 0.866 |
| ATC-B | 65 (0.65) | 55 (0.55) | 0.96 | 0.92 | 9 (0.9) | 0.698 |
| ATC-C | 92 (0.92) | 52 (0.52) | 0.95 | 0.92 | 10 (1) | 0.847 |
| ATC-D | 71 (0.71) | 70 (0.7) | 0.97 | 0.93 | 10 (1) | 0.7835 |
| ATC-E | 74 (0.74) | 65 (0.65) | 0.95 | 0.91 | 9 (0.9) | 0.752 |
| | W = (0.35, | 0.35, | 0.1, | 0.1, | 0.1) | |
| Ordered candidate set: {ATC-A, ATC-C, ATC-D, ATC-E, ATC-B} | | | | | | |

The QoS weight vector assigned the same weight to $Q_C$ and $Q_M$. After calculation, the optimized candidate model set could be obtained as {ATC-A, ATC-C, ATC-D, ATC-E, ATC-B}. The five ATC service models were assembled, respectively, to the same five airport operation control simulation systems, and the airliner instance was set to 100. The execution times of the five simulation applications operated by MS-CDS-2P are shown in Figure 10.

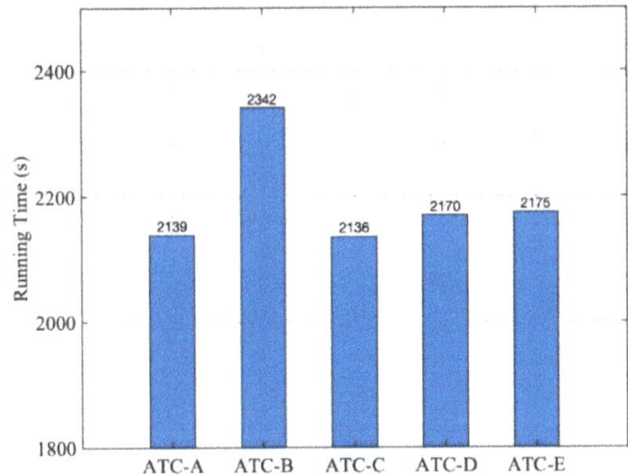


**Figure 10.** Running times of five simulation systems (2).

These applications could run effectively, and it can be seen that in MS-CDS-2P operation mode, the order of the ATC models corresponding to the execution efficiency of the five simulation applications was not completely consistent with the optimization order in the model candidate set. Because the QoS weight vector W was set merely according to the experimental architecture without detailed analysis, it was impossible to accurately quantify the extent to which the model performance $Q_M$ and network communication capability $Q_C$ affected the entire execution efficiency of the simulation application.

The experimental results show that the searched model can work together with other models and implement the simulation task, which verifies the effectiveness of the semantic search framework. Further, the proposed QoS-based simulation model selection method can support users in selecting the model which has the biggest target QoS index (Q) according to their QoS preference. However, it cannot always give the optimum solution that could optimize a certain performance (execution efficiency) of a complex simulation system.

## 5.2. Discussion

The experiment first tested the performance of the simulation application under three patterns: S-1P, TDS-2P, and CDS-2P. The results prove that, compared with TDS, CDS can also effectively improve the execution efficiency of the simulation application with little performance loss, which demonstrates the practicability of CDS. Experiment 2 encapsulated the models with large computational loads into shareable services in the cloud environment by the C-RUM specification proposed in this paper. Then, by comparing the performance of MS-CDS and Scatter-CDS, the results prove that the MS distribution mode is better than the traditional scatter distribution mode at improving the execution efficiency of complex system simulation. This demonstrates the feasibility of C-RUM specification in cloud networking architecture and the effectiveness of the method, making the models with large computational loads into shared services, proposed in this paper. In experiment 3, the required ATC models were found under the proposed semantic search framework by attributes or correlation searching in the model description knowledge graph. The experiment assembled the searched model into the simulation application of the case study and verified that it can work together with other models and effectively implement the simulation task, which verified the correctness of the semantic search framework. Then, the model ranking based on the QoS weighted selection method was compared with the ranking of the execution time of actual simulation systems assembled by the models in the candidate set. This proved that the proposed QoS weighted-based simulation model selection method can select simulation models according to users' customized requirements, but the solution is not always the optimum one that could optimize the performance of the complex system simulation.

## 6. Summary and Future Work

A service-oriented model encapsulation and selection method for complex system simulation was proposed in this paper. This method first promotes the original RUM specification and puts forward C-RUM specification, which solves the problem of network communication in RUM specification. Models with large computational requirements are encapsulated into shareable services in the cloud architecture. The experimental results showed that model servitization can effectively improve the execution efficiency of complex system simulation applications. Then, the model semantic search framework is built in the simulation model description knowledge graph, which increases the correlation search ability compared with other semantic search methods. The QoS weighted-based model selection method supports users in customizing the weight of QoS indices and obtaining the ordered candidate model set by weighted comparison. This mechanism can support the selection of the required simulation model that satisfies users' QoS preference under the cloud architecture.

Future work should further improve the QoS weighted-based simulation model selection method, considering the limitation that it cannot assist users in selecting the optimal simulation service model directed at a specific simulation application or its certain performance. Also, future research should confirm the metric of the model QoS indices and study how to assign the corresponding QoS index weights.

**Author Contributions:** Conceptualization, S.X. and F.Z.; methodology, S.X.; software, Y.Y.; validation, S.X. and F.Z.; formal analysis, S.X. and F.Z.; investigation, S.X.; resources, W.T.; data curation, S.X. and Y.X.; writing—original draft preparation, S.X. and F.Z.; writing—review and editing, S.X. and Y.X.; visualization, S.X.; supervision, Y.Y. and F.Z.; project administration, W.T. and F.Z.; funding acquisition, Y.Y. and F.Z.

**Funding:** This work was supported in part by the National Natural Science Foundation of China (no. 61903368).

**Conflicts of Interest:** The authors declare no conflict of interest.

## References

1. Angelica, S.; Emanuele, P.; Andrea, Z.P.S. The Role of Complex Analysis in Modelling Economic Growth. *Entropy* **2018**, *20*, 883.
2. Martin, I.; Jan, C. A Behavioural Analysis of Complexity in Socio-Technical Systems under Tension Modelled by Petri Nets. *Entropy* **2017**, *19*, 572.
3. Yiping, Y.; Gang, L. High-performance Simulation Computer for Large-scale System-of-Systems Simulation. Journal of System Simulation. *J. Syst. Simul.* **2011**, *23*, 1617–1623.

4. Calheiros, R.N.; Ranjan, R.; Beloglazov, A.; Rose, C.A.F.D.; Buyya, R. CloudSim: A toolkit for modelling and simulation of cloud computing environments and evaluation of resource provisioning algorithms. *Softw. Pract. Exp.* **2011**, *41*, 23–50. [CrossRef]

5. Simon, J.E.T.; Azam, K.; Katherine, L.M.; Andreas, T.; Levent, Y.; Justyna, Z.; Pieter, J.M. Grand challenges for modelling and simulation: Simulation everywhere—from cyber infrastructure to clouds to citizens. *Simulation* **2015**, *91*, 648–665.

6. Feng, Z.; Yiping, Y.; Huilong, C. Reusable Component Model Development Approach for Parallel and Distributed Simulation. *Sci. World J.* **2014**. [CrossRef]

7. Sheng, B.; Zhang, C. Common intelligent semantic matching engines of cloud manufacturing service based on OWL-S. *Int. J. Adv. Manuf. Technol.* **2016**, *84*, 103–118. [CrossRef]

8. Singh, S.; Chana, I. QRSF: QoS-aware resource scheduling framework in cloud computing. *J. Supercomput.* **2015**, *71*, 241–292. [CrossRef]

9. Pujara, J. Knowledge Graph Identification. *Int. Semant. Web Conf.* **2013**. [CrossRef]

10. Yang, S.; Hanquan, D.; Minghua, L. Research on Simulation Composability and Reusability Based on SOA. *J. Syst. Simul.* **2014**, *26*, 1522–1526.

11. Lee, H.; Yang, J.S.; Kang, K.C. Domain-oriented variability modeling for reuse of simulation models. *Simulation* **2014**, *90*, 438–459. [CrossRef]

12. Jianbo, L.; Yiping, Y. Research on the Development Approach for Reusable Model inParallel Discrete Event Simulation. *Discrete Dyn. Nat. Soc.* **2015**. [CrossRef]

13. Yiping, Y.; Feng, Z. A Reusable Simulation Model Development and Usage Method. China Patent 1 0353755.5, 14 August 2013.

14. Purohit, L.; Kumar, S. Web Service Selection using Semantic Matching. In Proceedings of the International Conference on Advances in Information Communication Technology & Computing, Bikaner, India, 12–13 August 2016.

15. Wu, M.C.; Gu, J.Z. OWL-S Semantic Extension in the Dynamic Combination of Web services. *Comput. Appl. Softw.* **2007**, *24*, 69–71.

16. Jiao, H.; Zhang, J.; Li, J.H.; Shi, J. Research on cloud manufacturing service discovery based on latent semantic preference about OWL-S. *Int. J. Comput. Integr. Manuf.* **2017**, *30*, 433–441. [CrossRef]

17. Zhang, T.; Liu, Y.; Zha, Y. Semantic Web-based approach to simulation services dynamic discovery. *Comput. Eng. Appl.* **2007**, *43*, 15–19.

18. Song, L.-L.; Li, Q. Research on simulation model description ontology and its matching model. *Comput. Eng. Appl.* **2008**, *44*, 6–12.

19. Li, T.; Li, B.H.; Chai, X.D. Layered simulation service description framework oriented to cloud simulation. *Comput. Integr. Manuf. Syst.* **2012**, *18*, 2091–2098.

20. Cheng, C.; Chen, A.Q. Study on Cloud Service Evaluation Index System Based on QoS. *Appl. Mech. Mater.* **2015**, *742*, 683–687.

21. Tong, Z.; Yunsheng, L.; Yabing, Z. Optimal Approach to QoS-Driven Simulation Services Composition. *J. Syst. Simul.* **2009**, *21*, 4990–4994.

22. Jiahang, L.; Junli, S.; Liqun, J. A QoS Evaluation Model for Cloud Computing. *Comput. Knowl. Technol.* **2010**, *6*, 8801–8803.

23. Li, B.H.; Chai, X.; Hou, B.; Li, T.; Zhang, Y.B.; Yu, H.Y.; Tang, Z. Networked Modeling & Simulation Platform Based on Concept of Cloud Computing—Cloud Simulation Platform. *J. Syst. Simul.* **2009**, *21*, 5292–5299.

24. Chen, T.; Chiu, M.C. Development of a cloud-based factory simulation system for enabling ubiquitous factory simulation. *Robot. Comput. Integr. Manuf.* **2017**, *45*, 133–143. [CrossRef]

25. Jeanjacques, M.C. *Web services Description Language (WSDL) Version 1.2*; World Wide Web Consortium (W3C): Cambridge, MA, USA, 2003.

26. Xu, Z.L.; Sheng, Y.P.; He, L.R.; Wang, Y.F. Review on Knowledge Graph Techniques. *J. Univ. Electron. Sci. Technol. China* **2016**, *45*, 589–606.

27. Organization, T.M. Resource Description Framework (RDF). In *Encyclopedia of Gis*; Shekar, S., Xiong, H., Eds.; Spring: Berlin, Germany, 2004; pp. 6–19.

28. Siqi, X.; Feng, Z.; Yiping, Y.; WenJie, T. A Description Method of Cloud Simulation Model Resources based on Knowledge Graph. In Proceedings of the 4th International Conference on Cloud Computing and Big Data Analytics, IEEE, Chengdu, China, 12–15 April 2019; pp. 655–663.

29. You, M.; Wang, S.; Hung, P.C.K. A Highly Accurate Prediction Algorithm for Unknown Web Service QoS Values. *IEEE Trans. Serv. Comput.* **2017**, *9*, 511–523.

30. Francis, N.; Green, A.; Guagliardo, P. Formal Semantics of the Language Cypher. *arXiv* **2018**, arXiv:1802.09984.

*Article*

# Minimum Memory-Based Sign Adjustment in Signed Social Networks

**Mingze Qi** [1,†], **Hongzhong Deng** [1,*,†] and **Yong Li** [2]

[1] College of Systems Engineering, National University of Defense Technology, Changsha 410073, China
[2] School of Economic and Management, Changsha University, Changsha 410022, China
* Correspondence: hongzhongdeng@gmail.com
† These authors contributed equally to this work.

Received: 11 June 2019; Accepted: 24 July 2019; Published: 25 July 2019

**Abstract:** In social networks comprised of positive (P) and negative (N) symmetric relations, individuals (nodes) will, under the stress of structural balance, alter their relations (links or edges) with their neighbours, either from positive to negative or vice versa. In the real world, individuals can only observe the influence of their adjustments upon the local balance of the network and take this into account when adjusting their relationships. Sometime, their local adjustments may only respond to their immediate neighbourhoods, or centre upon the most important neighbour. To study whether limited memory affects the convergence of signed social networks, we introduce a signed social network model, propose random and minimum memory-based sign adjustment rules, and analyze and compare the impacts of an initial ratio of positive links, rewire probability, network size, neighbor number, and randomness upon structural balance under these rules. The results show that, with an increase of the rewiring probability of the generated network and neighbour number, it is more likely for the networks to globally balance under the minimum memory-based adjustment. While the Newmann-Watts small world model (NW) network becomes dense, the counter-intuitive phenomena emerges that the network will be driven to a global balance, even under the minimum memory-based local sign adjustment, no matter the network size and initial ratio of positive links. This can help to manage and control huge networks with imited resources.

**Keywords:** structural balance; minimum memory based sign adjustment; social networks; NW network; convergence

## 1. Introduction

Structural balance theory has attracted many researchers from different fields to study signed social networks, which are composed of positive (P) and negative (N) edges defined on a set of $n$ individuals (nodes) [1–10]. The most interesting question is whether (and how) the signed social networks can evolve to a balanced and steady structure under individual stress-reducing adjustments. Pioneering research into classic structural balance theory and empirical studies can be found in [1,2,5,6,11–15].

Many different assumptions on stress reduction in signed networks under local adjustment have been designed to check whether the network will reach a global balance [5,6,9,13,16–21]. Rules which mix imbalance stress reduction with homophily and other bilateral pressures have also been proposed, such as those found in [8,15,22,23].

Other researchers, considering the amount of information that a node holds, proposed some "global" and "local" sign adjustments, including add and delete mechanisms [6,19]. Under a global adjustment

*Entropy* **2019**, *21*, 728; doi:10.3390/e21080728 

mechanism, individuals are all granted information about the whole network and can take this into account in their sign adjustments. However, under local mechanisms, individuals can only assay the factors in their local neighbourhood [13,24]. Local angle- or information-based sign adjustment mechanisms often cause global turbulence and imbalance in the network.

Alter mechanisms are based upon socio–psychological insights and claim to reflect individuals' real processes. With limitations in information access, the calculative ability of individuals becomes constrained to the cognition span of their neighbourhoods [20,25,26]. Montgomery [20] reformulated balance theory, by allowing actors to possess an incomplete awareness of the evaluations held by other actors and by adopting balance closure (modified to allow incomplete awareness) as an equilibrium concept. Their analysis revealed that an actor's "indirect awareness" of imbalance is necessary but not sufficient for that actor's ambivalence in the balance closure. Volstorf [25] proposed that, with increasing size of the interaction group, the memory becomes error-prone in the game and individuals may categorize partners into types to decrease memory effort. A memory test showed that 126 recruited participants from Berlin universities could remember rare partner types better than they remembered common ones. The authors also proposed an ecologically rational memory strategy in social interactions. Brashears [26] constructed a mathematical model of the evolution of relationships and explored the consequences of triadic interaction rule for the relation of nations and on the polarity configuration of a system of nations, and found that a special triadic interaction rule produced only two long term triadic configurations: unipolarity and bipolarity.

Individuals can only remember details of interactions with important friends and enemies accumulated in their life histories [27]. Limited information affects the decisions of an individual in many ways. Kottonau [28] presented an agent-based computational model, a memory model of new product diffusion within a consumer social network on the micro level, and discussed the effect of memory level on habit breaking and product adoption. Winke and Stevens [29] investigated the specificity of memory in co-operative contexts and found that memory accuracy is robust to differences in the cooperative context; however, the social network size did correlate with memory accuracy. Their findings suggested that the demands of interacting in a large social network may require excellent memory. Hassanibesheli [27] investigated how history (or memory) has global consequences on the evolution of a signed social system. They found that past relations surely impact on the evolution of the system and will prolong the time necessary to reach "balanced states", but do not change the dynamical attractors of the system.

With increases in network size, type, and scale, social networks have become more and more complex. One feasible way to study a social network is with a computer-generated network model. Experiments and simulation results on letter pass networks [30,31], actors collaborate networks [32,33], organize networks [34,35], co-authorship networks [36,37], telephone calls networks [38,39], and email networks [40] have showed that social networks are characterized as "small-world" and with "six degrees of separation". The computer-generated network model plays an important role in social network analysis.

Based upon these considerations, the paper proceeds as follows: in Section 1, we introduce the basic signed social network model. In Section 2, we analyze network fluctuations under randomly chosen node sign adjustment. In Section 3, we propose a minimum memory-based sign adjustment rule and study the impacts of the initial ratio of positive links $\alpha_0$, the rewire probability $p_r$, network size, neighbour number $K$, and randomness upon structural balance. Finally, the paper summarizes in Section 4.

## 2. The Network Model and Sign Adjustment Rules

### 2.1. The Signed Social Network Model

With the objective of eventually studying the balancing adjustment of signed social networks and considering (as mentioned above) that most social networks are small world networks, we initially built a signed social network Newman–Watts small world model (NW model) [41]. The growth (formation) process of this network model is as follows:

Signed NW network model:

1.  The network is assigned $n$ nodes and a regular ring lattice is constructed on the nodes, where each node is connected to a total of $K$ neighbours, each side with $K/2$ neighbours (where $K$ is even integer).
2.  Select all node pairs in turn.
3.  Add links between the selected node pairs with probability $p_r$, if no self-loops and link duplication. We name the probability $p_r$ as the rewiring probability.
4.  Each symmetric link is randomly set to a positive sign with a probability of $\alpha_0$ and to negative with a probability of $1 - \alpha_0$.

We use $T_i (i = 0, 1, 2, 3)$, to represent the numbers of the four kinds of three-cycles, where the subscript $i$ represents the number of positive links in the cycle. For example, $T_3$ is a balanced three-cycle in which all three edges are positive. According to classical structural balance theory, as showed in Figure 1, $T_0$ and $T_2$ are imbalanced triangles and $T_1$ and $T_3$ are balanced triangles.

The density of the NW network will increase with $p_r$ as

$$Density \approx (\frac{nK}{2} + (\frac{n(n-1) - nK}{2})p_r)/(\frac{n(n-1)}{2}) = (\frac{n - K - 1}{n - 1})p_r + \frac{K}{n - 1}. \tag{1}$$

So, according to the structure balance theory I, the initial expected balance of a structure is

$$\beta(3) = \frac{p_{T_1} + p_{T_3}}{p_{T_0} + p_{T_1} + p_{T_2} + p_{T_3}} = 3\alpha_0(1 - \alpha_0)^2 + \alpha_0^3. \tag{2}$$

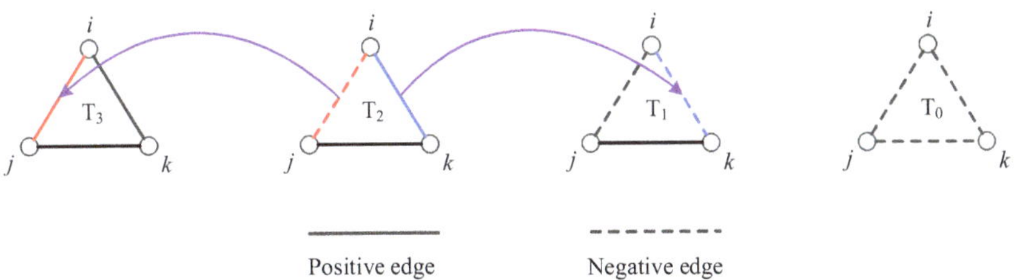


**Figure 1.** Four kinds of triangle: $T_0$, $T_1$, $T_2$, and $T_3$. The solid and dashed lines, respectively, represent the positive and negative links in triangles. $T_1$ and $T_3$ are balanced triangles. $T_0$ and $T_2$ are imbalanced triangles. For example, if the selected imbalanced cycle is a $T_2$ cycle in the middle and the selected duty node is $i$, then changing the sign of either the red link $i - j$ or blue link $i - k$ can change the imbalanced cycle $T_2$ to a balanced cycle $T_3$ or $T_1$.

## 2.2. Random Adjustment Rule

Following [17], individuals in the generated social network, under the stress from imbalanced three-cycles, may adjust the cycles in which they are located. Today, with information blast and the increase of network size, the ratio of information that individuals can hold become less. If individuals are extremely short-sighted, they will randomly and locally adjust only one selected three-cycle, as if they have zero memory about their neighbour. We call this rule the random adjustment rule. The iterative adjustment process is accomplished as follows:

1. Randomly select a three-cycle from the network.
2. If the selected cycle is balanced, then return to step 1.
3. If the cycle is imbalanced, select any one of its constituent nodes as the "duty node" and change the sign of any one of duty node's two links, in order to achieve balance in the cycle.

Note that all nodes, under this iterative adjustment rule, are selected with equal probability; none has priority. Of course, with the process of effective adjustment, there will be fewer imbalanced triads. According to the adjustment rule, the selection probability of all nodes are the same, but their subsequent actions are different. Nodes in an imbalanced triad may change their sign, but nodes in a balanced triad will be kept the same. In each step, as soon as the duty node has successfully changed one sign in its links, all other nodes in the network are refreshed with this information. This means that all adjustment is open and transparent. Obviously, this adjustment is entirely random and does not take any local information, let alone global information, into account. As the network is incomplete, each edge may belong to many three-cycles and, so, the sign adjustment of one edge may "infect" the balance of the other three-cycles to which this edge belongs. The random adjustment may generate new and more imbalanced 3-cycles in the process of solving the imbalance problem of the selected three-cycle.

## 2.3. Minimum Memory-Based Sign Adjustment Rules

Simulation results under random sign adjustment rules have showed that the network will become random and imbalanced. Thus, the question arises as to how the network will evolve in the case where an individual has little memory about their neighbours. In the real world, amongst large groups of friends, only several best friends keep in touch regularly and develop bonds which probably materially affect their behaviour (the memory in this paper is the number of neighbours that each node can remember and will take into account while the node adjusts its sign. This is different from the historic memory of relation and adjustment in [27]). Thus, one may assume that limited memory is more grounded in reality.

We firs, consider the simplest condition, where only the attitude of one important neighbour is taken into consideration. We name this the minimum memory-based sign adjustment rule. The sign adjustment rule is as follows:

1. Set all nodes to remember only one of their most important neighbours (regardless of whether it is friend or enemy). At the beginning of the simulation, each node randomly selects one neighbour from amongst all of its neighbours to compose its close neighbour set.
2. Select a three-cycle at random from the network.
3. If the cycle is balanced, then return to step 2.
4. If the cycle is imbalanced, randomly select one of its nodes as the duty node.
5. Change the sign of any one of the two edges which link the duty node in the cycle if the sign change can strictly increase the balance ratio of the duty node with his best neighbour.

Note that the relationship between node and its close neighbour can be negative. The close neighbour set of each node does not change over the whole life cycle.

## 3. Results and Discussion

The evolution of NW networks under the sign random adjustment are depicted in Figure 2.

The results in Figure 2 indicate that, under the random sign adjustment rule, the NW network became imbalanced and disorderly, no matter the initial ratio of positive links, the rewiring probability $p_r$, the neighbour number $K$, and the network size $n$. The numeric value of the balance ratio of social networks converged to $\beta(3) \approx 0.5$, $T_1, T_2 \approx 3/8$, and $T_0 = T_3 \approx 1/8$. The network was imbalanced, and nodes kept adjusting their signs randomly. The ratios of different three-cycles were statistically stable. The results verify the ordinary common-sense that local balance optimization can not let a network reach a global balance.

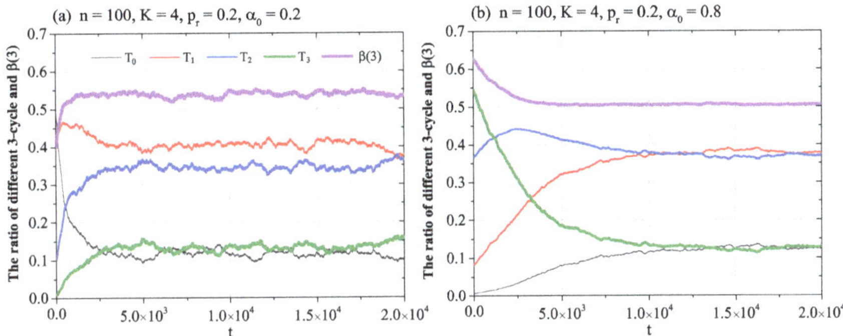



**Figure 2.** The evolution ratio of different three-cycles in the Newmann-Watts (NW) network under the sign random adjustment rule. The horizontal axis is the simulation time $t$. The vertical axis gives the ratio of three-cycles. The sampled plot in (**a**) gives the results for a NW network with parameters $n = 100$, $K = 4$, $p_r = 0.2$, and $\alpha_0 = 0.2$. In (**b**), the network parameters are $n = 100$, $K = 4$, $p_r = 0.8$, and $\alpha_0 = 0.8$.

Figure 3 depicts the simulation results in social networks.

The results in Figure 3 show that:

(1) While the rewire probability was lower ($p_r = 0.2$), as showed in the upper two panels of Figure 3a,b, the NW network could not reach a global balance $\beta(3) = 1$, but converged to an imbalanced state with a stable value $\beta(3) \neq 1$. The size of this stable value is jointly impacted by the other variables.

(2) With an increase of the rewiring probability $p_r$, as in the bottom Figure 3c,d, the NW network became denser. If the rewire probability was high enough (e.g., $p_r = 0.8$), under the minimum memory-based sign adjustment, the NW network reached a global balance $\beta(3) = 1$.

(3) A comparison of Figures 2 and 3 shows that random adjustment could not lead network to global balance; however, under the memory-based sign adjustment rule, even though it was a minimum memory with only one fixed neighbour, the network could (but not surely) converge to a global balance.

To see whether convergence to global balance was dependent on any other variables, such as network size $K$, rewire probability $p_r$, and initial ratio of positive links $\alpha_0$, we simulated the network evolution under the independent influence of these variables.

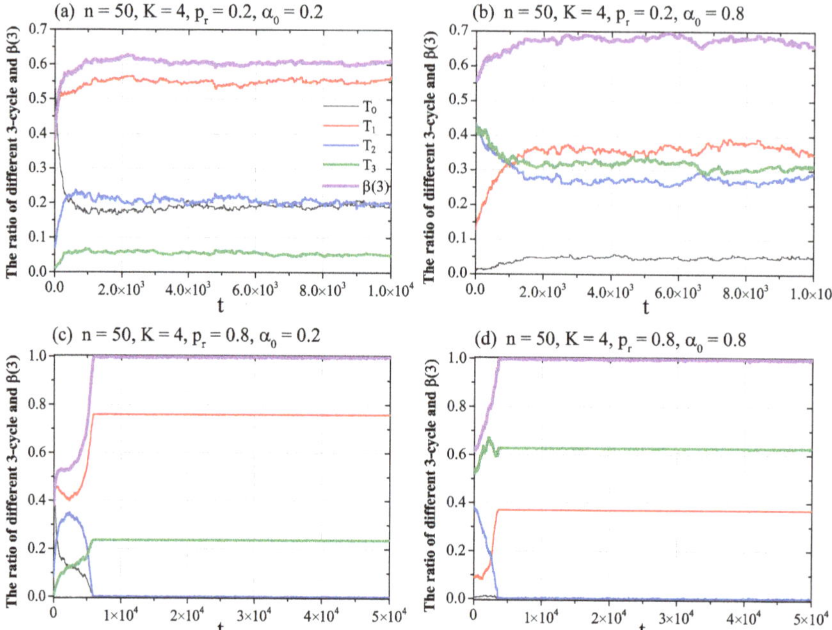

**Figure 3.** The evolution ratio of different 3-cycles in the NW network under the minimum memory-based sign adjustment rule. The horizontal axis is the simulation time $t$. The vertical axis gives the ratio of three-cycles. (**a**) shows the results for a NW network with parameters $n = 50$, $K = 4$, $p_r = 0.2$, and $\alpha_0 = 0.2$; (**b**) is a network with $n = 50$, $K = 4$, $p_r = 0.2$, and $\alpha_0 = 0.8$; (**c**) gives the results for a NW network with $n = 50$, $K = 4$, $p_r = 0.8$, and $\alpha_0 = 0.2$; and (**d**) is a network with $n = 50$, $K = 4$, $p_r = 0.8$, and $\alpha_0 = 0.8$.

The results in Figure 4 show that

(1) In Figure 4a, where the rewire probability was $p_r = 0.2$, if the initial ratio of positive links $\alpha_0$ was less than about 0.7, the NW network converged to an imbalanced steady state, $\beta(3) \approx 0.58$, and could not converge to a global balance $\beta(3) = 1$. If $\alpha_0 > 0.7$, the convergent value of $\beta(3)$ increased. The bigger the value of $\alpha_0$, the greater of balance ratio $\beta(3)$ was. The reason for this is that, while $\alpha_0 > 0.7$, the NW network initially had more positive links, and global balance was possibly close to the initial state.

(2) If the rewire probability was $p_r = 0.8$, as shown in Figure 4b, no matter the value of the initial ratio of positive links, the ratio of balanced 3-cycle converged to $\beta(3) \approx 0.99$, and the NW network nearly reached a global balance. (In Figure 4b, the steep drop in value of $\beta(3) \approx 0.955$ while $p_r = 0.5$ is an effect of the randomness). The trend of all nodes becoming a homophily could not be influenced by the initial ratio of positive links.

(3) In most conditions, except where the initial ratio of positive links was at its maximum value (1) or minimum value (0), the convergence of NW network and its convergent value of $\beta(3)$ were immune to the initial ratio of positive links $\alpha_0$.

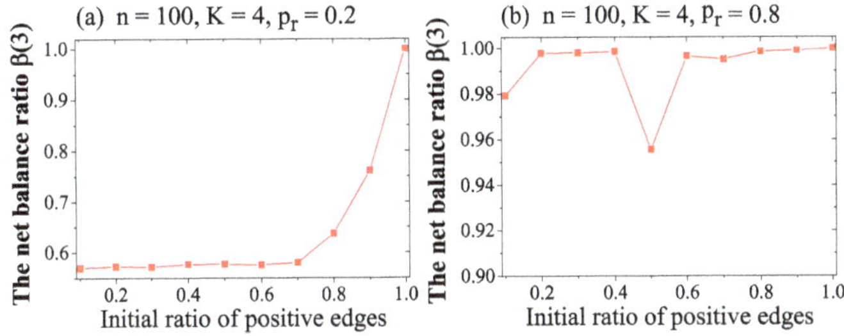

**Figure 4.** The impact of initial ratio of positive links $\alpha_0$ on NW network evolution under the minimum memory-based sign adjustment rule. The horizontal axis is the initial ratio of positive links $\alpha_0$. The vertical axis is the ratio of balanced 3-cycles. Each data plot is the average value of 10 simulations and each simulation lasted 5000 steps. The rewire probability was $p_r = 0.2$ in (**a**) and $p_r = 0.8$ in (**b**). The other variables were fixed at $n = 100$ and $K = 4$.

In Figure 4, the results indicate that $\beta(3)$ could be influenced by rewire probability $p_r$. To see how can rewire probability independently impacted on the NW network's convergence, we conducted another simulation, as follows:

These results in Figure 5 show that

(1) The curve in Figure 5a was similar to the curve in Figure 5b. Once again, this result verified the conclusion that the initial ratio of positive links had little influence on NW network's evolution, except for when it was close to zero or unity.

(2) While the rewire probability was small ($p_r < 0.5$), the NW network, under the minimum sign adjustment rule, could not converge to a global balance but, instead, converged to an imbalanced stable state $\beta(3) \approx 0.6$; no matter the value of the initial ratio of positive links.

(3) If the rewire probability was bigger than about 0.7, the NW network converged to a global balance $\beta(3) = 1$. This convergence was immune to the initial ratio of positive links $\alpha_0$.

(4) While the rewire probability was about $p_r \approx 0.6$, whether the network could converge to a global balance or not was random and may have been influenced by random factors in the NW network model and the sign adjust process.

(5) In Figure 5, a critical value (which also can be named as the chaotic area), of rewire probability $p_r \approx 0.6$ emerged clearly. If the rewire probability exceeded the critical value $p_r \approx 0.6$, the NW network converged to a global balance $\beta(3) \approx 1$; the NW network became a homophily group and was divided into two opposite subgroups, with positive links inside the subgroup and negative links among them. If the rewire probability was $p_r < 0.5$, the balance ratio of the NW network converged to about $\beta(3) = 0.58$, indicating an imbalanced network.

(6) With an increase of the rewire probability, the NW network became denser. While $p_r = 0.6$, the density of the network with 100 nodes was about 0.616, according to Equation (1) in Section 2; the influence of sign adjustment could spread to whole network and lead these networks to a global balance easily, even when the adjustment was based on minimum memory (only taking one fixed neighbour's attitude into account). When the rewire probability was lower ($p_r < 0.5$), the NW network was sparser, with fewer three-cycles. The adjustment of each imbalanced three-cycle had a weak influence and could not spread its influence to other three-cycles quickly. When the rewire probability was very small, there were only a few, maybe several, three-cycles in the NW network. Among these fewer three-cycles, if parts of an imbalanced three-cycle were not constituted by the node's close neighbour, then the adjustment of these imbalanced three-cycles had no influence on node's local balance ratio with

his close neighbour. Thus, these imbalanced three-cycles, although imbalanced, were kept unchanged throughout the whole simulation.

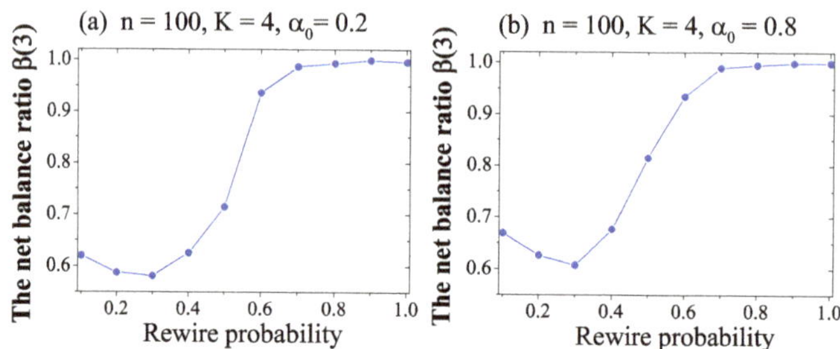

**Figure 5.** The impact of rewire probability $p_r$ on NW network evolution under the minimum memory-based sign adjustment rule. The horizontal axis is the rewire probability. The vertical axis is the ratio of balanced three-cycles. Each data plot is the average value of 10 simulations and each simulation lasted 50,000 steps. The initial ratio of positive links was $\alpha_0 = 0.2$ in (**a**) and $\alpha_0 = 0.8$ in (**b**). Other variables were fixed at $n = 100$ and $K = 4$.

From the results in Figures 4 and 5, we observed the emergence of a critical value of the rewire probability and that the convergence behaviour was independent of the initial ratio of positive links. To check whether the critical value and balance ratio were impacted by network size and neighbour $K$, we conducted another simulation, and obtained the following results:

The results in Figure 6 indicate that

(1) With an increase of $K$, the network became denser and the influence of each sign adjustment could affect more nodes. So, the convergent value of the ratio of balanced 3-cycle increased, as shown in Figure 6a. What we should emphasize here is that, when $K$ was big and the rewire probability was very small, although the network was dense, the network had few triangles and the balance ratio was hard to increase. Thus, while two networks were of the same density, the small $K$, big $p_r$ network was better than big $K$, small $p_r$ network, as more cycles contributed towards faster convergence to a global balance.

(2) The results in Figure 6b show that network size had little influence on the convergence trend of a network. The shape of the curve in Figure 6b is similar to these curves in former Figure 3 When the network size was smaller (e.g., $n = 20$, as shown in Figure 6b), even though the rewire probability was small, the ratio of balanced three-cycle could reach about 0.82. The reason for this is that, in a small network, each sign adjustment has a larger influence, relative to that in a huge network, and will spread its sign adjustment influence quicker and widely.

(4) Network size had great influence on the converge time. While the network was small, each sign adjustment had a comparatively greater influence on the whole network, and the network easily converged to a global balance.

(5) When the network became large, the number of required simulation iterations increased very quickly, as it is very hard for a huge network to converge to global balance. For example, if a complete connected network's size is 500, then each node has $C_{499}^2 = 124{,}251$ triangles, but only 498 related triangles contain both duty nodes and its only remembered neighbour, simultaneously. So, in each iteration, the probability of selecting one related triangle is $498/124{,}251 = 1/499 \approx 0.002$. Furthermore, not every related triangle is imbalanced and needs to adjust. Only imbalanced triangles need adjustment. The probability that the related triangle is imbalanced is $3(\alpha_0)^2(1 - \alpha_0) + (1 - \alpha_0)^3$. For example, if $\alpha_0 = 0.8$, the initial balance ratio of this complete connected network is about $3 \times 0.8 \times 0.2^2 + 0.8^3 = 0.608$.

The initial probability to select a useful imbalanced and related triangle in one iteration in a completely connected network is about $\frac{1}{499} \times 0.608 \approx 0.00122$. This value is very small, and will become even smaller with the evolution of the network, with the increase of $a_0$. This means that only in about one in about every 820 iterations can the network adjust to become little more balanced. Every useful sign adjustment of an imbalanced triangle can only enhance the balance ratio of whole network $\beta(3)$ by about $6/C_{500}^3 \approx 0.000000048$. To reach global balance $\beta(3) = 1$, then, the network will need somewhere around $\frac{(1-0.608) \times C_{500}^3}{6} \times \frac{499}{0.608} \approx 1.11 \times 10^9$ iterations, on average. Under the same conditions, a complete connected network of size 100 will only need $\frac{(1-0.608) \times C_{500}^3}{6} \times \frac{99}{0.608} \approx 1.72 \times 10^6$ iterations. The needed convergence time is $T(n) = O(n^4)$. So, the time needed for a huge network to converge to a global balance will increase quickly and the computation time become very very long for large networks.

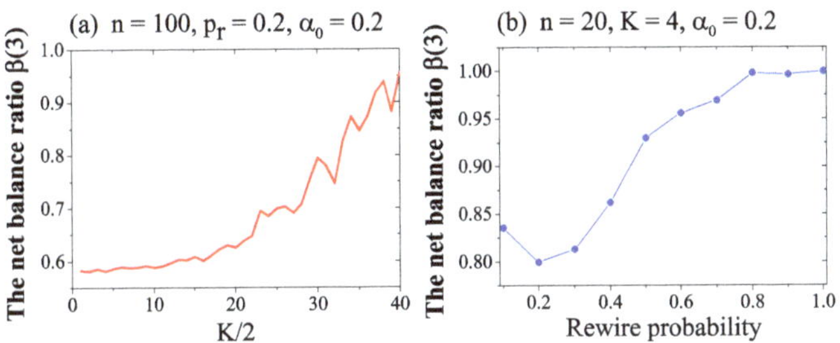

**Figure 6.** The independent influence of the parameter $K$ and network size $n$ on ratio of balanced 3-cycles. The horizontal axis is $K/2$ in (**a**) and rewire probability in (**b**). The vertical axis is the ratio of balanced 3-cycles. The network size was 100 in (**a**) and 20 in (**b**). Other variables were fixed at $p_r = 0.2$ and $\alpha_0 = 0.2$.

As the NW network was created by the NW model, the created NW networks were not absolutely the same, as there was randomness in the NW model. The sign adjustment process will also be impacted by the randomness of the node and cycle selection. To see whether this randomness affected the network convergence trends and values, we carried out a deviation analysis, as shown in Figure 7.

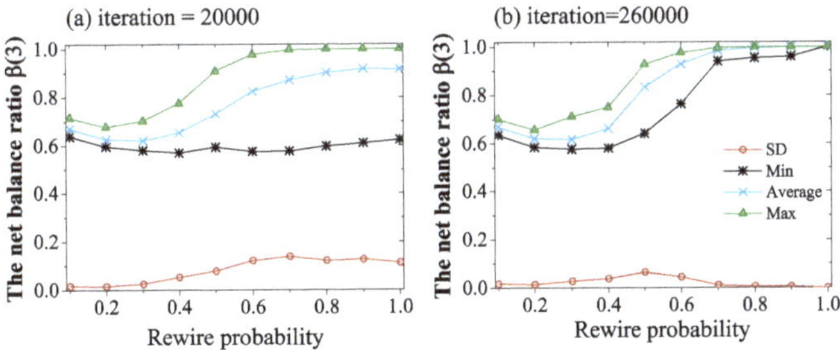

**Figure 7.** The standard deviation analysis of the ratio of balanced three-cycles. The horizontal axis is the rewiring probability. The vertical axis is the ratio of balanced 3-cycles. The blue line is the average value of 50 trials under the same conditions. The green and black lines are the maximum and minimum values among the 50 trials. The red line is the standard deviation of these 50 trials. Each data plot shows the evolution simulations of 20,000 iterations in (**a**) and 260,000 iterations in (**b**). The other variables were fixed at $n = 100$, $K = 4$, and $\alpha_0 = 0.2$.

From the results shown in Figure 7, we can see that

(1) The influence of random factors decreased with the length of the simulation time and number of iterations.

(2) When the simulation time was short and rewire probability small, as shown in Figure 7a, the network converged quickly, for there were fewer links and triangles in the network. The standard deviation was lower, and the maximum, average, and minimum values were very close. With increased rewire probability, the standard deviation enlarged and the gap between the maximum and minimum values expanded. With $p_r > 0.6$, some lucky networks had already reached a global balance $\beta(3) = 1$, as shown by the yellow dashed line; however, some unlucky networks were still imbalanced, and could not reach a global balance, even when the rewire probability was near unity. The randomness in the sign adjustment process could not ensure that every triangle became balanced.

(3) With a longer simulation time, as showed in Figure 7b, and high rewire probability, a denser network also consistently converged to a global balance. The standard deviation decreased with the number of simulation iterations. This result also, indirectly, verified our former conclusion in Figure 6b.

(4) When the rewire probability was close to $p_r = 0.5$, the above-mentioned chaotic area emerged, and the influence of randomness was greater. This result is similar to the results in [42,43].

## 4. Conclusions

We have studied the influence of random and minimum memory-based sign adjustment rules on the evolution of signed social networks and analyzed the impacts of the initial ratio of positive edges $\alpha_0$, the rewiring probability $p_r$, network size, neighbour number $K$, and randomness upon the balancing convergence value. We found that the minimum memory-based sign adjustment can lead to a network global balance if the rewire probability in the NW network exceeds a critical value. With larger rewire probability, the network is denser and it is easier for the influence of each sign adjust to spread to the whole network.

This discovery can help researchers to judge whether an opinion change will spread to the whole network and help network designers to manage and control large social networks. In this paper, we only studied two kinds of simple sign adjustment in the NW network model and the influence of some network characters, some other characters of the social network may also have an important influence on the network evolution. For example, the cluster coefficient, as an index of triangle density in the network, may restrict the influence area of each balance adjustment and, so, may affect the convergence or converge speed of the network. Thus, future research should include the impact of the network model, distribution of degree, cluster coefficient of social networks, and incorporate real adjustment rules, according to empirical observations of the people in and the evolution of real social networks.

**Author Contributions:** M.Q. and H.D. designed the study; M.Q., H.D. and Y.L. performed the research and wrote the paper. All authors have read and approved the final manuscript.

**Funding:** This work was supported by the National Natural Science Foundation of China (NSFC) [grant number 71771214, 71690233].

**Acknowledgments:** The authors are grateful to Peter Abell, Federico Varese, Ray Reagans, and Olav Sorenson for their help and work on this paper.

**Conflicts of Interest:** The authors declares no conflict of interest.

## References


1. Heider, F. Attitudes and cognitive organization. *J. Psychol.* **1946**, *21*, 107–112. [CrossRef] [PubMed]
2. Cartwright, D.; Harary, F. Structural balance: A generalization of Heider's theory. *Psychol. Rev.* **1956**, *63*, 277–293. [CrossRef] [PubMed]
3. Opp, K.D. Balance Theory: Progress and Stagnation of a Social Psychological Theory. *Philos. Soc. Sci.* **1984**, *14*, 27–49. [CrossRef]
4. Hallinan, M.T.; Hutchins, E.E. Structural Effects on Dyadic Change. *Soc. Forces* **1980**, *59*, 225–245. [CrossRef]
5. Deng, H.; Abell, P. A Study of Local Sign Change Adjustment in Balancing Structures. *J. Math. Sociol.* **2010**, *34*, 253–282. [CrossRef]
6. Deng, H.; Abell, P.; Li, J.; Wu, J. A study of sign adjustment in weighted signed networks. *Soc. Netw.* **2012**, *34*, 253–263. [CrossRef]
7. Lewis, K. How Networks Form: Homophily, Opportunity, and Balance. In *Emerging Trends in the Social and Behavioral Sciences: An Interdisciplinary, Searchable, and Linkable Resource*; Wiley Online Library: Hoboken, NJ, USA, 2015; pp. 1–14.
8. Deng, H.; Abell, P.; Engel, O.; Wu, J.; Tan, Y. The influence of structural balance and homophily/heterophobia on the adjustment of random complete signed networks. *Soc. Netw.* **2016**, *44*, 190–201. [CrossRef]
9. Rawlings, C.M.; Friedkin, N.E. The structural balance theory of sentiment networks: Elaboration and test. *Am. J. Sociol.* **2017**, *123*, 510–548. [CrossRef]
10. Kirkley, A.; Cantwell, G.T.; Newman, M.E.J. Balance in signed networks. *Phys. Rev. E* **2019**, *99*, 012320. [CrossRef] [PubMed]
11. Sørensen, A.B.; Hallinan, M.T. A stochastic model for change in group structure. *Sociol. Rev.* **1976**, *24*, 143–166. [CrossRef]
12. Doreian, P.; Krackhardt, D. Pre-transitive balance mechanisms for signed networks. *Math. Sociol.* **2001**, *25*, 43–67. [CrossRef]
13. Macy, M.W.; Willer, R. From factors to actors: Computational sociology and agent-based modeling. *Rev. Sociol.* **2002**, *28*, 143–166. [CrossRef]
14. Ilany, A.; Barocas, A.; Koren, L.; Kam, M.; Geffen, E. Structural balance in the social networks of a wild mammal. *Anim. Behav.* **2013**, *85*, 1397–1405. [CrossRef]
15. Yap, J.; Harrigan, N. Why does everybody hate me? Balance, status, and homophily: The triumvirate of signed tie formation. *Soc. Netw.* **2015**, *40*, 103–122. [CrossRef]
16. Hummon, N.P.; Doreian, P. Some dynamics of social balance processes: Bringing Heider back into balance theory. *Soc. Netw.* **2003**, *25*, 17–49. [CrossRef]
17. Antal, T.; Krapivsky, P.L.; Redner, S. Dynamics of social balance on networks. *Phys. Rev. E* **2005**, *72*, 36121. [CrossRef] [PubMed]
18. Kulakowski, K.; Gawronski, P.; Gronek, P. The Heider Balance: A Continuous Approach. *Int. J. Mod. Phys. C* **2005**, *16*, 707–716. [CrossRef]
19. Ludwig, M.; Abell, P. An evolutionary model of social networks. *Eur. Phys. J. B* **2007**, *58*, 97–105. [CrossRef]
20. Montgomery, J.D. Balance Theory with Incomplete Awareness. *J. Math. Sociol.* **2009**, *33*, 69–96. [CrossRef]
21. Marvel, S.A.; Kleinberg, J.M.; Kleinberg, R.D.; Strogatz, S.H. Continuous-time model of structural balance. *Proc. Natl. Acad. Sci. USA* **2011**, *108*, 1771–1776. [CrossRef]
22. Kossinets, G.; Watts, D.J. Origins of Homophily in an Evolving Social Network. *Am. J. Sociol.* **2009**, *115*, 405–450. [CrossRef]
23. Mei, W.; Cisneros-Velarde, P.; Friedkin, N.E.; Bullo, F. Dynamic Social Balance and Convergent Appraisals via Homophily and Influence Mechanisms. *arXiv* **2017**, arXiv:1710.09498.
24. Rijt, A.V.D. The Micro-Macro Link for the Theory of Structural Balance. *J. Math. Sociol.* **2011**, *35*, 94–113. [CrossRef]
25. Volstorf, J.; Rieskamp, J.; Stevens, J.R. The Good, the Bad, and the Rare: Memory for Partners in Social Interactions. *PLoS ONE* **2011**, *6*, e18945. [CrossRef] [PubMed]

26. Brashears, M.E.; Brashears, L.A. The Enemy of My Friend Is Easy to Remember: Balance as a Compression Heuristic. In *Advances in Group Processes*; Emerald Group Publishing Limited: Bingley, UK, 2016; Volume 33.

27. Hassanibesheli, F.; Hedayatifar, L.; Safdari, H.; Ausloos, M.; Jafari, G. Glassy States of Aging Social Networks. *Entropy* **2017**, *19*, 246. [CrossRef]

28. Kottonau, J.; Burse, J.; Pahl-Wostl, C. A consumer memory-based model of new product diffusion within a social network. In Proceedings of the 10th Meeting of the Annual Workshop on Computational and Mathematical Organisation Theory, CMOT, Computational Social Organisational Science Conference, CASOS, CMU, Pittsburgh, PA, USA, 21–24 July 2000.

29. Winke, T.; Stevens, J.R. Is cooperative memory special? The role of costly errors, context, and social network size when remembering cooperative actions. *Front. Robot. AI* **2017**, *4*, 52. [CrossRef]

30. Milgram, S. The small world problem. *Psychol. Today* **1967**, *2*, 60–67.

31. Guare, J. *Six Degrees of Separation: A Play*; Vintage: New York, NY, USA, 1990.

32. Adamic, L.A.; Huberman, B.A. Power-Law Distribution of the World Wide Web. *Science* **2000**, *287*, 2115. [CrossRef]

33. Amaral, L.A.N.; Scala, A.; Barthelemy, M.; Stanley, H.E. Classes of small-world networks. *Proc. Natl. Acad. Sci. USA* **2000**, *97*, 11149–11152. [CrossRef]

34. Davis, G.F.; Yoo, M.; Baker, W.E. The Small World of the American Corporate Elite, 1982–2001. *Strateg. Organ.* **2003**, *1*, 301–326. [CrossRef]

35. Davis, G.F.; Greve, H.R. Corporate Elite Networks and Governance Changes in the 1980s. *Am. J. Sociol.* **1997**, *103*, 1–37. [CrossRef]

36. Bordons, M.; Gomez, I. Collaboration networks in science. In *The Web of Knowledge: A Festschrift in Honor of Eugene Garfield*; Information Today Inc.: Medford, NJ, USA, 2000; pp. 197–213.

37. Barabasi, A.; Jeong, H.; Neda, Z.; Ravasz, E.; Schubert, A.; Vicsek, T. Evolution of the social network of scientific collaborations. *Phys. A-Stat. Mech. Appl.* **2002**, *311*, 590–614. [CrossRef]

38. Aiello, W.; Chung, F.R.K.; Lu, L. A random graph model for massive graphs. In Proceedings of the Thirty-Second Annual Acm Symposium on Theory of Computing, Portland, OR, USA, 21–23 May 2000; pp. 171–180.

39. Aiello, W.; Chung, F.R.K.; Lu, L. Random evolution in massive graphs. In Proceedings of the 42nd IEEE Symposium on Foundations of Computer Science, Newport Beach, CA, USA, 7 August 2002; pp. 97–122.

40. Ebel, H.; Mielsch, L.I.; Bornholdt, S. Scale-free topology of e-mail networks. *Phys. Rev. E* **2002**, *66*, 35103. [CrossRef] [PubMed]

41. Newman, M.; Watts, D. Renormalization Group Analysis of the Small-World Network Model. *Phys. Lett. A* **1999**, *263*, 341–346. [CrossRef]

42. Abell, P.; Ludwig, M. Structural Balance: A Dynamic Perspective. *J. Math. Sociol.* **2009**, *33*, 129–155. [CrossRef]

43. Deng, H.Z.; Abell, P.; Li, Y.; Wu, J.; Tan, Y.J. Network Size Impact upon Global Balance Structure in Small Complete Network. *Appl. Mech. Mater.* **2015**, *713* 2276–2279. [CrossRef]

*Article*

# A SOM-Based Membrane Optimization Algorithm for Community Detection

Chuang Liu *, Yingkui Du and Jiahao Lei

School of Information Engineering, Shenyang University, Liaoning 110044, China; yikui.du.cn@gmail.com (Y.D.); 18691661820@163.com (J.L.)
* Correspondence: chuang.liu@mail.dlut.edu.cn

Received: 26 April 2019; Accepted: 23 May 2019; Published: 25 May 2019

**Abstract:** The real world is full of rich and valuable complex networks. Community structure is an important feature in complex networks, which makes possible the discovery of some structure or hidden related information for an in-depth study of complex network structures and functional characteristics. Aimed at community detection in complex networks, this paper proposed a membrane algorithm based on a self-organizing map (SOM) network. Firstly, community detection was transformed as discrete optimization problems by selecting the optimization function. Secondly, three elements of the membrane algorithm, objects, reaction rules, and membrane structure were designed to analyze the properties and characteristics of the community structure. Thirdly, a SOM was employed to determine the number of membranes by learning and mining the structure of the current objects in the decision space, which is beneficial to guiding the local and global search of the proposed algorithm by constructing the neighborhood relationship. Finally, the simulation experiment was carried out on both synthetic benchmark networks and four real-world networks. The experiment proved that the proposed algorithm had higher accuracy, stability, and execution efficiency, compared with the results of other experimental algorithms.

**Keywords:** community detection; membrane algorithm; self-organizing map network; complex networks; optimization

---

## 1. Introduction

Many networks can be simulated by complex networks, such as social networks, biological networks, and the World Wide Web. The study of complex networks is increasingly attracting the attention of researchers from many different fields. These complex networks are represented by nodes and edges. In order to clearly understand the structural characteristics and functional characteristics of complex networks, finding the relationship between these nodes and edges is especially important for studying the composition of the network and understanding the functional characteristics of the network. As a method to revealing the relationship between nodes and edges in the network, community structure has become a hot research topic in network science. More and more researchers are paying attention to community detection problems in complex networks [1–3].

There are many algorithms for studying community detection, including the graph partitioning algorithm, hierarchical clustering, modularity optimization algorithm, label propagation algorithm, partition-based clustering algorithm, evolutionary algorithm, etc. [4]. Among many algorithms, evolutionary algorithms can solve the problems of community detection without prior knowledge. These problems need to be converted into optimization problems first, and then they can be solved by using evolutionary algorithms, such as the genetic algorithm (GA), particle swarm optimization (PSO), differential evolution (DE), etc. Such algorithms have the ability to automatically detect the number of communities when the number of communities in the network is unknown, and is more suitable for solving real network problems [1]. The application of evolutionary algorithms in complex networks has attracted the attention of many researchers. Tasgin et al. first proposed a genetic algorithm to solve these kinds of complex problems [5]. Pizzuti proposed a GA-based community detection algorithm, which introduced a genetic representation and the concept of community score as the fitness function to detect community structure in complex networks [6]. Pizzuti proposed a multiobjective genetic algorithm to find communities in complex networks. The method maximizes the intra-connections inside each community and minimizes inter-connections between different communities [7]. Gong et al. proposed a synergy of a genetic algorithm with a hill-climbing strategy as the local search procedure to optimize modularity destiny to explore the network at different resolutions [8]. Pizzuti et al. proposed a many-objective optimization algorithm for community detection in multi-layer networks [9]. Meo et al. proposed a scalable method to maximize modularity in large networks, which is a new clustering method that couples the accuracy of global approaches with the scalability of local methods [10]. Grass-Boada et al. proposed a multi-objective overlapping community detection algorithm, which is based on the Pareto-dominance based multi-objective evolutionary algorithmsand global and local approaches for discovering overlapping communities [11]. Berahmand et al. proposed a local approach by detecting and expanding core nodes through extended local similarity of nodes [12]. Shi et al. proposed a locally-biased spectral approximation approach to adapt the Lanczos method for local community detection, which apply a fast random walk, personalized PageRank, and heat kernel diffusion [13]. Moradi et al. proposed an extension genetic algorithm with a novel local search strategy for community detection [14].

In summary, the research results of community detection based on evolutionary algorithms mainly focus on network coding, group initialization, evolution rule design, and objective function selection. Although the above literature has obtained a wealth of research results, the accuracy and complexity of these algorithm still needs to be improved. In this paper, we proposed an evolutionary membrane community detection algorithm based on self-organizing maps (SOM), named EMCD-SOM. SOM, an unsupervised learning algorithm for clustering and high-dimensional visualization, is an artificial neural network developed by simulating the characteristics of the human brain's processing signals [15]. The proposed algorithm consists of objects, reaction rules, and membrane structure. An object presents a partition result of the complex network. Reaction rules include GA and DE. In the skin membrane, GA is utilized as reaction rules to evolve the objects. DE is introduced as reaction rules in the region of each membrane. In order to find the optimal membrane structure, SOM determines the number of membranes by learning the information of the objects. To evaluate the performance of EMCD-SOM, synthetic benchmark networks and four real-world networks were conducted by the proposed EMCD-SOM. The experimental results showed that the proposed method was more useful and effective than other state-of-the-art algorithms including FastNewman [16], LconDanon [17], GA-NET [6], CMM [18], and Meme-net [8] from the literature.

The main contributions of this paper are summarized below:

- The SOM neural network may learn and mine the structure of the current objects in the decision space, which is beneficial for guiding the local and global search of the proposed algorithm;
- The number of membranes of the proposed EMCD-SOM is determined according to the characteristics of SOM mapping similar data to adjacent neurons.
- GA and DE are employed as reaction rules to evolve the objects in the different region of membrane;
- The proposed EMCD-SOM can implement the balance of exploration and exploitation in four real world networks.

The rest of this paper is organized as follows. In Section 2, the description of the proposed EMCD-SOM is elaborated. In Section 3, the simulation results are evaluated on the benchmark test problems in comparison with some state-of-the-art evolutionary algorithms. Moreover, this section includes a sensitivity analysis for the proposed EMCD-SOM. Finally, Section 4 summarizes the concluding remarks of this paper.

## 2. The Proposed Approach

This section will explain the principles of the proposed EMCD-SOM based on a membrane system. Since the membrane system consists of three elements: object, reaction rule, and membrane structure, the proposed algorithm also has these elements. In the proposed EMCD-SOM, the focus is on how to achieve these three elements. The object as the first element in the region of membrane represents candidate solution for network partitioning. The second element is the reaction rule, which are designed to evolve objects in different region of membranes. The membrane structure is the last element, which helps to promote the exchange of information between membranes and enhance the diversity of objects. These features are very useful in developing a new evolutionary algorithm to improve its solving performance.

The pseudo-code of the proposed EMCD-SOM is given in the Algorithm 1.

---

**Algorithm 1** The pseudo-code of the proposed EMCD-SOM.

---

**Input:** The parameters of the proposed algorithm are initialized, including the number of objects in

each elementary membrane, each object within its boundaries.

**Output:** The best object is found from the different elementary membranes.

1: The objects are initialized from the search space.
2: The fitness of these objects is calculated according to the modularity density function in Equation (3).
3: **while** End Condition **do**
4:     Determining the number of membrane ($NC$) by invoking SOM
5:     **for** $i = 1; i < NC; i + +$ **do**
6:         Evolving the objects in the region of elementary membrane according to the DE-based

reaction rule.

7:     **end for**
8:     The objects from the region of elementary membrane are released into the region of skin membrane.
9:     All objects in the region of skin membrane are evolved according to the GA-based reaction rule.
10: **end while**

---

### 2.1. Object and Its Initialization

The object is encoded as a partition of community in the complex network. Depending on the number of network communities, each object can be represented as a set of real integer values. In the proposed algorithm, an object is defined as:

$$X = (x_1, x_2, \cdots, x_n) \tag{1}$$

where $n$ represents the number of the nodes in a complex network, and $x_i$ is the $i$-th node and is an integer change from 1 to $n$. A community consists of nodes with the same value. The graphical illustration of the object coding is shown in Figure 1. As can be seen from Figure 1, there are 14 nodes and a total of three communities represented by objects. It is worth mentioning that the number of communities is automatically determined by the proposed algorithm. In the worst case, a complex network with $n$ nodes can be divided into $n$ communities.

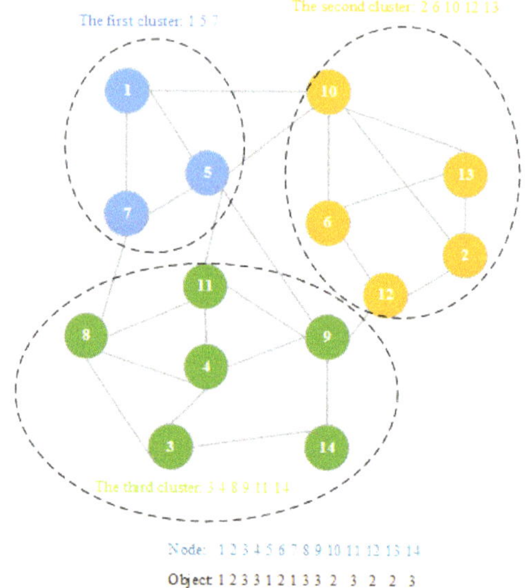

Figure 1. A generic illustration of the representation of a discrete object.

The object represents the result of network partitioning in the proposed EMCD-SOM. It is initialized according to Equation (2):

$$x_{i,j} = \lceil x_j^l + (x_j^u - x_j^l) \times r \rceil + 1 \tag{2}$$

where $1 \leq i \leq N$, $N$ is the number of objects in the region of all membranes. $1 \leq j \leq n$, $n$ represents the maximum value of the node identifier in a complex network. $x_{i,j}$ is the value of the $j$-th identifier in the $i$-th object, which is an integer value from 1 to $n$. $x_j^l$ represents the $j$-th lower limit of the identifier in the complex network, which has a value of 1, and $x_j^u$ represents the upper boundary value of the $j$-th identifier of the identifier in the complex network, which is $n$. $r$ can generate a random number on the interval $(0, 1)$. In the formula, the ceiling operations is utilized to ensure that $x_{i,j}$ is an integer value.

### 2.2. Objective Function

Among many objects in the region of membranes, how to determine which object is the best forthe best community partition requires the use of the objective function. The modularity density widely used in community detection problems [19], and its definition is given in Equation (3).

$$f = \sum_{i=1}^{N} \left( \frac{L(V_i, V_i) - L(V_i, \overline{V_i})}{|V_i|} \right) \tag{3}$$

where $L(V_1, V_2) = \sum_{i \in V_1, j \in V_2} A_{ij}$, and $L(V_1, \overline{V_2}) = \sum_{i \in V_1, j \in \overline{V_2}} A_{ij}$, and $\overline{V_2} = \Omega - V_2$, and $A$ is the adjacent matrix of the network, and $\Omega = V_1, V_2, \cdots, V_N$ is a partition.

The value of the objective function is one of the most critical steps that guides the object's search direction. The modularity density values are utilized to evaluate the quality of objects in all membranes. The higher modularity density value has, the better community structure is attained by the proposed algorithm. If the modularity density value is equal to 1, the network partition represents a very good community structure.

### 2.3. Membrane Structure

Since the proposed algorithm is based on a membrane system, it inherits the same network structure from the membrane system. In order to simplify the implementation of this structure, the proposed algorithm is defined as a structure containing only the elementary membrane. Each elementary membrane can be thought of as an evolutionary unit. In the experiment, we found that the number of membranes is difficult to set. To solve this problem, we used a self-organizing mapping network (SOM) to determine the number of elementary membranes, specifically using SOM to discover the structural information of the decision space of objects, and then determine the number of elementary membranes. The details of SOM are given below.

SOM, an unsupervised learning algorithm proposed by Kohonen for clustering and high-dimensional visualization, is an artificial neural network developed by simulating the characteristics of the human brain's processing signals. It is characterized by the ability to map high-dimensional distributions to low dimensions and maintain mapping invariance. In recent years, SOM have been applied to the solution of optimization problems. Jin et al. proposed a SOM with a novel learning rule to solve the traveling salesman problem (TSP) [20]. Villmann et al. proposed a hybrid system combining SOM and evolutionary algorithms to promote neighborhood cooperation [21]. Zhang et al. proposed a self-organizing multiobjective evolutionary algorithm. SOM is employed to establish the neighborhood relationship among current solutions [22]. Liang et al. proposed a multi-objective particle swarm optimization algorithm based on SOM, which mainly uses SOM to discover the structural information of population and the multi-objective Pareto solution set, and then guides the particle flight [23]. The topology of a two-dimensional SOM is shown in Figure 2.

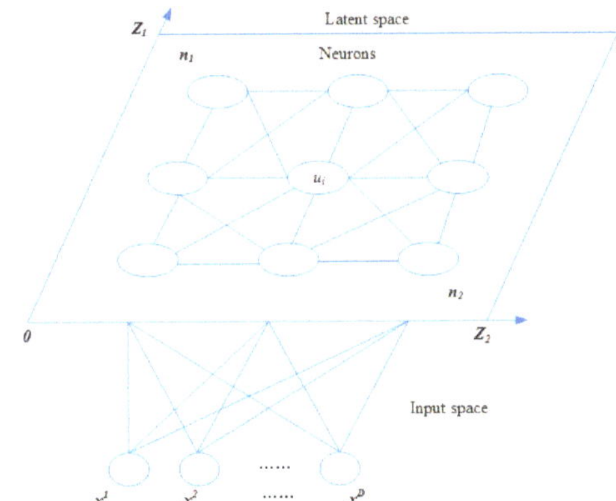


**Figure 2.** An illustration of a two-dimensional self-organizing map network (SOM).

As shown in the figure, SOM consists of an input layer and a competition layer (output layer). The number of input layer neurons is $D$, and the competition layer consists of a one-dimensional or two-dimensional planar array of $N = n_1 \times n_2$ neurons. Each neuron $u_i \in (1, 2, \cdots, N)$ has its own location information $Z^{u_i} = (z_1^{u_i}, z_2^{u_i})$ and weight information $W^{u_i} = (w_1^{u_i}, w_1^{u_i}, \cdots, w_D^{u_i})$. The network is fully connected, that is, each input node is connected to all output nodes. SOM consists of a training phase and a clustering phase. In the first stage, the training data is randomly selected, the winning neurons are selected according to the Euclidean distance, and the weights of the winning neurons and their neighboring neurons are updated. The second stage is mapping test data to neurons and similar data to neighboring neurons. The number of membranes of the proposed EMCD-SOM is determined according to the characteristics of SOM mapping similar data to adjacent neurons. Furthermore, the number of clusters in the SOM is used to determine the number of membranes in the proposed algorithm. The structure of

EMCD-SOM is conducive to improving search efficiency and is suitable for solving community detection problems.

In the proposed algorithm, the objects in the region of elementary membrane are evolved by the reaction rule according to the differential evolution algorithm. When objects from different membranes are evolved, they are released into the region of the skin membrane. These objects will continue to evolve by calling genetic algorithm-based reaction rules. Then, they are aggregated into several classes using SOM and these clustered objects are in turn sent to the region of elementary membrane and are evolved by invoking the reaction rule. After executing several generations, some good objects can be generated by executing reaction rules in the different elementary membranes. The best object can be found by comparing the modularity density values of these objects.

*2.4. Reaction Rules*

The reaction rule is inspired by the chemical reaction of the objects and the way of handling the compound. Reaction rules can be implemented through mechanisms that can develop objects into the direction of the global optimal partition of the network. According to "No Free Lunch", there is no single optimization algorithm to solve every optimization problem effectively and efficiently. In other words, different algorithms possess a different accuracy to solve the same optimization problem. The ensemble of state-of-the-art algorithms can obtain a better solution than using a single algorithm. Inspired by this, we employed the GA algorithm and the DE algorithm to evolve objects in both the skin membrane and elementary membrane.

GA is a computational model that simulates the natural evolution of Darwin's biological evolution theory and the biological evolution process of genetic mechanism. It is a method to search for optimal solutions by simulating natural evolutionary processes. In each generation, the optimal individual is selected based on the individual's adaptability in the problem domain, and new individuals are generated by crossover and mutation operations in the genetic operator. In the proposed algorithm, GA acts as a reaction rule in the skin membrane. More specifically, the individual in GA is represented by the object. The selection operation is used to select the parent population of mating in the GA. Here we used a wide range of deterministic tournament selection operators. The crossover operation was implemented by two-way crossing over operation in the literature [8]. In mutation, we randomly selected a object in the region of the skin membrane. A point mutation was employed, which randomly picked a dimension value on the object and then randomly changed the value to its neighbor's dimension value. GA facilitated global search by the proposed algorithm. The parameters of GA were given as follows: Crossover probability = 0.8, mutation probability = 0.2.

DE was employed as a reaction rule in elementary membranes. DE is an optimization algorithm based on differential and simple mutation operation and one-to-one competitive survival strategy, which reduces the complexity of genetic operations. It generates new individuals through differential mutation with some different strategies including DE/rand/1, DE/best/1, DE/best/2, DE/rand-to-best/1, etc. In order to improve the diversity of candidate solutions, DE introduces crossover to operate on target vectors and mutation vectors to generate new experimental vectors. In the proposed algorithm, DE/best/1 was utilized to evolve objects in the region of the elementary membrane. A modified binomial crossover was employed to assign the value of either dimension in an object to the value of the corresponding dimension in another object [24]. The parameters of DE were given as follows: $F = 0.9$ is called the differential weight. $CR = 0.3$ is called the crossover probability.

## 3. Experimental Evaluation

The performance of the proposed algorithm was validated in a series of experiments based on both synthetic benchmark networks and the four real-world networks by comparing it with state-of-art algorithms. Section 3.1 will discuss the details of these networks. Section 3.2 will describe the experimental condition in running the simulation. Section 3.3 will give several metrics of the experimental algorithms. Section 3.4 will give the simulation result of the LFR (Lancichinetti–Fortunato–Radicchi)benchmark network calculated by all experimental algorithms. Section 3.5 will discuss the experimental results based on the evaluation metrics of the experimental algorithm on different network datasets.

*3.1. Synthetic Benchmark Networks and Four Real-World Networks*

3.1.1. Description of Synthetic Benchmark Betworks

The first set of experiments is the LFR benchmark network presented by Lancichinetti and Radicchi in [25], which has power law degree distribution and variable sized communities. It is the most widely used benchmark network for testing the performance of algorithms in community detection. Compared with other synthetic networks, LFR networks can reflect some important features of complex real-world systems. In the simulation, the number of nodes in the LFR network was 1000, the average degree was 15, the maximum degree was 50, the mixing parameter was 0.1, the minimum planted community size was 20, and the maximum planted community size was 50.

3.1.2. Description of Four Real-World Networks

In the following experiments, four real-world networks were employed to test the performance of the proposed algorithm, including the Zachary's karate club network, American college football club network, Krebs America Political Book network, and Bottlenose dolphins network. The ground-truths of these networks has been known. More details about the definition of these network datasets can be discussed as follows. The Zachary's karate club network, constructed by Zachary, is a network of relations between 34 members of a karate club over a period of two years [26]. The karate club is split into two communities of almost the same size on account of disagreements between the administrator and the instructor of the club. The American college football network consists of 115 vertices and 613 edges, which is divided into 12 communities, which was first proposed by Girvan and Newman [27]. Vertices in the network represent teams which are identified by their college names, and edges represent the regular season games between the two teams they connect. This Krebs America political book network consists of 105 vertices and 441 edges between books purchased together during the 2004 presidential election, which was compiled by Krebs [28]. Bottlenose Dolphins network consists of 62 vertices and 60 edges based on social acquaintances, which is naturally divided into two large groups: The male group and the female one [29]. Each node represents a dolphin living over a period of 7 years in the bottlenose dolphins network. The related parameters of each real-world network are described in Table 1.

**Table 1.** Parameters of the real-world networks.

| Datasets | Nodes | Edges | Communities |
|---|---|---|---|
| Zachary's karate club network | 34 | 78 | 2 |
| American college football club network | 115 | 613 | 12 |
| Krebs America political book network | 105 | 441 | 3 |
| Bottlenose dolphins network | 62 | 60 | 2 |

*3.2. Experimental Conditions*

In the experiments, some related community detection algorithms were employed to compare with the proposed algorithm. These algorithms consist of Fast–Newman, Lcon-Danon, GA-net, Meme-net, and MOGA-net. Some of them, including GA-net and Meme-net, are single-objective algorithms, while the rest are non-evolutionary algorithms. They were run in Windows 7 enterprise version under the hardware environment of Intel Pentium dual-core 2.93 GHZ and 16 GB RAM. The proposed algorithm was implemented using Matlab2015.

Since the results of the community detection method based on evolutionary algorithm depend on the validity of the random search process, 30 repeated tests were performed independently on both synthetic benchmark networks and 4 real-world networks, and statistical results were calculated in order to evaluate the statistical performance of algorithms and reduce statistical errors. Moreover, 4 statistical metrics were designed, such as Mean, Std, Worst, and Best. These metrics were employed to evaluate the solving performance of these various algorithms.

### 3.3. Evaluation Measures

At present, there are many metrics for evaluating the effectiveness of community detection algorithms that detect the quality of network partitions of complex networks. Among these metrics, the normalized mutual information (NMI) are the most widely used in community detection of complex networks. In addition, to further evaluate the quality of the experimental results, some clustering indicators were introduced include the F-measure and Rand Index.

NMI is a similarity measure estimating the similarity between detected partitions and true ones. A higher NMI value represents a greater similarity between two partitions. If NMI takes its maximum value which is equal to 1, all communities obtained by the experimental algorithms are identical to all real communities. In the following experiment, NMI was used to evaluate the results between true partition and the partition obtained by experimental algorithms. The definition of NMI(A, B) is shown in Equation (4):

$$NMI(A,B) = \frac{-2\sum_{i=1}^{C_A}\sum_{j=1}^{C_B} D_{ij}\log(\frac{D_{ij}N}{D_i \cdot D_j})}{\sum_{i=1}^{C_A} D_i \cdot \log(\frac{D_i}{N}) + \sum_{j=1}^{C_B} D_j \cdot \log(\frac{D_j}{N})} \tag{4}$$

where $A$ and $B$ are partitions of a network, and $C_A$ represents the number of communities in $A$ while $C_B$ denotes that of $B$. $D$ is a confusion matrix, and $D_{i,j}$ stands for the number of nodes in community $i$ of $A$ that also appear in community $j$ of $B$. $N$ is the number of elements. $D_i$ is the sum over row $i$ of $D$ while $D_j$ is the sum of elements in column $j$.

F-measure is also called F-score, which is a weighted harmonic averaging of Precision and Recall. It is a commonly used evaluation standard in the clustering field and is often used to evaluate the quality of the classification model. The definition of F-measure is shown in Equation (5):

$$F = 2 \times \frac{PR}{P + R} \tag{5}$$

where $P$ is the precision and $R$ is the recall rate.

Rand Index(RI) is also called Rand measure, which is a measure of the similarity between two data clusterings. In the experiments, Rand Index is employed to measure the similarity between real partitions and the partitions obtained by experimental algorithms. The definition of Rand Index is shown in Equation (6):

$$RI = \frac{a}{a + b} \tag{6}$$

where $a$ can be considered as the number of agreements between real partitions and the partitions obtained by experimental algorithms, and $b$ as the number of disagreements between real partitions and the partitions obtained by experimental algorithms.

### 3.4. Experiments on Synthetic Benchmark Networks

In the following experiment, the LFR network consisted of a network of size 1000 with a mixing parameter fixed at 0.1. All experimental algorithms ran independently 30 times in the networks. The statistical results of the evaluation indicators with NMI, F-measure, and Rand Index were used to evaluate the performance of all experimental algorithms.

As shown in Table 2, the proposed EMCD-SOM achieved the best results on all indicators in comparison with other experimental algorithms. FastNewman had suboptimal results on the synthetic benchmark networks. Due to the fact that Meme-net runs for a long time and there is no calculation result, the statistical result was represented by '-'. In summary, compared with other experimental methods, the proposed algorithm was suitable for solving networks with a large number of nodes.

**Table 2.** The statistical values obtained by the experimental algorithms on the synthetic benchmark networks of size 1000 with a mixing parameter fixed at 0.1. GA-NET: GeneticAlgorithm-NET; CMM: Convexified Modularity Maximization; Meme-net: Memeticalgorithm-net; EMCD-SOM: The proposed algorithm; NMI: normalized mutual information; RI: Rand Index.

| Metrics | Statistics | FastNewman [16] | LconDanon [17] | GA-NET [6] | CMM [18] | Meme-net [8] | EMCD-SOM |
|---|---|---|---|---|---|---|---|
| NMI | Mean | 0.952684 | 0.945996 | 0.872757 | 0.939711 | - | 0.992237 |
| | Std | $5.64601 \times 10^{-16}$ | 0 | 0.0186498 | 0.0136735 | - | 0.0115922 |
| | Worst | 0.952684 | 0.945996 | 0.827308 | 0.915167 | - | 0.947601 |
| | Best | 0.952684 | 0.945996 | 0.899495 | 0.969452 | - | 1 |
| F-measure | Mean | 0.881533 | 0.943461 | 0.858099 | 0.86981 | - | 0.976459 |
| | Std | $3.38761 \times 10^{-16}$ | 0 | 0.0256338 | 0.0270183 | - | 0.0337187 |
| | Worst | 0.881533 | 0.943461 | 0.79845 | 0.825811 | - | 0.854329 |
| | Best | 0.881533 | 0.943461 | 0.898216 | 0.937594 | - | 1 |
| RI | Mean | 0.986993 | 0.992146 | 0.983747 | 0.975294 | - | 0.996954 |
| | Std | $3.38761 \times 10^{-16}$ | $4.51681 \times 10^{-16}$ | 0.00267911 | 0.00821687 | - | 0.00554023 |
| | Worst | 0.986993 | 0.992146 | 0.977668 | 0.960883 | - | 0.971924 |
| | Best | 0.986993 | 0.992146 | 0.988004 | 0.993564 | - | 1 |

## 3.5. Experiments on Real-World Networks

In this section, the proposed algorithms were compared with other algorithms for 4 real-world datasets with real partitions known in the following experiment. All experimental algorithms were run 30 times, independently. The statistical results of NMI, F-measure, and Rand Index were utilized to evaluate the performance of the experimental algorithms.

### 3.5.1. Display Network Partition

We visualized the community detection results obtained by the proposed algorithm on 4 real-world datasets with real partitions known. As shown in Figures 3–6, the community division was the best result from 30 runs, and almost every partition had a good community structure and was similar to the real division of the network. The results of Figure 3 show that the proposed algorithm can obtain different levels of community structure on Zachary's karate club network. The proposed algorithm could discover 2 communities, as shown in Figure 3, which is consistent with the real community structure in Table 1.

The community structure detected by the proposed algorithm on the American college football network is shown in Figure 4. It can be seen from Figure 4 that the proposed algorithm detected 11 partitions, but only a few nodes had community partitioning errors. The real network had 12 partitions in Table 1.

As seen Figure 5 in the US political book network, due to the complexity of the network structure, the proposed algorithm had a community structure with 4 communities, but the actual network partition was 3 in Table 1.

Lastly, Figure 6 shows the results of the community of the Bottlenose dolphins network obtained by the proposed algorithm. As shown in Figure 6, the number of the community obtained by the proposed algorithm was larger than the result of the real network in Table 1.

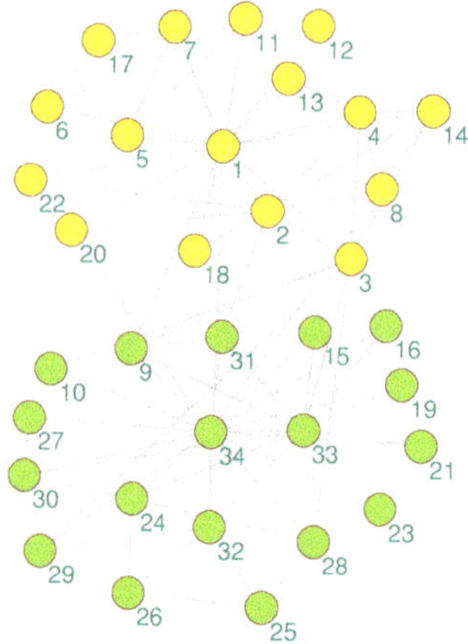

**Figure 3.** The community detection result of the proposed algorithm on Zachary's karate club network.

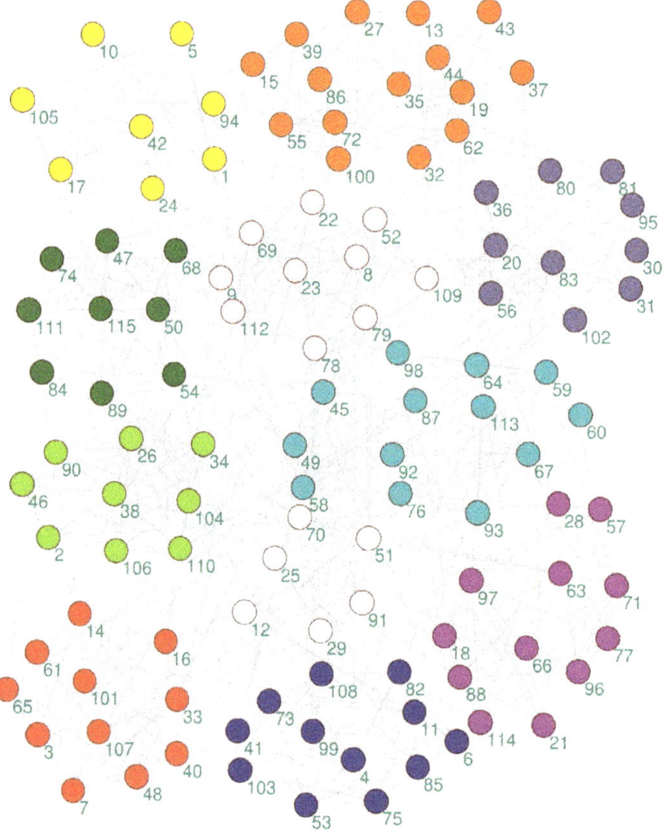

**Figure 4.** The community detection result of the proposed algorithm on the American college football club network.

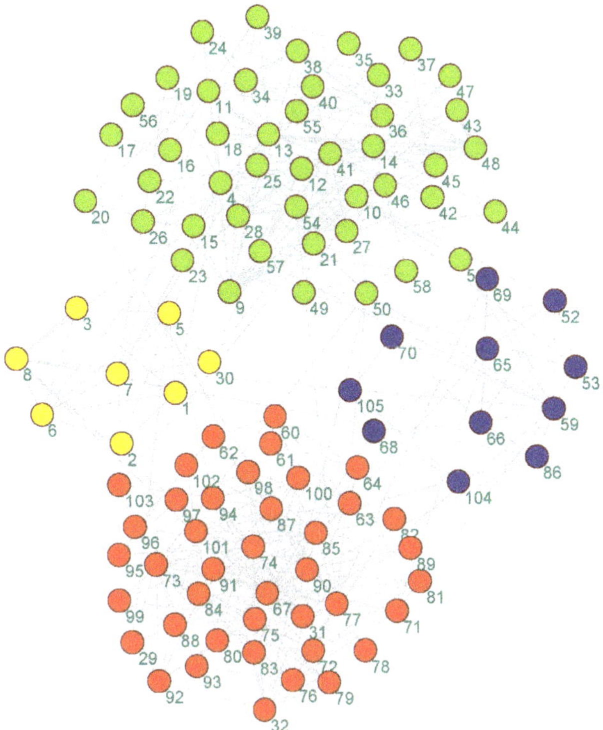

**Figure 5.** The community detection result of the proposed algorithm on the Krebs America political book network.

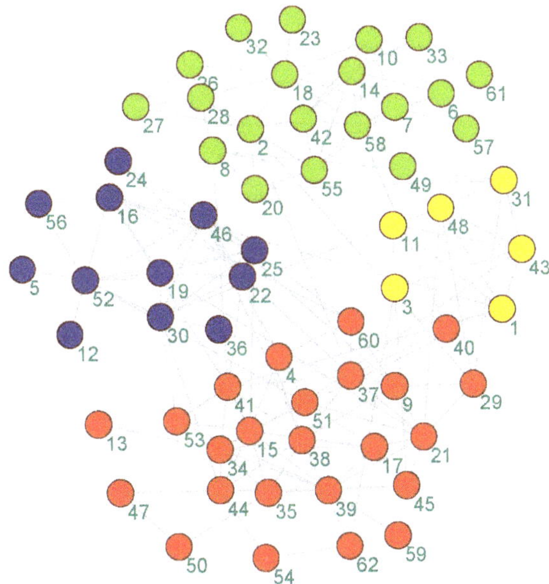

**Figure 6.** The community detection result of the proposed algorithm on the Bottlenose dolphins network.

### 3.5.2. Comparison of the Proposed Algorithm with Other Algorithms

In this section, Tables 3–5 show the community detection effect of the proposed algorithm and other experimental algorithms running 30 times with 3 evaluation indicators on 4 real networks. As shown in Tables 3–5, compared to other algorithms, the proposed algorithm had a good performance in community detection on 4 real-world networks.

The NMI values of all experimental algorithms are shown in Table 3. On Zachary's karate club network, the best results obtained by the proposed algorithm indicated that it can all converge to the global optimal $NMI = 1$. The result indicates that the community obtained by the proposed algorithm was exactly the same as the real community. This result can also be obtained from Figure 3. To illustrate the performance of the proposed algorithm, we sorted these algorithms according to the average of the NMI indicator as follows: CMM, Meme-net, EMCD-SOM, FastNewman, GA-NET, and LconDanon. Compared with Meme-net, the proposed algorithm obtained the suboptimal community partition result.

On the American college football club network, the proposed algorithm gained the best average NMI of 0.900987 in all experimental algorithms. CMM attained the second-best NMI average. The performance of these algorithms was sorted as follows: EMCD-SOM, CMM, Meme-net, LconDanon, FastNewman, and GA-NET.

On Krebs America political book network, the proposed algorithm found the second-best NMI average of 0.528597, which is not much different from FastNewman. The best result, out of the 30 times, belonged to the proposed EMCD-SOM. According to the average value of NMI, these algorithms were sorted as follows: FastNewman, EMCD-SOM, LconDanon, Meme-net, CMM, and GA-NET.

On the Bottlenose dolphins network, the proposed algorithm obtained the fourth average. These algorithms were sorted as follows: CMM, LconDanon, FastNewman, EMCD-SOM, Meme-net, and GA-NET.

Next, all experimental algorithms were evaluated by calculating the F-measure, which was conducted on the real-world networks. This indicator is often used to evaluate the quality of the classification model. The F-measure values obtained by the experimental algorithms on real-world networks are shown in Table 4.

As seen in Table 4, the proposed algorithm could obtain the best results for the F-measure indicator compared with all experimental algorithms on most of real-world networks. Compared with the proposed algorithm, CMM gained the best result on Dolphins, and Meme-net gained the best result on Karate Club, and FastNewman gained the best result on Political Book and Dolphins.

Finally, all experimental algorithms were evaluated according to the Rand Index indicator. This indicator is often used to measure the similarity between two data clusterings. The Rand Index values obtained by the experimental algorithms on the real-world networks are shown in Table 5.

As we can see, compared with the other 5 community detection methods for Rand Index on real networks, the proposed EMCD-SOM could get satisfactory results, especially in the American college football club network. For the karate network, Meme-net gained the best result. For Football club, the proposed algorithm gained the best result. FastNewman gained the best result on the Political book and Dolphins network in terms of the Rand Index. It is worth noting that the proposed algorithm was similar with FastNewman on the Political book network.

Finally, although the proposed algorithm was not optimal, the proposed algorithm showed stable results on different networks, which indicates that the proposed algorithm is suitable for solving community structure partitioning problems in complex networks.

**Table 3.** The NMI values obtained by the experimental algorithms on the real-world networks with real partitions known.

| Networks | NMI | FastNewman [16] | LconDanon [17] | GA-NET [6] | CMM [18] | Meme-net [8] | EMCD-SOM |
|---|---|---|---|---|---|---|---|
| Karate Club | Mean | 0.692467 | 0.530471 | 0.662719 | 1 | 0.759591 | 0.729539 |
| | Std | $2.25841e \times 10^{-16}$ | 0 | 0.041038 | 0 | 0.12226 | 0.0916947 |
| | Worst | 0.692467 | 0.530471 | 0.593038 | 1 | 0.699488 | 0.6895798 |
| | Best | 0.692467 | 0.530471 | 0.707135 | 1 | 1 | 1 |
| Football Club | Mean | 0.697732 | 0.72976 | 0.36438 | 0.900688 | 0.877428 | 0.900987 |
| | Std | $1.1292 \times 10^{-16}$ | $3.38761 \times 10^{-16}$ | 0.0326597 | 0.00603723 | 0.0338035 | 0.0128863 |
| | Worst | 0.697732 | 0.72976 | 0.287833 | 0.896274 | 0.757927 | 0.858186 |
| | Best | 0.697732 | 0.72976 | 0.432277 | 0.914376 | 0.924195 | 0.91137 |
| Political Book | Mean | 0.530814 | 0.522288 | 0.407465 | 0.454128 | 0.46474 | 0.528597 |
| | Std | $4.51681 \times 10^{-16}$ | $2.25841 \times 10^{-16}$ | 0.0204818 | $3.38761 \times 10^{-16}$ | 0.0283599 | 0.0190332 |
| | Worst | 0.530814 | 0.522288 | 0.361427 | 0.454128 | 0.425702 | 0.482507 |
| | Best | 0.530814 | 0.522288 | 0.449338 | 0.454128 | 0.522001 | 0.553662 |
| Dolphins | Mean | 0.5727 | 0.574277 | 0.431174 | 0.814113 | 0.52687 | 0.567711 |
| | Std | $1.1292 \times 10^{-16}$ | $2.25841 \times 10^{-16}$ | 0.0350064 | $1.1292 \times 10^{-16}$ | 0.0510336 | 0.0432212 |
| | Worst | 0.5727 | 0.574277 | 0.363285 | 0.814113 | 0.396634 | 0.501266 |
| | Best | 0.5727 | 0.574277 | 0.523461 | 0.814113 | 0.612508 | 0.660154 |

**Table 4.** The F-measure values obtained by the experimental algorithms on real-world networks with real partitions known.

| Networks | F-measure | FastNewman [16] | LconDanon [17] | GA-NET [6] | CMM [18] | Meme-net [8] | EMCD-SOM |
|---|---|---|---|---|---|---|---|
| Karate Club | Mean | 0.828011 | 0.758621 | 0.810516 | 0.812349 | 0.907227 | 0.89563 |
| | Std | $4.51681 \times 10^{-16}$ | $3.38761 \times 10^{-16}$ | 0.0345437 | 0.0292515 | 0.0471795 | 0.0353847 |
| | Worst | 0.828011 | 0.758621 | 0.761594 | 0.771371 | 0.884034 | 0.884034 |
| | Best | 0.828011 | 0.758621 | 0.846678 | 0.878937 | 1 | 1 |
| Football Club | Mean | 0.607997 | 0.624275 | 0.357385 | 0.888643 | 0.829276 | 0.881271 |
| | Std | $3.38761 \times 10^{-16}$ | $4.51681 \times 10^{-16}$ | 0.0259086 | 0.0102019 | 0.0593904 | 0.0222667 |
| | Worst | 0.607997 | 0.624275 | 0.304809 | 0.866702 | 0.654615 | 0.806481 |
| | Best | 0.607997 | 0.624275 | 0.415762 | 0.902567 | 0.914482 | 0.896491 |
| Political Book | Mean | 0.819664 | 0.792252 | 0.631611 | 0.778402 | 0.721159 | 0.810397 |
| | Std | $1.1292 \times 10^{-16}$ | $2.25841 \times 10^{-16}$ | 0.0476347 | $1.1292 \times 10^{-16}$ | 0.0532029 | 0.0256946 |
| | Worst | 0.819664 | 0.792252 | 0.541227 | 0.778402 | 0.617422 | 0.736497 |
| | Best | 0.819664 | 0.792252 | 0.700829 | 0.778402 | 0.806321 | 0.834708 |
| Dolphins | Mean | 0.786624 | 0.70509 | 0.549487 | 0.968117 | 0.671548 | 0.721252 |
| | Std | 0 | $3.38761 \times 10^{-16}$ | 0.056409 | 0 | 0.0584518 | 0.0520816 |
| | Worst | 0.786624 | 0.70509 | 0.444878 | 0.968117 | 0.567638 | 0.665973 |
| | Best | 0.786624 | 0.70509 | 0.753607 | 0.968117 | 0.778187 | 0.88149 |

**Table 5.** The Rand Index values obtained by the experimental algorithms on real-world networks with real partitions known.

| Networks | RI | FastNewman [16] | LconDanon [17] | GA-NET [6] | CMM [18] | Meme-net [8] | EMCD-SOM |
|---|---|---|---|---|---|---|---|
| Karate Club | Mean | 0.841355 | 0.707665 | 0.770291 | 0.762686 | 0.88164 | 0.866845 |
| | Std | $2.25841 \times 10^{-16}$ | $2.25841 \times 10^{-16}$ | 0.0276138 | 0.0295904 | 0.0601917 | 0.0451438 |
| | Worst | 0.841355 | 0.707665 | 0.730838 | 0.734403 | 0.85205 | 0.85205 |
| | Best | 0.841355 | 0.707665 | 0.802139 | 0.834225 | 1 | 1 |
| Football Club | Mean | 0.880702 | 0.887109 | 0.836476 | 0.971647 | 0.953755 | 0.973221 |
| | Std | $4.51681 \times 10^{-16}$ | $5.64601 \times 10^{-16}$ | 0.0252958 | 0.00177524 | 0.0241369 | 0.00652113 |
| | Worst | 0.880702 | 0.887109 | 0.762319 | 0.972387 | 0.886651 | 0.949352 |
| | Best | 0.880702 | 0.887109 | 0.88177 | 0.979863 | 0.984744 | 0.978032 |
| Political Book | Mean | 0.828205 | 0.804212 | 0.703199 | 0.759341 | 0.757045 | 0.820733 |
| | Std | $2.25841 \times 10^{-16}$ | $1.1292 \times 10^{-16}$ | 0.0192073 | $5.64601 \times 10^{-16}$ | 0.034364 | 0.0203903 |
| | Worst | 0.828205 | 0.804212 | 0.6663 | 0.759341 | 0.707692 | 0.764103 |
| | Best | 0.828205 | 0.804212 | 0.730403 | 0.759341 | 0.817216 | 0.843223 |
| Dolphins | Mean | 0.713908 | 0.684294 | 0.570739 | 0.936542 | 0.645672 | 0.679129 |
| | Std | $3.38761 \times 10^{-16}$ | $2.25841 \times 10^{-16}$ | 0.0295801 | 0 | 0.0288785 | 0.0398455 |
| | Worst | 0.713908 | 0.684294 | 0.52935 | 0.936542 | 0.597039 | 0.640402 |
| | Best | 0.713908 | 0.684294 | 0.700159 | 0.936542 | 0.718139 | 0.814384 |

## 4. Conclusions

This paper proposed a membrane algorithm based on a self-organizing map network named EMCD-SOM, which was used to solve complex network community detection problems. According to the characteristics of community detection, the proposed algorithm gave the realization principle of object, reaction rule, and membrane structure. The encoded object represented the partitioning result of community detection. Genetic algorithm and differential evolution were employed as two reaction rules to evolve objects in different regions of the membranes. The proposed algorithm used SOM to determine the number of elementary membranes and fully exploit neighborhood information. The effectiveness of the proposed algorithm was evaluated on four real-world networks. Compared with other algorithms, the results showed that our algorithm could achieve better performance, indicating that EMCD-SOM has great potential in solving community detection problems. In addition, because EMCD-SOM adopts modularity density as an objective function, it can effectively solve the resolution limitation problem of the modularity degree, and reasonably divide the network structure at different resolutions. In the future, EMCD-SOM will be improved so that it can effectively detect communities in overlapping networks, large-scale networks, and multi-level heterogeneous networks.

**Author Contributions:** Conceptualization, C.L.; methodology, C.L.; software, J.L; validation, J.L.; writing–original draft preparation, C.L.; writing—review and editing, Y.D.

**Funding:** This project was supported by the Startup Research Fund for Ph.D of Liaoning Province, China (Grant No. 20170520364), Key R&D Program Guidance Plan of Liaoning Province, China (Grant No.2018104013), the Technological Innovation Program for Young People of Shenyang City, China (Grant No. RC180338), and the Science and Technology Program of Shenyang City, China (Grant No. F16-155-9-00, and F17-180-9-00).

**Conflicts of Interest:** The authors declare no conflict of interest.

## References

1. Pizzuti, C. Evolutionary Computation for Community Detection in Networks: A Review. *IEEE Trans. Evol. Comput.* **2018**, *22*, 464–483. [CrossRef]

2. Dakiche, N.; Tayeb, F.B.S.; Slimani, Y.; Benatchba, K. Tracking community evolution in social networks: A survey. *Inf. Proc. Manag.* **2019**, *56*, 1084–1102. [CrossRef]

3. Liu, J.; Abbass, H.A.; Tan, K.C. Evolutionary Community Detection Algorithms. In *Evolutionary Computation and Complex Networks*; Springer International Publishing: Cham, Switzerland, 2019; pp. 77–115. [CrossRef]

4. Lancichinetti, A.; Fortunato, S. Community detection algorithms: A comparative analysis. *Phys. Rev. E Stat. Nonlinear Soft Matter Phys.* **2009**, *80*, 056117. [CrossRef]

5. Tasgin, M.; Herdagdelen, A.; Bingol, H. Community detection in complex networks using genetic algorithms. *arXiv* **2007**, arXiv:0711.0491.

6. Pizzuti, C. Ga-net: A genetic algorithm for community detection in social networks. In *Parallel Problem Solving from Nature–PPSN X*; Springer: Berlin, Germany, 2008; pp. 1081–1090.

7. Pizzuti, C. A multiobjective genetic algorithm to find communities in complex networks. *IEEE Trans. Evol. Comput.* **2012**, *16*, 418–430. [CrossRef]

8. Gong, M.; Fu, B.; Jiao, L.; Du, H. Memetic algorithm for community detection in networks. *Phys. Rev. E* **2011**, *84*, 056101. [CrossRef]

9. Pizzuti, C.; Socievole, A. Many-objective optimization for community detection in multi-layer networks. In Proceedings of the 2017 IEEE Congress on Evolutionary Computation (CEC), San Sebastian, Spain, 5–8 June 2017; pp. 411–418.

10. Meo, P.D.; Ferrara, E.; Fiumara, G.; Provetti, A. Mixing local and global information for community detection in large networks. *J. Comput. Syst. Sci.* **2014**, *80*, 72–87. [CrossRef]

11. Grass-Boada, D.H.; Pérez-Suárez, A.; Gago-Alonso, A.; Bello, R.; Rosete, A. Multi-objective Overlapping Community Detection by Global and Local Approaches. In *Progress in Pattern Recognition, Image Analysis, Computer Vision, and Applications*; Mendoza, M.; Velastín, S., Eds.; Springer International Publishing: Cham, Switzerland, 2018; pp. 272–280.

12. Berahmand, K.; Bouyer, A.; Vasighi, M. Community Detection in Complex Networks by Detecting and Expanding Core Nodes Through Extended Local Similarity of Nodes. *IEEE Trans. Comput. Social Syst.* **2018**, *5*, 1021–1033. [CrossRef]

13. Shi, P.; He, K.; Bindel, D.; Hopcroft, J.E. Locally-biased spectral approximation for community detection. *Knowl. Syst.* **2019**, *164*, 459–472. [CrossRef]

14. Moradi, M.; Parsa, S. An evolutionary method for community detection using a novel local search strategy. *Physica A Stat. Mech. Appl.* **2019**, *523*, 457–475. [CrossRef]

15. Kohonen, T. The self-organizing map. *Proc. IEEE* **1990**, *78*, 1464–1480. [CrossRef]

16. Newman, M.E. Fast algorithm for detecting community structure in networks. *Phys. Rev. E* **2004**, *69*, 066133. [CrossRef] [PubMed]

17. Danon, L.; Diaz-Guilera, A.; Duch, J.; Arenas, A. Comparing community structure identification. *J. Stat. Mech. Theor. Exp.* **2005**, *2005*, P09008. [CrossRef]

18. Chen, Y.; Li, X.; Xu, J. Convexified modularity maximization for degree-corrected stochastic block models. *Ann. Stat.* **2018**, *46*, 1573–1602. [CrossRef]

19. Li, Z.; Zhang, S.; Wang, R.S.; Zhang, X.S.; Chen, L. Quantitative function for community detection. *Phys. Rev. E* **2008**, *77*, 036109. [CrossRef]

20. Jin, H.D.; Leung, K.S.; Wong, M.L.; Xu, Z.B. An efficient self-organizing map designed by genetic algorithms for the traveling salesman problem. *IEEE Trans. Syst. Man. Cybernet.* **2003**, *33*, 877–888. [CrossRef]

21. Villmann, T.; Villmann, B.; Slowik, V. Evolutionary algorithms with neighborhood cooperativeness according to neural maps. *Neurocomputing* **2004**, *57*, 151–169. [CrossRef]

22. Zhang, H.; Zhou, A.; Song, S.; Zhang, Q.; Gao, X.Z.; Zhang, J. A Self-Organizing Multiobjective Evolutionary Algorithm. *IEEE Trans. Evol. Comput.* **2016**, *20*, 792–80. [CrossRef]

23. Liang, J.; Guo, Q.; Yue, C.; Qu, B.; Yu, K. A self-organizing multi-objective particle swarm optimization algorithm for multimodal multi-objective problems. In *Advances in Swarm Intelligence*; Tan, Y., Shi, Y., Tang, Q., Eds.; Springer International Publishing: Cham, Switzerland, 2018; pp. 550–560.

24. Jia, G.; Cai, Z.; Musolesi, M.; Wang, Y.; Tennant, D.A.; Weber, R.J.M.; Heath, J.K.; He, S. Community Detection in Social and Biological Networks Using Differential Evolution. In *Learning and Intelligent Optimization*; Hamadi, Y., Schoenauer, M., Eds.; Springer: Berlin/Heidelberg, Germany, 2012; pp. 71–85.

25. Lancichinetti, A.; Fortunato, S.; Radicchi, F. Benchmark graphs for testing community detection algorithms. *Phys. Rev. E* **2008**, *78*, 046110. [CrossRef]

26. Zachary, W.W. An information flow model for conflict and fission in small groups. *J. Anthropol. Res.* **1977**,*33*, 452–473. [CrossRef]

27. Girvan, M.; Newman, M.E. Community structure in social and biological networks. *PNAS* **2002**, *99*, 7821–7826. [CrossRef]

28. Newman, M. Mark Newman's Network Data Collection. Available online: http://www-personal.umich.edu/~mejn/netdata/ (accessed on 24 May 2019).

29. Lusseau, D.; Schneider, K.; Boisseau, O.J.; Haase, P.; Slooten, E.; Dawson, S.M. The bottlenose dolphin community of Doubtful Sound features a large proportion of long-lasting associations. *Behav. Ecol. Sociobiol.* **2003**, *54*, 396–405. [CrossRef]

*Article*

# Image Entropy for the Identification of Chimera States of Spatiotemporal Divergence in Complex Coupled Maps of Matrices

**Rasa Smidtaite** [1,2,*], **Guangqing Lu** [3,*] and **Minvydas Ragulskis** [1]

[1]  Center for Nonlinear Systems, Kaunas University of Technology, Studentu 50-146, LT-51368 Kaunas, Lithuania; minvydas.ragulskis@ktu.lt
[2]  Department of Applied Mathematics, Kaunas University of Technology, Studentu 50-318, LT-51368 Kaunas, Lithuania
[3]  School of Electrical and Information Engineering, Jinan University, 206 Qianshan Road, Zhuhai 519070, China
*  Correspondence: rasa.smidtaite@ktu.lt (R.S.); tgqluyp@jnu.edu.cn (G.L.); Tel.: +370-671-13831 (R.S.); +86-136-31257365 (G.L.)

Received: 29 March 2019; Accepted: 20 May 2019; Published: 24 May 2019

**Abstract:** Complex networks of coupled maps of matrices (NCMM) are investigated in this paper. It is shown that a NCMM can evolve into two different steady states—the quiet state or the state of divergence. It appears that chimera states of spatiotemporal divergence do exist in the regions around the boundary lines separating these two steady states. It is demonstrated that digital image entropy can be used as an effective measure for the visualization of these regions of chimera states in different networks (regular, feed-forward, random, and small-world NCMM).

**Keywords:** chimera states; coupled map lattice; nilpotent matrix

---

## 1. Introduction

Chimera state is a dynamical spatiotemporal behavior when structured patterns of coherence and incoherence occur. This phenomenon was first observed in a network of non-locally coupled identical oscillators [1]. The existence of chimera states has been investigated in theory [2–4] as well as it has been proved in several experiments [5–7].

Chimeras are observed in optical [7,8], chemical [9,10], neuronal systems [11,12]. Experimental verification of chimeras in the system of non-locally coupled Belousov-Zhabotinsky chemical oscillators in a two-dimensional array is reported in [10]. The relativistic quantum chimera state is uncovered in two-dimensional Dirac material systems where the manifestations of both integrable and chaotic dynamics may be controlled electrically [8]. The coexistence of coherent and incoherent states, known as chimeras, is particularly important for neuronal systems. These states have also been linked to Parkinson's disease, epileptic seizures, and even to schizophrenia [11]. The occurrence of chimera states in two-dimensional and three-dimensional networks of Hindmarsh-Rose oscillators representing realistic models of neuronal ensembles is identified in [12].

Initially it was thought that chimeras can be observed only in networks of non-locally coupled oscillators [1]. Later studies revealed that besides non-locally connected networks [3,4,13–16], these states can be found in local [17–19] as well as in global [6,20] coupling topologies. Chimera patterns are analyzed in networks of Logistic maps with hierarchical connectivities [21]. The robustness of chimera patterns to inhomogeneities in a lattice of identical FitzHugh-Nagumo oscillators with irregular coupling topologies is demonstrated in [22]. Besides these symmetric coupling topologies, chimera states are also observed in Erdős-Rényi [23], small-world [24], scale-free [25], heterogeneous [26] networks. The emergence of chimeras in a multiplex network with two non-identical interconnected layers is investigated in [27]. It is shown that the range of parameters displaying chimera states in the first homogeneous layer is affected by the changes in coupling of the same nodes in the second layer. Neural modular network is analyzed in [28] where neurons are assumed to be connected with electrical synapses within their communities and with chemical synapses across them—these two coupling types cause the formation of chimera-like states. To evaluate behavior of neurons measures of synchronization, metastability, and chimera-like states are estimated. The study of multiscale network [29] observes how the appearance of chimera states in global ring is influenced by the changes in topology of subnetworks.

The current study is focused on the dynamics of complex coupled maps of matrices. It is demonstrated that chimera states of spatiotemporal divergence do exist in the regions around the boundary lines separating the quiet state and the diverged state. That highlights the importance of this paper (chimera states have not been previously explored in coupled maps of matrices). Moreover, chimera states of spatiotemporal divergence are investigated in different types of networks, including random networks. The exploration of the effects induced by the network structure and the development of entropy-based visualization technique for chimera states of spatiotemporal divergence are the main objectives of this paper.

## 2. Preliminary Notes and the Objective

### 2.1. A Network of Coupled Maps

A paradigmatic model of a lattice of translational invariance with periodic boundary conditions, comprising $m$ real-valued, single-variable time-discrete maps that are coupled to their closest neighbors reads [14]:

$$x^{(t+1)}(i) = f\left(x^{(t)}(i), a\right) + \frac{\varepsilon}{2P} \sum_{j=i-P}^{i+P} \left(f\left(x^{(t)}(j), a\right) - f\left(x^{(t)}(i), a\right)\right) \tag{1}$$

where $i$ is the number of the node ($i = 1, 2, \ldots, m$); $t$ is discrete time ($t = 0, 1, 2, \ldots$); $x^{(t)}(i)$ is the scalar nodal variable; $\varepsilon$ is the coupling parameter within the interval $(0, 1)$; $P$ is a fixed number of nearest neighbors to either side ($P \geq 0$). The local dynamics of every element $i$ on the one-dimensional ring is described by the Logistic map:

$$f\left(x^{(t)}(i), a\right) = ax^{(t)}(i)\left(1 - x^{(t)}(i)\right) \tag{2}$$

where $0 < a \leq 4$ and the initial condition is bounded to $0 \leq x^{(0)}(i) \leq 1$ in order to ensure the mapping to the interval $x^{(t)}(i) \in [0, 1]$ [30]. Please note that all parameters $a$ of the Logistic map are identical for all nodes, but initial conditions $x^{(0)}(i)$ are randomly distributed in interval $[0, 1]$.

At $P = 1$ Equation (1) describes a standard coupled map lattice (CML). At $P \geq 2$ Equation (1) represents a regular network of coupled maps. The coupling radius $r$ is defined as $r = \frac{P}{m}$. Please note that $r = 0.5$ ($r = \frac{m-1}{2m}$ if $m$ is odd) corresponds to global coupling.

### 2.2. A Network of Coupled Map of Matrices

CMLs play an important role in modelling such complex phenomena as, spatiotemporal chaos, spatial bifurcations, global travelling waves [31–33]. A scalar iterative nodal variable at each node of a CML can be replaced by a matrix variable [34]. All scalar variables $x^{(t)}(i)$ are replaced by $2 \times 2$ matrices $\begin{bmatrix} x_{11}^{(t)}(i) & x_{12}^{(t)}(i) \\ x_{21}^{(t)}(i) & x_{22}^{(t)}(i) \end{bmatrix}$ in Equations (1) and (2). Such a transition from a scalar Logistic map (Equation (2)) to a single Logistic map of matrices is explained in detail in [35]. All square $2 \times 2$ matrices can be classified into idempotent and nilpotent matrices; however only nilpotent matrices can generate the effect of divergence in an isolated Logistic map of matrices when the absolute values of the matrix elements grow unbounded [35]. Therefore, all $2 \times 2$ matrices in this paper will be set as nilpotent matrices.

Please note that a $2 \times 2$ nilpotent matrix can be uniquely characterized by its single Eigenvalue $\lambda^{(t)}$ and a scalar nilpotent parameter $\mu^{(t)}$ [35]. Appropriate re-arrangements and the collection of terms do transform the CML described by Equation (1) and Equation (2) into a one-dimensional coupled map lattice of matrices (1D CMLM) [34]:

$$\lambda^{(t+1)}(i) \;=\; a\lambda^{(t)}(i)\left(1 - \lambda^{(t)}(i)\right), \tag{3}$$

$$\mu^{(t+1)}(i) \;=\; (1 - \varepsilon)\, a\mu^{(t)}(i)\left(1 - 2\lambda^{(t)}(i)\right) + \frac{\varepsilon}{2}\left(a\mu^{(t)}(i+1)\left(1 - 2\lambda^{(t)}(i+1)\right)\right.$$
$$\left. + a\mu^{(t)}(i-1)\left(1 - 2\lambda^{(t)}(i-1)\right)\right), \tag{4}$$

where $0 \le \lambda^{(0)}(i) \le 1$ is the single Eigenvalue of the initial nilpotent matrix at node $i$; $\mu^{(0)}(i) = 1$ $(i - 1, 2, \ldots, m)$ is the nilpotent parameter of the initial nilpotent matrix at node $i$. The nilpotent model of a 1D CMLM comprises two scalar maps—therefore the lattice parameters $\lambda^{(t)}(i)$ and $\mu^{(t)}(i)$ are computed directly instead of performing matrix computations on the 1D lattice [34]. Please note that the divergence of a node $i$ is represented by the unbounded growth of $\mu^{(t)}(i)$.

The main objective of this paper is to investigate the dynamics of a network of coupled maps where scalar map variables are replaced by matrix variables. The model of such networks of coupled maps of matrices (NCMM) follows from Equations (1) and (3):

$$\mu^{(t+1)}(i) = f\left(\mu^{(t)}(i), \lambda^{(t)}(i), a\right) + \frac{\varepsilon}{2P} \sum_{j=i-P}^{i+P} \left(f(\mu^{(t)}(j), \lambda^{(t)}(j), a) - f(\mu^{(t)}(i), \lambda^{(t)}(i), a)\right), \tag{5}$$

where

$$f(\mu^{(t)}(i), \lambda^{(t)}(i), a) = a\mu^{(t)}(i)\left(1 - 2\lambda^{(t)}(i)\right) \tag{6}$$

but Eigenvalues of nilpotent matrices are computed directly according to Equation (3). At $P = 1$ the NCMM reduces to a 1D CMLM which (as shown in [34]) can generate fractal patterns of $\mu^{(t)}(i)$ representing spatiotemporal divergence that can be controlled by the coupling parameter between the nodes.

In other words, the main objective of this paper is to investigate if NCMMs (at $P \ge 2$) can exhibit chimera states of spatiotemporal divergence. Such NCMMs will be called regular NCMMs due to the orderly connectivity of neighboring nodes.

## 3. Chimera States of Spatiotemporal Divergence in Regular NCMMs

### 3.1. Spatiotemporal Divergence in a Regular NCMM

A regular NCMM comprising 200 nodes is investigated in this section. The parameter of the Logistic map $a$ is set to 3.699956 (the onset of chaos); the coupling parameter $\varepsilon$ is set to 0.4. Initial Eigenvalues $\lambda^{(0)}(i)$; $i = 1, 2, \ldots, 200$ are randomly distributed in the interval $(0, 1)$. The regular NCMM is iterated in 1000 time-forward steps according to Equation (5). The evolution of the network at $P = 4$ $(r = 0.02)$; $P = 5$ $(r = 0.025)$ and $P = 6$ $(r = 0.03)$ is depicted in Figure 1 parts (a), (b) and (c) respectively.

The regular NCMM diverges after a turbulent transient process at $P = 4$ (absolute numerical values of $\mu^{(t)}(i)$ are truncated to 5 in Figure 1a for the clarity of presentation). However, the regular NCMM calms down (Figure 1c) when each node is connected to 12 adjacent neighbors ($P = 6$). It appears that the degree of connectivity can be used to control the divergence of the network.

It is interesting to observe that the evolution of the regular NCMM results into a complex pattern at $P = 5$ (Figure 1b). The nodes are grouped into clusters of temporary divergence, however they calm down and re-explode again during the turbulent evolution of the network in time (Figure 1b). As mentioned in the Introduction, chimera states describe a dynamical spatiotemporal behavior when structured patterns of coherence and incoherence occur [1]. The definition of chimera states is extended in this paper. Figure 1b depicts a spatiotemporal behavior when structured patterns of quiet states and diverging states occur. The quiet state of a node $i$ is defined as the state when $\mu^{(t)}(i)$ tends to zero. The spatiotemporal divergence of the node $i$ is defined as the state when the modulus of $\mu^{(t)}(i)$ exceeds a pre-set level (this level is set to 5 in all computational experiments in this paper). Such behavior of the network is described as chimera states of spatiotemporal divergence. Such a complex behavior of the regular NCMM raises a question about the global view of the dynamics of the network in the parameter plane $\varepsilon - r$.

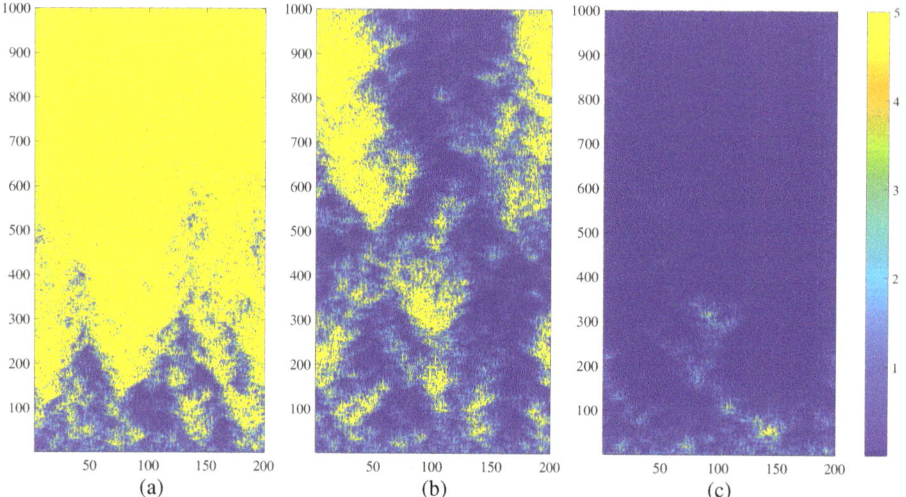

**Figure 1.** The transient dynamics of a regular NCMM comprising 200 nodes ($a = 3.699956$; $\varepsilon = 0.4$; $\lambda^{(0)}(i)$; $i = 1, 2, \ldots, 200$ are randomly distributed in the interval $(0, 1)$) represented by the variation of $\mu^{(t)}(i)$. The network diverges at $r = 0.02$ (part (**a**)); generates complex patterns at $r = 0.025$ (part (**b**)); and calms down at $r = 0.03$ (part (**c**)). Numerical values of $\mu^{(t)}(i)$ are truncated to 5 for the clarity of presentation.

As mentioned previously, the standard definition of chimera states is modified to the definition of chimera states of spatiotemporal divergence in this paper. In other words, structured patterns of coherence and incoherence are replaced by structured patterns of quiet and diverging states. It would be tempting to rename the diverging states as chaotic states. Also, it must be noted that spatiotemporal chaos is a well-explored phenomenon in cellular automata [36].

However, transitional states of temporary divergence cannot be defined as chaotic transients. By the definition, a chaotic attractor is bounded in the phase space. In our model, the evolution of a nodal variable $\mu^{(t)}(i)$ is not bounded. This is illustrated in Figure 2 where the evolution of 5 nodes ($\mu^{(t)}(1)$, $\mu^{(t)}(50)$, $\mu^{(t)}(100)$, $\mu^{(t)}(150)$ and $\mu^{(t)}(200)$) is visualized in time interval $500 \leq t \leq 1000$ at the set of system parameters corresponding to Figure 1b. The numerical values of $\mu^{(t)}(i)$ are cropped to 5 in Figure 2a—but the uncropped values of $\mu^{(t)}(i)$ are depicted in Figure 2b. It is clear from Figure 2 that the evolution of $\mu^{(t)}(i)$ cannot be described as the bursting chaos [37] (bounded in the phase space). Therefore, the definition of chimera states of spatiotemporal divergence is used in this paper.

Coherent states are represented as quiet states. However, diverging nodes evolve in radically different trajectories (Figure 2b)—what corresponds to the incoherent states.

The visualization of the transient dynamics of the NCMM at every point of the parameter plane $\varepsilon - r$ poses serious technical problems. Instead, the regular NCMM is evolved until the transient processes cease down and the steady-state evolution of the network is registered for 150 time-forward steps. That results into a grayscale digital image representing the values of $\mu^{(t)}(i)$; $1 \leq i \leq 200$; $0 \leq \mu^{(t)}(i) \leq 5$ (the size of the digital image in pixels is $200 \times 150$). Then, this digital grayscale image representing the steady-state evolution of the network is reduced into one single scalar number representing the entropy of that image (we use the standard MATLAB function *entropy*).

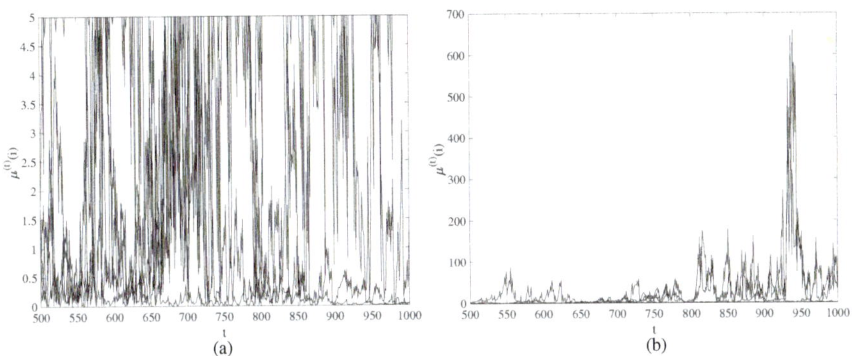

**Figure 2.** The evolution of $\mu^{(t)}(1)$, $\mu^{(t)}(50)$, $\mu^{(t)}(100)$, $\mu^{(t)}(150)$ and $\mu^{(t)}(200)$ in time interval $500 \leq t \leq 1000$ at the set of system parameters corresponding to Figure 1b. The numerical values of $\mu^{(t)}(i)$ are cropped to 5 in part (**a**) and are shown uncropped in part (**b**).

The schematic diagram representing this information reduction process is illustrated in Figure 3. The parameter $r$ is set to 0.05; all other system parameters (except $\varepsilon$) are kept the same. The coupling parameter $\varepsilon$ is varied from 0 to 1 and image entropy is computed for the steady-state evolution of the regular NCMM for each discrete value of $\varepsilon$ (Figure 3). Please note that the image entropy for the quiet network (Figure 3c) and the diverged network (Figure 3a,e) are all equal to zero. However, chimera-type states of spatiotemporal divergence yield entropies larger than zero (Figure 3b,d).

The relationship between the image entropy and the coupling parameter $\varepsilon$ yields two distinct peaks in Figure 3. Such behavior of the regular NCMM is very interesting. Initially, when the coupling parameter $\varepsilon$ is small, the network diverges (Figure 3). When the coupling parameter $\varepsilon$ exceeds a critical value (over 0.38), the network's final state is the quiet state (Figure 3). That corresponds well to the phenomenon observed in 1D CMLM—the effect of divergence can be controlled by increasing the coupling parameter $\varepsilon$ [34]. However, a completely unexpected behavior of the regular NCMM is observed when the coupling parameter $\varepsilon$ exceeds the upper threshold (around 0.83)—the network diverges again (Figure 3).

Such behavior of the regular NCMM reminds a coupled network of dendritic neurons [38]. A strongly coupled network of dendritic neurons tends to synchronize (what is dangerous to the functionality of brain). The well-known medical procedure known as "the gamma knife" can be used to eliminate synchronized tangles of dendritic neurons causing epileptic seizures. Simulation results in [38] show that the annihilation of too many synaptic links between neurons (caused by the overexposure of the network by a high dose of radiation therapy) leads to a synchronized state of the random network again. A similar effect can be observed in Figure 3—which shows an astonishing similarity (in terms of long-term behavior) between two networks of a completely different physical and mathematical origin.

Moreover, the regular NCMM exhibits a completely unique feature (compared to the network of dendritic neurons)—the dynamics of the network in the narrow region between the quiet mode and the divergence mode can be characterized by the existence of chimera states of spatiotemporal divergence (Figure 3b,d). Notably, image entropy detects the region of the existence of such chimera states very well (Figure 3).

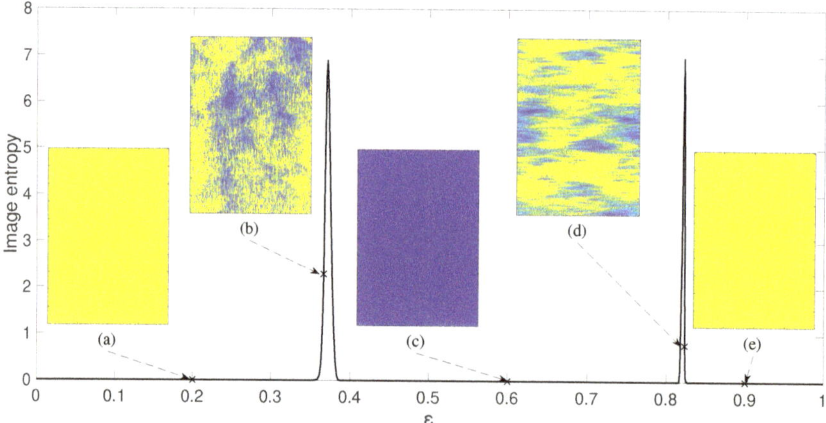

**Figure 3.** Image entropy of patterns is calculated for regular network when parameter $r$ is set to 0.05. Network coupling parameter $\varepsilon$ is set to 0.2, 0.366, 0.6, 0.823 and 0.9 in parts (**a**), (**b**), (**c**), (**d**) and (**e**). Image entropy is equal to 2.29 and 0.785 in parts (**b**) and (**d**) respectively.

Finally, chimera states of spatiotemporal divergence can be identified in the whole parameter plane $\varepsilon - r$ (Figure 4a). Chimera states of spatiotemporal divergence are located at the boundary between the quiet regime and the divergence regime (Figure 4a). The geometric shape of this boundary is very sensitive to the variation of $r$ when $r$ is small—but gets less sensitive when the regular network becomes denser (Figure 4a).

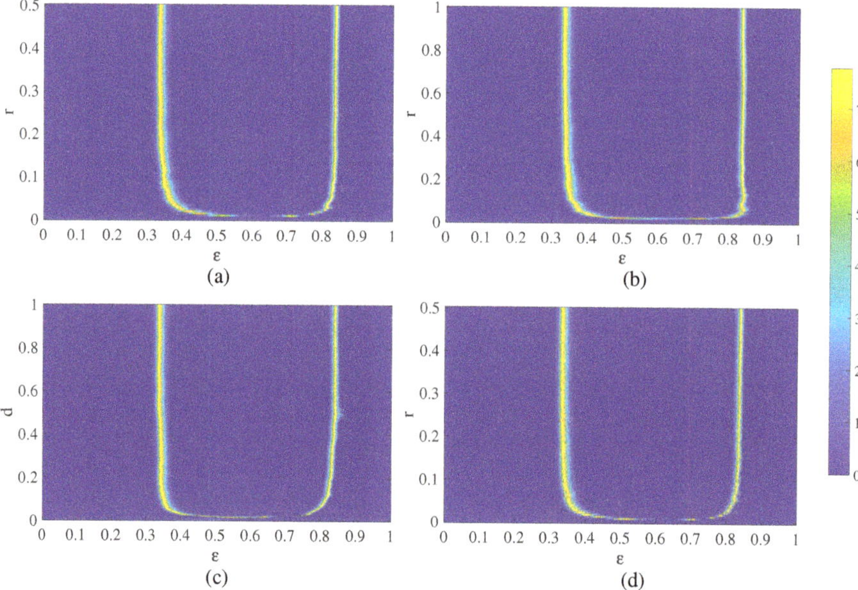

**Figure 4.** The visualization of chimera states of spatiotemporal divergence for networks of different structure: a regular NCMM (part (**a**), parameter plane $\varepsilon - r$); a regular unidirectional NCMM (part (**b**), parameter plane $\varepsilon - r$); the Erdős-Rényi NCMM (part (**c**), parameter plane $\varepsilon - d$); the small-world NCMM (part (**d**), parameter plane $\varepsilon - r$). The colorbar denotes numerical values of the entropy computed for steady-state evolution of the networks.

### 3.2. Chimera States of Spatiotemporal Divergence in a Regular Feed-Forward NCMM

Diffusive couplings between adjacent nodes is a paradigmatic choice for modelling neural networks which proves adequate in many cases [39]. However, feed-forward connectivity is also believed to play a significant role in a neuroscience context [40,41]. Each node is unidirectionally coupled to its successive neighbors in a feed-forward network:

$$\mu^{(t+1)}(i) = f\left(\mu^{(t)}(i), \lambda^{(t)}(i), a\right) + \frac{\varepsilon}{P}\sum_{j=i}^{i+P}\left(f\left(\mu^{(t)}(j), \lambda^{(t)}(j), a\right) - f\left(\mu^{(t)}(i), \lambda^{(t)}(i), a\right)\right) \quad (7)$$

Please note that the coupling radius $r = \frac{P}{m}$ now ranges from $r = \frac{1}{m}$ for a local feed-forward network to $r = \frac{m-1}{m}$ for global unidirectional coupling.

Computational experiments are continued with a regular feed-forward NCMM comprising 200 nodes ($a = 3.699956$; $\varepsilon = 0.4$; $\lambda^{(0)}(i)$ are randomly distributed in the interval $(0, 1)$). The evolution of the network at $P = 5$ ($r = 0.025$); $P = 7$ ($r = 0.035$) and $P = 9$ ($r = 0.045$) is depicted in Figure 5 parts a, b, and c respectively.

The regular feed-forward NCMM diverges at $P = 5$ (Figure 5a). The network exhibits chimera states of spatiotemporal divergence at $P = 7$ (Figure 5b) and completely calms down at $P = 9$ (Figure 5c). It is interesting to note that the feed-forward connectivity changes the shape of chimera states (Figure 5b)—the unidirectional coupling can be clearly identified from Figure 5.

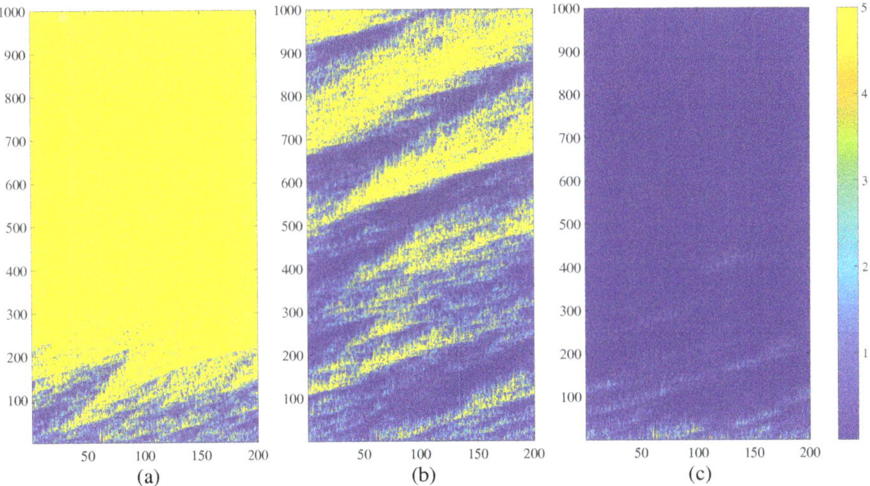

**Figure 5.** The transient dynamics of a regular directional NCMM comprising 200 nodes ($a = 3.699956$; $\varepsilon = 0.4$; $\lambda^{(0)}(i)$; $i = 1, 2, \ldots, 200$ are randomly distributed in the interval $(0, 1)$) represented by the variation of $\mu^{(t)}(i)$. The network diverges at $r = 0.025$ (part (**a**)); generates complex fractal-type patterns at $r = 0.035$ (part (**b**)); and calms down at $r = 0.045$ (part (**c**)). Numerical values of $\mu^{(t)}(i)$ are truncated to 5 for the clarity of presentation.

The location of chimera states of spatiotemporal divergence for the regular feed-forward NCMM are shown in parameter plane $\varepsilon - r$ in Figure 4b. Chimera states are located at the boundary between the quiet regime and the divergence regime—but a surprising is the fact that the geometric shape of this region is very similar to Figure 4a.

## 4. Chimera States of Spatiotemporal Divergence in a Complex NCMM

Most social, biological, and technological networks exhibit non-trivial topological features, with patterns of connection between their nodes that are neither purely regular nor purely random. Three relevant characteristics are usually employed to characterize a complex network—randomness, heterogeneity and modularity [42].

One extreme are regular networks. These are usually man-made networks that have the lowest heterogeneity and lowest randomness (as discussed in Sections 3.1 and 3.2). Another extreme is random Erdős-Rényi networks [43]. Such random networks have low heterogeneity and the degree distribution will be a Gaussian bell-shaped curve. The emergence and visualization of chimera states of spatiotemporal divergence in a random Erdős-Rényi NCMM is investigated in Section 4.1.

Most real-world networks, however, do not have homogeneous distribution of degree that regular or random networks have. The number of connections each node has in most real-world networks varies greatly and they are positioned somewhere between regular and random networks. A typical real-world network is proposed in [44] where the connections between the nodes in a regular graph are rewired with a certain probability. The resulting networks can be positioned between the regular and random networks according to their topological structure—and are referred to as small-world networks. The emergence and visualization of chimera states of spatiotemporal divergence in a small-world NCMM is investigated in Section 4.2.

### 4.1. Chimera States of Spatiotemporal Divergence in the Erdős-Rényi NCMM

The Erdős-Rényi NCMM network is generated by starting with a disconnected set of nodes that are then paired with a uniform probability. The coupling density of the Erdős-Rényi NCMM is defined as the ratio between the existing number of edges $n_r$ and the maximum number of edges in a complete network: $d = \frac{2n_r}{m(m-1)}$. Please note that $0 \leq d \leq 1$.

The model of the Erdős-Rényi network is adopted from [45]:

$$\mu^{(t+1)}(i) = (1-\varepsilon)f\left(\mu^{(t)}(i), \lambda^{(t)}(i), a\right) + \frac{\varepsilon}{k_i} \sum_{j=1}^{m} T_{i,j}(d)f(\mu^{(t)}(j), \lambda^{(t)}(j), a), \tag{8}$$

where the mapping function $f$ remains the same as in Equation (6); $\varepsilon$ is the coupling parameter; $i$ is the degree of the node $i$. The adjacency matrix $T_{i,j}$ represents the Erdős-Rényi random network where the average degree of node $i$ is set to $d$. The iterative relationship for $\lambda^{(t)}(i)$ also remains the same as in Equation (3).

The Erdős-Rényi NCMM diverges at $d = 0.031$ (Figure 6a). The network exhibits complex transient states of spatiotemporal divergence at $d = 0.033$ (Figure 6b) and completely calms down at $d = 0.035$ (Figure 6c).

It is well-known that the visualization of chimera states in a random network poses serious technical problems because adjacent nodes do not necessarily belong to the same chimera state [25]. In other words, the visualization of interpretable chimera states requires special and not always clearly defined node permutation algorithms [25].

Despite the before-mentioned problems with the visualization of chimera states, we continue with the digital image entropy-based algorithm without the node permutation (Figure 4c). The results are surprising. First of all, the geometric shape of the region of chimera states of spatiotemporal divergence is very similar to Figure 4b. Secondly, the boundaries of the region of chimera states are smooth—the random nature of the network does not substantially change the geometric shape of the region.

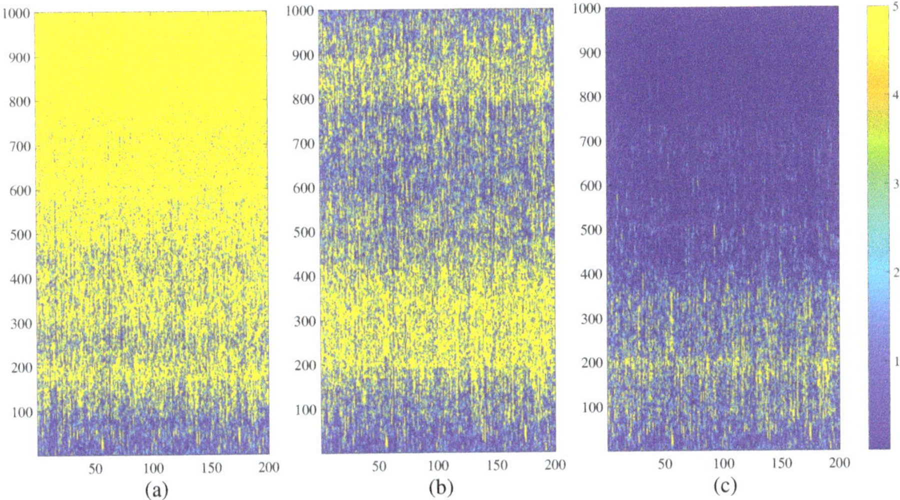

**Figure 6.** The transient dynamics of the Erdős-Rényi NCMM comprising 200 nodes ($a = 3.699956$; $\varepsilon = 0.4$; $\lambda^{(0)}\,(i)$; $i = 1, 2, \ldots, 200$ are randomly distributed in the interval $(0,\,1)$) represented by the variation of $\mu^{(t)}\,(i)$. The network diverges at $d = 0.031$ (part (**a**)); generates complex fractal-type patterns at $d = 0.033$ (part (**b**)); and calms down at $d = 0.035$ (part (**c**)). Numerical values of $\mu^{(t)}\,(i)$ are truncated to 5 for the clarity of presentation.

### 4.2. Chimera States of Spatiotemporal Divergence in the Small-World NCMM

Computational experiments are continued with the small-world NCMM. To obtain a small-world network the Watts-Strogatz model is considered [44]. Watts-Strogatz network is constructed starting from a ring lattice with $m$ nodes and $k$ edges per node. Each pair of nodes is rewired with probability $\beta$. Please note that a regular network is generated at $\beta = 0$. However, when all edges are rewired ($\beta = 1$) a ring lattice is transformed into a random graph.

The implementation of the small-world network of CMM is similar to Equation (8) except that the adjacency matrix is computed according to the Watts-Strogatz model [44].

As a starting point a ring lattice with $P$ nearest neighbors (Equation (1)) is considered—which results in the construction of undirected networks. The probability $\beta$ to rewire the target node is set to 0.2 in all calculations.

The Watts-Strogatz NCMM diverges at $P = 3$ ($r = 0.015$) in Figure 7 part (a). The network experiences transient processes of spatiotemporal divergence at $P = 4$ ($r = 0.02$) in Figure 7 part (b) and completely calms down at $P = 5$ ($r = 0.025$) in Figure 7 part (c).

Chimera states of spatiotemporal divergence for the small-world NCMM in the $(r, \varepsilon)$ parameter plane are shown in Figure 4d. Surprisingly, the shape of the highlighted region is very similar to Figure 4a–c—even though the network topology is completely different.

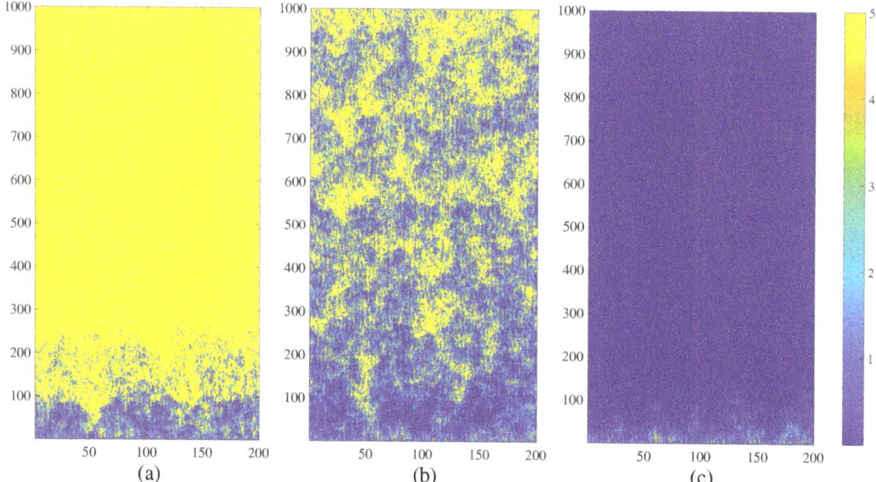

**Figure 7.** The transient dynamics of the small-world network NCMM comprising 200 nodes ($a = 3.699956$; $\varepsilon = 0.4$; $\lambda^{(0)}$ ($i$); $i = 1, 2, \ldots, 200$ are randomly distributed in the interval $(0, 1)$; $\beta = 0.2$) represented by the variation of $\mu^{(t)}$ ($i$). The network diverges at $r = 0.015$ (part (**a**)); generates complex patterns of spatiotemporal divergence at $r = 0.02$ (part (**b**)); and calms down at $r = 0.025$ (part (**c**)). Numerical values of $\mu^{(t)}$ ($i$) are truncated to 5 for the clarity of presentation.

## 5. Concluding Remarks

The visualization of chimera states in a regular one-dimensional lattice does not cause much difficulties because these chimera states are represented by compact time-varying clusters of synchronized nodes. However, the concept of the space is lost in complex networks, which makes it not straightforward to define a chimera state [25]. To enhance the view of chimera states, the rearrangement of nodes can be done. The node with the highest degree is labelled to be the first, then other nodes are arranged according to their distance from the first node [25].

The visualization scheme for chimera states in this manuscript is not based on the rearrangement on nodes. Moreover, chimera states in NCMM are not states of spatiotemporal synchronization between the neurons or other types of nonlinear oscillators. Chimera states in NCMM do exist in the regions around the boundary lines separating the quiet state or the state of divergence. These chimera states represent the self-organization of nodes into spatiotemporal clusters of divergence. It appears that image entropy is an effective measure for the visualization of the regions of chimera states in NCMM. Moreover, the proposed techniques work well with different topology networks (regular, feed-forward, random, and small-world NCMM). The network structure has a strong impact to the geometrical shape of chimera states of spatiotemporal divergence (compare Figure 1b, Figure 5b, Figure 6b, Figure 7b). However, it appears that the boundary line separating the quiet states and the diverged states is not strongly affected by the structure of the network—which is a completely counter-intuitive result. This robustness of the geometric shape of boundary lines against the network structure has important implications for different potential applications—desynchronization of complex coupled maps of matrices, temporary control of divergence in coupled maps of matrices, etc. These applications remain clear objectives of future research.

The existence (and appropriate visualization) of chimera states of spatiotemporal divergence is already an interesting result in nonlinear dynamics of complex CMLs of matrices. The sensitivity of these chimera states to different perturbations, the potential of chimera states to embed and to transmit secret visual information—these are important questions falling out of the scope of this paper—but remaining a definite objective of future research.

**Author Contributions:** Conceptualization, M.R.; methodology, R.S. and M.R.; software, R.S. validation, R.S. and G.L.; formal analysis, R.S.; investigation, R.S.; resources, G.L. and M.R.; data curation, R.S. and G.L.; writing–original draft preparation, M.R.; writing–review and editing, R.S., G.L. and M.R.; visualization, R.S.; supervision, M.R.; project administration, G.L. and M.R.; funding acquisition, R.S.

**Funding:** This research was supported by the Business Support Fund of Kaunas University of Technology (AlgebMIS).

**Conflicts of Interest:** The authors declare no conflict of interest.

## References

1. Kuramoto, Y.; Battogtokh, D. Coexistence of coherence and incoherence in nonlocally coupled phase oscillators. *Nonlinear Phenom. Complex Syst.* **2002**, *5*, 380–385.
2. Zakharova, A.; Kapeller, M.; Schöll, E. Chimera Death: Symmetry Breaking in Dynamical Networks. *Phys. Rev. Lett.* **2014**, *112*, 154101. [CrossRef]
3. Panaggio, M.J.; Abrams, D.M. Chimera states: Coexistence of coherence and incoherence in networks of coupled oscillators. *Nonlinearity* **2015**, *28*, 67–87. [CrossRef]
4. Bukh, A.; Strelkova, G.; Anishchenko, V. Spiral wave patterns in a two-dimensional lattice of nonlocally coupled maps modeling neural activity. *Chaos Soliton Fractals* **2019**, *120*, 75–82. [CrossRef]
5. Martens, E.A.; Thutupalli, S.; Fourrière, A.; Hallatschek, O. Chimera states in mechanical oscillator networks. *Proc. Natl. Acad. Sci. USA* **2013**, *110*, 10563–10567. [CrossRef] [PubMed]
6. Hart, J.D.; Bansal, K.; Murphy, T.E.; Roy, R. Experimental observation of chimera and cluster states in a minimal globally coupled network. *Chaos* **2016**, *26*, 094801. [CrossRef] [PubMed]
7. Li, X.W.; Bi, R.; Sun, Y.X.; Zhang, S.; Song, Q.Q. Chimera states in Gaussian coupled map lattices. *Front. Phys.* **2018**, *13*, 130502. [CrossRef]
8. Xu, H.Y.; Wang, G.L.; Huang, L.; Lai, Y.C. Chaos in Dirac electron optics: Emergence of a relativistic quantum chimera. *Phys. Rev. Lett.* **2018**, *120*, 124101. [CrossRef]
9. Nkomo, S.; Tinsley, M.R.; Showalter, K. Chimera States in Populations of Nonlocally Coupled Chemical Oscillators. *Phys. Rev. Lett.* **2013**, *110*, 244102. [CrossRef]
10. Totz, J.F.; Rode, J.; Tinsley, M.R.; Showalter, K.; Engel, H. Spiral wave chimera states in large populations of coupled chemical oscillators. *Nat. Phys.* **2018**, *14*, 282–286. [CrossRef]
11. Majhi, S.; Bera, B.K.; Ghosh, D.; Perc, M. Chimera states in neuronal networks: A review. *Phys. Life Rev.* **2018**. [CrossRef]
12. Hizanidis, J.; Kanas, V.G.; Bezerianos, A.; Bountis, T. Chimera states in networks of nonlocally coupled Hindmarsh-Rose neuron models. *Int. J. Bifurc. Chaos* **2014**, *24*, 1450030. [CrossRef]
13. Shepelev, I.A.; Bukh, A.V.; Strelkova, G.I.; Vadivasova, T.E. Chimera states in ensembles of bistable elements with regular and chaotic dynamics. *Nonlinear Dyn.* **2017**, *90*, 2317–2330. [CrossRef]
14. Malchow, A.K.; Omelchenko, I.; Schöll, E.; Hövel, P. Robustness of chimera states in nonlocally coupled networks of nonidentical logistic maps. *Phys. Rev. E* **2018**, *98*, 012217. [CrossRef]
15. Bogomolov, S.A.; Slepnev, A.V.; Strelkova, G.I.; Schöll, E.; Anishchenko, V.S. Mechanisms of appearance of amplitude and phase chimera states in ensembles of nonlocally coupled chaotic systems. *Commun. Nonlinear Sci. Numer. Simul.* **2017**, *43*, 25–36. [CrossRef]
16. Bukh, A.; Rybalova, E.; Semenova, N.; Strelkova, G.; Anishchenko, V. New type of chimera and mutual synchronization of spatiotemporal structures in two coupled ensembles of nonlocally interacting chaotic maps. *Chaos Interdiscipl. J. Nonlin. Sci.* **2017**, *27*, 111102. [CrossRef] [PubMed]
17. Laing, C.R. Chimeras in networks with purely local coupling. *Phys. Rev. E* **2015**, *92*, 050904(R). [CrossRef] [PubMed]
18. Clerc, M.G.; Coulibaly, S.; Ferré, M.A.; García-Ñustes, M.A.; Rojas, R.G. Chimera-type states induced by local coupling. *Phys. Rev. E* **2016**, *93*, 052204. [CrossRef]
19. Kundu, S.; Majhi, S.; Bera, B.K.; Ghosh, D.; Lakshmanan, M. Chimera states in two-dimensional networks of locally coupled oscillators. *Phys. Rev. E* **2018**, *97*, 022201. [CrossRef]

20. Schmidt, L.; Krischer, K. Clustering as a Prerequisite for Chimera States in Globally Coupled Systems. *Phys. Rev. Lett.* **2015**, *114*, 034101. [CrossRef]

21. zur Bonsen, A.; Omelchenko, I.; Zakharova, A.; Schöll, E. Chimera states in networks of logistic maps with hierarchical connectivities. *Eur. Phys. J. B* **2018**, *91*, 65.

22. Omelchenko, I.; Provata, A.; Hizanidis, J.; Schöll, E.; Hövel, P. Robustness of chimera states for coupled FitzHugh-Nagumo oscillators. *Phys. Rev. E* **2015**, *91*, 022917. [CrossRef] [PubMed]

23. Lopes, M.A.; Goltsev, A.V. Distinct dynamical behavior in Erdős-Rényi networks, regular random networks, ring lattices, and all-to-all neuronal networks. *Phys. Rev. E* **2019**, *99*, 022303. [CrossRef] [PubMed]

24. Sawicki, J.; Omelchenko, I.; Zakharova, A.; Schöll, E. Chimera states in complex networks: Interplay of fractal topology and delay. *Eur. Phys. J. Spec. Top.* **2017**, *226*, 1883–1892. [CrossRef]

25. Zhu, Y.; Zheng, Z.; Yang, J. Chimera states on complex networks. *Phys. Rev. E* **2014**, *89*, 022914. [CrossRef]

26. Li, B.; Saad, D. Chimera-like states in structured heterogeneous networks. *Chaos* **2017**, *27*, 043109. [CrossRef]

27. Ghosh, S.; Zakharova, A.; Jalan, S. Non-identical multiplexing promotes chimera states. *Chaos Soliton Fractals* **2018**, *106*, 56–60. [CrossRef]

28. Hizanidis, J.; Kouvaris, N.E.; Zamora-López, G.; Díaz-Guilera, A.; Antonopoulos, C.G. Chimera-like States in Modular Neural Networks. *Sci. Rep.* **2016**, *6*, 19845. [CrossRef] [PubMed]

29. Makarov, V.V.; Kundu, S.; Kirsanov, D.V.; Frolov, N.S.; Maksimenko, V.A.; Ghosh, D.; Dana, S.K.; Hramov, A.E. Multiscale interaction promotes chimera states in complex networks. *Commun. Nonlinear Sci.* **2019**, *71*, 118–129. [CrossRef]

30. May, R.M. Simple mathematical models with very complicated dynamics. *Nature* **1976**, *261*, 459–467. [CrossRef]

31. Zhang, Y.Q.; He, Y.; Wang, X.Y. Spatiotemporal chaos in mixed linear-nonlinear two-dimensional coupled logistic map lattice. *Phys. A* **2018**, *490*, 148–160. [CrossRef]

32. Huang, T.; Zhang, H. Bifurcation, chaos and pattern formation in a space-and time-discrete predator-prey system. *Chaos Soliton Fractals* **2016**, *91*, 92–107. [CrossRef]

33. Fernandez, B. Selective chaos of travelling waves in feedforward chains of bistable maps. *arXiv* **2018**, arXiv:1811.08310.

34. Guangqing, L.; Smidtaite, R.; Navickas, Z.; Ragulskis, M. The effect of explosive divergence in a coupled map lattice of matrices. *Chaos Soliton Fractals* **2018**, *113*, 308–313.

35. Navickas, Z.; Smidtaite, R.; Vainoras, A.; Ragulskis, M. The logistic map of matrices. *Discret. Cont. Dyn. B* **2011**, *3*, 927–944.

36. Miranda, G.H.B.; Machicao, J.; Bruno, O.M. Exploring spatio-temporal dynamics of cellular automata for pattern recognition in networks. *Sci. Rep.* **2016**, *6*, 37329. [CrossRef]

37. Zheng, Y.H.; Lu, Q.S. Spatiotemporal patterns and chaotic burst synchronization in a small-world neuronal network. *Phys. A Stat. Mech. Its Appl.* **2008**, *387*, 3719–3728. [CrossRef]

38. Sakyte, E.; Ragulskis, M. Self-calming of a random network of dendritic neurons. *Neurocomputing* **2011**, *74*, 3912–3920. [CrossRef]

39. Cross, M.C.; Hohenberg, P.C. Pattern formation outside of equilibrium. *Rev. Mod. Phys.* **1993**, *65*, 851. [CrossRef]

40. Goldman, M.S. Memory without feedback in a neural network. *Neuron.* **2009**, *61*, 621–634. [CrossRef] [PubMed]

41. Zankoc, C.; Fanelli, D.; Ginelli, F.; Livi, R. Desynchronization and pattern formation in a noisy feed-forward oscillator network. *Phys. Rev. E* **2019**, *99*, 012303. [CrossRef] [PubMed]

42. Solé, R.V.; Valverde, S. Information theory of complex networks: On evolution and architectural constraints. In *Complex Networks*; Springer: Berlin/Heidelberg, Germany, 2004; pp. 189–207.

43. Erdős, P.; Rényi, A. On Random Graphs. *Publ. Math.* **1959**, *6*, 290–297.

44. Watts, D.J.; Strogatz, S.H. Collective dynamics of 'small-world'networks. *Nature* **1998**, *393*, 440. [CrossRef]

45. Shinoda, K., Kaneko, K. Chaotic Griffiths Phase with Anomalous Lyapunov Spectra in Coupled Map Networks. *Phys. Rev. Lett.* **2016**, *117*, 254101. [CrossRef] [PubMed]

*Article*

# Evolution Model of Spatial Interaction Network in Online Social Networking Services

Jian Dong [1], Bin Chen [1,*], Pengfei Zhang [1], Chuan Ai [1], Fang Zhang [1], Danhuai Guo [2,3] and Xiaogang Qiu [1]

[1] College of System Engineering, National University of Defense Technology, Changsha 410073, China; jiandong.nudt@foxmail.com (J.D.); hncszpf@163.com (P.Z.); rogeraichuan@gmail.com (C.A.); fangzhang.nudt@foxmail.com (F.Z.); 13874934509@139.com (X.Q.)

[2] Computer Network Information Center, Chinese Academy of Sciences, 4th South Fourth Road Zhongguancun, Beijing 100190, China; guodanhuai@cnic.cn

[3] University of Chinese Academy of Sciences, 19th Yuquan Road, Beijing 100049, China

* Correspondence: nudtcb9372@gmail.com; Tel.: +86-0731-8457-4332

Received: 22 March 2019; Accepted: 23 April 2019; Published: 24 April 2019

**Abstract:** The development of online social networking services provides a rich source of data of social networks including geospatial information. More and more research has shown that geographical space is an important factor in the interactions of users in social networks. In this paper, we construct the spatial interaction network from the city level, which is called the city interaction network, and study the evolution mechanism of the city interaction network formed in the process of information dissemination in social networks. A network evolution model for interactions among cities is established. The evolution model consists of two core processes: the edge arrival and the preferential attachment of the edge. The edge arrival model arranges the arrival time of each edge; the model of preferential attachment of the edge determines the source node and the target node of each arriving edge. Six preferential attachment models (Random-Random, Random-Degree, Degree-Random, Geographical distance, Degree-Degree, Degree-Degree-Geographical distance) are built, and the maximum likelihood approach is used to do the comparison. We find that the degree of the node and the geographic distance of the edge are the key factors affecting the evolution of the city interaction network. Finally, the evolution experiments using the optimal model DDG are conducted, and the experiment results are compared with the real city interaction network extracted from the information dissemination data of the WeChat web page. The results indicate that the model can not only capture the attributes of the real city interaction network, but also reflect the actual characteristics of the interactions among cities.

**Keywords:** city interaction network; evolution model; preferential attachment; WeChat; maximum likelihood

## 1. Introduction

With the rapid development of the Internet, smart phones, and information technology, online social networking services such as Facebook, Twitter, Sina Weibo, and WeChat have developed rapidly. These platforms facilitate the interactions among users and accelerate the dissemination of emotions and opinions contained in the information. Meanwhile, these platforms provide a rich source of social media including geospatial information for the research of social networks [1–4]. The interactions of users in social networks usually manifest as the viewing and forwarding of information. More and more research shows that geographical space, which seems to be a bridge between online and offline, affects the interactions of users in social networks [5–7].

Spatial interaction is the process whereby entities at different points in physical space make contacts, demand/supply decisions, or locational choices [8]; for example, trade in goods among different countries or regions, human migration among cities or countries, and people in different cities communicating with each other by phone or social media software. In social networks, spatial interactions are formed by users who belong to different spatial locations through viewing and forwarding information. Naturally, spatial interactions can be described by complex network [9], where nodes represent spatial locations, which can be cities, provinces, or countries, and edges represent interactions of entities in different spatial locations. The research on the characteristics of the spatial interaction network in social networks and their evolutionary mechanisms is of great significance for providing location-based business services, planning and managing communication network facilities, and formulating regional economic development policies. In addition, the results also can be used to improve the performances of several types of applications in various fields, such as social network analysis [10] and affective computing [11–13].

The existing network evolution models mainly include the random graph models (RGM) [14–16], generated network models (GNM) [17,18], and data-driven network models (DDNM) [19–21]. Random graph models, such as Poisson random graphs and generalized random graphs, attempt to apply the connecting probability and changing strategy of the edge to a certain number of nodes to generate a random network that meets specific statistical characteristics (such as average degree, degree distribution, joint degree distribution, and degree-degree correlation). Generated network models, such as preferential attachment models and their variants, try to generate a network that reflects certain characteristics of the real network (such as a power-law distribution, small-world characteristics, and homogeneity) through certain node-adding, edge-adding, and edge-changing rules from simple graphs (regular graphs). These two widely-used models can usually generate networks with some characteristics of the real network, but they cannot satisfy multiple characteristics at the same time. Moreover, these models usually do not consider the geospatial characteristics of networks, making it difficult to describe the evolution process of spatial interaction network.

Generally, distance and location are the two important factors of geospatial characteristics. On the one hand, it is found that the interaction frequency among users has a distance decay effect. People tend to communicate more with friends who are close to them geographically, while users who are far away from each other are less likely to interact [22–25]. On the other hand, the behaviors of people living in similar geographical locations, such as the same city, often show similarities, while people in different geographical locations will have different behavior patterns due to economic and cultural differences, thus affecting the information interactions among regions [22].

Gravity laws are commonly found in spatial interaction networks such as crowd flow networks, population migration networks, and commodity trade networks. Thus, a gravity model for spatial interaction is proposed by analogy with the law of universal gravitation. The gravity model provides an estimate of the traffic between two or more regions (such as the number of trips and the quantity of commodity trade). In a spatial interaction network, the gravity model can be interpreted as the frequency of interactions between two nodes. The frequency is proportional to the strength of the two nodes and inversely proportional to the power of the distance between the two nodes. The gravity model has become a classic model for interpreting and predicting the interactions of spatial networks and is widely used in many fields including transportation planning [26], population migration [27,28], international trade [29,30], and disease transmission [31]. Although the gravity model is simple, intuitive, easy to calculate, and involves geographical factors, it lacks a rigorous theoretical foundation. In addition, the gravity model is deterministic and cannot explain the fluctuation of the interaction between two nodes in the spatial interaction network [32]. Therefore, this kind of static estimation is not suitable for describing the evolution of spatial networks.

This paper proposes a spatial interaction network at the city level, which is called the city interaction network. We study the evolution mechanism of the city interaction network formed in the process of information dissemination in social networks, where nodes represent cities and edges represent interactions among cities. We consider the evolution model of the city interaction network from the perspective of the edge, that is how each edge is added to the city interaction network. A evolution model for describing the interactions among cities is established. The evolution model consists of two core processes: the edge arrival and the preferential attachment of the edge. The edge arrival model arranges the arrival time of each edge; the model of preferential attachment of the edge determines the source node and the target node of each arriving edge. Six preferential attachment models (Random-Random, Random-Degree, Degree-Random, Geographical distance, Degree-Degree, Degree-Degree-Geographical distance) are built, and the maximum likelihood approach is used to do the comparison. Finally, the evolution experiments using the optimal model (Degree-Degree-Geographical distance) are conducted, and the experiment results are compared with the real city interaction network extracted from the information dissemination data of the WeChat web page.

Preferential attachment of edges: The preferential attachment model assumes that when a new node joins the network, it creates a constant number of edges, where the selection of the target node for each edge is proportional to the degree of the node [33]. In addition to degree, the node age and geographic distance of the edge can be applied to the preferential attachment model [34]. This paper considers the evolution of the network from the perspective of the edge. Therefore, when an edge is added to the network, the source node and the target node are selected according to preferential attachment of edges.

Evaluation by the maximum likelihood: The maximum likelihood approach is usually used to compare a series of models numerically and select the best model (and parameters) to interpret the data [35]. As our understanding of real-world networks improves, likelihood remains unchanged, while the generative models improve to incorporate the new understanding. Success in modeling can therefore be effectively tracked [34]. The maximum likelihood approach is widely used to estimate network model parameters [35–37] and select the optional model [34,38]. Therefore, this paper uses the maximum likelihood approach to evaluate and compare different network evolution models based on empirical data.

WeChat: WeChat is one of the most popular social networking platforms in China. As of the second quarter of 2016, WeChat has covered more than 94% smart phones in China, with 0.8 billion monthly active users. WeChat has powerful social functions and a large number of users, and WeChat has integrated almost all aspects of people's lives, including payment, location-based services, shopping, games, and entertainment. Therefore, WeChat is an appropriate system to study the characteristics and evolution mechanism of the spatial interaction network in social networks.

The rest of this paper is organized as follows: the second section introduces the dissemination data of the WeChat web page and constructs the city interaction network. The third section introduces the evolution model of the city interaction network. In the fourth section, the maximum likelihood method is used to evaluate the six preferential attachment models and to select the optimal model and parameters. In the fifth section, the optimal model is used for network evolution, and the obtained evolutionary network is compared with the real city interaction network. The potential biases and model extension are discussed in the sixth section, and the seventh section is the conclusion.

## 2. Preliminaries

### 2.1. Dataset

WeChat provides three basic functions: instant messaging (including single and group chat), moments (where users publish, comment, and forward information), and official accounts (including subscription accounts and service accounts). Users can interact with their friends by posting text, voice, pictures, emoticons, location, video, web links, and other information. This paper studies the dissemination data of the WeChat web page (HTML5) collected by third-party service companies.

The recording process of the WeChat web page data can be described as: when a web page with a certain theme is created and published by the creator through the official accounts, the content of this web page can be viewed by other users. Users who view the web page can send it to their moments or WeChat friends, or not forward it. Thus, the users who view (or forward) and the users who are viewed (or forwarded) are recorded.

The dissemination data of WeChat web page were obtained, and the time span of the data was from 2–8 July 2016. There were 622,637 records in total, and each record can be represented by a six-tuple <pageID, sourceID, targetID, type, time, ip>, where pageID represents the unique identity of the web page, sourceID and targetID represent the unique identity of the user, type represents the behavior type of target, including viewing and forwarding, time represents the time when the behavior of targetID occurs, and ip represents the IP address of targetID. In order to protect the privacy of users, web page identity and user identity were anonymized.

*2.2. City Interaction Network*

Most of the researches related to geography use self-reported data to identify the location of users, which is often inaccurate. By locating users with IP addresses, the errors of self-reported data can be avoided. Song et al. analyzed several large IP address databases, including the Chunzhen IP address database, the Taobao IP address database, the Sina IP address database, and the Baidu IP address database [39]. They found that the four IP address databases were quite different, and when the administrative division level was lower, the coverage rate and coincidence rate of IP address databases would decrease, while the data availability would also decrease. However, considering the coverage rate and coincidence rate of the four IP address databases, they believed that the credibility of the Taobao IP database was the highest. Therefore, the Taobao IP address database was used in our work to locate the IP address in the data to the corresponding cities in China. Finally, the IP address in the data was located in 34 provincial divisions of China (including 23 provinces, 4 municipalities, 5 autonomous regions, and 2 special administrative regions), a total of 372 cities. The number of cities corresponding to each provincial division is shown in Table 1.

**Table 1.** City distribution of 34 provincial divisions in China. China has 34 provincial divisions, including 23 provinces, 4 municipalities, 5 autonomous regions, and 2 special administrative regions.

| Province | Number of Cities | Province | Number of Cities | Province | Number of Cities |
|---|---|---|---|---|---|
| Beijing | 1 | Tianjin | 1 | Hebei | 11 |
| Inner Mongolia | 12 | Liaoning | 14 | Jilin | 9 |
| Shanghai | 1 | Jiangsu | 13 | Zhejiang | 21 |
| Fujian | 16 | Jiangxi | 9 | Shandong | 11 |
| Hubei | 18 | Hunan | 17 | Guangdong | 14 |
| Hainan | 18 | Chongqing | 1 | Sichuan | 21 |
| Yunnan | 16 | Xizang | 7 | Shannxi | 10 |
| Qinghai | 8 | Ningxia | 5 | Xinjiang | 15 |
| Shanxi | 11 | Heilongjiang | 13 | Anhui | 11 |
| Henan | 17 | Guangxi | 14 | Guizhou | 9 |
| Gansu | 14 | Hong Kong | 1 | Macao | 1 |
| Taiwan | 12 | | | | |

Figure 1 shows the active frequency of users in each provincial division. The active frequency of a province is the number of users located in that province. The active frequency was more in the east and less in the west. The top three provincial divisions with the highest frequency were Shandong, Henan, and Guangdong, and the active frequency of Xizang, Xinjiang, and Taiwan was low. This fully reflects that information interaction is affected by political, economic, cultural, geographical, and demographic factors.

**Figure 1.** The active frequency of users in 34 provincial divisions of China. The transition of colors from red to yellow indicates the reduction of active frequency, and the corresponding data of each color are given by the color bar in the lower left corner.

Based on the data of the web page dissemination in WeChat, the city interaction network $G_t = (V, E_t, W_t)$ can be constructed. $G_t$ is a dynamic directed network, $V = \{v_1, v_2, v_3, \cdots, v_N\}$ is the set of nodes in the network, representing cities of China, and the number of nodes is $N$; $E_t = \{e_1, e_2, e_3, \cdots, e_{M_t}\}$ is the set of edges of the network from Time 0–$t$, representing the interactions among cities, and the number of edges is $M_t$; $W_t = \{w_1, w_2, w_3, \cdots, w_{M_t}\}$ is the weight set of edges in the network from Time 0–$t$, representing the number of interactions among cities. The dynamics of the city interaction network $G_t$ is reflected in the changes of the edge and weight. We took the cities in Shandong province as an example to elaborate the construction process of the city interaction network. At $t = 0$, $G_t$ is a network containing only 17 isolated nodes (the number of cities in Shandong province). When a WeChat web page is published by a user in Jinan and users in Dezhou view or forward this web page, then a directed edge from Jinan to Dezhou is established. The weight of the directed edge is the number of Dezhou users viewing the web page. With the dissemination of the web page, it was assumed that the interaction network one day later is as shown in Figure 2. At this time, the number of nodes in the interaction network was $N = 17$, and the number of edges was $M_t = 22$ (bidirectional edges are denoted as two edges), where $t = 1$ (day). The city interaction network in this paper allows self-connected edges, which represents the interactions in the same city.

**Figure 2.** Schematic diagram of the city interaction network in Shandong province. Red dots represent the nodes of the network, and black arrows represent the directed edges of the network. The arrows start from the source node and point to the target node. The bidirectional arrow indicates that the two nodes are source and target nodes of each other.

Take the starting time of data (2 July 2016 00:00) as the time $t = 0$, and construct the city interaction network. The time span of the network is $T$. Table 2 lists the basic properties of the network $G_T$, including the number of nodes, number of edges, number of self-connected edges, average degree of nodes, density, average clustering coefficient, and average shortest path length.

**Table 2.** Basic properties of the city interaction network $G_T$. $N$ represents the number of nodes, $M_T$ the number of edges, $M_T^{se}$ the number of self-connected edges, $k_T^{avg}$ the average degree of nodes, $\rho_T$ the density, and $L_T$ the average length of the shortest path.

| $T$ | $N$ | $M_T$ | $M_T^{se}$ | $k_T^{avg}$ | $\rho_T$ | $L_T$ |
|---|---|---|---|---|---|---|
| 2–8 July 2016 | 372 | 30,438 | 353 | 163.65 | 0.22 | 1.73 |

According to the basic properties of the network $G_T$ listed in Table 2, an overall understanding of the interaction among cities was obtained through the dissemination of WeChat web page. The network involved 372 nodes and 30,438 edges, which indicates that not every two nodes had connected edges. On average, each node only had connections with 163.65 nodes, and the density of the network was only 0.22. It can be seen that although WeChat has a large number of users in China and covers all cities, each city will not interact with all other cities in the short term. The average shortest path length of the network was 1.73, which means that the average hop from one node to another node was 1.73. There were 353 self-connected edges in the network, and only 19 nodes had no self-connected edges. A total of 622,637 interaction records were recorded, among which, 350,578 records were the interactions in the same city, accounting for 56%. It can be seen that users were more inclined to interact with users in the same city.

Figure 3 shows the number of non-isolated nodes and the number of edges in the city interaction network as a function of time. Figure 3a shows the number of non-isolated nodes in the city interaction network as a function of time. Non-isolated nodes represent the nodes that have interacted with other nodes. In the initial stage, the number of non-isolated nodes grew rapidly, and the growth became slow until the number of nodes was close to $N$. Figure 3b shows the number of edges in the city interaction network as a function of time. The number of edges in the network kept increasing, but due to the limitation of the number of nodes, the growth of the number of edges gradually slowed down. In the case where the number of non-isolated nodes in the network was almost constant, the number of edges still kept growing. This also reflects the limitations of the evolution of the city interaction network from the perspective of nodes.

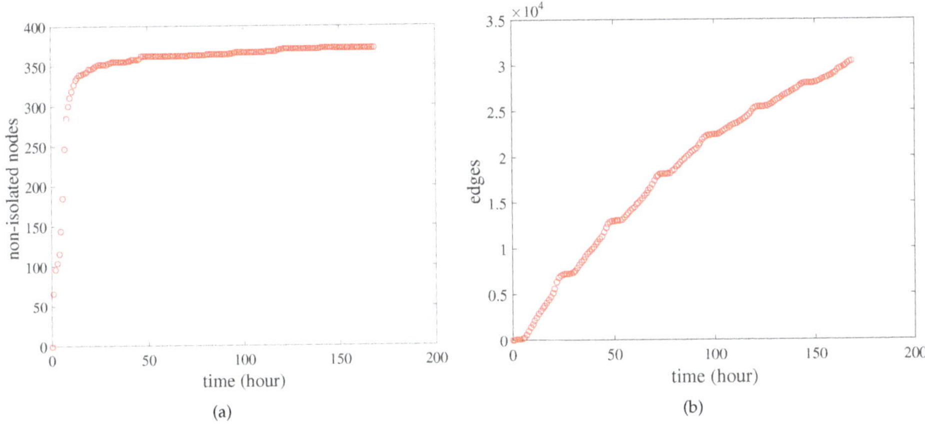

**Figure 3.** The number of non-isolated nodes and the number of edges in the city interaction network as a function of time. (**a**) The number of non-isolated nodes in the city interaction network as a function of time. (**b**) The number of edges in the city interaction network as a function of time. Each data point in the figure represents the number of non-isolated nodes (or edges) in the city interaction network from $t = 0$ to the current time. The time interval between two data points is one hour.

### 2.3. Notation

Let $Z$ denote the set of edges to be added to the network, $t(z), z \in Z$ the time when an edge $z$ is added to the network, and $z_{u,v}^t$ an edge $z$ added to the network at time $t$, and its source node and target node are connected to node $u$ and node $v$ respectively. Let $k_t(v)$ denote the degree of node $v$ at time $t$ and $d(u,v)$ denote the geography distance between node $u$ and node $v$.

### 3. Evolution Model

We consider the evolution model of the city interaction network from the perspective of the edge. The model consists of two core processes: the edge arrival and the preferential attachment of the edge. The edge arrival determines the arrival time of each edge; the preferential attachment of the edge determines the source node and the target node of each arriving edge.

For an edge $z$, it is composed of a node pair:

$$z = (u,v), u, v \in V, \tag{1}$$

where $V$ represents the node set and does not change with the network evolution. Assuming that the arrival time of the edges is a function of time in $\Delta t$, then the arrival time of each edge in $\Delta t$ will be arranged, and all edges can be expressed in the time sequence according to the arrival time:

$$Z = z^{t_1}, z^{t_2}, \cdots, z^{t_C}, \tag{2}$$

$$t_1 \leqslant t_2 \leqslant \cdots \leqslant t_C, \tag{3}$$

where $C$ is the length of the sequence $Z$, and Formula (3) guarantees the time-ordered arrival of the edges.

Select the source node $u$ and the target node $v$ from node set $V$ according to a certain preferential attachment for the edge arriving at time $t$:

$$P(z^t_{u,v}) \sim X(\Theta), \tag{4}$$

where $X(\Theta)$ represents a distribution function and $\Theta$ is the parameter of the distribution function. Finally, the network evolution is realized by updating the edge and weight. The edge arrival and preferential attachment of the edge are described in detail below.

*3.1. Edge Arrival*

Figure 4 shows the interaction quantity among cities of the data (each record represents an interaction) as a function of time. In the figure, each data point represents the interaction quantity among cities from time $t = 0$ to the current time, and the red line is the fitting of the function. It can be seen from the figure that the interaction quantity was a linear function of time, which satisfies $f(t) = 4025t - 4.51e4$, and the time unit is hours. Since each edge represents the interaction among nodes, $f(t)$ can be used to describe the number of arriving edges. Thus, the number of edges added to the network per unit time is a constant $\varepsilon = 4025$, and the time interval for each arriving edge is $t_i - t_{i-1} = 1/\varepsilon, i = 2, 3, \cdots, C$. Let the time of the first arrived edge be $t_1 = 0$, so that the time of each arriving edge is determined.

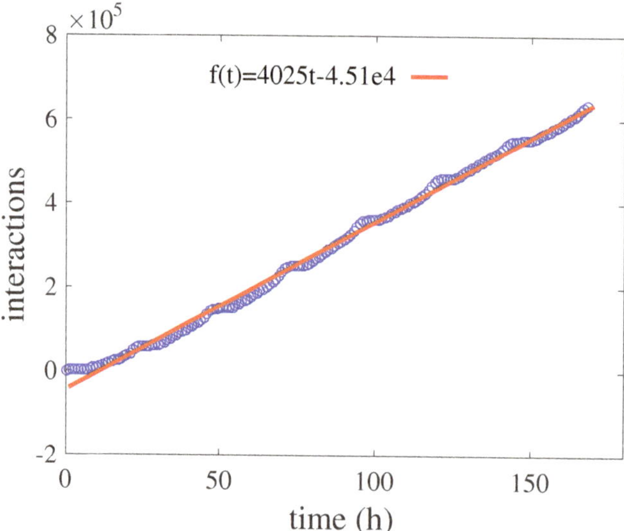


**Figure 4.** The interaction quantity among cities of the data as a function of time. Each data point represents the interaction quantity among cities from time $t = 0$ to the current time, and the red line is the fitting for the function; the fitting expression is given in the figure.

*3.2. Preferential Attachment of the Edge*

In this paper, the evolution of the city interaction network is considered from the perspective of the edge. Therefore, when an edge is added to the network, its source node and the target node will be selected according to a certain mechanism. This selection mechanism is called preferential attachment of the edge. Here, six different preferential attachment models are considered in this paper:

**Random-Random (RR)**: for the arrived edge at time $t$, two nodes are randomly selected from the node set $V$ as its source node and the target node, respectively:

$$P_{RR}(z_{u,v}^t) = \frac{1}{N_t^2}.$$

(5)

**Random-Degree (RD)**: for the arriving edge at time $t$, a node is randomly selected from the node set $V$ as its source node, and the selection of its target node is proportional to the degree of nodes in the network:

$$P_{RD}(z_{u,v}^t) = \frac{[k_t(v)]^\alpha}{N \sum_{i \in V}[k_t(i)]^\alpha}.$$

(6)

**Degree-Random (DR)**: for the arrived edge at time $t$, a node is randomly selected from the node set $V$ as its target node, and the selection of its source node is proportional to the degree of nodes in the network:

$$P_{DR}(z_{u,v}^t) = \frac{[k_t(u)]^\beta}{N \sum_{i \in V}[k_t(i)]^\beta}.$$

(7)

**Geographical distance (G)**: for the arrived edge at time $t$, the selection of its source node and target node is proportional to the geographical distance between the two nodes:

$$P_G(z_{u,v}^t) = \frac{[d(u,v)]^\gamma}{\sum_{i,j \in V}[d(i,j)]^\gamma}.$$

(8)

**Degree-Degree (DD)**: for the arrived edge at time $t$, the selection of its source node and target node is proportional to the degree of the nodes in the network. The degree index for the source node is $\alpha$, and the degree index for the target node is $\beta$:

$$P_{DD}(z_{u,v}^t) = \frac{[k_t(v)]^\alpha [k_t(u)]^\beta}{\sum_{i,j \in V}[k_t(i)]^\alpha [k_t(j)]^\beta}.$$

(9)

**Degree-Degree-Geographical distance (DDG)**: for the arrived edge at time $t$, the selection of its source node and target node is proportional to the degree of the nodes in the network and to the geographical distance between the source node and the target node. The degree index for the source node is $\alpha$; the degree index for the target node is $\beta$; and the distance index is $\gamma$:

$$P_{DDG}(z_{u,v}^t) = \frac{[k_t(v)]^\alpha [k_t(u)]^\beta [d(u,v)]^\gamma}{\sum_{i,j \in V}[k_t(i)]^\alpha [k_t(j)]^\beta [d(i,j)]^\gamma}.$$

(10)

## 4. Evaluation

In this section, a quantitative approach is applied to compare the accuracies of different preferential attachment models. The network is often considered to be the result of an evolutionary random process that drives its growth, including new nodes and new edges [35]. Given real data about network evolution, the extent to which the assumptions of a model are supported by the data using the maximum likelihood approach can be tested. The maximum likelihood approach is usually used to compare a series of models numerically and to select the best model (and parameters) to interpret the data. Estimating the likelihood of a preferential attachment model $M$ involves considering each arriving edge $z^t$ and computing the likelihood $P_M(z_{u,v}^t)$ that the edge $z^t$ selects the actual source node $u$ and the actual target node $v$ according to the model $M$. Therefore, the likelihood of network $G_T$ generated by model $M$ can be expressed as:

$$P_M(G_T) = \prod_{t \in T} P_M(z_{u,v}^t).$$

(11)

To obtain better numerical accuracy, the log-likelihood is used in this paper:

$$log(\prod_t P_M(z_{u,v}^t)) = \sum_t log(P_M(z_{u,v}^t)). \tag{12}$$

Since the city interaction network had self-connecting edges, which represents the interaction in the same city, we assumed that the distance of self-connecting edges was 20 kilometers (consider each city contour as a circle, and 20 kilometers is the approximate average of the radius of all cities). Figure 5 shows the relationship between the log-likelihood of models and different parameters. The RR model had no parameters, and its log-likelihood was a constant $-3{,}185{,}899$. In addition to the RR model, the log-likelihoods of the other five models were all convex functions of the model parameters, so the maximum likelihood of each model can be found to estimate the best parameters of the model. Table 3 lists the maximum log-likelihood of different preferential attachment models and the optimal parameters under the maximum log-likelihood. It can be seen from Figure 5 that, under the same parameter, the log-likelihood of the RD model and DR model was approximately equal. This reflects that the RD model and DR model had similar effects on the network evolution, and the selection of the source node and the target node was equal. Figure 5d also reflects this point. Figure 5c shows the relationship between the log-likelihood and parameter $\gamma$ of G model, and its maximum log-likelihood was significantly higher than that of the RR model, RD model, DR model, and DD model, indicating that the distance played an important role in the evolution of the city interaction network. The DDG model considered both the node degree and the geography distance among nodes in the network evolution process. It can be seen that the maximum log-likelihood of DDG model was the highest, which was 22% higher than that of the DD model and 11% higher than that of the G model. In addition, in the DD model, when $\alpha = 1.0$, $\beta = 1.0$, its log-likelihood was the maximum. In the G model, when $\gamma = -1.6$, its log-likelihood was the maximum. The DDG model, which considered the node degree and the geography distance, obtained the maximum likelihood when $\alpha = 0.6$, $\beta = 0.6$, $\gamma = -1.5$. This indicates that the distance made the degree of the node less important. Then, we applied the DDG preferential attachment model with parameters $\alpha = 0.6$, $\beta = 0.6$, $\gamma = -1.5$ to the evolution of the city interaction network.

## 5. Network Evolution

In order to verify the city interaction network model and the evolution process of the network, network evolution experiments were conducted. We considered the real network $G_{3T/4}$ from 2–4 July 2016 and evolved it from $t = \frac{3}{4}T$ until $t = T$. Specifically, the edge arrival model was used to determine the edges arriving at time $t \in [\frac{3}{4}T, T]$. For each arriving edge, the DDG preferential attachment model was used to select its source node and target node. Finally, the evolutionary network $G_T'$ with the same time length as the real network $G_T$ was obtained. $G_T$ and $G_T'$ were analyzed by the comparison of the statistical characteristics and community structure of the network.

Figure 6 shows the statistical characteristics of real network $G_T$ and evolutionary network $G_T'$. Figure 6a,b are considered from the edge properties. Figure 6a shows the weight distribution of the edges. It can be seen that the weight distributions of the real network and the evolutionary network followed the power-law distribution. The weight distribution of real network $G_T$ was fitted as shown in the dotted black line. The power exponents of the weight distributions of real network and evolutionary network were 1.92 and 1.99, respectively (the weight distributions of the real network and evolutionary network approximately overlapped, so the fit line of the weight distribution of the evolutionary network is not drawn). The weight of the edge represents the interaction among cities, and the power-law distribution of the weight distribution reflects that only a few cities had frequent interactions, while the interactions among most cities was very small. Figure 6b shows the geographical distance distribution of edges. The geographical distance distribution of edges is a property that connects the network with geographical space. Most of the interactive distances among cities were about 100 km. As the distance continued to increase, the probability of interaction became smaller. In addition, 20 km was also the high-frequency distance of city interaction (the distance was denoted as 20 km if the interaction occurred in the same city), indicating that the interaction in the same city occupied a large proportion.

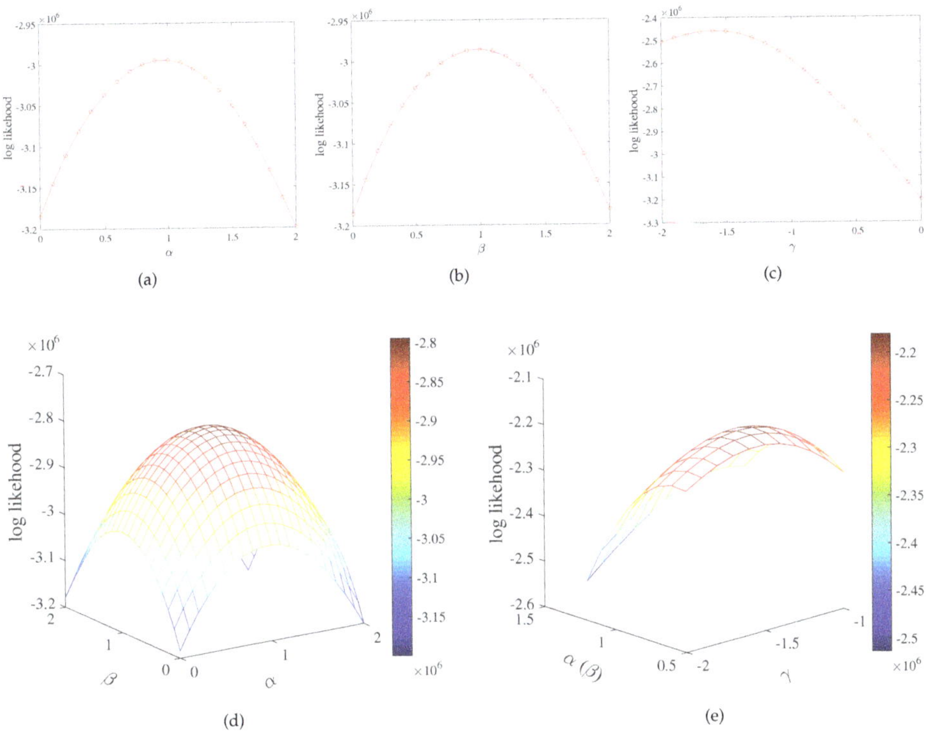

**Figure 5.** The relationship between log-likelihood of models and different parameters. (**a**) The relationship between the log-likelihood of the Random-Degree (RD) model and parameter $\alpha$. (**b**) The relationship between the log-likelihood of the Degree-Random (DR) model and parameter $\beta$. (**c**) The relationship between the log-likelihood of the Geographical distance (G) model and parameter $\gamma$. (**d**) The relationship between the log-likelihood of the Degree-Degree (DD) model and parameters $\alpha$ and $\beta$. (**e**) The relationship between the log-likelihood of the Degree-Degree-Geographical distance (DDG) model and parameters $\alpha(\beta)$ and $\gamma$.

**Table 3.** The maximum log-likelihood of different preferential attachment models and the optimal parameters under the maximum log-likelihood.

| Model | Parameter | The Maximal Log-Likelihood |
|---|---|---|
| RR | - | $-3{,}185{,}899$ |
| RD | $\alpha = 1.0$ | $-2{,}994{,}407$ |
| DR | $\beta = 1.0$ | $-2{,}985{,}583$ |
| G | $\gamma = -1.6$ | $-2{,}456{,}443$ |
| DD | $\alpha = 1.0$<br>$\beta = 1.0$ | $-2{,}794{,}647$ |
| DDG | $\alpha = 0.6$<br>$\beta = 0.6$<br>$\gamma = -1.5$ | $-2{,}180{,}441$ |

Figure 6c–f are considered from the perspective of node properties. Figure 6c shows the node weight distribution; the horizontal ordinate is the node number, and the numbering order is arranged in descending order of node weight. The node weight of a node is the sum of all the weights of edges connected with the node, which reflects the interactions between the node and its neighbor nodes. Figure 6d shows the betweenness centrality distribution of nodes; the horizontal ordinate is the node number, and the numbering order is arranged in descending order of the betweenness centrality of nodes. The betweenness centrality is to measure the importance of a node to connect with other nodes. By comparing the real network $G_T$ with the evolutionary network $G'_T$, it can be found that the node weight and betweenness centrality of some nodes in the evolutionary network were obviously higher or lower than the real network, but the overall trend was consistent with the real network. The provincial capital is the economic, political, and cultural center of a province, which is also reflected in the city interaction network. In the real network shown in Figure 6c,d, provincial capitals have relatively high node weight and betweenness centrality, such as Beijing, Shanghai, Guangzhou, Suzhou, Tianjin, and Hangzhou, which can also be reflected in the evolutionary network. Figure 6e shows the relationship between node degree and node weight. Figure 6f shows the relationship between node degree and node betweenness centrality. The greater the degree of nodes, the greater the node weight and the betweenness centrality.

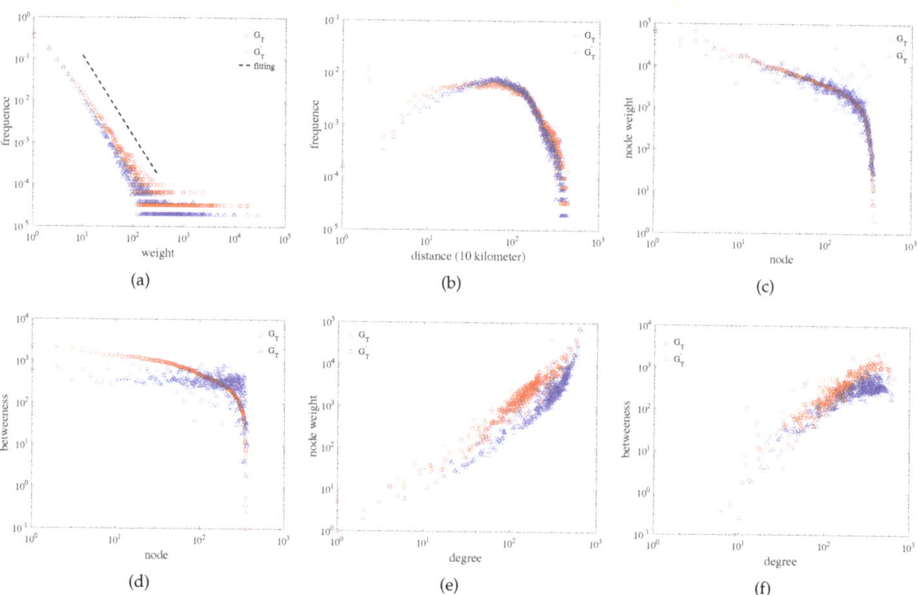

**Figure 6.** Statistical characteristics of real network $G_T$ and evolutionary network $G'_T$. (**a**) The weight distribution of edges. The weight distribution of real network $G_T$ is fitted as shown in the dotted black line. (**b**) The geography distance distribution of edges. The distance is in units of 10 kilometers. (**c**) The node weight distribution. The horizontal ordinate is the node number, and the numbering order is arranged in descending order of node weight. (**d**) The betweenness centrality distribution of nodes. The horizontal ordinate is the node number, and the numbering order is arranged in descending order of the betweenness centrality of nodes. (**e**) The relationship between node degree and node weight. (**f**) The relationship between node degree and node betweenness centrality. In the figure, the red circle marks represent the statistical characteristics of the real network $G_T$, and the blue triangle marks represent the statistical characteristics of the evolutionary network $G'_T$. All subgraphs are plotted on log-log coordinates.

For the real network $G_T$ and evolutionary network $G_T'$, two community detection methods, Louvain [40] and Infomap [41], were used to extract the community structure of the network, and the Normalized Mutual Information (NMI) was used to evaluate the results of community detection. The evaluation results are shown in Table 4. $G_T - PAD$ represents the comparison between the community structure of real networks and the provincial administrative divisions in China; $G_T - G_T'$ represents the comparison between the community structure of the real network and that of the evolutionary network. It can be found that the community structure of the real network was consistent with the administrative division to a certain extent, and it also shows the influence of the distance factor on the interactions among cities. In addition, the community structure of evolutionary network and real network was also similar, which indicates that the preferential attachment model in this paper can describe the emergence of community to a certain extent. This is mainly because the distance factor was considered in the model, so that cities in the same province were easily connected and formed communities. In general, the evolutionary network can be well matched with the real network, which reflects that the model can not only capture the properties of the real city interaction network, but also reflect the geographical characteristics of the interactions among cities.

**Table 4.** Evaluation results of community detection in undirected networks. $G_T$ represents the real network, $PAD$ represents Provincial Administrative Divisions in China, and $G_T'$ represents the evolutionary network.

| Comparison | Louvain | Infomap |
|---|---|---|
| $G_T - PAD$ | 0.738 | 0.831 |
| $G_T - G_T'$ | 0.715 | 0.850 |

## 6. Discussion

### 6.1. Potential Biases

In this paper, the evolution of the city interaction network was modeled and analyzed by using the interactive data formed in the process of information dissemination. There is no doubt that the use of one dataset to explain the results is not complete enough. Since our model was data-driven, the edge arrival model and maximum likelihood method were data-dependent. For the edge arrival model, different spatial interactive data may have different situations. The selection of model parameters in this paper was based on the method of maximum likelihood. The optimal parameters of the model can be found using real data. Therefore, different datasets will lead to different optimal parameters of the model. The evolution model was evaluated by comparing the structure characteristics of the evolutionary network and the real network. From the results, the model can capture the properties of the real city interaction network, but this is only limited to the city interaction network formed in the process of information dissemination. In the process of information dissemination, the interaction of information enables people to express their emotions and opinions. It is helpful to understand people's emotional tendency by considering the semantic characteristics of interactive information in the spatial interaction network.

Moreover, compared with cities in other countries, Chinese cities have some specificities. (1) China is a vast country, and the distance between cities is relatively large, making distance factors play an important role in the interactions of cities. (2) The distribution of Chinese cities shows a convergent pattern, which is different from Western countries. As a result, China has many large cities with large populations, such as Beijing, Shanghai, and Guangzhou. (3) The provincial administrative divisions in China are established around large cities, and the cities within the province are more likely to interact. The higher the level of political and economic development of the city, the more obvious the interaction. (4) China has a large population and a high Internet penetration rate, which makes information spread rapidly and widely. The results of this paper were obtained in this context. However, if the background were changed to some countries with a relatively small scale and the development levels of cities within the country were similar to each other, the influence of the distance factor on the interactions among cities may not be well reflected. Therefore, different countries have influence on the settings of the model.

*6.2. Model Extension*

The preferential attachment model in this paper belongs to a link prediction model based on the similarity of the network structure. Essentially speaking, a model for link prediction makes a guess about the factors resulting in the existence of links, which is actually what an evolving model wants to show. Up to now, the studies of link prediction overwhelmingly emphasized undirected networks. However, the study of link prediction in directed networks is inadequate [42].

The current common method for extending the technology applied to undirected networks to directed networks is to divide the degrees into outdegree and indegree, such as community detection [43–46]. According to this ideas, our model can be extended to directed networks. Take the DDG model as an example: the model can be extended to a directed network: Directed-Degree-Degree-Geographical distance (DiDDG): for the arriving edge at time $t$, the selection of its source node is proportional to the out-degree of the nodes in the network; the selection of its target node is proportional to the in-degree of the nodes; meanwhile, the selection of its source node and target node is proportional to the geographical distance between the source node and the target node. The degree index for the source node is $\alpha$; the degree index for the target node is $\beta$; and the distance index is $\gamma$:

$$P_{DiDDG}(z_{u,v}^t) = \frac{[k_t^{out}(v)]^\alpha [k_t^{in}(u)]^\beta [d(u,v)]^\gamma}{\sum_{i,j \in V} [k_t^{out}(i)]^\alpha [k_t^{in}(j)]^\beta [d(i,j)]^\gamma}. \tag{13}$$

In the modified model, the degree is divided into the out-degree and in-degree for consideration, so that the probability of connecting an edge between node $u$ and node $v$ will vary depending on the direction of the edge.

## 7. Conclusions

This paper studied the evolution mechanism of the city interaction network formed in the process of information dissemination in social networks, where nodes represent cities and edges represent interactions among cities. We considered the evolution model of the city interaction network from the perspective of the edge. In the model, the nodes were fixed, and the evolution process of the edge consisted of two core processes: the edge arrival and the preferential attachment of the edge. The model of edge arrival determines the arrival time of each edge; the model of preferential attachment of the edge determines the source node and the target node of each arriving edge. Six preferential attachment models were considered, and the comparison was done by the maximum likelihood approach. We found that the degree of the node and the geographic distance of the edge were the key factors affecting the evolution of the city interaction network. The DDG preferential attachment model, which considered both the node degree and the geographical distance among nodes in the network evolution process, was the best of the six models. Finally, we conducted the evolution experiments using the most optimal model and compared it with the real city interaction network extracted from the information dissemination data of the WeChat web page. By comparing the weight, geographical distance, node weight, and betweenness centrality of the real network and the evolutionary network, it was found that the evolutionary network could be well matched to the real network, which reflects that the model can describe the actual characteristics of the interactions among cities. Our research is of great significance for providing location-based business services, planning and managing communication network facilities, and formulating regional economic development policies.

However, there are still some limitations in our work. On the one hand, the evolution process of the city interaction network is affected by a variety of factors, such as politics, economy, population, etc. A comprehensive comparative analysis of the effects of these factors plays a significant role in the evolution model. These factors should be considered in the evolution model in future work. On the other hand, our work was verified by the real dissemination data of the WeChat web page; whether the model is applicable to the evolution of other spatial interaction networks still needs to be further verified.

**Author Contributions:** Conceptualization, J.D. and C.A.; methodology, J.D.; software, J.D.; validation, J.D., B.C., and C.A.; formal analysis, J.D. and F.Z.; investigation, F.Z. and P.Z.; resources, B.C., D.G., and X.Q.; data curation, D.G., F.Z., and P.Z.; writing, original draft preparation, J.D.; writing, review and editing, J.D., B.C., C.A., and P.Z.; visualization, J.D. and P.Z.; supervision, J.D., B.C., P.Z., D.G., and X.Q.; project administration, J.D.; funding acquisition, B.C.

**Funding:** This study is supported by the National Key Research & Development (R & D) Plan under Grant No. 2017YFC1200300, the National Natural Science Foundation of China under Grant Nos. 71673292 and 71673294, the National Social Science Foundation of China under Grant No. 17CGL047, the Beijing National Science Foundation of China under Grant No. 91224006, and the Guangdong Key Laboratory for Big Data Analysis and Simulation of Public Opinion.

**Conflicts of Interest:** The authors declare no conflict of interest. The funders had no role in the design of the study; in the collection, analyses, or interpretation of data; in the writing of the manuscript; nor in the decision to publish the results.

## References


1. Kietzmann, J.H.; Hermkens, K.; McCarthy, I.P.; Silvestre, B.S. Social media? Get serious! Understanding the functional building blocks of social media. *Bus. Horiz.* **2011**, *54*, 241–251. [CrossRef]
2. Wolfe, A.W. Social network analysis: Methods and applications by Stanley Wasserman; Katherine Faust. *Am. Ethnol.* **1997**, *24*, 219–220. [CrossRef]
3. Guille, A. Information diffusion in online social networks. In Proceedings of the 2013 SIGMOD/PODS Ph.D. Symposium, New York, NY, USA, 23 June 2013; pp. 31–36. [CrossRef]
4. Liu, L.; Qu, B.; Chen, B.; Hanjalic, A.; Wang, H. Modelling of information diffusion on social networks with applications to WeChat. *Phys. A* **2018**, *496*, 318–329. [CrossRef]
5. Laniado, D.; Volkovich, Y.; Scellato, S.; Mascolo, C.; Kaltenbrunner, A. The impact of geographic distance on online social interactions. *Inf. Syst. Front.* **2018**, *20*, 1203–1218. [CrossRef]
6. Deville, P.; Song, C.; Eagle, N.; Blondel, V.D.; Barabási, A.L.; Wang, D. Scaling identity connects human mobility and social interactions. *Proc. Natl. Acad. Sci. USA* **2016**, *113*, 7047–7052. [CrossRef] [PubMed]
7. Barthelemy, M. Spatial Networks. In *Encyclopedia of GIS*; Springer: New York, NY, USA, 2014; Chapter 2, pp. 1967–1976. [CrossRef]
8. Roy, J.R.; Thill, J.C. Spatial interaction modelling. *Pap. Reg. Sci.* **2003**, *83*, 339–361. [CrossRef]
9. Dejon, B. Spatial interaction network flow models. In *Vorträge der Jahrestagung 1977 / Papers of the Annual Meeting 1977 DGOR*; Brockhoff, K., Dinkelbach, W., Kall, P., Pressmar, D.B., Spicher, K., Eds.; Physica-Verlag HD: Heidelberg, Germany, 1978; pp. 377–386.
10. Chiancone, A.; Franzoni, V.; Li, Y.; Markov, K.; Milani, A. Leveraging zero tail in neighbourhood for link prediction. In Proceedings of the 2015 IEEE/WIC/ACM International Conference on Web Intelligence and Intelligent Agent Technology (WI-IAT), Singapore, 6–9 December 2015; Volume 3, pp. 135–139. [CrossRef]
11. Franzoni, V.; Milani, A.; Biondi, G. SEMO: A semantic model for emotion recognition in web objects. In Proceedings of the International Conference on Web Intelligence, Leipzig, Germany, 23–26 August 2017; pp. 953–958. [CrossRef]
12. Franzoni, V.; Milani, A.; Vallverdu, J. Emotional affordances in human-machine interactive planning and negotiation. In Proceedings of the International Conference on Web Intelligence, Leipzig, Germany, 23–26 August 2017; pp. 924–930. [CrossRef]
13. Franzoni, V.; Milani, A.; Nardi, D.; Vallverdú, J. Emotional machines: The next revolution. *WI* **2019**, *17*, 1–7. [CrossRef]
14. Erdős, P.; Rényi, A. On the strength of connectedness of a random graph. *Acta Biochim. Biophys. Acad. Sci. Hung.* **1964**, *12*, 261–267. [CrossRef]
15. Molloy, M.; Reed, B. A critical point for random graphs with a given degree sequence. *Random Struct. Algorithms* **1995**, *6*, 161–179. [CrossRef]

16. Newman, M.E.J.; Strogatz, S.H.; Watts, D.J. Random graphs with arbitrary degree distributions and their applications. *Phys. Rev. E Stat. Nonlinear Soft Matter Phys.* **2001**, *64*, 026118. [CrossRef] [PubMed]
17. de Solla Price, D.J. Networks of scientific papers. *Science* **1965**, *149*, 510–515. [CrossRef]
18. Newman, M.E.J. Prediction of highly cited papers. *Europhys. Lett.* **2014**, *105*, 28002. [CrossRef]
19. Maslov, S.; Sneppen, K. Specificity and stability in topology of protein networks. *Science* **2002**, *296*, 910–913. [CrossRef]
20. Maslov, S.; Sneppen, K.; Zaliznyak, A. Detection of topological patterns in complex networks: Correlation profile of the internet. *Phys. A* **2004**, *333*, 529–540. [CrossRef]
21. Robins, G.; Pattison, P.; Kalish, Y.; Lusher, D. An introduction to exponential random graph (p*) models for social networks. *Soc. Netw.* **2007**, *29*, 173–191. [CrossRef]
22. Cho, E.; Myers, S.A.; Leskovec, J. Friendship and mobility: User movement in location-based social networks. In Proceedings of the 17th ACM SIGKDD International Conference on Knowledge Discovery and Data Mining, San Diego, CA, USA, 21–24 August 2011; pp. 1082–1090. [CrossRef]
23. Illenberger, J.; Nagel, K.; Flötteröd, G. The role of spatial interaction in social networks. *Netw. Spat. Econ.* **2013**, *13*, 255–282. [CrossRef]
24. Scellato, S.; Mascolo, C.; Musolesi, M.; Latora, V. Distance matters: Geo-social metrics for online social networks. In Proceedings of the 3rd Wonference on Online Social Networks, Boston, MA, USA, 22–25 June 2010; p. 8.
25. Goldenberg, J.; Levy, M. Distance is not dead: Social interaction and geographical distance in the Internet era. *arXiv e-prints* **2009**, arXiv:cs.CY/0906.3202.
26. Khadaroo, J.; Seetanah, B. The role of transport infrastructure in international tourism development: A gravity model approach. *Tour. Manag.* **2008**, *29*, 831–840. [CrossRef]
27. Davis, K.F.; D'Odorico, P.; Laio, F.; Ridolfi, L. Global spatio-temporal patterns in human migration: A complex network perspective. *PLoS ONE* **2013**, *8*, e53723. [CrossRef]
28. Lewer, J.J.; den Berg, H.V. A gravity model of immigration. *Econ. Lett.* **2008**, *99*, 164–167. [CrossRef]
29. Dueñas, M.; Fagiolo, G. Modeling the international-trade network: A gravity approach. *J. Econ. Int. Coord.* **2013**, *8*, 155–178. [CrossRef]
30. Carrère, C. Revisiting the effects of regional trade agreements on trade flows with proper specification of the gravity model. *Eur. Econ. Rev.* **2006**, *50*, 223–247. [CrossRef]
31. Xia, Y.; Bjørnstad, O.N.; Grenfell, B.T. Measles metapopulation dynamics: A gravity model for epidemiological coupling and dynamics. *Am. Nat.* **2004**, *164*, 267–281. [CrossRef] [PubMed]
32. Simini, F.; González, M.C.; Maritan, A.; Barabási, A.L. A universal model for mobility and migration patterns. *Nature* **2012**, *484*, 96–100. [CrossRef]
33. Barabasi, A.L.; Albert, R. Emergence of scaling in random networks. *Science* **1999**, *286*, 509–512. [CrossRef] [PubMed]
34. Leskovec, J.; Backstrom, L.; Kumar, R.; Tomkins, A. Microscopic evolution of social networks. In Proceedings of the 14th ACM SIGKDD International Conference on Knowledge Discovery and Data Mining, Las Vegas, NV, USA, 24–27 August 2008; pp. 462–470. [CrossRef]
35. Wiuf, C.; Brameier, M.; Hagberg, O.; Stumpf, M.P.H. A likelihood approach to analysis of network data. *Proc. Natl. Acad. Sci. USA* **2006**, *103*, 7566–7570. [CrossRef] [PubMed]
36. Leskovec, J.; Faloutsos, C. Scalable modeling of real graphs using Kronecker multiplication. In Proceedings of the 24th International Conference on Machine Learning, Corvalis, OR, USA, 20–24 June 2007; pp. 497–504. [CrossRef]
37. Wasserman, S.; Pattison, P. Logit models and logistic regressions for social networks: I. An introduction to Markov graphs andp. *Psychometrika* **1996**, *61*, 401–425. [CrossRef]
38. Bezáková, I.; Kalai, A.; Santhanam, R. Graph model selection using maximum likelihood. In Proceedings of the 23rd International Conference on Machine Learning, Pittsburgh, PA, USA, 25–29 June 2006; pp. 105–112. [CrossRef]
39. Song, J.; Xu, K.; Song, M.; Zhan, X. Credibility evaluation method of domestic IP address database. *J. Comput. Appl.* **2014**, *34*, 4–6.
40. Blondel, V.D.; Guillaume, J.L.; Lambiotte, R.; Lefebvre, E. Fast unfolding of communities in large networks. *J. Stat. Mech. Theory Exp.* **2008**, *2008*, P10008. [CrossRef]

41. Edler, D.; Guedes, T.; Zizka, A.; Rosvall, M.; Antonelli, A. Infomap bioregions: Interactive mapping of biogeographical regions from Species Distributions. *Syst. Biol.* **2017**, *66*, 197–204. [CrossRef]
42. Lü, L.; Zhou, T. Link prediction in complex networks: A survey. *Phys. A* **2011**, *390*, 1150–1170. [CrossRef]
43. Su, C.; Guan, X.; Du, Y.; Wang, Q.; Wang, F. A fast multi-level algorithm for community detection in directed online social networks. *J. Inf. Sci.* **2017**. [CrossRef]
44. Chang, C.; Lee, D.; Liou, L.; Lu, S.; Wu, M. A probabilistic framework for structural analysis and community detection in directed networks. *IEEE/ACM Trans. Network.* **2018**, *26*, 31–46. [CrossRef]
45. Agreste, S.; De Meo, P.; Fiumara, G.; Piccione, G.; Piccolo, S.; Rosaci, D.; Sarné, G.M.L.; Vasilakos, A.V. An empirical comparison of algorithms to find communities in directed graphs and their application in web data analytics. *IEEE Trans. Big Data* **2017**, *3*, 289–306. [CrossRef]
46. Yang, L.; Silva, J.C.; Papageorgiou, L.G.; Tsoka, S. Community structure detection for directed networks through modularity optimisation. *Algorithms* **2016**, *9*, 73. [CrossRef]

MDPI
St. Alban-Anlage 66
4052 Basel
Switzerland
Tel. +41 61 683 77 34
Fax +41 61 302 89 18
www.mdpi.com

*Entropy* Editorial Office
E-mail: entropy@mdpi.com
www.mdpi.com/journal/entropy